Springer Series in
Surface Sciences

2

Springer Series in **Surface Sciences**

Editors: Gerhard Ertl and Robert Gomer

Volume 1: **Physisorption Kinetics**
By H. J. Kreuzer, Z. W. Gortel

Volume 2: **The Structure of Surfaces**
Editors: M. A. Van Hove, S. Y. Tong

Volume 3: **Dynamical Phenomena at Surfaces, Interfaces and Superlattices**
Editors: F. Nizzoli, K.-H. Rieder, R. F. Willis

Volume 4: **Desorption Induced by Electronic Transition, DIET II**
Editors: W. Brenig, D. Menzel

The Structure of Surfaces

Editors: M.A. Van Hove and S.Y. Tong

With 223 Figures

Springer-Verlag
Berlin Heidelberg New York Tokyo

Dr. Michel A. Van Hove

Department of Chemistry, University of California
Berkeley, CA 94720, USA

Professor S. Y. Tong

The University of Wisconsin-Milwaukee, Department of Physics and
The Laboratory for Surface Studies
Milwaukee, WI 53201, USA

Series Editors:

Professor Dr. Gerhard Ertl

Institut für Physikalische Chemie, Universität München, Sophienstraße 11
D-8000 München 2, Fed. Rep. of Germany

Professor Robert Gomer

The James Franck Institute, The University of Chicago, 5640 Ellis Avenue,
Chicago, IL 60637, USA

ISBN 3-540-15410-8 Springer-Verlag Berlin Heidelberg New York Tokyo
ISBN 0-387-15410-8 Springer-Verlag New York Heidelberg Berlin Tokyo

Library of Congress Cataloging in Publication Data. Main entry under title: The Structure of surfaces. (Springer series in surface sciences ; 2) Selected papers presented at the First International Conference on the Structure of Surfaces, held Aug. 13–16, 1984 at the University of California, Berkeley. Includes indexes. 1. Surface chemistry–Congresses. I. Van Hove, M. A. (Michel A.), 1947-. II. Tong, S. Y. III. International Conference on the Structure of Surfaces (1st : 1984 : University of California, Berkeley) IV. Series.
QD501.A1S77 1985 541.3'453 85-4789

Offset printing: Beltz Offsetdruck, 6944 Hemsbach/Bergstr. Bookbinding: J. Schäffer OHG, 6718 Grünstadt
2153/3130-543210

Preface

This book is a collection of selected papers presented at the First International Conference on the Structure of Surfaces (ICSOS-1). ICSOS-1 was held on the Berkeley campus of the University of California during August 13-16, 1984. The International Organizing Committee members were: S.Y. Tong (Chairman), M.A. Van Hove (Vice-Chairman), D.A. King (Secretary), D.J. Chadi (Treasurer), D.L. Adams, A.M. Bradshaw, M.J. Cardillo, J.E. Demuth, J. Eckert, G. Ertl, B.I. Lundqvist, J.B. Pendry, Y. Petroff, M. Simonetta, J.R. Smith, G.A. Somorjai, J. Stöhr, R. Ueda, and X.D. Xie.

The series of ICSOS meetings was initiated to assess the status of surface-structural determination and the relationship between surface or interface structures and physical or chemical properties of interest. The subject matter includes solid and adsorbate-covered surfaces, well-established and promising new surface-sensitive techniques, and results of experimental and theoretical studies.

The physical and chemical properties of a surface or interface are often critically determined by its atomic-scale structure. A variety of techniques has been developed to study this structure and its connection with the surface or interface properties of single crystals and of imperfect and amorphous interfaces.

The papers in this book cover the theory of surface structure, new analytical techniques for surface structure, new developments in established structural techniques, recent structural results, defect structures, and phase transitions at surfaces.

Berkeley, Milwaukee
October 1984

M.A. Van Hove
S.Y. Tong

Acknowledgments

We wish to acknowledge the many organizations and individuals whose contributions made possible the First International Conference on the Structure of Surfaces and these Proceedings. We express our deep appreciation to our many sponsors: the Air Force Office of Scientific Research, the American Physical Society, the A.T.&T. Bell Laboratories, the American Vacuum Society, the Exxon Research and Engineering Corporation, the General Motors Corporation, the International Business Machines Corporation, the International Union for Pure and Applied Physics, the International Union for Vacuum Science, Technique and Applications, the Lawrence Berkeley Laboratory, the Northern California Chapter of the American Vacuum Society, and the Xerox Corporation. Particular thanks go to all the individuals who contributed much to the well-being of both the Conference and the Proceedings: A.A. Maradudin, A. Searcy, E. Kozak, S.W. Wang, A. Kahn, T. Woodward, C. Coolahan, N. Su, D. Arbuckle, P. Little, P. Marlin, M. Hilton, B. Naasz, B.E. Bent, J. Carrazza, D.F. Ogletree, G. Blackman, E. Andersen, C.M. Mate, and D. Kelly. An important element was of course the contribution from the International Advisory Committee members: D. Aberdam, J.C. Bertolini, M. Cardona, G. Comsa, L.C. Feldman, F. Garcia Moliner, D.R. Hamann, D. Haneman, A.A. Lucas, T.E. Madey, K. Müller, S. Nakamura, A.G. Naumovetz, P.R. Norton, G. Rovida, W.E. Spicer, J.F. van der Veen, A. van Oostrom, and R.F. Willis. Finally, we are grateful to the many authors and referees who worked under severe time constraints on the Proceedings articles.

Contents

Introduction 1

Part I Theory of Surface Structure

I.1 General Discussion 4

1. Theory of Surface Reconstruction. By M.L. Cohen (With 1 Figure) .. 4

2. Electronic and Magnetic Properties of Transition-Metal Surfaces, Interfaces and Overlayers. By L.M. Falicov, R.H. Victora, and J. Tersoff 12

3. The Binding of Adsorbates to Metal Surfaces By S. Holloway and J.K. Nørskov (With 7 Figures) 18

I.2 Specific Applications 29

4. Energy Minimization Calculations for Diamond (111) Surface Reconstructions. By D. Vanderbilt and S.G. Louie (With 3 Figures) . 29

5. Total Energies and Atom Locations at Solid Surfaces By R. Richter, J.R. Smith, and J.G. Gay (With 1 Figure) 35

6. Theory of Hydrogen on Metal Surfaces By M.S. Daw and S.M. Foiles (With 2 Figures) 41

Part II New Surface Structure Techniques

II.1 Techniques Based on Electrons 48

7. The Surface Topography of a Pd(100) Single Crystal and Glassy $Pd_{81}Si_{19}$ Studied by Scanning Tunneling Microscopy. By M. Ringger, H.R. Hidber, R. Schlögl, P. Oelhafen, H.J. Güntherodt, K. Wandelt, and G. Ertl (With 4 Figures) 48

8. Theory of the Scanning Tunneling Microscope. By J. Tersoff 54

9. Reflection Electron Microscopy Studies of Crystal Lattice Termination at Surfaces. By T. Hsu and J.M. Cowley (With 6 Figures) 55

II.2 Techniques Based on Photons and Other Probes 60

10. Surface Structure by X-Ray Diffraction
By I.K Robinson (With 4 Figures) 60

11. Optical Transitions and Surface Structure
By P. Chiaradia, A. Cricenti, G. Chiarotti, F. Ciccacci, and
S. Selci (With 3 Figures) 66

12. High-Resolution Infrared Spectroscopy and Surface Structure
By Y.J. Chabal (With 4 Figures) 70

13. Optical Second Harmonic Generation for Surface Studies
By Y.R. Shen (With 6 Figures) 77

14. NMR and Surface Structure. By C.P. Slichter (With 4 Figures) 84

Part III Developments in Existing Techniques

III.1 LEED and Electron Propagation 92

15. Determination of Surface Structure by LEED
By F. Jona, J.A. Strozier, Jr., and P.M. Marcus (With 2 Figures) . 92

16. Structure Determination of Molecular Adsorbates with Dynamical LEED
and HREELS. By M.A. Van Hove 100

17. Computer Controlled LEED Intensity and Spot Profile Determination
By K. Müller and K. Heinz (With 6 Figures) 105

18. On the Role of Space Inhomogeneity of Electron Damping in LEED
By I. Bartoš and J. Koukal (With 2 Figures) 113

19. Attenuation of Isotropically Emitted Electron Beams
By R. Mayol, F. Salvat, and J. Parellada-Sabata (With 2 Figures) . 117

III.2 Diffuse LEED, NEXAFS/XANES and SEXAFS 124

20. LEED, XANES and the Structure of Disordered Surfaces
By J.B. Pendry (With 3 Figures) 124

21. The Structure of Organic Adsorbates from Elastic Diffuse LEED
By D.K. Saldin, D.D. Vvedensky, and J.B. Pendry (With 2 Figures) . 131

22. Multiple Scattering Effects in Near-Edge X-Ray Absorption Spectra
By D.D. Vvedensky, D.K. Saldin, and J.B. Pendry (With 1 Figure) .. 135

23. NEXAFS and SEXAFS Studies of Chemisorbed Molecules: Bonding,
Structure and Chemical Transformations
By J. Stöhr (With 3 Figures) 140

24. Current Status and New Applications of SEXAFS: Reactive
Chemisorption and Clean Surfaces. By P.H. Citrin (With 5 Figures) 149

III.3 High-Resolution Electron Energy Loss Spectroscopy (HREELS) 156

25. Electron-Phonon Scattering and Structure Analysis
By M. Rocca, H. Ibach, S. Lehwald, M.-L. Xu, B.M. Hall, and S.Y. Tong (With 6 Figures) 156

26. Shape Resonances in OH Groups Chemisorbed on the (100) Surface of Ge-Si Alloys. By H.H. Farrell, J.A. Schaefer, J.Q. Broughton, and J.C. Bean 163

27. Structure and Temperature-Dependent Polaron Shifts on Si(111) (2x1)
By C.D. Chen, A. Selloni, and E. Tosatti (With 4 Figures) 170

III.4 Atom and Ion Scattering 176

28. Surface Structure Analysis by Atomic Beam Diffraction
By J. Lapujoulade, B. Salanon, and D. Gorse (With 6 Figures) 176

29. Structure Analysis of a Semiconductor Surface by Impact Collision Ion Scattering Spectroscopy (ICISS): Si(111) $\sqrt{3}\times\sqrt{3}$ R30°Ag
By M. Aono, R. Souda, C. Oshima, and Y. Ishizawa (With 4 Figures) 187

III.5 Photoemission 191

30. Surface Structure Determination with ARPEFS
By J.J. Barton, S.W. Robey, C.C. Bahr, and D.A. Shirley (With 4 Figures) 191

31. Angle Resolved XPS of the Epitaxial Growth of Cu on Ni(100)
By W.F. Egelhoff, Jr. (With 2 Figures) 199

32. Evidence for Diffusion at 80K of Gold Atoms Through Thin, Defective Oxide Layers. By S. Ferrer, C. Ocal, and N. Garcia (With 4 Figures) 204

III.6 Neutron Scattering 210

33. Surface Characterization by the Inelastic Scattering of Neutrons from Adsorbates. By C.J. Wright (With 3 Figures) 210

34. Infrared and Neutron-Scattering Studies of Ethene Adsorbed onto Partially Exchanged Zinc A Zeolite. By J. Howard, J.M. Nicol, and J. Eckert (With 3 Figures) 219

Part IV Clean and Adsorbate-Covered Metals

IV.1 Clean Metal Surfaces 226

35. Theoretical Study of the Structural Stability of the Reconstructed (110) Surfaces of Ir, Pt and Au. By H.-J. Brocksch and K.H. Bennemann (With 1 Figure) 226

36. The Structure and Surface Energy of Au(110) Studied by Monte Carlo Method. By T. Halicioğlu, T. Takai, and W.A. Tiller (With 1 Figure) 231

37. Long- and Short-Range Order Fluctuations in the H/W(100) System By R.F. Willis (With 6 Figures) 237

IV.2 Atomic Adsorption on Metal Surfaces 246

38. Synchrotron X-Ray Scattering Study of a Chemisorption System: Oxygen on Cu(110) Surface. By K.S. Liang, P.H. Fuoss, G.J. Hughes, and P. Eisenberger (With 3 Figures) 246

39. Helium Diffraction from Oxygen-Covered Nickel Surfaces By I.P. Batra, T. Engel, and K.H. Rieder (With 2 Figures) 251

40. Competing Reconstruction Mechanisms in H/Ni(110) By R.J. Behm, K. Christmann, G. Ertl, V. Penka, and R. Schwankner (With 3 Figures) 257

IV.3 Molecular Adsorption on Metal Surfaces 264

41. The Uses and Limitations of ESDIAD for Determining the Structure of Surface Molecules. By T.E. Madey 264

42. The Study of Simple Reactions at Surfaces by High-Resolution Electron Energy Loss Spectroscopy By N.V. Richardson, C.D. Lackey, and M. Surman (With 6 Figures) .. 269

Part V Semiconductors

V.1 Elemental Semiconductors 278

43. Triangle-Dimer Stacking-Fault Model of the Si(111)7x7 Surface Bonding Configuration. By E.G. McRae (With 6 Figures) 278

44. Structure of the Si(111)2x1 Surface By I.P. Batra, F.J. Himpsel, P.M. Marcus, R.M. Tromp, M.R. Cook, F. Jona, and H. Liu (With 3 Figures) 285

45. Refinement of the Buckled-Dimer Model for Si(001)2x1 By Y.S. Shu, W.S. Yang, F. Jona, and P.M. Marcus (With 4 Figures) 293

46. Surface Relaxation and Vibrational Excitations on the Si(001)2x1 Surface. By D.C. Allan and E.J. Mele (With 2 Figures) 298

V.2 Compound Semiconductors 303

47. The Geometric Structure of the (2x2)GaAs(111) Surface By G. Xu, S.Y. Tong, and W.N. Mei (With 4 Figures) 303

48. A Comparison Between the Electronic Properties of GaAs(111) and GaAs($\bar{1}\bar{1}\bar{1}$). By R.D. Bringans and R.Z. Bachrach (With 4 Figures) ... 308

49. X-Ray Diffraction from the (3x3) Reconstructed ($\bar{1}\bar{1}\bar{1}$)B Surface of InSb. By R.L. Johnson, J.H. Fock, I.K. Robinson, J. Bohr, R. Feidenhans'l, J. Als-Nielsen, M. Nielsen, and M. Toney (With 3 Figures) 313

V.3 Adsorbate-Covered Semiconductors 317

50. Atomic and Electronic Structure of p(1x1) Overlayers of Sb on the the (110) Surfaces of III-V Semiconductors. By C.B. Duke, C. Mailhiot, A. Paton, K. Li, C. Bonapace, and A. Kahn (With 2 Figures) 317

51. Models for Si(111) Surface upon Ge Adsorption By S.B. Zhan, J.E. Northrup, and M.L. Cohen (With 4 Figures) 321

52. The Graphite (0001)-(2x2)K Surface Intercalated Structure By N.J. Wu and A. Ignatiev (With 4 Figures) 326

Part VI Defects and Phase Transitions

VI.1 Theoretical Aspects 334

53. Engergetics of the Incommensurate Phase of Krypton on Graphite: A Computer Simulation Study. By M. Schöbinger and F.F. Abraham (With 3 Figures) 334

54. Theory of Commensurate-Incommensurate Phase Transitions on W(001) By S.C. Ying and G.Y. Hu (With 3 Figures) 341

55. Molecular Dynamics Investigation of Dislocation-Depinning Transitions in Mismatched Overlayers. By K.M. Martini, S. Burdick, M. El-Batanouny, and G. Kirczenow (With 2 Figures) 347

56. Quantitative Analysis of LEED Spot Profiles By M. Henzler (With 1 Figure) 351

57. Measurement of the Specific Heat Critical Exponent Using LEED By N.C. Bartelt, T.L. Einstein, and L.D. Roelofs (With 3 Figures) 357

58. Short Range Correlations in Imperfect Surfaces and Overlayers By J.M. Pimbley and T.-M. Lu (With 3 Figures) 361

59. Domain Size Determination in Heteroepitaxial Systems from LEED Angular Profiles. By D. Saloner and M.G. Lagally (With 4 Figures) 366

VI.2 Experimental Studies 375

60. Diffusion and Interaction of Adatoms By G. Ehrlich (With 19 Figures) 375

61. Atom-Probe and Field Ion Microscope Studies of the Atomic Structure and Composition of Overlayers on Metal Surfaces. By T.T. Tsong and M. Ahmad (With 7 Figures) 389

62. LEED Studies of Physisorbed Noble Gases on Metals and Interadatom Interactions. By M.B. Webb and E.R. Moog (With 3 Figures) 397

63. Phases and Phase Transitions in Two Dimensional Systems with Competing Interactions. By R.J. Birgeneau, P.M. Horn, and D.E. Moncton (With 5 Figures) 404

64. Diffraction from Pinwheel and Herringbone Structures of Nitrogen and Carbon Monoxide on Graphite. By S.C. Fain, Jr. and H. You (With 2 Figures) 413

65. X-Ray Scattering Studies: The Structure and Melting of Pb on Cu(110) Surfaces. By S. Brennan, P.H. Fuoss, and P. Eisenberger (With 3 Figures) 421

Index of Contributors 427

Subject Index 429

Introduction

The structure of a solid's surface influences many of its essential properties, among them the chemical, electronic, and vibrational properties. To fully understand surface properties, one must often first determine to an atomistic accuracy the spatial coordinates of atoms in the surface region. Normally, an accuracy of a few percent of the interatomic distances is needed to fix certain electronic and vibrational characteristics. The interlayer spacings, bond lengths, bond angles, and bonding configurations need to be accurately determined and the results understood. Much work in this field has focused on single-crystal surfaces, but techniques have also been introduced which yield structural information at imperfect, amorphous, and other kinds of surfaces or interfaces. The contributions to these proceedings reflect the great diversity and complexity of surface science, covering theory, techniques, and structural results. They have been grouped according to the following topics.

In the first chapter, the theory of surface structure is treated. The purpose of the articles in this chapter is to use theory to predict stable structures. The ability to predict surface structures using theoretical methods is an ultimate goal of theory and it requires a highly sophisticated understanding of surface forces. Towards this goal, theory has made significant gains.

The second chapter groups together articles that deal with the new analytical techniques for surface structure. The vitality of this area is demonstrated by the large number of new methods presented here. These include the adaptation of established techniques developed elsewhere to the study of surfaces. New techniques can complement existing techniques and provide a useful cross-check. Some new techniques give first-time information on new types of surfaces (e.g., amorphous surfaces and solid-liquid or solid-solid interfaces), while others yield qualitatively different structural information (e.g., scanning tunneling microscopy and nuclear magnetic resonance).

New developments in established structural techniques are covered in Chap.3. This chapter presents studies wherein various extensions, improvements, and simplifications have led to more diverse applications, increased accuracies, or greater efficiencies in surface-structure determination. It also reflects the many new ideas that continue to emerge in this area.

Chapter 4 presents recent structural results that have been obtained on clean metal surfaces as well as on atomic and molecular adsorbates on those surfaces. The articles cover, among other topics, clean-surface and adsorbate-induced reconstructions, and molecular adsorption geometries. Results in these areas are significant for the atomic-scale understanding of oxidation, catalysis, and other surface phenomena.

In Chap.5, corresponding advances in the determination of semiconductor surface structures are presented. Progress has been achieved in particular on several reconstructed elemental and compound semiconductor surfaces. In addition, metal-semiconductor interfaces and intercalated structures point to new directions of interest.

The mutually related topics of defects and phase transitions are covered in Chap.6. Defects and adsorbate-adsorbate interactions often play important roles in surface reactions and crystal growth. Their study has progressed markedly in recent years, thanks to experimental and theoretical advances in direct imaging, diffraction, and the understanding of critical phenomena. The articles in this final chapter reflect the growing interest in these issues.

Part I

Theory of Surface Structure

I.1 General Discussion

1. Theory of Surface Reconstruction

Marvin L. Cohen

Department of Physics, University of California, and Materials and Molecular Research Division, Lawrence Berkeley Laboratory, Berkeley, CA 94720, USA

A review is presented of theoretical calculations used to determine surface structure. The focus is on applications of the pseudopotential total energy approach.

1.1 Introduction

I have been asked to give a general, introductory review describing the status of theoretical calculations of surface structure with emphasis on the pseudopotential method. The impetus for research in this field has come from experiments, with the largest fraction of the data being from LEED [1.1]. New experimental techniques and refinements of older approaches have added significant motivation for theoretical results. Among these are the scanning tunneling microscope, transmission electron microscopy, ion scattering, atomic scattering and diffraction, variations of EXAFS, optical and X-ray probes, and a variety of other approaches many of which were discussed at this conference. In these experiments, electrons, photons, neutrons, positrons, atoms, and ions are used to bombard surfaces, and the appropriate response functions are measured and analyzed. Some significant input has been made by theorists analyzing these data. In addition to providing the theoretical basis for evaluating the response functions, structural models are suggested to interpret the results. The symmetries of the observed spectra often rule out classes of models, and these analyses are very helpful for consolidating the data and putting constraints on the properties of the structures. Despite the large effort expended by talented researchers in this area, very few surface reconstructions have been "pinned down."

In the case of metals, reconstructions are usually not as dramatic as for semiconductors. Atoms at metal surfaces generally tend to pull back or relax, and the electrons at the surface adjust without much effort. We expect more dramatic changes in cases where the electrons are localized in bonds such as the covalent and partially covalent bonds in semiconductors. This would also be the case for electrons in the d orbitals of transition metals. Therefore, most research in this area has focused on these systems. This review will concentrate on semiconductor reconstructions.

What can theory say about these reconstructions? Can surface structure be predicted without input from experiment? These and other questions related to the use of experiments designed to probe electronic properties for structural determination have been considered in recent years by theorists working in the general area of electronic structure or band theory. In what follows, I will highlight pseudopotential calculations and, after a brief history and survey, describe some recent results.

1.2 Background

Accurate calculations of the electronic structure of bulk crystals became available in the 1960s. The pseudopotential [1.2] approach allowed a determination of the energy band spectrum for dozens of solids. These studies allowed detailed interpretation of optical and photoelectron spectra. In the early 1970's, the wave functions for electrons in the bulk were used to produce electron density or charge density plots. Theory was at a stage where one could say, "Tell me where the atoms are, and I'll tell you where the electrons are." However, if one asked about surface electrons using this approach, the procedure would be to treat the surface as the ends of a perfect solid. No reconstructions, surface states, or redistribution of the electronic charge were possible within the early models [1.3].

To allow for charge redistribution and the option to reconstruct the surface, two principal methods were developed [1.4,5]. In one approach [1.4], a matching of decaying orbitals to the propagating bulk electronic wave functions allowed for a description of surface states and charge redistribution. This method is a direct solution to the breaking of the translational symmetry caused by the surface. The second approach [1.5] involved the use of supercells [1.6] to accommodate localized geometries. An artificial supercell could model a surface by assuming a slab geometry containing about a dozen layers of Si atoms with a vacuum region on both sides. The slabs are repeated infinitely, and the top and bottom layers of the slab are associated with the solid surfaces. By treating the electronic charge self-consistently [1.4-6], the rearrangement of charge at the surface could be accounted for.

In this approach, the geometry of the surface is used as input. In the case of Si(111)-(2×1), it was believed at that time that the buckling model [1.7] was correct. This geometry was *assumed* for the atoms near the surface of the slab, and a selfconsistent calculation of the electronic states was performed. The results [1.5] yielded electron density maps, local density of states curves which give the density of states layer by layer into the solid, optical constants, band structures, and charge densities for the surface states, etc. The agreement with experiment appeared to be good especially for the optical data [1.8]. This work was extended to other surfaces, Schottky barriers, and heterojunctions [1.9]. However, it should be emphasized that the surface or interfacial structure was always taken from experiment as *input* to these calculations. Although the self-consistent pseudopotential calculations did answer some questions related to surface reconstruction, the approach was indirect. In particular, comparison of the calculated and measured electronic structure showed the superiority of some structures. In the main, the value of this approach was its contribution to our understanding of the electronic structure of surfaces and the consequences of surface reconstruction for electronic behavior.

1.3 Total Energy

Around 1980, the situation for structural calculations related to bulk properties improved dramatically. By using ab initio pseudopotential [1.10-15] and a momentum space formalism [1.16] for calculating total energies, it was possible to compute the total energy for different arrangements of atoms and find the lowest energy structure for a given volume [1.17]. This approach was applied to a number of problems [1.17]. It allowed the determination of a specific structure from a subset of candidates and gave accurate values for lattice constants, bulk moduli, cohesive energies, and properties of structural phase transitions with only the atomic number as input to generate the pseudopotentials. If the atomic mass was also used, phonon spectra and Gruneisen parameters could also be evaluated.

The extension to surfaces [1.18-20] came through the use of supercells and calculation of the total energy for a given surface geometry. This method made possible the extension to surfaces of total energy techniques designed for bulk properties. The first application was to Si(111) [1.18]. By computing the total energy as a function of relaxation, the relaxation energy could be computed. Minimizing the energy with respect to the relaxation gave an estimate of the equilibrium relaxation. In another early calculation for Si(001)-(2×1), it was possible to test [1.19] various models and demonstrate the lowering of the energy for the asymmetric dimer geometry compared to the symmetric dimer case. *Chadi* [1.21] had shown this using tight-binding models.

The tight-binding approach has also had a large impact on theoretical work in this area. In particular, the scheme used by *Chadi* [1.21] has been very successful. Most of the fully self-consistent pseudopotential approaches are based on a plane-wave expansion. Since semiconductor bonds are fairly localized, they require a large basis of the order of 100 plane waves to reproduce the electron states. In the tight-binding approach, the Hamiltonian matrix is about an order of magnitude smaller and therefore easier and less expensive to diagonalize. The matrix elements can be estimated and some scaling can be introduced to mimic the effects of self-consistency. This approach is not considered as "ab initio" as the fully self-consistent approach, but because it is flexible and relatively inexpensive, it has been applied to a wider variety of systems. In some cases, applications were made to fairly complex structures like the Si(111)-(7×7) geometry. For cases where comparison has been made with fully self-consistent calculations, the results are excellent.

As described above, the plane-wave pseudopotential approach is limited mainly by the cost of the computations. A fixed number of plane waves per atom in a cell is required to calculate the total energy of a specific geometry. Since the more complex geometries have larger cells and larger numbers of atoms in the unit cell, a larger matrix results. The cost of diagonalizing a matrix is proportional to n^x, where n is the matrix size and $2 \leq x \leq 3$.

It is possible to reduce the matrix size by choosing a set of basis functions which is a combination of plane waves and localized functions. This mixed basis scheme [1.22] has been used primarily for transition metals, but it can be used for highly localized covalent bonds such as those associated with C atoms. Recently, a first-principles linear combination of atomic orbitals approach using pseudopotentials has been developed [1.23]. This scheme can be used to compute total energies, and it has been applied to examine the diamond (111) surface [1.24]. The approach has some of the virtues of a small matrix scheme such as the tight-binding model, and it approximates well the results of plane-wave ab initio approaches for all cases tested.

In the fully self-consistent schemes, the electron-electron interactions are computed using a local density functional approach [1.25,26]. This method is designed to give a good account of the ground state, but excited state properties like band gaps are outside the scope of the original formulation. Many extensions have been suggested, but at present, there is no general agreement on a superior scheme. Band gaps calculated within the standard approach are usually underestimated by 30% to 50%, but ground-state energies are given accurately. These energies can be computed knowing only the electron density as a function of position. For calculations of surface reconstructions it is therefore necessary to recalculate the total energy every time the geometry is changed. For example, to compute the total energy as a function of relaxation requires the determination of the electron-electron potentials for each relaxed geometry, and this must be done self-consistently, i.e. the input electron density used to construct the electron-electron potential must be the same as the square of the resulting eigenfunctions obtained by diagonalizing the Hamiltonian matrix.

To determine a surface geometry using this approach, the energy is computed as the geometry is changed, and the minimum energy configuration is assumed to be the correct reconstruction. Unfortunately, it is possible to obtain "local minima" and hence not find the minimum energy structure.

Another related approach is to compute the Hellmann-Feynman forces on the surface atoms [1.27]. This scheme is similar to the total energy procedure described above, but often it is more useful. The starting point is the ideal or relaxed (1×1) geometry. With the density fixed, the components of the forces on each surface atom in the slab supercell are computed. The next step is to move the atoms in the direction of the forces. Because this changes the electron density, new forces are generated. Again, atoms are moved in the directions of the residual forces. This procedure is continued until the forces are zero or very small. The resulting zero-force geometry corresponds to the minimum energy structure. Unfortunately, just as in the case of total energy calculations, the results obtained may correspond to a local minimum.

1.4 Applications

The first detailed application of the pseudopotential energy-minimization force scheme considered [1.20] the Si(001)-(2×1) surface. Chadi's asymmetric dimer model was shown to have the lowest energy of the structures tested, but the resulting dimer bond length was found to be shorter than Chadi's value. More recently, using the microscopic parameters computed from the pseudopotential calculation, *Ihm* et al. [1.28] applied a position-space renormalization-group technique to study the structural phase transitions on this surface as a function of temperature. They found two families of reconstructed geometries and two critical temperatures representing the order-disorder transitions. *Pandey* in Ref. [1.28] has proposed another model which has some features in common with a buckling geometry but involves some π bonding between the broken bonds. Hence, at present, a Si(001)-(2×1) reconstruction geometry has not been agreed upon. However, it is gratifying to see that pseudopotential total energy schemes are playing a central role along with experimental probes in the search.

The most studied surface of the group IV materials is the (111) surface. The (1×1), (2×1), and (7×7) modifications have received the most attention. Because of the large number of atoms in the (7×7) unit cell, this geometry is considered to be too complex for a complete study with the plane-wave pseudopotential approach. However, smaller subunits like the Si(111)-(2×2) adatom and pyramidal-cluster geometries have been studied [1.29] to test (7×7) models [1.30,31].

As was discussed earlier, the ideal or (1×1) geometry was believed to undergo a (2×1) buckling distortion with a surface cell containing an "up" atom and a "down" atom. This distortion was thought to be necessary to produce a gap at the Fermi energy for the surface state and produce a semiconducting (2×1) reconstruction of the metallic (1×1) geometry. However, when the Hellmann-Feynman forces were calculated [1.32] for a (1×1) structure subject to a buckling distortion, the surface resisted this distortion. Forces developed to return the surface to the (1×1) symmetry. Since data [1.33] on laser-annealed surfaces appeared to give (1×1) LEED patterns and the (1×1) atomic geometry was stable with respect to buckling distortions, it was suggested [1.34-36] that the (1×1) surface might be stable. Self-consistent spin-polarized pseudopotential calculations [1.36] showed that the (2×1) electronic distortion caused by antiferromagnetic ordering would lower the energy and yield a semiconductor surface even though the surface atomic structure was (1×1). The energies of the (2×1) antiferromagnetic and the (1×1) paramagnetic states are very close and dependent on the surface relaxation. Because the calculated results are sensitive to the functionals used for exchange and correlation, a definitive prediction is difficult. To my knowledge, at present there is no experimental support for a Si(111)-(1×1) ordering on cleaved surfaces.

Theoretical results for Ge(111) [1.36] also raise the possibility of a stable antiferromagnetic state. There is some experimental evidence [1.37,38] for a (1×1) LEED pattern after cleavage at liquid-helium temperatures. In addition, the surface appears to be nonmetallic. Several researchers have suggested that the low-temperature Ge(111) surface may be disordered and the (1×1) pattern could arise from contributions from bulk states.

Even though the *Haneman* [1.7] buckling model increases the total energy, it can be shown that some (2×1) structural distortions give a lower energy than the (1×1) state discussed above. Considerable effort has been expended to study a variety of (2×1) reconstructions of the surface atoms. Among these are the *Seiwatz* chain [1.39] and the *Chadi* π-bonded molecule model [1.40]. The most successful candidate appears to be the π-bonded chain model proposed for Si(111)-(2×1) by *Pandey* [1.41]. At first, this model was greeted with scepticism because it was agreed that too many bonds would have to be broken in the reconstruction processes and this would require a large energy. However, it was demonstrated [1.42] that a low-energy path exists with an energy barrier of less than 0.1 eV. This energy could be supplied in the process of cleaving the sample.

Using the minimum energy geometry, the surface state band structure or energy as a function of wave vector E(k) could be computed. The results were found to be in good agreement with angular-resolved photoemission measurements [1.42-46]. Considering the fact that the theoretical input is the atomic number and a subset of geometries, the agreement between experiment and theory is encouraging. Although the π-bonded chain model is likely to be correct, there is no proof that a lower energy geometry does not exist. One word of caution is that at this point the agreement with LEED measurements is marginal.

The situation is even less certain for Ge(111)-(2×1). Theory [1.40] predicts that the chain model has the lowest energy. Unlike the Si cases, there is not yet agreement between the experimental groups [1.47-49] doing angular-resolved photoemission experiments. Recent polarized optical measurements by Olmstead and Amer add considerable support to the chain model prediction [in Ref.1.28].

For diamond (111), LEED measurements are unable to distinguish between true (2×2) and disordered (2×1) domains. The total energy study [1.24] suggests that an undimerized π-bonded chain geometry has the lowest energy.

Other studies of the surfaces of group IV materials involve the determination of adsorption sites. Two examples are Al (*Northrup* in Ref. [1.28]) and Ge [1.50] on Si(111). By comparing the total energy for supercells with adsorbate atoms on different sites, the preferred sites are found.

1.5 Discussion and Conclusions

The major point to be emphasized is that theoretical total energy calculations are contributing on a level with many experimental probes to determine the reconstruction of surfaces. This review concentrated on group IV semiconductors, but pseudopotential theory has also been applied to study the surfaces of other semiconductors, simple metals, and transition metals. In particular, considerable research has been done on clean and adsorbate-covered III-V semiconductor surfaces. At this point, GaAs(110) is probably the best understood semiconductor surface. A good example of the use of the total energy pseudopotential approach to determine adsorbate sites is the study of Al on GaAs(110) [1.51].

Because of the predictive power of the new theoretical methods, the standard axiom that "decisions in physics are made by experiment" may not be universally true. I'm reminded of a figure (Fig.1.1) drawn by H. Kamamura of a torii, the gateway of a Shinto temple. The top of the torii representing truth is reached through experimental and theoretical physics. However, a crossbar is now added—computational physics. The impact of computational physics has been increasing, and this figure is symbolic of the strengthening of both theory and experiment by computation. We must beware of those cases where we find theory and experiment in agreement because of computation since all are incomplete or not thoroughly studied. In these cases, we have only reached the crossbar and have mistaken it for the "truth" level. However, I feel that these situations will be rare, and I am optimistic about future collaborative efforts between experimentalists and theorists and the role of computational physics.

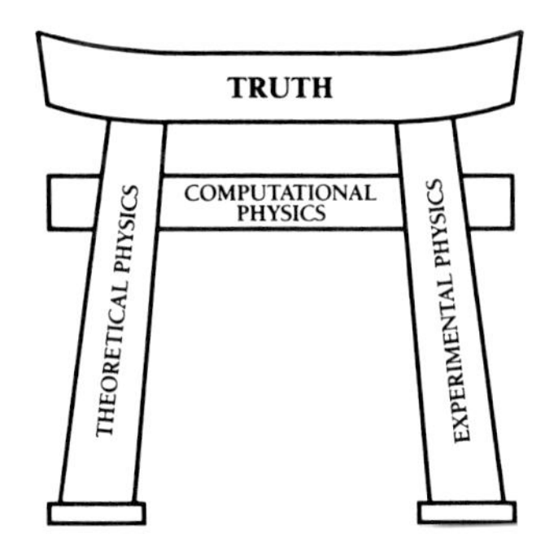

Fig.1.1

Acknowledgments. This work was supported by National Science Foundation Grant No. DMR 8319024 and by the Director, Office of Energy Research, Office of Basic Energy Sciences, Materials Sciences Division of the U.S. Department of Energy under Contract No. DE-AC03-76SF00098.

References

1.1 M.A. Van Hove, S.Y. Tong: *Surface Crystallography by LEED*, Springer Ser. Chem. Phys., Vol.2 (Springer, Berlin, Heidelberg 1979)
1.2 M.L. Cohen, V. Heine: *Solid State Physics* **24**, 37 (1970)
1.3 J.P. Walter, M.L. Cohen: Phys. Rev. Lett. **26**, 17 (1971); M.L. Cohen: Science **179**, 1189 (1973)
1.4 J.A. Appelbaum, D.R. Hamann: Rev. Mod. Phys. **48**, 3 (1976)
1.5 M. Schlüter, J.R. Chelikowsky, S.G. Louie, M.L. Cohen: Phys. Rev. B**12**, 4200 (1975)
1.6 M.L. Cohen, M. Schlüter, J.R. Chelikowsky, S.G. Louie: Phys. Rev. B**12**, 5575 (1975)
1.7 D. Haneman: Phys. Rev. **121**, 1093 (1961)
1.8 G. Chiarotti, S. Nannarone, R. Pastore, P. Chiaradia: Phys. Rev. B**4**, 3398 (1971)
1.9 M.L. Cohen: "Electrons at interfaces", in *Advances in Electronics and Electron Physics*, Vol.51, ed. by L. Marton, C. Marton (Academic, New York 1980) pp.1-62
1.10 T. Starkloff, J.D. Joannopoulos: Phys. Rev. B**16**, 5212 (1977)
1.11 A. Zunger, M.L. Cohen: Phys. Rev. B**18**, 5449 (1978)
1.12 D.R. Hamann, M. Schlüter, C. Chiang: Phys. Rev. Lett. **43**, 1494 (1979)
1.13 G. Kerker: J. Phys. C**13**, L189 (1980)
1.14 M.T. Yin, M.L. Cohen: Phys. Rev. B**25**, 7403 (1982)
1.15 S.G. Louie, S. Froyen, M.L. Cohen: Phys. Rev. B**26**, 1738 (1982)
1.16 J. Ihm, A. Zunger, M.L. Cohen: J. Phys. C**12**, 4401 (1979)
1.17 M.L. Cohen: Physica Scripta T**1**, 5 (1982)
1.18 J. Ihm, M.L. Cohen: Solid State Comm. **29**, 711 (1979)
1.19 J. Ihm, D.J. Chadi, M.L. Cohen: Phys. Rev. B**21**, 4592 (1980)
1.20 M.T. Yin, M.L. Cohen: Phys. Rev. B**24**, 2303 (1981)
1.21 D.J. Chadi: Phys. Rev. Lett. **43**, 43 (1979); J. Vac. Sci. Technol. **16**, 1290 (1979)
1.22 S.G. Louie, K.M. Ho, M.L. Cohen: Phys. Rev. B**19**, 1774 (1974)
1.23 J.R. Chelikowsky, S.G. Louie: Phys. Rev. B**29**, 3470 (1984)
1.24 D. Vanderbilt, S.G. Louie: This Volume, p.29
1.25 P. Hohenberg, W. Kohn: Phys. Rev. **136**, B864 (1964)
1.26 W. Kohn, L. Sham: Phys. Rev. **140**, A1333 (1965)
1.27 J. Ihm, M.T. Yin, M.L. Cohen: Solid State Comm. **37**, 491 (1981)
1.28 J. Ihm, D.H. Lee, J.D. Joannopoulos, J.J. Xiong: In *Proceedings of the 17th International Conference on the Physics of Semiconductors*, San Francisco, 1984 (Springer, Berlin, Heidelberg 1984)
1.29 J.E. Northrup, M.L. Cohen: Phys. Rev. B**29**, 1966 (1984)
1.30 W.A. Harrison: Surf. Sci. **55**, 1 (1976)
1.31 G. Binnig, H. Rohrer, C. Gerber, E. Weibel: Phys. Rev. Lett. **50**, 120 (1983)
1.32 J.E. Northrup, J. Ihm, M.L. Cohen: Phys. Rev. Lett. **47**, 1910 (1981)
1.33 D. Zehner, C.W. White, P. Heimann, B. Reihl, F.J. Himpsel, D.E. Eastman: Phys. Rev. B**24**, 4875 (1981)
1.34 R. Del Sole, D.J. Chadi: Phys. Rev. B**24**, 7430 (1981)
1.35 C.B. Duke, W.K. Ford: Surf. Sci. **111**, L685 (1981)
1.36 J.E. Northrup, M.L. Cohen: Phys. Rev. B**29**, 5944 (1984)
1.37 V.Yu. Artistov, N.I. Golovko, V.A. Grazhulis, Y.A. Ossipyan, V.I. Talyanskii: Surf. Sci. **177**, 204 (1982)
1.38 D. Haneman, R.Z. Bachrach: J. Vac. Sci. Technol. **21**, 337 (1982)
1.39 R. Seiwatz: Surf. Sci. **2**, 473 (1964)
1.40 D.J. Chadi: Phys. Rev. B**26**, 4762 (1982)
1.41 K.C. Pandey: Phys. Rev. Lett. **47**, 1913 (1981)
1.42 J.E. Northrup, M.L. Cohen: Phys. Rev. Lett. **49**, 1349 (1982)
1.43 F.J. Himpsel, P. Heimann, D.E. Eastman: Phys. Rev. B**24**, 2003 (1981)

1.44 R.I.G. Uhrberg, G.V. Hansson, J.M. Nicholls, S.A. Flodstrom: Phys. Rev. Lett. **48**, 1032 (1982)
1.45 F. Houzay, G. Guichar, R. Pinchaux, G. Jezequel, F. Solal, A. Barsky, P. Steiner, Y. Petroff: Surf. Sci. **132**, 40 (1983)
1.46 J.E. Northrup, M.L. Cohen: Phys. Rev. B**27**, 6553 (1983)
1.47 J.M. Nicholls, G.V. Hansson, R.I.G. Uhrberg, S.A. Flodstrom: Phys. Rev. B**27**, 2594 (1983)
1.48 J.M. Nicholls, G.V. Hansson, U.O. Karlsson, R.I.G. Uhrberg, R. Engelhardt, K. Seki, S.A. Flodstrom, E.E. Koch: Phys. Rev. Lett. **52**, 1555 (1984)
1.49 F. Solal, G. Jezequel, A. Barsky, P. Steiner, R. Pinchaux, Y. Petroff: Phys. Rev. Lett. **52**, 360 (1984)
1.50 S.B. Zhang, J.E. Northrup, M.L. Cohen: This Volume, p.321
1.51 J. Ihm, J.D. Joannopoulos: Phys. Rev. Lett. **47**, 679 (1981)

2. Electronic and Magnetic Properties of Transition-Metal Surfaces, Interfaces and Overlayers

L.M. Falicov and R.H. Victora

Department of Physics, University of California, and Materials and Molecular Research Division, Lawrence Berkeley Laboratory, Berkeley, CA 93720, USA

J. Tersoff

I.B.M. Thomas J. Watson Research Center Yorktown Heights, NY 10598, USA

Results of calculations of the electronic and magnetic properties of transition-metal surfaces, interfaces and overlayers are presented for a variety of systems. They involve Ni, Co, Fe and Cr in a diversity of forms, including alloys, metastable configurations, and overlayers on nonmagnetic metals. The overall behavior of these systems can be interpreted in terms of four qualitative rules which are presented, analyzed, and illustrated.

2.1 Introduction

There is considerable current interest in the magnetic and related electronic properties of 3d magnetic transition-metal surfaces and overlayers. These metals exhibit itinerant magnetism: their magnetization derives from the spin polarization of the itinerant d electrons. In moving down the periodic table from Ni, there is a decrease in the number of these d electrons (an increase in the number of d holes), and a consequent increase in the bulk magnetization [2.1] from 0.61 Bohr magnetons in Ni, to 1.72 in Co, and 2.22 in Fe. Beyond Fe lie the more complicated magnetic structures of Mn and Cr. In particular Cr has an antiferromagnetic ground state [2.2] in which at the maximum of an incommensurable spin density wave there is a magnetization of 0.59 Bohr magnetons. In all these elements, the itinerant nature of the d electrons makes the magnetic properties a sensitive function of local environment. Consequently the presence of a dissimilar neighbor, as found in an interface, or the absence of some neighbors, as found at a surface, may cause considerable changes in the local magnetic properties.

We have calculated the electronic and magnetic properties for many surface and overlayer systems [2.3-8]. We used a Slater-Koster parametrized tight-binding scheme in which the one- and two-center integrals are fitted to the bulk band structures of the elements; 4s, 4p and 3d electrons are included. The electron-electron interaction consists of single-site contributions and is sufficiently general to allow for realistic effects such as nonrigid exchange splitting. The interaction is treated self-consistently in the Hartree-Fock approximation. Our scheme has been tested against experimental data [2.9,10] and against state-of-the-art first-principles calculations [2.11,12] on several occasions, and has produced consistently excellent agreement [2.5-7,13].

In this contribution we use our theoretical results, combined with experimental information, to develop systematically some qualitative rules for predicting the magnetization and density-of-state effects of these complicated systems. In addition we examine in more detail several systems where unusual or unexpected phenomena occur.

2.2 Surfaces

A particularly important system is bcc Fe and its surfaces. The experimentally observed bulk spin polarization [2.7] (twice the magnetization divided by the g factor) is 2.12. We have calculated the spin polarization of the (110) surface to be 2.55 and that of the (100) surface to be 2.90. These results are easily understood by considering the simple Stoner theory [2.14] which suggests that the magnetization of a ferromagnet increases with the electron-electron interaction and decreases with the bandwidth. An iron atom at the (100) surface has four missing nearest neighbors as compared with a bulk atom. As a consequence the projected density-of-states bandwidth in such a surface atom is much narrower. The surface atom has an enhanced spin polarization relative to the bulk. An iron atom at the (110) surface has only two missing nearest neighbors; its projected bandwidth is intermediate between the bulk and the (100) surface atoms, and so is its magnetization. The conclusion to be drawn is that elemental surfaces increase the magnetic moment of atoms as compared with the bulk: the more missing neighbors, the higher the local magnetization.

The validity of the preceding argument can be tested by examining the behavior of other transition elements, e.g., Ni, Co and Cr. Nickel has a bulk spin polarization [2.5] of 0.56 and we have calculated the surface spin polarization to be 0.74 for the (100) surface, and 0.65 for the (111) surface. These results agree with the nearest-neighbor argument because the fcc (100) surface atom has four of the twelve bulk neighbors missing, while there are only three missing neighbors in the (111) surface. However, it is clear that the magnetization changes and the concomitant effects in the electronic spectrum, such as the exchange splitting, are smaller in Ni than in Fe. On the other hand, we find [2.8] that the (100) surface of Cr has a spin polarization of 3.00, an increase by a factor of 4.4 from the antiferromagnetic bulk value. The magnetization of Cr is enhanced at the surface considerably more than that of Fe. We find [2.6] that for Co the influence of the surfaces is small, as in Ni.

A proper explanation of these disparate effects of the surfaces on these materials lies in the fact that two conditions are necessary for the existence of d electron magnetism: a sufficiently strong electron-electron interaction, as compared to the bandwidth, and an availability of holes (unoccupied d states). Bulk Ni and Co are near saturation: almost all available holes are in the minority spin bands. Iron, on the other hand, has almost one additional unmagnetized d hole. Chromium has four additional unmagnetized holes. Consequently an enhanced electron-electron interaction to bandwidth ratio, as provided by the surfaces, can influence only those magnetic materials which have not reached saturation, i.e., there is a considerable increase of the magnetization at the surfaces of Fe and Cr, but small effects in Ni and Co.

The preceding discussion is oversimplified: hybridization between the d and the sp electrons somewhat blurs the angular momentum character of the electronic states. This means that there is no true value of the spin polarization which could be called "saturation". It just becomes progressively harder to increase the magnetization beyond a given value. This value is essentially reached in the bulk at the end of the series, and may be achieved for the other elements in other environments. In these "saturated" situations the available empty states of the majority spin do not have a large density at the Fermi level, i.e., they are essentially of sp character. Despite these complexities, the basic idea is that the closer an element is to magnetic saturation, the less the effect that surfaces have on its magnetic properties.

The "healing length", i.e., the distance over which a given disturbance disappears, is also a strong function of the saturation of the spin polarization. This effect is clearly exemplified in our calculations of the magnetic properties of the (100) surface of bcc Cr. Here [2.8] the strongly enhanced local magnetization penetrates several layers into the bulk. The surface layer has a polarization of 3.00; the second layer has a polarization of 1.56; the third, 1.00. The whole structure is antiferromagnetic. The effect is a consequence of the extensive hybridization between states centered in neighboring atoms. In particular, the exchange splitting of a given atom is considerably enhanced by a larger exchange splitting of a nearest neighbor. This effect, striking in Cr, is observed to a lesser extent in Fe, but is negligible in Ni surfaces where saturation makes increases in polarization very difficult. (Third layers in Ni structures are almost completely "healed".)

A particularly interesting application of these concepts occurs in the ordered FeCo alloy and its surfaces. Ordered FeCo has a magnetization [2.7] of 4.85 per two-atom unit cell. This value is considerably higher than the sum of the two magnetizations of the constituent atoms: $1.72+2.22=3.94$. Our calculations, in agreement with neutron diffraction data, find that almost all the increased moment occurs in the Fe atom. This is because the strong electron-electron interaction of Co helps increase the exchange splitting of the upper Fe bands and, since Fe is an unsaturated magnet, this perturbation increases the Fe magnetization to a value of approximately 3.0. The effect of Fe on Co, on the other hand, is much weaker, because the polarization of the d holes in Co is essentially saturated. Consequently, the magnetization of Co decreases only slightly from what would be its bulk bcc value.

The saturation of both the Fe and Co spin polarizations in the ordered bulk FeCo alloy should make the magnetic moment relatively insensitive to the presence of surfaces. Our calculations show this to be the case: a Co atom at the (100) surface increases its spin polarization by only 0.25 to a value of 2.03, whereas an Fe atom increases its by 0.34, to a value of 2.95. Effects at the (110) surface are even smaller: the increases are 0.08 for Co and 0.09 for Fe. These are much smaller increases than found for pure iron – 0.78 for the (100) surface – and clearly support our general arguments.

2.3 Interfaces

Overlayers introduce complexities beyond that of the simple surfaces because of the effects of the film-substrate interface. To understand the magnetic properties of these overlayers we first calculated the electronic and magnetic properties of some interfaces [2.3,5]. In particular, we examined [2.5] the nickel-copper (100) and (111) interfaces. We found that the sp electrons of Cu hybridize considerably with the Ni d electrons. This effect reduces the interface-projected density of states near the Fermi level and makes it difficult for the interface Ni atoms to saturate. Consequently, the spin polarization of Ni at both the (100) and (111) interfaces is found to be 0.38, a considerable reduction from the 0.56 bulk value. We also found that if the Ni-substrate coupling is increased above its Ni-Cu value, as should be the case for simple metals like lead and aluminum, then the interface layer is unmagnetized for the (100) case also. These results point out that the effect of a nonmagnetic substrate such as Cu is to reduce the magnetic moment of the transition metal in direct contact with it to a value below that of the bulk.

2.4 Environmental Changes in the Magnetization: Qualitative Rules

We have thus far introduced four important observations which directly resulted from our calculations and experimental results. We may call these observations qualitative rules.

1. The removal of nearest neighbors of its own kind reduces the projected bandwidth of a magnetic transition metal atom and thus increases the electron-electron interaction to bandwidth ratio. This effect, most evident at surfaces, tends to enhance magnetism.

2. Magnetization enhancement is sizeable only in those elements where the bulk magnetization is not close to saturation, i.e., where holes in the d band exist which can still be polarized. Considerable enhancement is therefore expected for Cr and Fe; the effect is small for Co and Ni. It is also small for the surface enhancement of Fe in the FeCo alloy, where the alloying effect has already produced "saturation" of the Fe magnetization.

3. The presence of a strongly magnetized atom with a large exchange splitting near a weakly magnetic but polarizable atom with a smaller splitting considerably enhances the magnetization of the latter.

4. The presence of a nonmagnetic unpolarizable atom next to—and coupled to—a magnetic transition-element atom tends to decrease or fully destroy the magnetization of the latter.

2.5 Overlayers

The above rules are conceptually very important but unfortunately only qualitative. In systems like overlayers, where all four rules apply simultaneously and where they act in opposite directions, only a full self consistent calculation can yield the final result: no a priori prediction is possible.

Monoatomic overlayers of Ni on Cu(100) and Cu(111) surfaces provide a clear example of this point. As discussed previously, the effect of the Cu interface is to decrease the Ni magnetization. On the other hand, the effect of the free surface is to enhance it. Our calculations [2.5] show that the magnetization of the (111) Ni monolayer is nearly zero, whereas the (100) monolayer has essentially bulk magnetization. These results correspond well with our previous arguments that fcc (100) surfaces have higher magnetization than (111) surfaces.

Comparison with the interface results is, however, not so straightforward. As previously noted, the spin polarization of both interfaces is 0.38, a decrease by a factor of 1.47 from the bulk value. Clearly the exchange splitting of the bulk atoms helps to maintain a sizeable magnetization at the interface. On the other hand, the presence of a surface ("the other side" of a monolayer) also tends to enhance the magnetization. Which effect is more important—hybridization with the strongly magnetized bulk atoms or the enhancement caused by the free surface—is clearly a sensitive function of environmental variables such as surface orientation and chemical composition, and not susceptible to simple qualitative arguments.

The extent of this sensitivity is demonstrated by our calculations [2.6] for Co overlayers on the Cu(111) surface. Here the monolayer has a spin polarization of 1.63, greater than the values for the inner atoms of the dilayer: 1.58. In other words, the surface enhancement of the magnetization is more important in this case than the enhancement caused by the nearest-neighbor exchange splitting. It is probably the result of Co having more holes than Ni, and a monolayer moment close to the bulk value, a case similar to the monolayer of Ni on Cu(100).

A more predictable system is that consisting of an Fe monolayer on the (110) surface of the ordered FeCo alloy. Here the Fe magnetization is expected to be higher than its ordinary value at the (110) surface of bcc Fe: the substrate has a larger exchange splitting in the alloy than in pure Fe. Our calculation [2.7] finds the additional enhancement to be 0.08 and 0.12, depending on the Fe atom position, relative to the spin polarization of 2.55 found at the Fe(110) free surface.

Another result of considerable interest [2.5] is the fact that in the extreme strong coupling limit, when the magnetic transition metal hybridizes infinitely strongly to the conduction states of the substrate, both a monolayer and a dilayer of Ni(100) show no magnetization whatsoever — two "dead" magnetic layers — whereas a triatomic layer shows considerable spin polarization (0.61 at the surface, 0.45 for the intermediate atoms) even though the interface Ni atoms are magnetically dead.

So far the discussion has centered on the local values of the spin polarization. In fact many other interesting properties involving, inter alia, the electronic density of states, the total electronic energies and the spatial distribution of charge and spin polarization can be investigated at the same time. The system consisting of a monolayer of Co on the Cu(111) surface provides a particularly interesting example [2.6] because of the richness of its low-energy configurations. In particular, an antiferromagnetic two-atom unit surface cell is found to have a total electronic energy 2.04 eV per surface atom higher than the ferromagnetic ground state. Even more interesting is the existence of a spatially modulated two-atom surface cell state, with an energy only 0.41 eV per surface atom higher than the ferromagnetic ground state. The spatially modulated state, which should be easily accessible, consists of equal charge and magnetization in the two atoms of the unit cell, but different distribution of the magnetization among the various d orbitals. It possesses a surface projected density of states very similar to that of the ferromagnetic ground state.

The type of calculation discussed here is not necessarily restricted to smooth, uniform surfaces. By enlarging the surface unit cell it is possible, albeit computationally expensive, to include surface defects such as steps, terraces, kinks and partial overlayers, all structures of considerable importance in heterogeneous catalysis [2.15-20]. Our calculation [2.21] of the electronic properties of a paramagnetic partial layer (two-third coverage) of either Cu or Ni on paramagnetic Ni proves that the calculations are feasible and their results provide useful information on chemical as well as electronic properties of these interesting and practical systems.

Acknowledgments. One of us (R.H.V.) would like to acknowledge an A.T.&T. Bell Laboratories predoctoral fellowship, under whose sponsorship this work was done.

This work was supported at the Lawrence Berkeley Laboratory by the Director, Office of Energy Research, Office of Basic Energy Sciences, Materials Science Division of the Department of Energy under Contract Number DE-AC03-76SF0098.

References

2.1 C. Kittel: *Introduction to Solid State Physics*, 5th ed. (Wiley, New York 1976) p.465
2.2 H.L. Skriver: J. Phys. F**11**, 97 (1981)
2.3 J. Tersoff, L.M. Falicov: Phys. Rev. B**25**, 2959 (1982)
2.4 J. Tersoff, L.M. Falicov: Phys. Rev. B**26**, 459 (1982)
2.5 J. Tersoff, L.M. Falicov: Phys. Rev. B**26**, 6186 (1982)
2.6 R.H. Victora, L.M. Falicov: Phys. Rev. B**28**, 5232 (1983)
2.7 R.H. Victora, L.M. Falicov, S. Ishida: Phys. Rev. B**30**, 3896 (1984)
2.8 R.H. Victora, L.M. Falicov: To be published
2.9 U. Gradmann, G. Waller, R. Feder, E. Tamura: J. Magn. Magn. Mater. **31-34**, 883 (1983)
2.10 C. Rau, S. Eichner: Phys. Rev. Lett. **47**, 939 (1981)
2.11 D.S. Wang, A.J. Freeman, H. Krakauer: Phys. Rev. B**26**, 1340 (1982)
2.12 S. Ohnishi, A.J. Freeman, M. Weinert: Phys. Rev. B**28**, 6741 (1983)
2.13 R.H. Victora, L.M. Falicov: Phys. Rev. B**30**, 259 (1984)
2.14 J.M. Ziman: *Principles of the Theory of Solids*, 2nd ed. (Cambridge University Press, Cambridge 1972) p.339
2.15 G.A. Somorjai, R.W. Joyner, B. Lang: Proc. R. Soc. London, Ser. A**331**, 335 (1972)
2.16 B. Lang, R.W. Joyner, G.A. Somorjai: J. Catal. **27**, 405 (1972)
2.17 W. Erley, H. Wagner: Surf. Sci. **74**, 333 (1978)
2.18 M.A. Chesters, G.A. Somorjai: Surf. Sci. **52**, 21 (1975)
2.19 M. Boudart: Adv. Catal. **20**, 153 (1969)
2.20 J.W.A. Sachtler, M.A. Van Hove, J.P. Bibérian, G.A. Somorjai: Phys. Rev. Lett. **45**, 1601 (1980)
2.21 J. Tersoff, L.M. Falicov: Phys. Rev. B**24**, 754 (1981)

3. The Binding of Adsorbates to Metal Surfaces

S. Holloway

The Donnan Laboratories, University of Liverpool
Liverpool L69 3X, Great Britain

J.K. Nørskov

NORDITA, Blegdamsvej 17, DK-2100 Copenhagen, Denmark

The total energy of an adsorbate as a function of position outside a metal surface determines, within the adiabatic approximation, the equilibrium position, the chemisorption energy, the vibrational spectrum and the activation energies for adsorption, diffusion and further reactions on the surface. Results obtained using the effective medium approach to calculate the total interaction energy are reviewed. Due to the simplicity of the approach the full potential energy surface has been calculated for a number of adsorption systems, and it is possible to relate the properties of the interaction potential to the parameters describing the atom and surface in question. Specific topics to be discussed are hydrogen and oxygen chemisorption on transition metals, quantum diffusion of chemisorbed hydrogen, and the oxygen incorporation and initial oxidation of metals.

3.1 Introduction

It is seductive to contemplate that once the ground-state adiabatic potential energy hypersurface for a particular adsorbate-metal surface combination has been established, the gates to understanding both the static and dynamic properties of adsorption will open [3.1,2]. The calculation, however, of the total energy of the combined system as a function of the set of nuclear positions is indeed an awesome task. Even with very efficient methods, such as the local density approximation to treat electron-electron correlations [3.3], the low symmetry combination of a surface with an adsorbate present makes a direct calculation extremely demanding. Consequently, relatively few examples exist in the literature. One scheme has been to extend the well-tried methods of quantum chemistry and represent the surface as a relatively small cluster of atoms [3.4]. At the other end of the scale, conventional band-structure methods have been modified to include the presence of a surface by considering thin slabs of substrate atoms upon which arrays of the adsorbate are chemisorbed [3.5,6]. Lying between these two extremes is perhaps the most commonly used model for adsorption systems, employing the jellium model to describe the metal [3.7]. Here the metal ion cores have been smeared out to form a homogeneous positive background which terminates abruptly at a plane. Although this is a rather idealized model, much of our basic understanding regarding adsorption on metallic systems has been derived from extensive calculations on this substrate.

An alternative approach to the problem is to make judicious approximations in the total energy calculation, so enabling low-symmetry problems to be treated. This short review contains a statement of the strengths and weaknesses of one of these techniques, the effective-medium theory [3.8-11]. Whilst the theoretical framework of the method is presented and discussed, an attempt is made to provide a conceptual picture of the interactions involved in chemisorption and indicate the important substrate and adsorbate parameters which determine the topology of the potential energy hypersurface.

Section 3.2 contains a description of the effective-medium theory and how it stands in relation to ab initio calculations. Using potential energy hypersurfaces for hydrogen and oxygen adsorption on transition metals as prototypes, bond energies, bond lengths, vibrational properties and adsorption are discussed in Sects.3.3-5. Finally in Sect.3.6 conclusions are presented.

3.2 Effective Medium Approximation

The effective medium approach is based upon the premise that changes occurring when embedding a localized impurity into relatively extended hosts are restricted to occur within a small radius of the impurity. For the united system, space is then partitioned into two regions, "a" close to the impurity where the total potential is dominated by the contribution of the impurity potential, and the remainder, where the host potential wins. Thus the technique inherently estimates DIFFERENCES between various hosts and not ABSOLUTES. Until now, it has been used extensively to estimate changes in binding energies of atoms and molecules both outside and inside metal surfaces. It has been shown that the above scheme, when evaluated to first order in potential in both regions, yields for the change in total energy

$$\delta E = \int_a \delta\phi_0^{-a}(\mathbf{r})\Delta n(\mathbf{r})d\mathbf{r} + \delta\left\{\int_{-\infty}^{\varepsilon_F} \varepsilon\Delta n(\varepsilon)d\varepsilon\right\} \quad . \tag{3.1}$$

Here the first, purely electrostatic term involves the difference $\delta\phi_0^{-a}(\mathbf{r})$ in electrostatic potential between the new and old hosts and the atom-induced charge density $\Delta n(\mathbf{r})$ in the old host. The integral is over the atomic region "a" only and the superscript -a on $\delta\phi_0$ signifies that only the charge outside "a" must be included. In the second term $\Delta n(\varepsilon)$ is the atom-induced density of states and the term involves the difference in the sum of atom-induced one-electron energies evaluated for a rigidly shifted potential inside region "a". Although the result (3.1) is derived within the local density approximation, a similar result can be obtained for a nonlocal exchange-correlation energy. Both the derivation and the result closely resemble the force calculations of *Anderson* et al. [3.12].

Although it is possible to use (3.1) in a number of modes, in this short review attention is focused exclusively on its application to atomic species interacting with metal surfaces. For this problem the essence of the scheme is in finding a suitably tractable (high symmetry) host which can be used to approximate the low-symmetry problem of interest. To this end it has become practice to use the homogeneous electron gas as an effective medium. By performing a weighted average of the true surface electron density $\bar{n}_0(r)$ over the atom-induced electrostatic potential, it has been shown that the atom-surface interaction energy ΔE for an atom located at position $\mathbf{R}$ is given by [3.11]

$$\begin{aligned}\Delta E(\mathbf{R}) &= E^{hom}(\bar{n}_0(\mathbf{R})) + \int_a \delta\phi_0^{-a}(\mathbf{r})\Delta\bar{n}(\mathbf{r})d\mathbf{r} + \delta\left\{\int_{\infty}^{\varepsilon_F} \varepsilon\Delta n(\varepsilon)d\varepsilon\right\} \\ &= \Delta E_{eff}^{hom}(\bar{n}_0(\mathbf{R})) + \Delta E_{cov}(\mathbf{R}) \quad ,\end{aligned} \tag{3.2}$$

where $\Delta E_{eff}^{hom}(\bar{n}_0(\mathbf{R}))$ is given by the energy $\Delta E^{hom}(\bar{n}_0)$ of an atom in a homogeneous electron gas. This term also includes the electrostatic term from

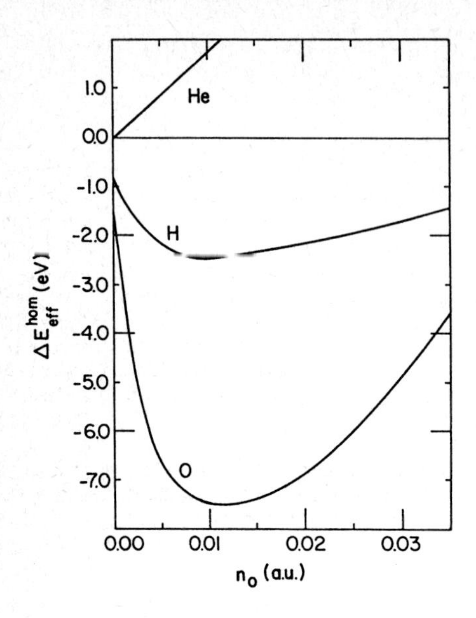

Fig.3.1. Interaction energy $\Delta E_{eff}^{hom}(n_0)$ of He, H and O with a homogeneous electron gas as a function of electron gas density n_0 [3.13]

(3.1). The second term ΔE_{cov}, which may be thought of as a covalent interaction, is the one-electron term in (3.1).

As previously mentioned, the tractability of the problem of an atom embedded in a homogeneous electron gas makes it an attractive choice for an "effective medium". There have been a number of independent calculations of $\Delta E_{eff}^{hom}(\bar{n}_0)$ for a variety of atoms, illustrative examples of which are shown in Fig.3.1. It should be stressed that part of the beauty of the scheme is that once the functional $\Delta E_{eff}^{hom}(\bar{n}_0)$ has been obtained, there is no need to repeat the calculation when changing systems, a fact which results in a considerable reduction of effort and computing time.

When considering classes of adsorbates it is useful to divide the atomic species into two generic classes, the rare gases and the rest. Rare gas atoms, as epitomized by He in Fig.3.1, have a repulsive interaction for all densities. This is nothing more than the cost in kinetic energy of orthogonalizing an increasing number of electron gas states to the deep-lying atomic states. The linearity of the functional is the basis for the assumption often employed when analyzing He-scattering experiments, that the interaction energy is proportional to the surface electron density [3.14]. The second class of adsorbates considered here is the reactive gases. Figure 3.1 shows the functional forms for hydrogen and oxygen. At zero electron density, the limiting value of ΔE_{eff}^{hom} corresponds to the negative of the atomic affinity. For very low densities the bandwidth and work function of an electron gas goes to zero. Consequently, for species with a positive electron affinity, it will always be favorable to transfer an electron on to this level and gain energy, thus giving the incorrect limit infinitely far away from a surface. From a practical point of view, the asymptotic behavior far into the vacuum is of little concern since this portion of the potential energy hypersurface is not important in determining interesting dynamical effects. As the density is increased, ΔE_{eff}^{hom} passes through a minimum which reflects the propensity for such gases to form chemical bonds. The lowering of the total energy can be envisaged within a solid-state description to be due to a screening of the electron-electron repulsion in the negative ion. At even higher densities, the kinetic energy repulsion, as described previously for He, again dominates the total interaction energy.

Up until this point it has tacitly been assumed that the electron charge density exterior to a particular metal surface of interest is readily available: with only a few exceptions this unfortunately is not generally the case. Although first-principle calculations of clean metal surfaces are appearing more frequently in the literature, the complexity of performing, for example, a LAPW calculation, has resulted in a very limited data base. As a viable alternative, the use of a superposition of atomic densities has been suggested [3.15]. At least in one case it has been shown that at the level of the approximation scheme used in deriving (3.2), the subtle differences in using the self-consistent charge density over the superposition are of little or no consequence [3.16].

So ΔE_{cov} reflects the difference in the one-electron spectrum of going from the homogeneous electron gas to the metal surface. If it is a free-electron-like metal, these changes are expected to be small. To first order in $(n_0(\mathbf{r})-\bar{n}_0)$, then [3.11]

$$\Delta E^{(1)}_{cov}(\mathbf{R}) = \int_a (n_0(\mathbf{r}) - \bar{n}_0)\Delta V(\mathbf{r} - \mathbf{R})d\mathbf{r} \quad , \tag{3.3}$$

where ΔV is the atom-induced effective potential in the homogeneous electron gas. For transition metals, the interaction with the d bands must be treated to higher order in the adsorbate-metal d-interaction V_{ad}. The starting point is, however, not the free atom, but a renormalized atom embedded in an electron gas. As discussed above, such atoms can be regarded as screened negative ions and they have filled valence states well below the Fermi energy. The renormalized atom thus has a "closed shell" well below the metal d bands and thus the interaction is fairly weak and may be calculated using the resonant level model.

Hydrogen chemisorption upon Ni(100) represents one of the "benchmark" systems, and to place the effective medium approach in perspective it is illustrative to compare its performance with existing calculations. There exist two ab initio treatments of this problem, each representing different ends of the spectrum. *Upton* and *Goddard* [3.4] used a 28 atom nickel cluster to model the surface and employed a generalized valence bond description to treat the hydrogen-nickel interaction. *Umrigar* and *Wilkins* [3.6] treated the interaction of a complete $p(1\times1)$ monolayer of hydrogen on a 5 layer thick nickel slab using a local density approximation for exchange and correlation in an LAPW framework. Table 3.1 compares the results from these calculations with both the effective medium result and selected experimental results. The general agreement is seen to be very good. Results indicate an absolute accuracy of the effective medium theory to be approximately 0.5 eV with a greater confidence in energy differences.

3.3 Atomic Chemisorption

Beginning with (3.2), the atomic superposition approximation and the energy functional for $\Delta E^{hom}_{eff}(\bar{n}_0)$, it is possible to make a number of useful comments regarding the interaction of an atom with a metal surface. Incorporating the effects of interactions with metal d states via ΔE_{cov} enables a whole new

Table 3.1. Comparisons between results of different calculations for hydrogen chemisorbed in the fourfold coordinated center site outside a Ni(100) surface; the chemisorption energy ΔE, perpendicular vibrational frequency, $\omega_\perp$, and the Ni-H bond length d_{Ni-H} are included; also, the experimental values are shown when found

	H/Ni(100) center		
	ΔE [eV]	$\omega_\perp$ [mev]	d_{Ni-H} [Å]
Upton and Goddard [3.4][a]	-3.0	73	1.78
Umrigar and Wilkins [3.6]	-3.4	90	1.80
Present work	-2.7	76	1.91
Experimental	-2.7	74	-

[a]Upton and Goddard treat a single H atom on a Ni cluster, whereas Umrigar and Wilkins consider a p(1 × 1) H layer on a 5 layer Ni slab. The present work evaluates the interaction of an H atom on a semi-infinite Ni substrate. Thus although the systems treated are rather different, the results show a remarkable similarity considering the differences between the approaches. The values for Umrigar and Wilkins are only preliminary.

range of physical properties to be discussed quantitatively. This section is divided into subsections which cover particular properties of adsorption systems as would, for example, be sampled by different surface sensitive probes.

3.3.1 The Chemisorption Energy

Figure 3.2 compares the calculated chemisorption energies for hydrogen and oxygen for each of the 3d elements with experimental values. For hydrogen,

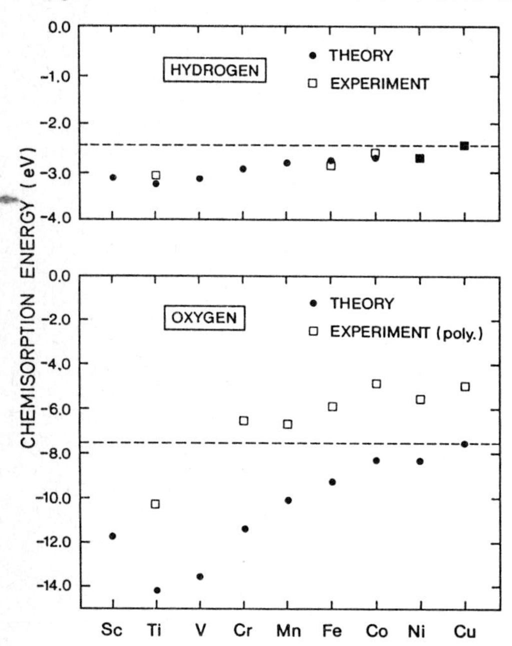

Fig.3.2. Comparison between calculated and experimental chemisorption energies for hydrogen and oxygen on the most close-packed surface of each of the 3d transition metals. The horizontal dashed line indicates the contribution to the chemisorption energy from $\Delta E^{hom}_{eff}(\bar{n}_0)$ in (3.2). This contribution is the same for all of the metals, since the optimum surface electron density where $\Delta E^{hom}_{eff}(\bar{n}_0)$ has its minimum value can always be found outside a surface. The trends along the series are given by ΔE_{cov}, (3.2). In the calculated values, no correction for the zero-point energy has been included [3.17]

experimental results correspond to single crystal studies whereas for oxygen the results are for polycrystalline samples. Outside a surface, the optimum value of the density, where $\Delta E^{hom}_{eff}(\bar{n}_0)$ has its minimum, can always be found, and thus irrespective of the surface this term contributes the same amount to the chemisorption energy, indicated in Fig.3.2 by the dashed line. The trends, therefore, must arise from differences in the interaction with the metal d electrons via ΔE_{cov} in the notation of this section. The increasing binding strength towards the left in the series can simply be related to a decrease in the occupancy of the antibonding adsorbate-metal d levels as the d band occupancy decreases. To second order in the interaction V_{ad}, the interaction is simply $-2(1-f)V^2_{ad}/(C_d-\varepsilon_a)$, where C_d is the center of the d bands, ε_a the renormalized adsorbate atomic level, and f the degree of filling of the d band [3.11,17]. The (1-f) behavior is very pronounced in Fig.3.2. While the absolute magnitude appears more than satisfactory for hydrogen, the results for oxygen appear depressed in energy by a few eV with respect to the experiment. This is probably due to shortcomings in the description of localized bonding using the local density approximation, a subject which has recently been discussed in some detail by *Jones* [3.18]. Generically it might be worth noting that for any other electronegative adsorbate which chemisorbs in an effective closed-shell configuration, the arguments presented above concerning the binding energy will apply equally well.

3.3.2 Bond Lengths

Whereas ΔE_{cov} determines the trends in binding energies along the transition metal series, it varies very weakly with distance outside a surface, Fig. 3.3. The bond length is thus determined mainly by the requirement that the surface electron density is equal to the optimum value. This, for instance, means that the bond length increases rapidly with coordination number, as illustrated in Table 3.2, where bond lengths and vibrational frequencies for hydrogen and oxygen on Ni(100) and (111) are shown.

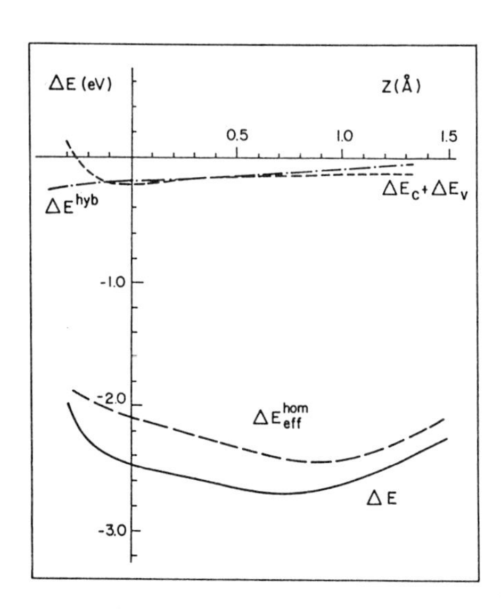

Fig.3.3. Binding energy of a hydrogen atom as a function of distance outside the fourfold coordinate center site on a Ni(100) surface. Both the total binding energy ΔE_{tot} and the various components in (3.2) are shown. ΔE_{cov} is split up into first ($\Delta E_c + E_v$, see also [3.18] and higher order (ΔE^{hyb}) terms [3.17]

Table 3.2. Comparison of theoretical and experimental bond lengths and perpendicular frequencies (harmonic approx.) for H and O chemisorbed on Ni(100) and (111)

	Coordination number	Bond length [Å]		Vibrational frequency [meV]	
	Theory	Theory	Exp.	Theory	Exp
H/Ni(100)	4	1.91		76	74
H/Ni(111)	3	1.81	1.84	131	139
O/Ni(100)[a]	4	1.83	1.98	30	(36)
O/Ni(111)	3	1.76	1.88	62	72

[a]For O/Ni(100), the experimental frequency shown is the "static lattice" value [3.20,21]. For experimental references, see [3.17,20].

3.4 Vibrational Excitations

Based upon the traditional interpretation of IR vibrational studies, the use of harmonic oscillator models to describe surface vibrations is now accepted in the surface vibrational literature [3.22]. While for studies involving only the observation of a fundamental excitation the use of either a harmonic or anharmonic pair interaction potential is rather arbitrary, when overtone spectroscopy becomes possible a better description of the gas-surface potential-energy hypersurface will become necessary [3.23]. Examining the form of the functional shown for hydrogen, for example, in Fig.3.1 it is clear that even for an atomic superposition of charge densities, a pair-potential form for $\Delta E(\mathbf{R})$ would be singularly inappropriate. Of course, in analyses of vibrational spectra, pair potentials have been very useful, particularly in the diagnostic mode, to determine whether low or high coordination sites are the stable bonding positions. To this end it is therefore desirable to formulate a similar "fingerprint" scheme from the more general potential-energy surfaces discussed above.

From Fig.3.4 it is clear that the pure perpendicular motion is determined by $\Delta E_{eff}^{hom}(\bar{n}_0)$. Within the harmonic approximation it is straightforward to show that the perpendicular vibrational frequency is given approximately by [3.17]

$$\omega \simeq \frac{1}{M}\sqrt{\frac{d^2\Delta E_{eff}^{hom}(n_0)}{d\bar{n}_0^2}}\;\frac{d\bar{n}_0}{dz} \quad , \tag{3.4}$$

where M is the atomic (reduced) mass and z denotes the direction perpendicular to the surface. Thus, when a true separation of perpendicular from parallel motion is possible (i.e., a separable harmonic potential), ω is determined principally by the normal derivative of the surface electron density. Further, assuming that $\bar{n}_0(r)$ can be described by a superposition of exponentially decaying atomic densities

$$\bar{n}_0(\mathbf{r}) = \sum_{\mathbf{R}} \bar{n}_0^a \exp(-\beta|\mathbf{r} - \mathbf{R}|) \quad , \tag{3.5}$$

and considering only high-symmetry sites, then

$$\omega \simeq \omega_0\beta \cos\theta \quad , \tag{3.6}$$

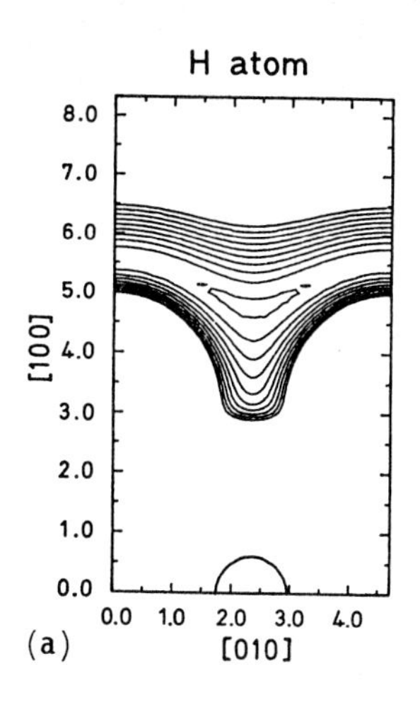

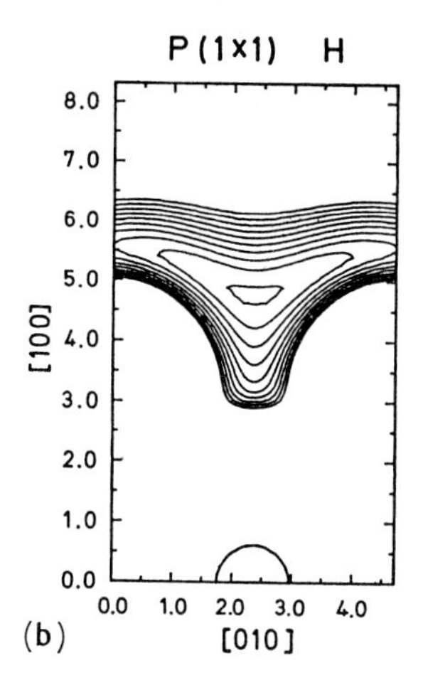

Fig.3.4a,b. Contours of constant potential energy for hydrogen outside a Ni(100) surface. The cut shown is perpendicular to the surface through the bridge and center sites. (**a**) For a single H atom; (**b**) for a p(1 × 1) monolayer. Contours are shown in steps of 0.05 eV, the lowest value being -2.7 eV

where θ is the bond angle with the surface normal and

$$\omega_0 = \frac{1}{\sqrt{M}} \sqrt{\frac{d^2 \Delta E_{eff}^{hom}(\bar{n}_0)}{d\bar{n}_0^2}}\, \bar{n}_0 \Bigg|_{\bar{n}_0 = \text{optimum}}$$

is a constant depending on the atom only. Since the density decay constant β changes only slowly from surface to surface, ω_0 is given basically by the bond angle and thereby by the coordination number and the substrate geometry.

Since $\Delta E_{eff}^{hom}(\bar{n}_0)$ simply modulates the spatial location of the chemisorption minimum, any lateral energetic changes must arise from ΔE_{cov}. For hydrogen this effect tends to be small and the resulting potential energy surface is characterized by low diffusion barriers and soft parallel vibrational modes. Figure 3.4 shows the ground-state potential energy surface for H/Ni(100). On the length scale of typical zero-point motion there is a great deal of anharmonicity in the potential which will manifest itself in a very significant coupling between perpendicular and parallel motion. The first 3 vibrational wave functions for H/Ni(100) are shown in Fig.3.5 [3.24]. The wave functions virtually fill out the potential energy contour corresponding to their eigenvalue, and this may be taken as a sign that the motion is chaotic in nature. Since the diffusion barriers are low, the higher energy states may tunnel onto neighboring sites, giving rise to the formation of energy bands. While the bandwidth is quite substantial for the more energetic states, this could be of less significance for higher hydrogen coverages [3.25]. To make contact with previous harmonic analyses it is instructive to break down the results of Fig.3.5, and discuss them in terms of the conventional wisdom of localized oscillators. Figure 3.6 shows a possible scheme corresponding to the $\bar{\Gamma}$ point of the 2-D SBZ. Harmonic oscillator eigenfunctions have been grouped corresponding to the classification of perpendicular and parallel modes of vibration. Whereas perpendicular modes all have A_1 symmetry, parallel modes have either E or A_1 symmetry depending upon whether their occupation number is even or odd. This in turn gives rise to the possibility of hybridization between pure perpendicular modes and selected pure parallel overtones, their interaction depending upon their energetic proximity and the strength of their coupling, which is simply a matrix element of the true

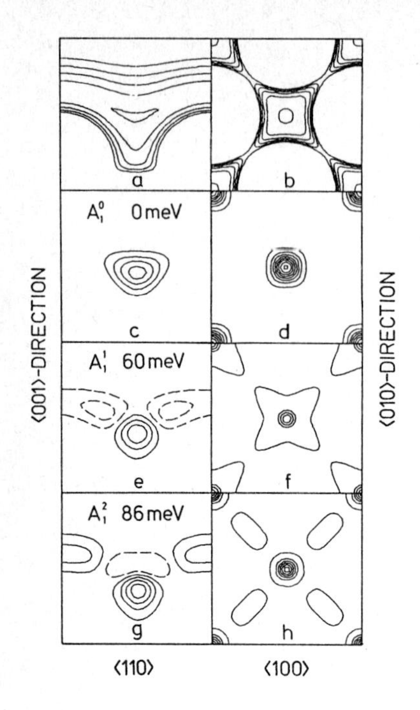

Fig.3.5a-h. Contours of constant potential energy for a single hydrogen atom outside a Ni(100) surface. Two cuts are shown: (**a**) perpendicular to the surface along the diagonal in (b) and (**b**) parallel to the surface through the absolute minimum. The contours are shown in steps of 0.05 eV the lowest value being -2.7 eV. Underneath (**c-h**) the corresponding hydrogen wave functions (*left*) and densities (*right*) are shown for the ground state and two excited states at ($k_{11}=0$)

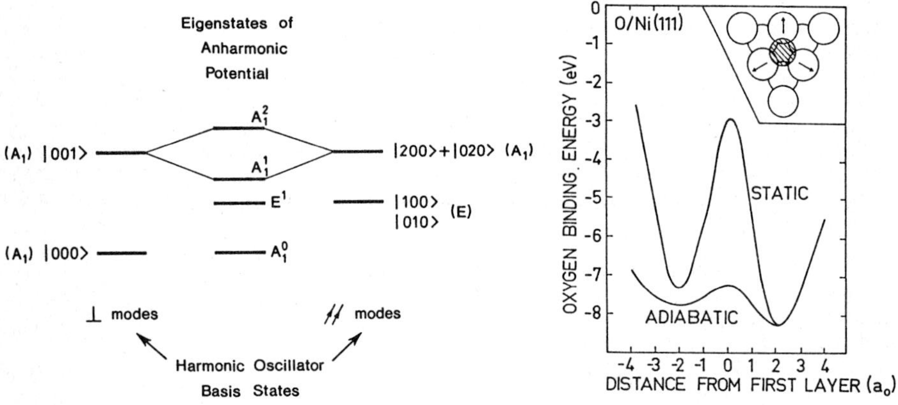

Fig.3.6

Fig.3.7

Fig.3.6. Schematic view of the eigenstates of the full anharmonic potential for a H atom outside a Ni (100) surface in terms of interacting harmonic solutions

Fig.3.7. Energy of oxygen outside a Ni(111) surface as a function of distance from the first layer. Both the static lattice potential and one where the Ni atoms have been allowed to relax to give the minimum energy ("adiabatic") are shown. The Ni-Ni interactions are described by a pair potential [3.20]

anharmonic potential. The central panel in Fig.3.6 shows the resulting eigenvalue scheme after allowing modes to interact, which should be compared to the $\bar{\Gamma}$ point spectrum shown in Fig.3.5. Such interactions are in fact well known in the study of vibrational motion and correspond to Fermi resonances. In addition to shifts in frequency, Fermi resonances are accompanied by "intensity borrowing" and it would be of great interest to see if this persists for the adsorption case.

Figure 3.4b shows the effect of increasing the hydrogen coverage on the potential energy surface. When each center site is occupied by a hydrogen atom ($p(1\times 1)$) the diffusion barrier begins to increase and thereby localizes the hydrogen to a greater degree. Nevertheless, it is still not possible to contemplate a harmonic description of the motion. A detailed analysis of the coverage dependence of vibrational properties will be presented elsewhere [3.26]. It should be said that concepts such as anharmonicity and delocalized eigenstates, while very pertinent for light atoms, are not expected to be of such importance for heavier adsorbates where separable potentials are more appropriate.

3.5 Absorption

The uptake of an adsorbed species into the bulk of a metal is again largely dominated by $\Delta E_{eff}^{hom}(\bar{n}_0)$ or, equivalently, by the kinetic energy repulsion. In Fig.3.7 an example of the potential energy variation for an oxygen atom moving into an Ni(111) surface is shown. The sizeable activation barrier is a consequence of the large electron density arising from the rigid first metal layer. If, however, the metal atoms are allowed to relax from the oxygen atom, the electron density decreases and the barrier follows. Since the interstitial electron density inside a transition metal is nominally larger than the optimum value in Fig.3.1, the energy of the oxygen (or hydrogen) atom inside the metal is therefore linearly dependent on the interstitial electron density [3.11]. As seen in Fig.3.7, the energy can be reduced considerable by relaxing the nearest-neighbor metal atoms outwards from the oxygen (hydrogen) atom. The cost of relaxing a surface atom is smaller than that for a bulk atom and therefore a subsurface site is typically more stable than one deeper into the bulk.

It should be mentioned that the outward relaxations of the metal lattice induced by the oxygen atoms inside the metal mediates to a very strong attractive oxygen-oxygen interaction [3.20]. Oxygen atoms close to each other can share the cost of relaxing the nearest-neighbor metal atoms. This is consistent with the strong tendency for island formation observed in oxidation reactions [3.27].

3.6 Conclusions

The effective medium theory, which is a method for comparing total interaction energies of an atom or molecule with different solids, has been described. For atomic chemisorption, detailed potential energy surfaces can be calculated very easily starting from the atom in a homogeneous electron gas. Furthermore, the properties of the interaction potential can be related to the properties of the surface and adsorbate in question.

References

3.1 J.R. Schrieffer: J. Vac. Sci. Technol. **13**, 335 (1976)
3.2 J.K. Nørskov: J. Vac. Sci. Technol. **18**, 420 (1981)
3.3 For a recent review, see *Theory of the Inhomogeneous Electron Gas*, ed. by S. Lundqvist, N.H. March (Plenum, New York 1983)
3.4 T.H. Upton, W.A. Goddard: Phys. Rev. Lett. **42**, 472 (1979)
3.5 P.J. Feibelman, D.R. Hamann: Solid State Commun. **34**, 215 (1980)
3.6 C. Umrigar, J.W. Wilkins: To be published
3.7 N.D. Lang, A.R. Williams: Phys. Rev. B**18**, 616 (1978)
3.8 B.I. Lundqvist, O. Gunnarsson, H. Hjelmberg, J.K. Nørskov: Surf. Sci. **89**, 196 (1979)
3.9 J.K. Nørskov, A. Houmøller, P. Johansson, B.I. Lundqvist: Phys. Rev. Lett. **46**, 257 (1981)
3.10 J.K. Nørskov, N.D. Lang: Phys. Rev. B**21**, 2136 (1980); M.J. Stott, E. Zaremba: Phys. Rev. B**22**, 1564 (1980)
3.11 J.K. Nørskov: Phys. Rev. B**26**, 2875 (1982)
3.12 O.K. Andersen, H.L. Skriver, H. Nohl, B. Johansson: Pure Appl. Chem. **52**, 93 (1979); O.K. Andersen: In *The Electronic Structure of Complex Systems*, NATO Advanced Study Institute, ed. by W. Temmerman, P. Phariseau (Plenum, New York 1982)
3.13 The most comprehensive calculations are those of M.J. Puska, R.M. Nieminen, M. Manninen: Phys. Rev. B**24**, 3037 (1980)
3.14 In a scattering calculation it would be necessary to match the decaying potential to a potential which ensures that the correct physical limit is obtained at large distances
3.15 C. Umrigar, M. Manninen, J.K. Nørskov: To be published
3.16 J.K. Nørskov: To be published
3.17 P. Nordlander, S. Holloway, J.K. Nørskov: Surf. Sci. **136**, 59 (1984)
3.18 R.O. Jones: In *Proc. Santa Barbara Workshop on Many Body Effects at Surfaces* (Academic, New York 1984)
3.19 Close to the metal atoms there is a third contribution to the interaction energy, not included in (3.2), describing the interaction with the metal cores [3.17]. This term can in some cases contribute to energy differences along the surface, in particular on top of a surface atom. In most other cases it is negligible [3.17,20]
3.20 B. Chakraborty, S. Holloway, J.K. Nørskov: Proc. ECOSS 6 York 1984, Surf. Sci. (in press)
3.21 S. Andersson, P.-A. Karlsson, M. Persson: Phys. Rev. Lett. **51**, 1278 (1983); and private communcation
3.22 H. Ibach, D.L. Mills: *Electron Loss Spectroscopy and Surface Vibrations* (Academic, New York 1982)
3.23 E. Heller, M.J. Davis: J. Phys. Chem. **86**, 2118 (1982)
3.24 M.J. Puska, R.M. Nieminen, M. Manninen, B. Chakraborty, S. Holloway, J.K. Nørskov: Phys. Rev. Lett. **51**, 1081 (1983)
3.25 The present description neglects self-trapping arising from lattice distortions, which is valid if (as assumed here) the presence of the hydrogen does not significantly alter the metal-metal interactions. If such changes should prove important, the surface diffusion would be significantly hampered. The motion would then be better described in terms of small-polaron hopping [3.20]; C.P. Flynn, A.M. Stoneham: Phys. Rev. B**10**, 3966 (1970); D. Emin, M.I. Baskes, W.D. Wilson: Phys. Rev. Lett. **42**, 791 (1979); B. Chakraborty, S. Holloway, J.K. Nørskov: To be published
3.26 J.K. Nørskov, S. Holloway, B. Chakraborty: To be published
3.27 S. Andersson, P.-A. Karlsson, M. Persson: Phys. Rev. Lett. **51**, 1278 (1983); and private communication

I.2 Specific Applications

4. Energy Minimization Calculations for Diamond (111) Surface Reconstructions

David Vanderbilt and Steven G. Louie

Department of Physics, University of California and Lawrence Berkeley Laboratory, Berkeley, CA 94720, USA

Total energies are calculated for a variety of structural models for the reconstructed 2×1 diamond (111) surface. An ab initio LCAO approach to local density theory, which incorporates a self-consistent treatment of interatomic charge transfer, is used. Among the structural models considered are the Haneman buckled, the Pandey π-bonded chain, the Chadi π-bonded molecule, and the Seiwatz single-chain models. The results strongly favor the π-bonded chain model; the others are shown to be implausible. When fully relaxed, the π-bonded chain model has an energy of ~0.3 eV/surface-atom *lower* than that of the relaxed 1×1 surface. *No* dimerization of the surface chain is found to occur.

4.1 Introduction

A remarkable variety of surface reconstructions occur on the (111) surfaces of the tetrahedral elements C, Si and Ge [4.1]. A possible common denominator may be the occurrence of a similar 2×1 reconstruction on all three elemental surfaces. While clear 2×1 LEED patterns are observed for Si and Ge(111) surfaces, LEED cannot distinguish between a true 2×2 or disordered domains of 2×1 for the diamond (111) surface [4.2]. However, the similarity of the angle-resolved photoemission (ARUPS) results for C [4.3], Si [4.4], and Ge [4.5] suggests that a common 2×1 structure may be responsible. The 2×1 structure disappears upon annealing for Si and Ge but appears upon annealing for C, indicating that it may be thermodynamically stable only for C. Thus the study of the diamond $2\times2/2\times1$ surface is of particular interest.

The π-bonded chain model proposed by *Pandey* [4.6] has attracted much attention as a possible candidate for the 2×1 structure. Energy-minimization calculations identify the Pandey chain structure as the lowest in energy of those tested for Si [4.7] and Ge [4.8], and the calculated dispersion of the occupied surface bands is in good agreement with the ARUPS data [4.4,5]. Ion backscattering [4.9] and optical [4.10] measurements appear to support this identification. However, contrary indications from LEED [4.11] and, most recently, photoemission [4.12] experiments have insured a continued controversy over this assignment.

Less experimental work has been done on the diamond $2\times2/2\times1$ surface [4.3,13-15]. A comparison of ARUPS results [4.3] with the calculated surface state dispersion provides indirect evidence for the Pandey π-bonded chain model [4.6,16], possibly with some dimerization along the chain [4.6]. However, discrepancies in the location and dispersion of the surface state persist, and the model remains controversal. The *Haneman* buckled model [4.17], the *Seiwatz* single chain model [4.18], and the *Chadi* π-bonded molecule model [4.19] are possible alternatives. No energy minimization calculations have previously been done for diamond.

Here, we report direct energy minimization calculations for these models. A first-principles linear combination of atomic orbitals approach has been used to calculate total energies in the pseudopotential [4.20] and local density (LDA) [4.21] approximations. The method is a generalization of the approach of *Chelikowsky* and *Louie* [4.22] to cases for which interatomic charge transfer must be treated self-consistently [4.23]. The calculations have been carried out using a slab geometry with 12 (10) atomic layers for 1×1 (2×1) cases, respectively.

4.2 Results

The energies of the various models are compared in Table 4.1. We begin with the *Haneman* buckling model [4.17]. Buckling of the ideal 1×1 surface is found to raise the energy. A similar result has been found for Si and Ge [4.7,8]; if anything, such a buckling distortion should be even less likely in diamond because of the Coulomb repulsion associated with charge transfer into the highly localized carbon dangling bond.

Table 4.1. Calculated total energies of diamond (111) 1×1 and 2×1 surface reconstruction models

Surface model	Energy [eV/surface-atom]
Ideal 1×1	0.00
Buckled[a] ($\Delta z = \pm0.26$ Å)	0.35
Chadi π-bonded molecule[b]	0.28
Seiwatz single chain[c]	1.30
Ideal Pandey π-bonded chain[d]	-0.05
Relaxed 1×1	-0.37
Relaxed Pandey π-bonded chain	-0.68
ΔE_{tot}, ±2% dimerization	+0.01
ΔE_{tot}, ±4% dimerization	+0.04
ΔE_{tot}, ±6% dimerization	+0.09

[a][4.17]; [b][4.19]; [c][4.18]; [d][4.6].

We find the ideal *Pandey* model [4.6], defined as having all bulk bond lengths (except for graphite-length bonds along the surface chains), to have an energy slightly lower than that of the ideal 1×1 surface. As we shall see, it is also lower than that of any of the other unrelaxed models.

The *Chadi* π-bonded molecule model [4.19] was tested and found to be higher in energy than the unrelaxed Pandey chain model, even though the tested geometry included relaxations as determined by Chadi using a tight-binding energy-minimization approach. Furthermore, the calculated surface state dispersion [4.16] is in very poor agreement with the ARUPS data [4.3]. Thus, we feel the model can be ruled out, and we have not tried to relax it further.

Finally, the unrelaxed Seiwatz chain model is found to have a total energy 1.30 eV above the ideal 1×1. We also tested a "relaxed" geometry provided by *Chadi* [4.24], again based upon tight-binding energy minimization; the energy was reduced to only 0.97 eV. Moreover, the position of the calculated surface bands [4.16] is in error by more than 2 eV. The model therefore appears untenable.

Thus the Pandey chain model is the only promising 2×1 model of those we considered. We have calculated the relaxations in some detail for this model. This was done by adjusting, one by one, the four surface-most bond lengths d_1 to d_4 in Fig.4.1, minimizing the energy for each one while the others were held constant. The procedure was repeated a second time to allow each bond to relax in a more fully relaxed environment. Next, we minimized with respect to buckling of the surface chain ($\phi \neq 0$ in Fig.4.1). Finally, the geometry was relaxed further using a Keating force-constant model [4.25] to direct the relaxation of the subsurface atoms in the middle of the slab, and the LDA total energy was calculated again. The final total energy for this fully relaxed geometry was found to be -0.68 eV/surface-atom. By comparison, energy minimization for the 1×1 case was also carried out, and led to a relaxed geometry with an energy of -0.37 eV/surface-atom. Thus, the π-bonded chain model appears to be a very promising candidate for the stable surface.

The relaxed chain geometry is shown in Fig.4.1 and Table 4.2, where the corresponding results for Si and Ge [4.7,8] are shown for comparison. Several interesting trends emerge. The relaxations of the surface chain bonds and others in the surface layer are similar for the three elements. Surprisingly, the subsurface interlayer bond d_4 was found to lengthen by a large amount, ~8% in diamond. This can be understood as being due to the highly directional nature of the bonding in C; the bond angle strains in the subsurface layer weaken this bond, thereby lengthening it. The tilting of the chain clearly

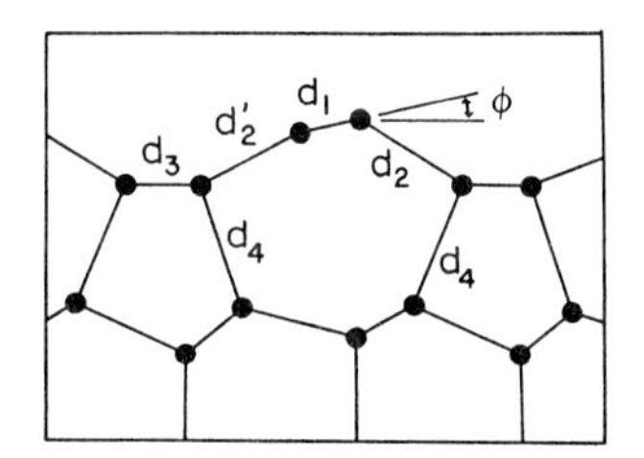

Fig.4.1. Side view of Pandey π-bonded chain model, with definition of parameters used to describe relaxed geometry in Table 4.2

Table 4.2. Relaxed geometries of π-bonded chain model for several elements. See Fig.4.1 for definition of parameters

	C	Si[a]	Ge[b]
Δd_1	-4%	-5%	-5%
Δd_2	1%	-1%	-1%
$\Delta d_2'$	1%	-3%	-3%
Δd_3	1%	0%	0%
Δd_4	8%	2%	0%
ϕ	3	9	13
E_{tot} eV	-0.68	-0.36	-0.34

[a][4.7]; [b][4.8].

increases from C to Si to Ge. The greater stiffness of bond angle restoring forces in C may be responsible. Finally, the total energy compared to the unrelaxed 1×1 is lowest for C, although it should be remembered that the scale of bonding energies is larger for C.

Dimerization of the relaxed chain geometry was tested. The k-point sample was carefully chosen to resolve any Peierls band splitting at the $\bar{J}\bar{K}$ zone boundary. Table 4.1 shows that the energy rises monotonically with the dimerization parameter. We have also tried smaller and larger dimerizations than those shown in Table 4.1, including a model proposed by *Chadi* [4.24] in which a radically large dimerization occurs. In all cases the dimerization energy was found to be positive. We conclude that dimerization does not occur.

In Fig.4.2 we show the calculated surface band structure in the gap region for the relaxed chain model. The dispersion of the occupied surface band along $\overline{\Gamma J}$ has been greatly reduced from that of *Pandey* [4.6], who found a difference of $\gtrsim 3$ eV between $E(\bar{\Gamma})$ and $E(\bar{J})$. This was already reduced to ~2.3 eV in our previous calculation on the ideal structure [4.16], and has now been reduced further to ~1.7 eV due to relaxations. The dispersion is thus in good agreement with experiment without the need for dimerization. The calculated band is too high by a rigid shift of ~1 eV, but this is also true (by ~0.3 eV and ~0.8 eV respectively) for Si and Ge [4.7,8].

Finally, Fig.4.3 shows the charge density for the chain geometry. It is evident from the total charge density in Fig.4.3a that the bond charge is

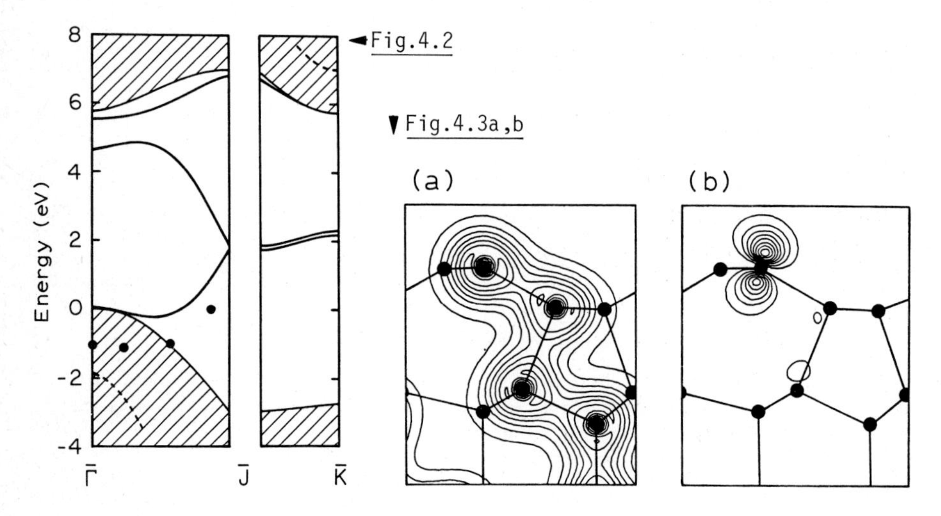

Fig.4.2. Calculated surface bands (*solid lines*) and resonances (*dashed lines*) for fully relaxed Pandey chain model. The bulk projected band structure (*shaded*) and the experimental ARUPS data of [4.3] (*black dots*) are shown for comparison

Fig.4.3a,b. Charge density contour plots in a plane perpendicular to the chain. Atom positions are indicated by filled circles; only half are in the plane of the plot. (**a**) Total charge density. (**b**) Charge density of the occupied surface band in the gap somewhat reduced along the weakened bond d_4. The charge density in the occupied surface band is shown in Fig.4.3b, clearly indicating the highly localized dangling-bond nature of this state. The dangling bonds can be seen to be nearly vertical, so that the π interactions are expected to be strong.

4.3 Discussion

The driving force for the reconstruction is the presence of an energetically unfavorable dangling bond containing an unpaired electron on each surface atom. The reconstruction models attempt to pair electrons via charge transfer or π-bonding, but must pay a price in Coulomb repulsion or elastic strain. Our total energy calculations indicate that the costs outweigh the gains in all cases except the π-bonded chain model.

Both the π-bonding and elastic energies are evidently larger for the π-bonded model for C, compared with those for Si and Ge. Since π bonding is more common in carbon chemistry, it would suggest stronger π bonding in C. The elastic energies are also expected to be larger, because the principal strains are due to bond bending, and the bond-angle force constants are relatively larger in C [4.25]. Energy minimization calculations for Si indicate [4.7] that the ideal Pandey chain model is already at -0.22 eV with respect to the ideal 1×1, while relaxations only lower this to -0.36 eV; for C the corresponding numbers are -0.05 eV and -0.68 eV, respectively. Thus considering relaxations is even more important for C than for Si or Ge.

The idea that the π-bonded chain might dimerize [4.6] was natural, given the need to reduce the dispersion of the occupied surface band and open a gap, and the analogy to polyacetylene. As we showed in the previous section, however, the dispersion has been corrected without the need for dimerization. Moreover, the analogy with $(CH)_x$ is only approximate; the elastic restoring forces must be stronger here because of the subsurface bonding, and the π interactions are weaker here because the dangling bonds are not entirely parallel.

Our surface band structure in Fig.4.2 is nominally metallic, whereas experiments do indicate an insulating surface with a gap [4.14]. In the absence of dimerization, antiferromagnetic ordering could open a gap along the $\overline{JK}$ zone boundary and lower the occupied surface bands [4.26,27]. This would tend to improve the agreement with the ARUPS data, and we think it likely that some such antiferromagnetic ordering occurs.

The discrepancy in the location of the occupied surface band, which is calculated to be ~1 eV higher than indicated by the ARUPS data [4.3], may partly be due to antiferromagnetic ordering. It may also partly be explained by the experimental difficulty in locating the Fermi level precisely [4.3]. Finally, it should be remembered that the LDA eigenvalues have no physical meaning as electron removal energies; the measured removal energy contains an electron-hole correlation which is not correctly included in LDA. Qualitatively, a correction for this effect would lower the theoretical occupied surface band, improving the agreement with experiment.

4.4 Summary

We have calculated the total energy of several proposed 2×1 reconstruction models for the C(111) surface. The *undimerized* Pandey π-bonded chain model is found to have the lowest energy, ~0.3 eV lower than that of a relaxed

1×1 structure. The dispersion of the calculated surface band is found to be in good agreement with experiment, without the need for dimerization. The other models appear implausible on the basis of total energy and surface state dispersion. Unfortunately, the relative paucity of experimental information on the diamond (111) $2 \times 2/2 \times 1$ surface makes our identification tentative at present; we hope to stimulate further experimental work in this area.

Acknowledgments. Support for this work was provided by NSF Grant No. DMR8319024 and by a program development fund from the Director of the Lawrence Berkeley Laboratory. CRAY computer time was provided by the Office of Energy Research of the Department of Energy. We (D.V. and S.G.L.) should also like to acknowledge receipt of Miller Institute and Alfred P. Sloane Foundation fellowships, respectively.

References

4.1 For a recent review, see D. Haneman: Adv. Phys. **31**, 165 (1982)
4.2 J.J. Lander, J. Morrison: Surf. Sci. **4**, 241 (1966)
4.3 F.J. Himpsel, D.E. Eastman, P. Heimann, J.F. van der Veen: Phys. Rev. B**24**, 7270 (1981)
4.4 F. Houzay et al.: Surf. Sci. **132**, 40 (1983) and references therein
4.5 J.M. Nicholls, G.V. Hansson, R.I.G. Uhrberg, S.A. Flodstrom: Phys. Rev. B**27**, 2594 (1983)
4.6 K.C. Pandey: Phys. Rev. Lett. **47**, 1913 (1981); Phys. Rev. B**25**, 4338 (1982)
4.7 J.E. Northrup, M.L. Cohen: J. Vac. Sci. Technol. **21**, 333 (1982)
4.8 J.E. Northrup, M.L. Cohen: Phys. Rev. B**27**, 6553 (1983)
4.9 R.M. Tromp, L. Smit, J.F. van der Veen: Phys. Rev. Lett. **51**, 1672 (1983)
4.10 P. Chiaradia et al.: Phys. Rev. Lett. **52**, 1145 (1984); M.A. Olmstead, N. Amer: Phys. Rev. Lett. **52**, 1148 (1984)
4.11 R. Feder, W. Mönch, P.P. Auer: J. Phys. C**12**, L179 (1979)
4.12 F. Solal et al.: Phys. Rev. Lett. **52**, 360 (1984)
4.13 B.B. Pate et al.: J. Vac. Sci. Technol. **19**, 349 (1981)
4.14 S.V. Pepper: Surf. Sci. **12**, 47 (1982)
4.15 W.S. Yang, F. Jona: Phys. Rev. B**29**, 899 (1984)
4.16 D. Vanderbilt, S.G. Louie: J. Vac. Sci. Technol. B**1**, 725 (1983)
4.17 D. Haneman: Phys. Rev. **121**, 1093 (1961)
4.18 R. Seiwatz: Surf. Sci. **2**, 473 (1964)
4.19 D.J. Chadi: Phys. Rev. B**26**, 4762 (1982)
4.20 D.R. Hamann, M. Schlüter, C. Chiang: Phys. Rev. Lett. **43**, 1494 (1979)
4.21 L. Hedin, B.I. Lundqvist: J. Phys. C**4**, 2064 (1971)
4.22 J.R. Chelikowsky, S.G. Louie: Phys. Rev. B**29**, 3470 (1984)
4.23 D. Vanderbilt, S.G. Louie: Phys. Rev. B**30**, 6118 (1984)
4.24 D.J. Chadi: Private communication; see also in *Proceedings of the IXth International Vacuum Congress and Vth International Conference on Solid Surfaces*, ed. by J.L. de Segovia (Imprenta Moderna, Madrid, Spain 1983) pp.80-88; J. Vac. Sci. Technol. B**2**, 948 (1984)
4.25 R. Tubino, L. Piseri, G. Zerbi: J. Chem. Phys. **56**, 1022 (1972)
4.26 M. Lannoo, G. Allan: Solid State Commun. **47**, 153 (1983)
4.27 A. Martin-Rodero et al.: Phys. Rev. B**29**, 476 (1984)

5. Total Energies and Atom Locations at Solid Surfaces

R. Richter, J.R. Smith, and J.G. Gay

Physics Department, General Motors Research Laboratories
Warren, MI 48090-9055, USA

With the advent of modern computers it is now possible to compute total energies at surfaces from first principles. However, the kinds of problems that can currently be dealt with are quite restrictive. A brief overview of such calculations, as well as the recently discovered universality in binding energy relations will be given. We show that it is now possible to calculate surface or cleavage energies of transition metals from first principles. This is done via our self-consistent local orbital (SCLO) method. The total energies of metal films of many thicknesses, for example of three, five, seven, and nine layers, are first computed; a bulk and surface energy are then derived from a least-squares linear plot of total energy versus film thickness. In the examples of copper(100) and silver(100), the deviation from the line is less than 0.1 eV. This shows the accuracy of our method since the slope of the line (the bulk energy) is four orders of magnitude larger than the intercept (twice the surface energy). Good agreement with experiment is obtained for all metals considered.

5.1 Introduction

With the application of modern bulk electronic structure methods to surfaces and the availability of today's computers, it is now a reasonable goal to calculate the equilibrium positions of individual atoms at surfaces. If one directly compares the energy levels for different geometries with spectroscopic data, one can sometimes run into difficulty in attempting to find an unambiguous geometry [5.1]. A more direct method is to calculate the energy of differing geometries and pick the one with the lowest total energy.

Progress has been made in the case of simple metal surfaces. *Barnett* et al. [5.2] have used pseudopotential perturbation theory to compute multilayer relaxation at simple or nearly-free-electron surfaces. *Bohnen* [5.3] has determined multilayer relaxation for these surfaces beyond the perturbation approximation.

Bimetallic interfaces [5.4] between simple metals and simple metal-semiconductor interfaces [5.5] have also been treated with some success. Equilibrium separations between surfaces have been determined by total energy minimization.

We know of no attempts to locate transition metal atoms at surfaces by calculating total energies for differing geometries, although recently there has been work on locating gas atoms chemisorbed on transition metal surfaces [5.6]. Transition metals are intrinsically more difficult to treat because of the localized d orbitals. However, it has been recently found [5.4] that simple and transition metals belong to the same universality class in total energy versus interatomic spacing [5.7]. It was found that the entire binding energy curve can be determined from the equilibrium binding energy and equilibrium vibrational frequency. For surfaces, it was found to apply, for example, to chemisorption [5.8] and adhesion [5.4]. This universal relation can be closely approximated by a Rydberg function [5.8].

The equilibrium binding energy for clean surface energetics is the surface energy. Thus the surface energy is a key variable in the computations of equilibrium positions of atoms at clean surfaces. In the following we discuss results of our first-principles calculations of the surface energies for Ag, Cu, Ni, and Fe.

Metals divide into two groups, nearly-free-electron (simple) and transition metals.

Nearly-free-electron metals generally have a low surface energy. When cleaved, only nondirectional bonds must be broken. The surface energy of a jellium slab has been calculated in various approximations [5.9]. The agreement with experiment has been good. Calculating the surface energy is straightforward: since the bulk energy of a jellium slab is known analytically, one merely subtracts it from the total energy of the truncated jellium and the remainder yields the surface energy.

Transition metals generally have a high surface energy. When cleaved, highly directional bonds must be broken. This means more energy per unit cell must be added. Unlike simple metals, transition metal surface energies have not in general been treated fully self-consistently [5.10,11]. We calculate the surface energy as follows. Given the total energy for slabs of different thicknesses n, we fit a straight line

$$E_n = nE_b + 2E_s \quad , \tag{5.1}$$

where E_n is the energy for a slab of thickness n, E_b is the energy per plane (bulk energy) and E_s is the surface energy. Thus the intercept of this straight line is twice the surface energy.

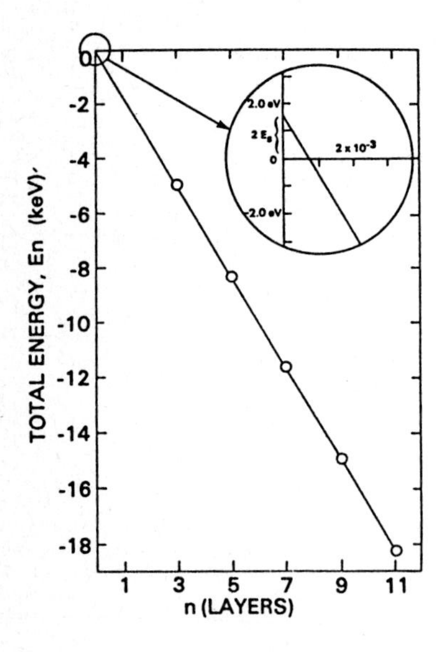

Fig.5.1. The total energy per unit cell versus film thickness for a Cu(100) film is shown. The inset shows the closeness of the intercept to zero. The intercept, 1.7 eV, is twice the surface energy of the Cu(100) surface

Of course, the problem is not so simple. The slope of the straight line is several orders of magnitude larger than the intercept. In our case, the bulk energy per plane is about a thousand eV and the intercept about one eV. A plot of energy versus thickness is shown in Fig.5.1. Thus one needs greater than the usual ten percent accuracy in the energy calculation to get 0.1 eV accuracy in the surface energy. In particular, the same approximation must be used to calculate the energy of the slabs of different thicknesses so that errors in the total energy are intensive. That is, they are absorbed into the energy per plane E_b.

5.2 Present Calculations

Here we use the self-consistent local orbital (SCLO) method [5.12] to calculate the energy of a thin slab of atoms.

5.2.1 Total Energy Formulation

The output from this calculation includes a charge density ρ and a set of energy eigenvalues ε. Using these, we calculate the total energy E of the configuration in the local density approximation [5.13] using (5.1):

$$E = \frac{1}{N}\left\{\sum_i \varepsilon_i - \frac{1}{2}\int \frac{\rho_v(\mathbf{r})\rho_v(\mathbf{r}')}{|\mathbf{r}-\mathbf{r}'|}\,d\mathbf{r}\,d\mathbf{r}' + \frac{1}{2}\sum_{i\neq j}\frac{Z_{c_i}Z_{c_j}}{|\mathbf{R}_i - \mathbf{R}_j|}\right.$$
$$\left. + \int [\rho(\mathbf{r})\varepsilon_{xc}(\mathbf{r}) - \rho_v(\mathbf{r})V_{xc}(\mathbf{r})]d\mathbf{r}\right\} \quad . \tag{5.2}$$

Details of the calculation will be given elsewhere [5.14]. Taken alone, the sum over eigenvalues $\sum_i\varepsilon_i$ does not give a good measure of the surface energy. For example, since the value of the work function ϕ varies slightly as the thickness of the film varies [5.15], this rigid shift of the bulk eigenvalues introduces essentially random noise into the calculation when comparing films of different thicknesses. However, the bulk part of the electrostatic double counting term [second term on the right-hand side of (5.2)] is shifted by a compensating amount $N\phi$ that will cancel out this fluctuation exactly, so that the sum of these two and not the individual pieces should be examined for meaning.

Why we can calculate to good enough accuracy. We compute matrix elements due to the potential from overlapping atoms exactly; the difference potential due to self-consistency is represented by a discrete Fourier series. This difference is found on a rather coarse mesh in real space (spacing about a half bohr, or about ten points between atoms in plane), our main numerical approximation. The largest errors we make in representing the difference potential are in the core region. The Fourier transform cannot mimic the fast variation near the core. Since the charge density in the core region of the surface atoms is very close to that of the bulk atoms [5.16], we will be making an error approximately proportional to the number of planes, which by (5.1) does not contribute to the surface energy. Further, the potential is a much smoother function of position than the charge density and hence is less sensitive to mesh spacing. This is tested by halving the mesh spacing. It is found that while the energy per plane changes dramatically, the surface energy is constant to 0.01 eV. Hence our error is moved into E_b, and not into the surface energy E_s.

We find that the density does not have to be very near to self-consistency for the total energy to have a stable value. The extremum property of the self-consistent charge density works in our favor here.

5.2.2 Copper, Silver, Nickel, and Iron

We have calculated the surface energy of the (100) surfaces of noble metals Cu and Ag and of the paramagnetic forms of Ni and Fe. The results are shown in the following table. The experimental values, taken from [5.17], were determined by taking the zero-temperature limit of the measured bulk modulus.

Table 5.1

Metal	E_s (Theory)	E_s (Theory)	E_s (Exp)
Cu	0.85 [eV]	2087 [erg/cm^2]	2016 [erg/cm^2]
Ag	0.89	1720	1543
Ni	1.16	2998	2664
Fe	1.47	2880	2452

The agreement with experiment is within 10% except in the case of iron. The experimental values were all derived by the same method, since the values derived by other measurements may differ by as much as 50%.

We expect our answers to be higher than experiment. Relaxation perpendicular to the surface has not been included; neither has the possibility of a lateral reconstruction. In fact, the more accurately we model the surface, the better an upper bound to the surface energy we expect our answer to be. Iron is known to have a complex relaxation perpendicular to the surface [5.18]. It has been shown that relaxation effects can significantly affect the value of the surface energy: for the (110) face of aluminum, including relaxation perpendicular to the surface, they reportedly lower the surface energy by 0.4 eV per unit cell [5.2]. It should be pointed out, however, that the difference between theory and experiment is too small to make definite conclusions about relaxation here. This is particularly true when one considers that there are no experimental values for single-crystal planes of these metals. The surface energy for the first iteration, that is, the output surface energy from an input potential computed from a charge density that is a simple superposition of atomic charge densities, is much higher than that of the self-consistent charge density. The difference is an order of magnitude for Cu; the surface energy for overlapping atoms is over 10 eV.

The influence of the exchange-correlation potential on our work should be mentioned. The Wigner correlation form was used. It is found from jellium work [5.19] that the use of the more accurate Ceperley-Alder correlation potential [5.20] will give a higher surface energy. In aluminum, for example, the Wigner correlation gives a value 100 erg/cm^2 lower than Ceperley-Alder correlation.

The cleavage energy per unit cell correlates well with the number of surface bonds broken. For Cu and Ag, very similar metals, the cleavage energies are identical within our error when viewed on a per unit cell basis. Since the lattice constant of Ag is larger than that of Cu, its cleavage energy per unit area is lower. Nickel, which has a lower d band filling, has a higher cleavage energy; Fe is higher still.

One measure of the accuracy of our total energy formulation is how well these data fall on a straight line. What is tabulated above is just twice the intercepts. The table below shows the deviations of each energy from the fitted line for copper and silver.

Table 5.2

Deviation [eV]				
Planes → Metal ↓	3	5	7	9
Cu	-0.04	-0.01	+0.11	+0.08
Ag	-0.03	+0.07	-0.03	

Since the values for copper and silver were so close to the line, we only used one pair of planes to get the surface energy of nickel or iron.

5.3 Summary

We have seen that given fixed atom locations for a transition metal surface, it is now possible to compute the surface energy from first principles. However, rearrangements or relaxations of the surface atomic layers may significantly reduce the energy needed to cleave a bulk crystal. As yet, the costs involved in calculating any but straightforward geometries are prohibitive. Self-consistency is essential, and the Hamiltonian must be determined quite accurately to obtain reasonable surface energies. It appears that even for slabs three atomic layers thick the surface energy is very close to the value for thicker films. Our results for the four metals suggest a correlation between band filling and surface energy per surface atom.

For simply cleaved surfaces the surface energy at low temperatures may be calculated to greater accuracy than can be found presently by experiment.

References

5.1 P.J. Feibelman, D.R. Hamann, F.J. Himpsel: Phys. Rev. B**22**, 1734 (1980); M.J. Cardillo, G.E. Becker, D.R. Hamann, J.A. Serri, L. Whitman, L.F. Mattheiss: Phys. Rev. B**28**, 494 (1983)
5.2 R.N. Barnett, U. Landman, C.L. Cleveland: Phys. Rev. B**28**, 1685 (1983)
5.3 K.-P. Bohnen: Phys. Rev. B**29**, 1045 (1984)
5.4 J.H. Rose, J.R. Smith, J. Ferrante: Phys. Rev. B**28**, 1835 (1983); J. Ferrante, J.R. Smith: Phys. Rev. B (to be published)
5.5 I.P. Batra: Phys. Rev. B**29**, 7108 (1984)
5.6 C. Umrigar, J.W. Wilkins: Bull. Am. Phys. Soc. **28**, 538 (1983)
5.7 In fact, nucleons and electron-hole liquids appear to belong to the same universality class as simple and transition metals. See J.H. Rose, J.P. Vary, J.R. Smith: Phys. Rev. Lett. **53**, 344 (1984)
5.8 J.P. Perdew, J.R. Smith: Surf. Sci. **141**, L295 (1984)
5.9 K.-P. Bohnen, S.C. Ying: Phys. Rev. B**22**, 1806 (1980); V. Sahni, J.P. Perdew, J. Gruenebaum: Phys. Rev. B**23**, 6512 (1981)
5.10 B. Delley, D.E. Ellis, A.J. Freeman, E.J. Baerends, D. Post: Phys. Rev. B**27**, 2132 (1983). These authors do not compute cleavage energies, but have computed surface tensions for clusters self-consistently.
5.11 J.A. Appelbaum, D.R. Hamann: Solid State Commun. **27**, 881 (1978). These authors stated a number for the surface energy of Cu(111), but did not

describe how they computed the total energy, although the electronic structure was computed self-consistently
5.12 J.R. Smith, J.G. Gay, F.J. Arlinghaus: Phys. Rev. B**21**, 2201 (1981)
5.13 See, for example, the total energy formulation in
J. Harris, R.O. Jones, J.E. Mueller: J. Chem. Phys. **75**, 3904 (1981)
5.14 J.R. Smith, J.G. Gay, R. Richter: To be published
5.15 P.J. Feibelman: Phys. Rev. B**27**, 1991 (1983); and
P.J. Feibelman, D.R. Hamann: Phys. Rev. B**29**, 6463 (1984). Here the work function ϕ can vary by 0.2 eV for slabs thicker than 3 planes
5.16 U. von Barth, C.D. Gelatt: Phys. Rev. B**21**, 2232 (1981)
5.17 H. Wawre: Z. Metallde **66**, 375 (1974)
5.18 J. Sokolov, F. Jona, P.M. Marcus: Solid State Commun. **49**, 307 (1984)
5.19 V. Sahni: Private communication
5.20 S.H. Vosko, L. Wilk, M. Nusair: Can. J. Phys. **58**, 1700 (1980) from a calculation by D.M. Ceperley, B.J. Alder: Phys. Rev. Lett. **45**, 566 (1980)

6. Theory of Hydrogen on Metal Surfaces

M.S. Daw and S.M. Foiles

Sandia National Laboratories, Livermore, CA 94550, USA

The embedded atom method [6.1,2] is used to study the classical behavior of hydrogen at all coverages on the Ni(110) and Pd(111) surfaces. For both surfaces, the hydrogen adsorption site is predicted to be the threefold site in agreement with the experimental observations. For Ni(110) the lowest energy ordered structure at $\Theta = 1$ is computed to be the (2×1). Molecular dynamics simulations of this system show a critical temperature for the order-disorder transition to be between 150 K and 200 K. For the Pd(111) surface, two structures $(\sqrt{3} \times \sqrt{3})R30^\circ$ and $(\sqrt{12} \times \sqrt{12})R30^\circ$ are predicted to be very close in energy. Quantum corrections are expected to make the $(\sqrt{3} \times \sqrt{3})R30^\circ$ lower in energy. Monte Carlo simulations at $\Theta = 1/3$ and 2/3 indicate that the order-disorder transition occurs near 125 K at both of these coverages with lower critical temperatures away from these ideal coverages. These predictions agree well with experimental observations.

6.1 Introduction

There is a great deal of interest in the behavior of H on transition metal surfaces because of the interesting ordering phenomena they exhibit. LEED diffraction experiments for H on Ni(110) have shown the existence of an ordered $c(2 \times 1)$ structure at $\Theta = 1$ which disorders at temperatures around 180 K [6.3,4]. Helium scattering results suggest that the H occupies a zigzag configuration of pseudo-threefold sites forming the (2×1) pattern [6.5]. For H-Pd(111), a recent LEED experiment has shown the existence of $(\sqrt{3} \times \sqrt{3})R30^\circ$ diffraction spots for coverages near $\Theta = 1/3$ and $\Theta = 2/3$ [6.6]. The intensity of these spots as a function of coverage shows maxima at these two specific coverages with a minimum for intermediate values. At the higher coverage the system disorders at $T_c = 105$ K, and at the lower, at $T_c \leqq 85$ K. The results on Pd(111) are in interesting contrast to previous work for H/Ni(111), where the H forms an ordered graphitic structure showing (2×2) spots with critical temperature around 270 K [6.7].

We report here calculations of the properties of H on Pd(111) and Ni(110) using the recently developed embedded atom method [6.1,2]. Adsorption sites and energies as well as ordered phases and critical temperatures are predicted. The results agree excellently with experiment.

6.2 Theory

In the Embedded Atom Method (EAM), the total energy of a system of atoms is given by [6.1,2]

$$E_{tot} = \sum_i F_i(\rho_{h,i}) + 1/2 \sum_{i,j} \phi_{ij}(R_{ij}) \quad . \tag{6.1}$$

In this expression, $\rho_{h,i}$ is the total electron density at atom i due to the rest of the atoms in the system, and F_i is the embedding energy to place an atom into that electron density. Finally, ϕ_{ij} is a short-range pair interaction representing the core-core repulsion of atoms i. This expression, with atoms i and j separated by a distance R_{ij}, for the total energy can be obtained from density functional theory by viewing each atoms as being embedded in a sea of electrons created by all the other atoms. Assuming that the host electron density is suitably uniform at each atom, the embedding energy is then a function of the electron density at the site. In this framework the embedding energy provides most of the cohesive energy of the solid, viewed as a binding between each atom and the electron gas. There is an embedding function for each element in the system (F_{Pd} and F_H) which is assumed to be universal in that the function depends only on the type of atom and the electron density but is independent of the other atomic types present in the system. The core-core repulsions are short-ranged and relatively weak.

The expression in (6.1) is useful provided one has a description of the host electron density at each atom (i.e., electron density established by all the other atoms). In this work, we make the approximation that the density in the solid is given as a superposition of atomic electron densities:

$$\rho_{h,i} = \sum_{j \neq i} \rho^a(R_{ij}) \quad . \tag{6.2}$$

Here, the total energy is a simple function of the positions of the atoms.

For any assemblage of H and Pd atoms, we then require the following functions to calculate the total energy: $F_{Pd}(\rho)$, $F_H(\rho)$, $\rho^a{}_{Pd}(r)$, $\rho^a{}_H(r)$, $\phi_{Pd-Pd}(r)$, $\phi_{Pd-H}(r)$, and $\phi_{H-H}(r)$. The function $F_H(\rho)$ has been determined for hydrogen in jellium calculations. The atomic densities are, of course, available from atomic calculations. The remaining functions must, for practical reasons, be determined semiempirically. This has been done by *Daw* and *Baskes* [6.1,2] by fitting to the following bulk properties of Ni and Pd: lattice constant, elastic constants, vacancy formation energy, sublimation energy in fcc and bcc phases, and the H heats of solution. The complete set of functions has been described in detail by *Daw* and *Baskes*, and we have used those functions unchanged for the present work. It is to be emphasized that the functions were fitted to bulk properties, and there has been no fitting to the surface properties.

It is important to note that the EAM (6.1,2) gives a framework for treating both adsorbate and substrate atoms uniformly, unlike the effective medium method [6.8], which calculates only the adsorbate energy. This allows calculations of both adsorbate and substrate relaxations, as well as adsorbate-adsorbate interactions mediated by the substrate. Also the computational simplicity of the method allows for the treatment of large unit cells needed to study disordered systems. Thus, both ordered and random states of the system can be treated on the same footing.

Because the EAM gives the potential energy of a solid as a function of the positions of the atoms, fully quantum-mechanical [6.7] solutions can be obtained straightforwardly [6.9]. However, we present here only classical solutions, and discuss the quantum effects in the conclusions.

There are three types of classical calculations that can be performed easily using (6.1,2): energy minimization, Monte Carlo, and molecular dynamics. In all cases, the calculations use a slab of metal atoms with the hydrogen applied to one side of the slab. Periodic boundary conditions are used for the two directions perpendicular to the exposed surfaces of the slab.

The energy minimizations and molecular dynamics calculations allow for the motion of all atoms, metal and H. The Monte Carlo calculations consider only the hydrogen motion to save computer time. This is not, however, a fundamental restriction of the EAM.

6.3 Results

The results of the EAM for clean metal surfaces as well as the single hydrogen adsorption energies have been reported earlier [6.2]. In summary, the method predicts the surface relaxations of the clean surface in agreement with experiment. It also predicts that the H adsorption sites for these two surfaces are the threefold site as it seen experimentally. Adsorption energies relative to the isolated atoms are within 5% of the experimental adsorption energies.

One of the major advantages of the EAM is the ability to treat high coverages of adsorbates on the surface without the restriction to ordered structures with small unit cells. Thus we have considered the ordering phenomena on these two surfaces. For Ni(110), energy minimization calculations were performed for various possible ordered structures at $\Theta = 1$. It was found that the lowest energy structure corresponds to a $c(2\times 1)$ symmetry with atomic positions in agreement with the helium diffraction experiments of *Engel* and *Rieder* [6.5].

Molecular dynamics simulations of the hydrogen motion on the Ni(110) surface were then performed to determine roughly the ordering temperature and to gain insight into the nature of the atomic motion. The degree of ordering is measured by the amplitude of the geometric structure factor S of the hydrogen:

$$S(k) = \langle \sum_{i,j} \exp[ik(R_i - R_j)] \rangle / N \quad , \tag{6.3}$$

where R_i are the atomic positions of the hydrogen, N is the number of H, k is the wave vector of interest, and the average is taken over the configurations generated by the molecular dynamics after the system has reached equilibrium. Figure 6.1 presents the value of the structure factor at the (2×1) spot computed as a function of temperature. Note that the EAM predicts an order-disorder transition for this coverage at a temperature between 150 K and 200 K. Experimentally, the system is known to disorder at 180 K, which agrees well with our calculations.

For hydrogen on Pd(111), the excess enthalpies per surface metal atom (enthalpy relative to the linear interpolation between the clean surface and a full monolayer coverage) has been computed for the following ideal ordered structures:

$(\Theta = 1)$ $\quad (1\times 1)$

$(\Theta = 1/3,\ 2/3)$ $\quad (\sqrt{3}\times\sqrt{3})R30^\circ$, $(\sqrt{12}\times\sqrt{12})R30^\circ$

$(\Theta = 1/2)$ $\quad (2\times 1)$, (2×2), (4×4)

$(\Theta = 1/4)$ $\quad (2\times 2)$, (4×4) .

To determine the zero-temperature phase equilibria, it is important to consider the energies of these structures away from ideal stoichiometry. This is done by assigning to each site in the unit cell a probability of occupation. These results are shown in Fig.6.2 for structures of lowest enthalpy:

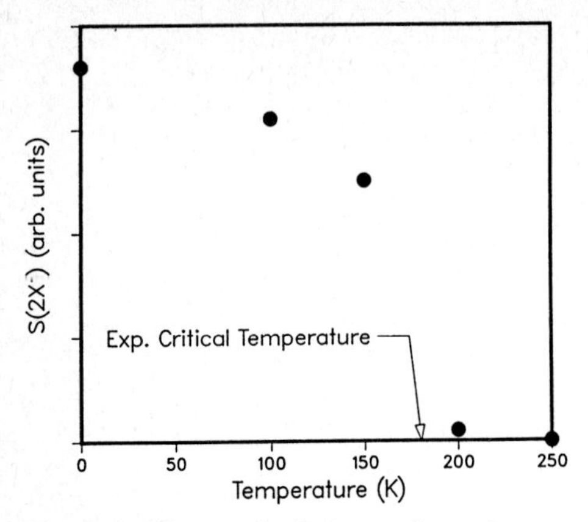

Fig.6.1. Theoretical intensity of (2×1) reflection vs temperature for H monolayer on Ni(110)

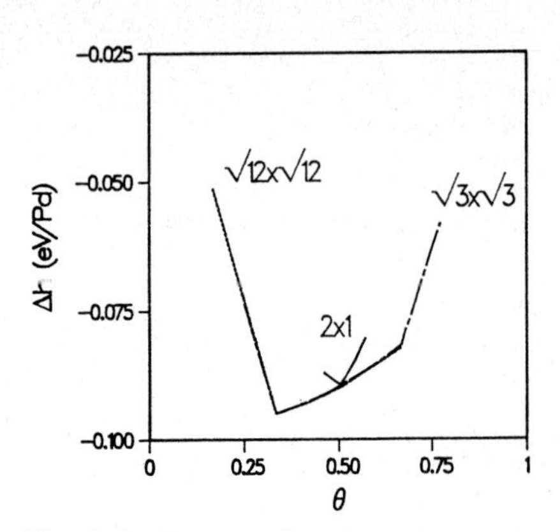

Fig.6.2. Excess enthalpy for H-Pd (111) as a function of coverage for structures described in the text

$(\sqrt{3}\times\sqrt{3})R30^\circ$, $(\sqrt{12}\times\sqrt{12})R30^\circ$, and (2×1) symmetries. Several conclusions are evident from the figure. First, the $\sqrt{3}$ and $\sqrt{12}$ structures are very close in energy over almost the whole coverage range. Also, the excess enthalpies of the $\sqrt{3}$ and $\sqrt{12}$ structures are concave upward for the coverage $1/3<\Theta<2/3$, implying that in this range the structure is actually a single phase between $\Theta=1/3$ and $\Theta=2/3$, rather than a coexistence of the ideal coverages. The enthalpy of the (2×1) structure is within theoretical error of the $\sqrt{3}$ (and $\sqrt{12}$) enthalpy at $\Theta=1/2$.

To estimate the critical temperatures, Monte Carlo simulations of the hydrogen positions were performed at coverages of 1/3, 1/2, and 2/3. The simulations showed that both the ordered $(\sqrt{3}\times\sqrt{3})R30^\circ$ and $(\sqrt{12}\times\sqrt{12})R30^\circ$ structures at $\Theta=1/3$ and 2/3 are stable for temperatures less than about 125 K. Above this temperature, the simulations produced site distributions without any long-range order. There is a definite short-range order in this temperature range: for $\Theta=2/3$, there are no 1st nearest neighbors and for $\Theta=1/3$ there are no 1st, 2nd, or 3rd nearest neighbors. At $\Theta=1/2$, the (2×1) structure was found to be stable for temperatures less than about 50 K, while the off-stoichiometry $(\sqrt{3}\times\sqrt{3})R30^\circ$ structure disordered for temperatures greater than 25 K. Thus, the order-disorder temperature at this coverage is very low.

6.4 Conclusions

The embedded atom method has been applied to calculate ordered and disordered structures of H on Ni(110) and Pd(111). Semiempirical functions determined previously from bulk properties have been applied without adjustment to surface calculations. These results indicate that the embedded atom method can accurately describe the energetics and critical temperatures of hydrogen on transition metal surfaces.

For H/Ni(110), the lowest energy structure is correctly predicted to be the (2×1). The critical temperature for the ordered structure of H/Ni(100) is calculated using molecular dynamics to be between 150 K and 200 K and is in excellent agreement with experiment.

For H/Pd(111) using a classical calculation, two structures, the $\sqrt{3}$ and $\sqrt{12}$, are found to have essentially degenerate energies. Quantum effects [6.7,9] may in fact cause a significant splitting between the two structures. The assumption that the two threefold hollows on Pd(111) are equivalent is true as far as the classical minima are concerned. However, because the two sites differ in the position of the second layer of Pd, the shapes of the potential energy curves are in fact different. The quantum zero-point motion of the protons sample the region around the minima, thus making the two threefold hollows inequivalent in the quantum regime. The $\sqrt{3}$ structure places hydrogens on one type of site, but the $\sqrt{12}$ structure is formed from both types of sites. Thus, the lowest energy structure has to be $\sqrt{3}$, favored by the difference in zero-point energies. We are presently calculating the zero-point energies and the results will be reported elsewhere. Using Monte Carlo methods, both the $\sqrt{3}$ and $\sqrt{12}$ structures are found to disorder at 125 K, within 40 K of the experimental values.

References

6.1 M.S. Daw, M.I. Baskes: Phys. Rev. Lett. **50**, 1285 (1983)
6.2 M.S. Daw, M.I. Baskes: Phys. Rev. B**29**, 6443 (1984)
6.3 T.N. Taylor, P.J. Estrup: J. Vac. Sci. Tech. **11**, 244 (1974)
6.4 R.J. Behm, K. Christmann, G. Ertl, V. Penka, R. Schwankner: To be published
6.5 T. Engel, K.H. Rieder: Surf. Sci. **109**, 140 (1981)
6.6 T.E. Felter, S.M. Foiles, M.S. Daw, R.H. Stulen: In preparation
6.7 K. Christmann, R.J. Behm, G. Ertl, M.A. Van Hove, W.H. Weinberg: J. Chem. Phys. **70**, 4168 (1979)
6.8 P. Nordlander, S. Holloway, J.K. Nørskov: Surf. Sci. **136**, 59 (1984)
6.9 M.J. Puska, R.M. Nieminen, M. Manninen, B. Chakraborty, S. Holloway, J.K. Nørskov: Phys. Rev. Lett. **51**, 1081 (1983)

Part II

New Surface Structure Techniques

II.1 Techniques Based on Electrons

7. The Surface Topography of a Pd(100) Single Crystal and Glassy $Pd_{81}Si_{19}$ Studied by Scanning Tunneling Microscopy

M. Ringger, H.R. Hidber, R. Schlögl, P. Oelhafen, H.J. Güntherodt, K. Wandelt* and G. Ertl*
Institut für Physik der Universität Basel
Klingelbergstraße 82, CH-4056 Basel, Switzerland
*Institut für Physikalische Chemie der Universität München
Sophienstraße 11, D-8000 München 2, FRG

7.1 Introduction

The methods available to study surfaces [7.1-3] may be divided into techniques which supply spatially resolved information (on the submicron scale) and into the others which yield integral information. Scanning Auger spectroscopy and electron microscopy used as SEM, TEM and STEM are perhaps the most powerful methods among the first group. A principal resolution limit of these techniques is given by the volume of interaction between the electrons of the "light source" and the specimen. This volume will always be larger than the volume of a single atom. Therefore a new microscopic technique which does not require such an interaction zone had to be found in order to reach the ultimate goal of atomic resolution in imaging surfaces. Such a technique is Scanning Tunneling Microscopy (STM) recently developed by *Binnig* and *Rohrer* and collaborators [7.4,5]. We have built a microscope similar to the published design principles.

After a short introduction into the mode of operation of this technique first results will be presented: we compare images of a Pd(100) surface and a $Pd_{81}Si_{19}$ glassy alloy surface obtained from STM and high-resolution SEM. The surface sensitivity of the STM technique is discussed using XPS and UPS data obtained from the specimen used in the microscopes.

7.2 Experimental

7.2.1 The Scanning Tunneling Microscope

The physical principle of vacuum tunneling has long been known [7.6]. A metal tip (here W) is brought very close (here ca. 10 Å) to the surface of a metallic specimen. A tunnel current depending in its strength on material constants and geometric factors results when a voltage (here ca. 0.1 V) is applied. The important property of the tunnel current is its exponential dependence [7.5] on the specimen-tip distance.

Experimentally, the tip is mounted on a tripod of three piezodrives. Two of them (the x and y drives) are controlled by a microcomputer and scan the tip reproducibly over the surface. The scan parameters determine the lateral resolution of the image, no magnification is involved. The third piezodrive is energized by the output of a P-I control unit which keeps the tunnel current constant. Assuming a constant work function of both tip and specimen, the controller output reflects the profile of the specimen. This output is recorded against the scan signal on an x-y recorder.

The approach of the sample to the tip is of crucial importance for the operation: therefore the sample is mounted on a piezoelectrically driven platform (the "louse") which allows positioning the sample without a mechanical contact.

The whole system is based on a vibration-free platform: vibrations are suppressed by a two-stage spring suspension with eddy current damping. This setup is placed in a vacuum of—at present—10^{-8} Torr; a UHV system with appropriate facilities for sample manipulation and surface cleaning is under construction.

7.2.2 Scanning Electron Microscopy

Most of the images were obtained in a JEOL CX 200 TEMSCAN microscope equipped with single-tilt side entry stage and X-ray analysis facilities. Specimens were mounted on large flakes of natural graphite to ensure good electrical contact and held in the standard graphite specimen holder. No surface treatments were employed. The beam conditions were usually 200 kV, 80 μA emission, smallest spot size and smallest condensor aperture. Under these conditions and with a liquid nitrogen cold trap at the sample location no contamination artifacts were observed. Some experiments were performed on a CAMSCAN S 4 instrument operating at 15 kV and with 150 μA emission. Specimens were mounted on standard holders with conductive silver paint.

7.2.3 Electron Spectroscopy

After examination under the microscopes, the small specimens were transferred under air onto a variable temperature sample holder. For core-level spectra molybdenum supports were used, valence band spectra were obtained from samples mounted on grafoil. A Leybold Heraeus EA 10 instrument equipped with Mg X-ray source, UPS and sputtering facilities was used. The base pressure was 5×10^{-11} Torr.

7.3 Results

7.3.1 The Pd(100) Surface

The sample was cut slightly off the (100) plane (vicinal angle 3.7 degrees) which should give rise to widespread surface steps. The crystal was characterized by various techniques including xenon adsorption [7.7]. From these results it emerged that the monoatomic surface steps were not aligned parallel to (100) or (110). The step height was estimated from the lattice constant fo Pd ($a = 3.88$ Å) to be between 3.6 Å and 3.7 Å. SEM images at medium resolution show that the surface contains large rupture zones with a pronounced step structure. The fracture zones occur with an average distance of ca. 50 micrometers in the optically smooth surface. One step was resolved to the image of Fig.7.1. A substructure is seen, consisting of sharp steps predominantly parallel to each other. The contrast problem is due to the size of the interaction zone between the electron beam and the object which here approaches the size of the structures imaged: the final limit of resolution for the beam characteristics of the SEM used is reached. Figure 7.2 presents a STM survey image with a resolution similar to the scale of Fig.7.1: the same step structure occurs modulated by some wide irregular dips. Note the clearer reproduction of details in the STM image compared to that of the SEM photograph. This is a consequence of the low-energy mode of operation of the STM.

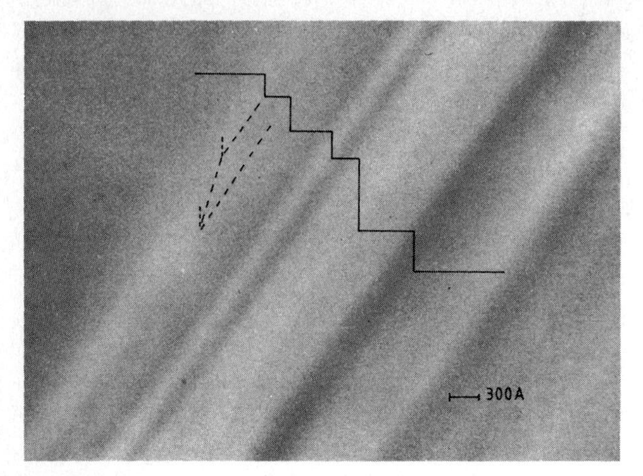

Fig.7.1. A high resolution SEM of a single step. The steps are indicated vertically; nonvertical slopes could possibly not be resolved. The height of the steps is overestimated due to shadowing effects

Fig.7.2 ↓ Fig.7.3 →

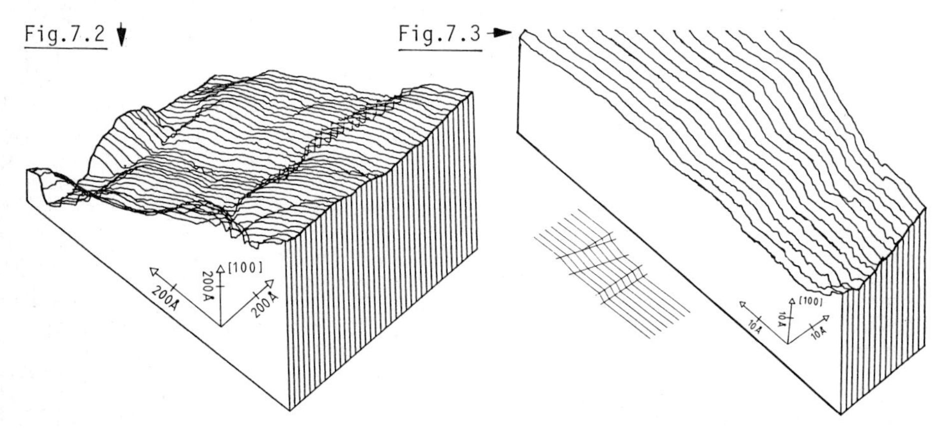

Fig.7.2. A survey STM over the Pd(100) surface. The orientation is only approximate

Fig.7.3. Resolution of mono- and double-atomic steps on Pd(100). The inset indicates that the steps are nonparallel and how a new terrace emerges from a point in the slope of the double atomic step

Examination of a flat area of Fig.7.2 at 20 times higher resolution leads to images as displayed in Fig.7.1. Regular mono- and double atomic steps with heights of 3.5 Å and 7.0 Å can be recognized. Since the STM imaging procedure does not rely on periodic structures, it is possible to detect nonparallel (Fig.7.3 inset) and even nonperiodic surface structures.

7.3.2 Glass $Pd_{81}Si_{19}$

Glassy alloys exhibit a quite different surface morphology which could be imaged on $Pd_{81}Si_{19}$ samples for the first time [7.8]. It reminds the viewer of a structure obtained by flattening droplets of a highly viscous liquid on a glass plate. This topography is much in line with the notion one gets from the preparation procedure [7.9] in which a droplet of liquid alloy is

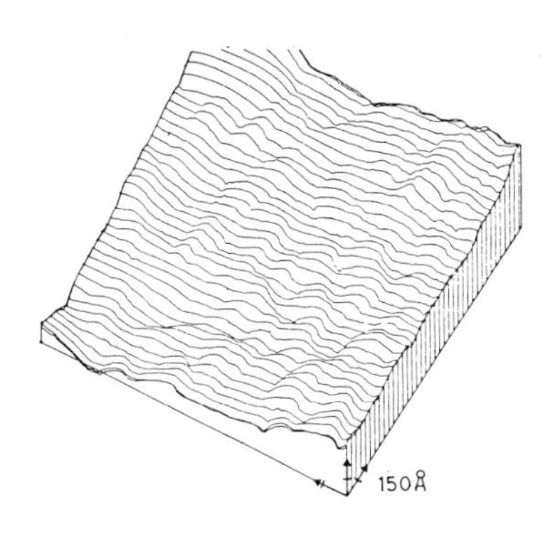

Fig.7.4. A STM survey scan over a glassy $Pd_{81}Si_{19}$ surface

squeezed between two copper disks. At the highest resolution obtained with this material the surface discontinuities exhibit streak-line contours much different from the sharp steps seen in crystalline surfaces.

Figure 7.4 presents a low-resolution scan with the STM. A general agreement between the topographies imaged with SEM and STM can be stated.

Characteristic are the flat areas which occur in two levels of height and the streak-like, nonparallel and nonperiodic edges. It can be seen that the two levels of height exhibit some fine structure and that the edges are sharp, i.e., not terraced within a resolution of ca. 50 Å. Again, the results of STM and SEM agree with each other.

7.3.3 Chemical Constitution of the Surface

Since the imaging method of STM operates with electrons of about 100 meV energy and has the resolving power to image single atoms, it is of interest to know the chemical constitution of the surface under examination. For the corresponding SEM images the composition of the first few monolayers is less important because of the greater mean free path length of the secondary electrons used here for the image formation. Table 7.1 reports the nominal surface composition measured by XPS for the two samples and a polycrystalline Pd foil which is permanently kept in UHV. Only carbon, oxygen, palladium and for the amorphous alloy silicon were present in concentrations larger than 0.2%.

Table 7.1. Surface compositions: values are given in atomic percent

Sample	% Pd	% C	% O	% Si
Pd(100)	33	26	41	-
Pd polycryst.	95	2	3	-
$Pd_{81}Si_{19}$ alloy	71	5	23	1

From the table the following conclusions emerge. The Pd(100) surface was massively contaminated with carbon-oxygen containing material. Even the reference was contaminated after prolonged sputtering and ca. 10 hours residence in the analyzer chamber in the upper 10^{-11} Torr regime. This result underlines again the necessity to work under real UHV conditions if the true metal surface is expected to be accessible to STM experiments for any length of time. The surface composition of the amorphous alloy deviates significantly from the bulk values. An SiO_2 film is present, which allows stable chemisorption of water (as shown by thermal desorption) which in turn forms at room temperature the specimen-vacuum interface.

The Pd samples were chosen partly because of their expected stability against surface oxidation in the non-UHV conditions of STM. The Pd 3d core level spectra confirm the identical and metallic nature of the Pd as probed by XPS.

Using XPS and He I UPS the density of states near the Fermi level was probed at different information depths. As already expected from the values of Table 7.1, both samples exhibit typically nonmetallic shapes of the valence band edges not only at the outer surface (UPS) but within the whole information depth of XPS.

In summary, it is obvious that the STM images here do not represent the genuine metal surface, provided the tip does not penetrate into the insulating overlayer. Furthermore it becomes clear that STM is, contrary to SEM, a really surface sensitive technique and that electron spectroscopy can help in the interpretation of the STM images.

7.4 Discussion

Our present STM version has revealed some hitherto unknown information about the surface topography of both crystalline Pd and glassy $Pd_{81}Si_{19}$ on a scale of ca. 10 Å resolution. It is assumed that at medium resolution it is insignificant whether the specimen is atomically clean or covered with contaminations a few monolayers thick.

Whereas crystalline Pd exhibits sharp contours of a terraced surface, the amorphous $Pd_{81}Si_{19}$ surface may be regarded as a flat plane with splashes of the supercooled liquid of the alloy irregularly distributed over the plane.

To interpret the STM images at higher resolution, the following hypothesis is put forward. At the low voltage used (ca. 0.1 V) it is unlikely that electrons from the metal surface tunnel both through the metal-overlayer barrier and through the whole contamination film [7.10]. Should the tip, however, penetrate into the overlayer to a depth comparable to the information depth of the valence band XPS, then the tunneling probability would be greatly enhanced and the barrier would be determined again by the characteristics of tip and specimen [7.11]. Provided the scan speed is sufficiently slow, the tip should then reproduce the surface topography at atomic resolution as exemplified in Fig.7.3. The smoothness of the structures seen is then not only due to a nonideally shaped tip [7.5,11] but also due to the presence of contamination.

Nevertheless, the STM demonstrated even under these conditions its unique ability to image nonperiodic details of the specimen (Fig.7.3). If the topography, however, contains sudden changes in height, e.g., at steep steps, the contamination hinders the free movement of the tip and either collisions with the specimen or damages of the delicate atomic arrangement at the surface of the tip give rise to the ripples seen in all the high-resolution STM images.

7.5 Summary

By comparing images of surfaces of Pd(100) and glassy $Pd_{81}Si_{19}$ obtained with STM and SEM respectively, it is demonstrated that the two methods agree well in their results at a resolution of ca. 30 Å. The principal limitations of the SEM technique and the advantages of the STM in the overlapping range of resolution of the two methods become apparent. The STM further gives first insight into the topographies of the two sample surfaces at a resolution in height of ca. 1 Å. As a consequence of the low energies used in STM, this method is highly surface sensitive and the images are affected by adsorbed overlayers. At all levels of resolution a clear difference in the topography of a crystalline and an amorphous solid surface was found.

Acknowledgments. We are grateful to P. Abt, H. Breitenstein, D. Holliger, E. Krattiger and A. Nassenstein for skillful technical support. We would like to thank P. Reimann for the preparation of the amorphous alloys and A. Kavanagh for assistance with the CAMSCAN SEM. Financial support of the Swiss National Science Foundation is gratefully acknowledged. Finally, we would like to thank Prof. J.M. Thomas (Department of Physical Chemistry, University of Cambridge) for giving us generous access to his electron microscopy facilities.

References

7.1 R. Gomer (ed): *Interactions on Metal Surfaces*, Topics in Applied Physics, Vol.4 (Springer, Berlin, Heidelber, New York 1975)
7.2 M.W. Roberts, C.S. McKee: *Chemistry of the Metal-Gas Interface* (Clarendon, Oxford 1978)
7.3 J.M. Thomas, R.L. Lambert: *Characterisation of Catalyst* (Wiley, New York 1980)
7.4 G. Binnig, H. Rohrer: Helv. Phys. Acta **55**, 55 (1982)
7.5 G. Binnig, H. Rohrer, C. Gerber, E. Weibel: Phys. Rev. Lett. **49**, 57 (1982)
7.6 J. Frenkel: Phys. Rev. **36**, 1604 (1930)
7.7 R. Mirinda, S. Daiser, K. Wandelt, G. Ertl: Surf. Sci. **131**, 61 (1983)
7.8 R. Schlögl, H.J. Güntherodt: J. Non Cryst. Solids (to be published)
7.9 F.E. Luborsky (ed.): *Amorphous Metallic Alloys* (Butterworths, London 1983) pp.8,26
7.10 M. Ringger, H. Hidber, R. Schlögl, H.J. Güntherodt: Appl. Phys. Lett. (to be published)
7.11 N. Garcia, C. Ocal, F. Flores: Phys. Rev. Lett. **50**, 2002 (1983)

8. Theory of the Scanning Tunneling Microscope

J. Tersoff

IBM T.J. Watson Research Center, P.O. Box 218
Yorktwon Heights, NY 10598, USA

The recent development of the "scanning tunneling microscope" (STM) by *Binnig* et al. [8.1-5] has made possible the direct real-space imaging of surface topography. In this technique, a metal tip is scanned along the surface while adjusting its height to maintain constant vacuum tunneling current. The result is essentially a contour map of the surface. This contribution reviews the theory [8.6-8] of STM, with illustrative examples. Because the microscopic structure of the tip is unknown, the tip wave functions are modeled as s-wave functions in the present approach [8.6,7]. This approximation works best for small effective tip size. The tunneling current is found to be proportional to the surface local density of states (at the Fermi level), evaluated at the position of the tip. The effective resolution is roughly $[2\text{Å}(R+d)]^{\frac{1}{2}}$, where R is the effective tip radius and d is the gap distance. When applied to the 2×1 and 3×1 reconstructions of the Au(110) surface, the theory gives excellent agreement with experiment [8.4] if a 9 Å tip radius is assumed. For dealing with more complex or aperiodic surfaces, a crude but convenient calculational technique based on atom charge superposition is introduced; it reproduces the Au(110) results reasonably well. This method is used to test the structure-sensitivity of STM. The Au(110) image is found to be rather insensitive to the position of atoms beyond the first atomic layer.

Finally, to illustrate the role of electronic structure, theoretical "images" for degenerately doped n and p type cleaved GaAs(110) are compared. The two images are strikingly different; for n type, the image does not reflect the actual surface topography because of electronic-structure effects. In the future, STM might be used to study surface electronic structure as well as topography.

References

8.1 G. Binnig, H. Rohrer, C. Gerber, E. Weibel: Appl. Phys. Lett. **40**, 178 (1982)
8.2 G. Binnig, H. Rohrer, C. Gerber, E. Weibel: Phys. Rev. Lett. **49**, 57 (1982);
G. Binnig, H. Rohrer: Surf. Sci. **126**, 236 (1983)
8.3 G. Binnig, H. Rohrer, C. Gerber, E. Weibel: Phys. Rev. Lett. **50**, 120 (1983)
8.4 G. Binnig, H. Rohrer, C. Gerber, E. Weibel: Surf. Sci. **131**, L379 (1983)
8.5 A.M. Baro, G. Binnig, H. Rohrer, C. Gerber, E. Stoll, A. Baratoff, F. Salvan: Phys. Rev. Lett. **52**, 1304 (1984)
8.6 J. Tersoff, D.R. Hamann: Phys. Rev. Lett. **50**, 1998 (1983)
8.7 J. Tersoff, D.R. Hamann: To be published
8.8 N. Garcia, C. Ocal, F. Flores: Phys. Rev. Lett. **50**, 2002 (1983)

9. Reflection Electron Microscopy Studies of Crystal Lattice Termination at Surfaces

Tung Hsu and J.M. Cowley

Department of Physics, Arizona State University, Tempe, AZ 85287, USA

Reflection Electron Microscopy (REM) is applied to the imaging of stacking faults immediately beneath the surface of bulk crystals. Strong contrast is obtained on stacking fault ribbons in graphite. The contrast on Pt(111) surfaces is also attributed to the stacking sequence change of the topmost layer of atoms.

9.1 Introduction

Using Bragg-reflected electrons to form images, reflection electron microscopy (REM) [9.1-5] is capable of imaging atomic steps, dislocations, reconstructions, etc., on the surface of bulk crystals (Fig.9.1). REM contrast due to stacking faults intersecting the surface has been calculated [9.6] but has not been observed experimentally. Variations in surface structure can arise from stacking faults which are parallel to the surface, that is, when the topmost layer(s) of atoms are stacked in a sequence different from that in the bulk. A similar situation is a crystal of large unit cell terminating (at the surface) at different nonequivalent layers of atoms. *Iijima* and *Hsu* have reported that for $TiS_{1.5}$, regions of different contrast separated by surface steps could be attributed to termination of the crystal at the different levels within the unit cell [9.7].

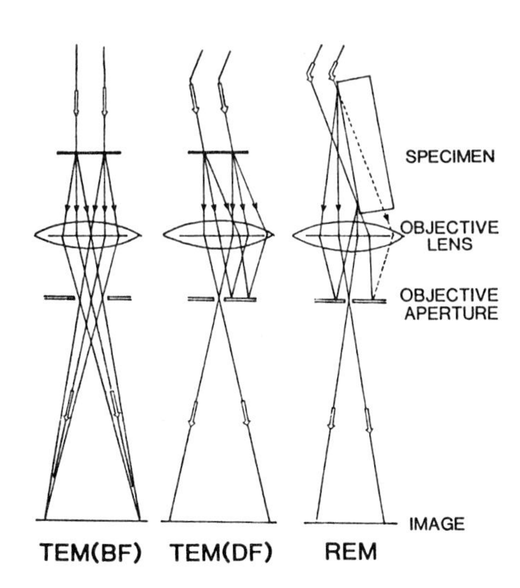

Fig.9.1. These ray diagrams compare Transmission Electron Microscopy (TEM) in Bright Field (BF) and Dark Field (DF) and Reflection Electron Microscopy (REM). REM is similar to TEM in DF except for the specimen part

The different terminations can enhance or suppress particular diffraction spots. Consequently, by selecting the proper diffraction spot to form the REM image, it is possible to resolve domains of stacking faults or domains of different atomic structures on the crystal surface.

9.2 Experimental and Results

Graphite. Extensive studies by electron microscopy and other methods have shown that graphite has basal dislocations in pairs and the region between the two dislocations is a stacking fault (Fig.9.2) [9.8]. These stacking faults form ribbons with widths of the order of 1000 Å and run parallel to the basal planes of the crystal. When these stacking fault ribbons are no more than a few atomic layers beneath the cleavage surface, they become detectable by the REM method.

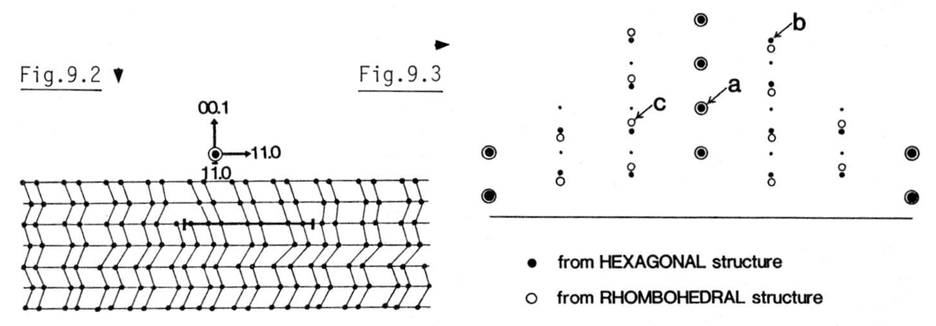

Fig.9.2. Cross-sectional view of the atomic structure of a stacking fault ribbon in graphite. The stacking fault ribbon runs in the direction perpendicular to the diagram. Its cross-section is indicated by the dark line

Fig.9.3. Reflection High Energy Electron Diffraction (RHEED) patterns of hexagonal graphite and rhombohedral graphite superimposed. Sizes of the spots give a rough indication of their relative intensities. Spots *a*, *b*, and *c* are used to form images in Fig.9.4a,b, and c, respectively. This drawing is based on diffraction patterns from the bulk crystals

The stacking sequence, be it hexagonal or rhombohedral, has very little effect on the intensity of the specularly reflected spots (0 0 . 2n). But along the nonspecular rods, the reflected intensities depend on whether there are hexagonal or rhombohedral stacking sequences (Fig.9.3) and imaging with these reflections can be used to image the stacking faults with high contrast.

In Fig.9.4a the REM image formed by using a (0 0 . 2n) spot shows little or no contrast between the ribbon and the perfect crystal. Only the dislocations on each side of the ribbon appear dark, due to the diffraction contrast originating from the lattice distortion [9.2]. However, when the crystal is tilted to excite nonspecular spots, the stacking fault ribbon appears dark or bright depending on the diffraction spot selected (Fig.9.4b,c).

Platinum. Previously reported REM images of Pt(111) surfaces [9.3] show an obvious difference in contrast between areas separated by one-atom-high steps. Figure 9.5 shows such an area. A possible origin of this contrast is that the topmost layer of atoms may be stacked in a sequence different from that of the bulk, resulting in a stacking fault parallel to, and one atom

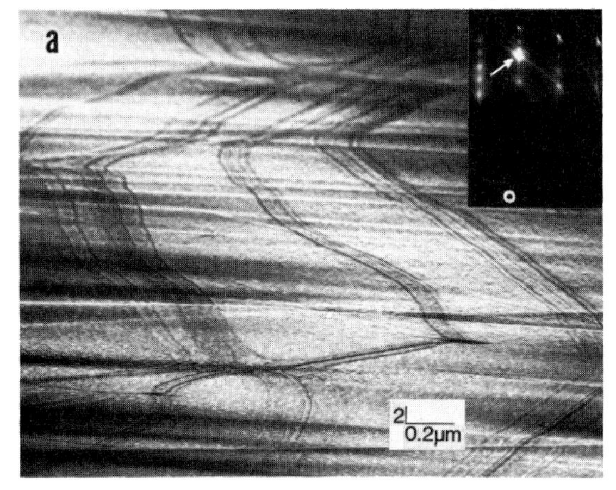

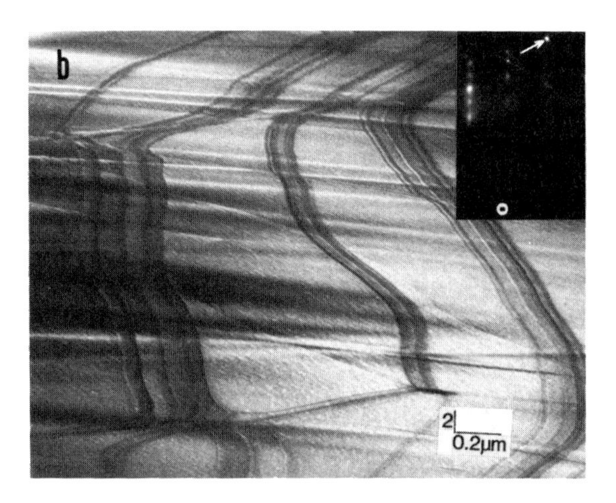

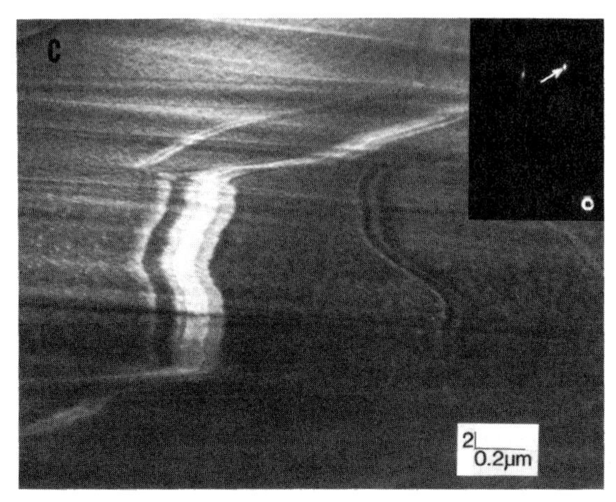

Fig.9.4a-c. REM images of stacking fault ribbons in graphite. Diffraction patterns are shown in the insets. Spots at arrows are used for imaging and they correspond to spots **a**, **b**, and **c**, respectively, in Fig. 9.3. Small circles indicate positions of (0 0 . 0) spots

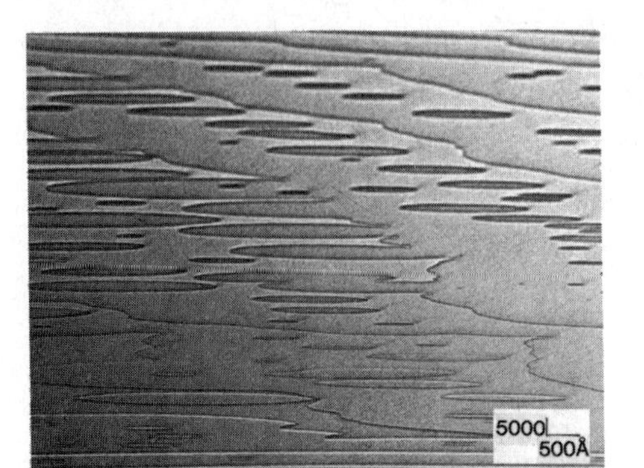

◄ Fig.9.5

Fig.9.6 ▼

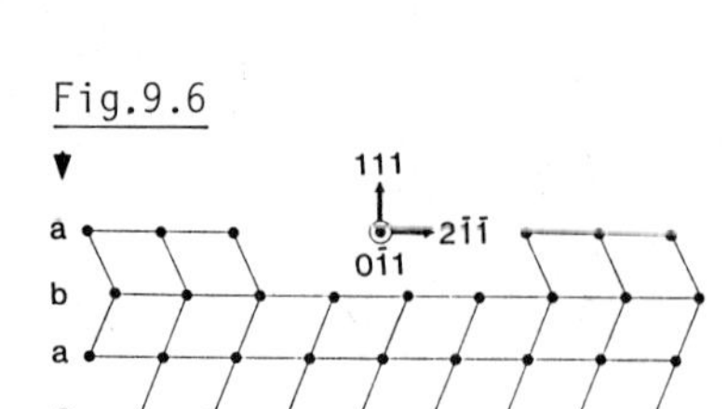

Fig.9.5. A REM image of Pt(111) showing different levels of contrast from areas separated by atomic steps

Fig.9.6. A proposed atomic structure of stacking fault parallel to Pt(111). The topmost layer of atoms are stacked differently. Evaporation created depressions, within which the crystal structure is perfect

beneath, the surface. The many circular loops, which appears as narrow ellipses due to the foreshortening in the electron beam direction, have been shown to be one-atom-deep depressions [9.3] and are very possibly caused by evaporation of surface atoms. If this is the case, the stacking fault model along a cross-section of the depression must be the one shown in Fig.9.6. Again, this stacking fault may be expected to have little effect on the specularly reflected (h,h,h) beams and more effect on the nonspecular beams. However, qualitative observations are not sufficient to allow interpretation of the differences between images recorded from specular and nonspecular reflections. For heavy metals such as Pt the many beam dynamical diffraction effects can produce contrast of this sort even from specular reflections if nonspecular reflections are present in the diffraction pattern. Detailed dynamical calculations are required to resolve the question unambiguously.

9.3 Conclusion

This work has shown that REM is sensitive to stacking faults parallel to the crystal surface and is capable of determining the structure of the stacking fault. It has not been determined whether these stacking fault ribbons in graphite are on the surface or several atomic layers beneath the surface. Comparison with computed intensity values is needed to resolve such questions. Similarly in the case of the Pt surface layers, the presence of surface stacking faults provides an obvious explanation for the observed contrast but a stronger theoretical basis is required for the quantitative interpretation.

Acknowledgment. This work is supported by NSF Grant DMR-7926460 and made use of the Facility for High Resolution Electron Microscopy at ASU supported by NSF Grant DMR-830651 and Arizona State University. The authors also thank Dr. A.W. Moore of Union Carbide Corp. for supplying the graphite specimen. One of the authors (TH) is supported by an IBM Fellowship.

References

9.1 P. Højlund-Nielsen, J.M. Cowley: Surf. Sci. **54**, 340 (1976)
9.2 K. Yagi, K. Takayanagi, G. Honjo: In *Crystals. Growth, Properties and Applications*, Vol.7 (Springer, Berlin, Heidelberg 1982)
9.3 T. Hsu, J.M. Cowley: Ultramicroscopy **11**, 239 (1983)
9.4 T. Hsu: Ultramicroscopy **11**, 167 (1983)
9.5 T. Hsu, S. Iijima, J.M. Cowley: Surf. Sci. **137**, 551 (1984)
9.6 H. Shuman: Ultramicroscopy **2**, 361 (1977)
9.7 S. Iijima, T. Hsu: In Procs. 10th Int. Congress of Electron Microscopy, Hamburg 1982, Vol.II, p.293
9.8 P. Delavignette, S. Amelinckx: J. Nucl. Mater. **5**, 17 (1962)

II.2 Techniques Based on Photons and Other Probes

10. Surface Structure by X-Ray Diffraction

I.K. Robinson

AT&T Bell Laboratories, Murray Hill, NJ 07974, USA

The diffraction of X-rays by crystalline matter has been well understood for 60 years. In that period a vast amount of practical expertise, in the form of X-ray crystallography, has been developed for the determination of crystal structures on the atomic scale. Low-energy electron diffraction (LEED) has extended crystallography to include crystal surfaces, but, as some of the accompanying papers have shown, data analysis is not so straightforward because of multiple scattering and the need for accurate atomic models. Neither of these is a serious problem in X-ray structure analysis.

The X-ray intensity scattered from a crystal is easily calculated by summing the contribution of every electron in the sample using the Thomson formula [10.1]. When applied to a monolayer at the surface of a crystal (or, equivalently, a reconstructed surface or other 2D arrays of atoms) with typical sources and instruments, the numbers in Table 10.1 result. The first lesson is that X-ray experiments with single monolayers are indeed practical. It is clear that under favorable choice of system, a conventional laboratory source is sufficient but for more general problems synchrotron radiation is needed.

Table 10.1. Calculated signal rates for the two reconstructed surfaces of interest to this paper. Approximate primary beam intensities (monochromatic) delivered to a typical sample are compared for conventional and synchrotron sources. The calculated diffraction signals assume a perfect 2D array of atoms, one per unit cell

Source	Primary beam Photons/s	Diffracted from Au(110) 2×1 Photons/s	Diffracted from Si(111) 7×7 Photons/s
60 kW Rotating anode	10^8	10	10^{-4}
SSRL bending magnet	10^{11}	10^4	10^{-1}
SSRL wiggler magnet	10^{12}	10^5	1
SSRL 54-pole wiggler	10^{13}	10^6	10

In proceeding to find the optimum experimental geometry, we must consider the function of an X-ray diffractometer. Conventionally, the detector and source direction lie in a "diffraction plane" fixed in space. The X rays are highly collimated within the plane but poorly normal to it. This gives rise to a resolution function which is highly asymmetric with the long direction normal to the diffraction plane [10.2]. The diffraction pattern of the 2D sample is an array of "rods", sharp in the direction parallel to the surface plane, but diffuse normal to it. Clearly the largest convolution of this pattern with the resolution function is when both long directions are aligned. Thus the surface plane is parallel to the diffraction plane with the incident and diffracted rays at glancing angle. This "glancing incidence" geometry [10.3] has another useful feature: the depth of penetration into the bulk is drastically reduced and so, therefore, is the thermal diffuse background upon which the surface signal sits.

Experimentally, a UHV preparation apparatus, a full 4-circle X-ray diffractometer and high-power X-ray source are required. Specific technical problems are the design of UHV-compatible X-ray windows and a means of precision (0.001°) sample manipulation [10.4]. Published work using the technique in recent years includes Ge(100) [10.5] and Au(110) [10.6] reconstructed surfaces as well as Pb/Cu(110) [10.7] and Xe/Graphite [10.8] adsorption systems. The high-resolution capabilities and the ability to determine atomic structures have been demonstrated. Much work is in progress and a large number of groups are adopting the technique. Some prospective areas of interest are crystal/crystal and crystal/liquid interfaces, 2D melting and phase transitions, nucleation and growth of overlayers and magnetic ordering at surfaces. The remainder of this paper is devoted to recent results on surface reconstructions that will serve as examples.

10.1 Observation of Domain Wall Structures in Au(110)

The structure of the reconstructed Au(110) surface has been previously reported [10.6,9]. Superlattice reflections were found close to half order positions along (100). A 2×1 missing row structure, in which every second row of atoms in the surface layer was absent, was derived by analysis of relative intensities. Pairwise displacements of 0.12 Å were found in the second layer, and this was confirmed by analysis of ion scattering data [10.10]. There was a systematic displacement of the peaks parallel to (100) which varied substantially from sample to sample (Fig.10.1), and was explained in terms of a distribution of domain walls that interrupted the coherence of the surface layer along (100). We proposed monoatomic steps in the surface as a suitable form of walls, consistent with scanning tunneling electron images of Au(110) [10.11].

We have recently succeeded in explaining the observed diffraction line shapes in Fig.10.1 with a suitably "random" distribution of these step domain walls. The absence of wall-wall interaction energy implies the probability of finding a wall is independent of whether it follows another wall or not. We model the Au(110) surface (following [10.9]) by assuming that the predominant structural units are monatomic steps with probability p and 2×1 missing row units with probability (1-p), as shown in Fig.10.2c. This is transformed into a 1-dimensional lattice model by indexing the structure with one site every *half* unit cell as shown in Fig.10.2d and writing the probability of a nearest-neighbor atom m sites away as P_m with

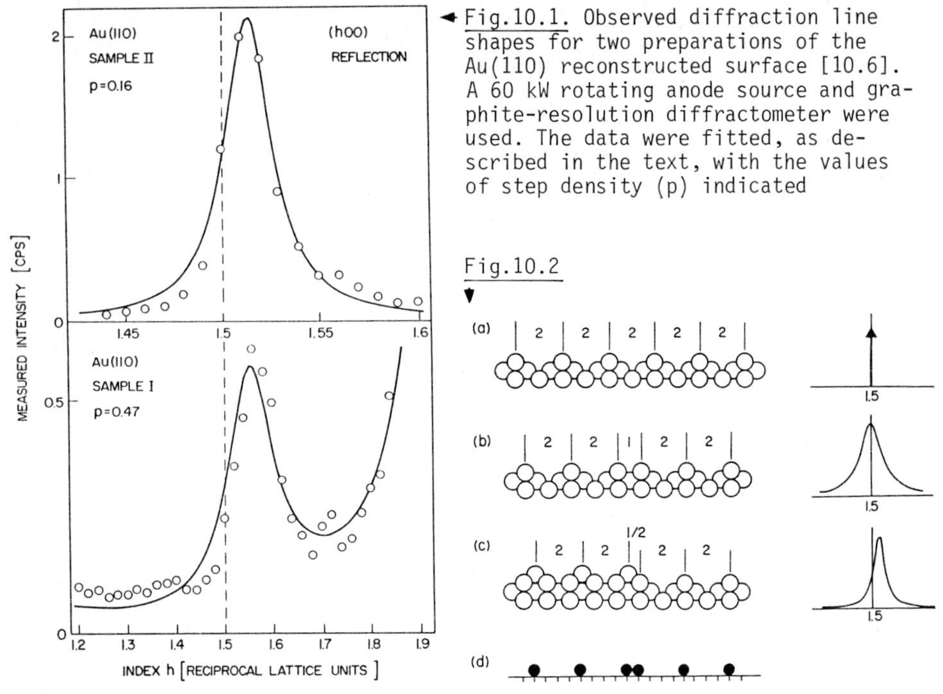

Fig.10.1. Observed diffraction line shapes for two preparations of the Au(110) reconstructed surface [10.6]. A 60 kW rotating anode source and graphite-resolution diffractometer were used. The data were fitted, as described in the text, with the values of step density (p) indicated

Fig.10.2a-d. Calculated radial diffraction profiles near the (3/2,0,0) superlattice reflection for models of Au(110) surfaces. Repeat spacings are indicated in units of the Au lattice constant, 4.0 Å. (**a**) Perfect missing-row structure. (**b**) Missing-row structure with random antiphase boundaries. (**c**) Missing-row structure with random monatomic steps. (**d**) Abstraction of structure in (c) to a 1D lattice model, with the allowed sites marked by ticks. The model does not specify whether the steps are up or down, so their density is not constrained externally

$$P_1 = p \quad ,$$

$$P_2 = P_3 = P_{m>4} = 0 \quad ,$$

$$P_4 = (1-p) \quad .$$

This model obeys the definition of "random" above.

If there is an atom at site $m=0$ then the cumulative probability of finding an atom at site m is

$$Q_m = \frac{1}{m!}\frac{\partial^m}{\partial x^m}\frac{1}{1-px-(1-p)x^4}\bigg|_{x=0} \quad ,$$

using the generator formalism. The structure factor is then simply the Fourier transform of Q_m which reduces exactly to

$$F(q) = \sum_{m=0}^{\infty} Q_m e^{iqm} = \frac{1}{1 - p e^{iq} - (1-p)e^{4iq}} \quad . \tag{10.1}$$

This functional form, which is similar to that calculated for random stacking faults in crystals [10.12], and its analogues for other types of domain walls have been evaluated and compared with observation. As found in [10.9], only the step domain wall of Fig.10.2c explains the observed diffraction peak positions. A counterexample is shown in Fig.10.2b.

Fits of (10.1) to the diffraction data are shown in Fig.10.1. In Figure 10.1 (upper), only the step density p and a scale factor were adjusted for agreement; in Fig.10.1 (lower) the range of momentum transfer was so large that thermal diffuse scattering (Δq^{-2}) around the (200) bulk Bragg peak had to be included as well. The resulting values of step density were $p = 0.16$ and 0.47, corresponding to average interstep spacings of 52 Å and 18 Å respectively. The slight systematic deviation of the calculated curve from the data in Fig.10.1b is probably due to repulsive interaction between the domain walls.

10.2 First Observation of X-ray Diffraction from Si(111)7 × 7

This is the prototypical example of the phenomenon of surface reconstruction, and was historically the first known case [10.13]. Because of quadratic dependence of the scattering intensity both on the atomic number of the scattering atom and on the unit cell area, the expected signal rates here are 10^5 times smaller than for Au(110). This necessitates the use of a storage ring source of X-rays (Table 10.1).

In collaboration with P. Fuoss and J. Stark of AT&T Bell Labs and P. Bennett of Arizona State University, we report diffraction from this surface seen at the Stanford Synchrotron Radiation Laboratory (SSRL). The surface was prepared by resistive heating to 1450 K in 10^{-10}Torr and was studied using the glancing incidence X-ray geometry on the focused wiggler beam line (VII-2).

Under parasitic running conditions, we surveyed a number of the 1/7th order reflections characteristic of the 7 × 7 reconstruction. The most intense one found was at the (1,3/7) position using the 2D hexagonal nomenclature of LEED or (34/21,-26/21,-8/21) in Miller index notation; this yielded 40 counts per second. The in-plane diffraction profiles (Fig.10.3, insert) were sharp, but still 50% broader than resolution. The surface coherence length was about 6000 Å, in agreement with other observations of the annealed surface [10.14,15].

Structural information normal to the surface is contained in the "rod profiles" of the superlattice reflections. These were measured by tilting the sample an angle χ about a direction bisecting the incident and diffracted beams [10.6], thereby adding a perpendicular component ($q_\perp$) to the in-plane momentum transfer ($q_\parallel$). A typical profile is shown in Fig.10.3. The solid line in Fig.10.3 is the dependence of the profile expected from resolution considerations mentioned in the introduction: the instrumental resolution function is approximated as a prolate ellipsoid with a 200:1 axial ratio (from the collimation angles of the diffractometer) which is aligned with the rod only at $q_\perp = 0$. As $q_\perp$ increases, the rod tilts with respect to the ellipsoid and the intersected length σ varies as

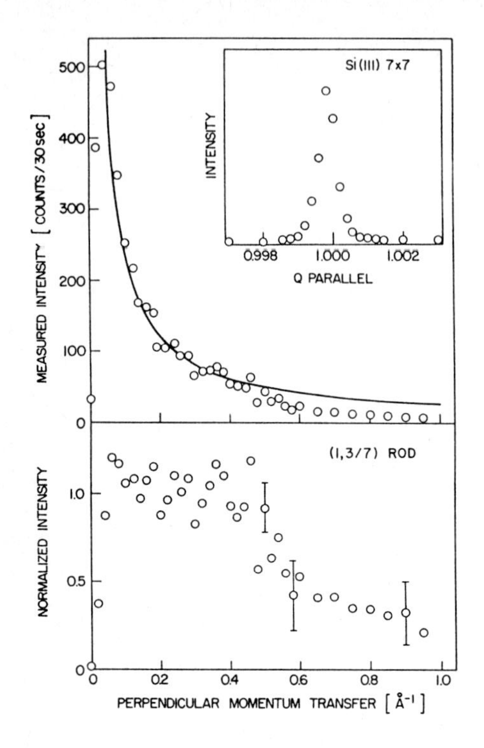

Fig.10.3. *Above*: Diffraction profile of the (1,3/7) superlattice reflection from the Si(111)7 × 7 surface perpendicular to the surface. The solid line is the variation of the width of the resolution function, $\sigma(q_\perp)$ from (10.2). The parallel component of momentum transfer for this reflection is $q_0 = 2.40$ Å^{-1}, so the sample tilts from $\chi = 0$ to $\chi = 22°$ over the range of the scan. *Below*: the same data normalized to $\sigma(q_\perp)$. Insert: radial diffraction profile of the (1,3/7) reflection. The ordinate is in units of q_0

$$\frac{1}{\sigma^2} = \frac{1}{a^2} + \sin^2\chi\left(\frac{1}{b^2} - \frac{1}{a^2}\right) \quad ,$$

$$\tan\chi = \frac{q_\perp}{q_\parallel} \quad , \qquad (10.2)$$

where $a = 2°$ and $b = 0.01°$ are the major and minor axes of the ellipsoid. Figure 10.3 shows that the observed scattering falls below this form at large $q_\perp$ values, and when normalized to it (Fig.10.3 lower) shows a smooth variation. The depth of the reconstruction, which is inversely related to the half-width of the rod, is estimated from the normalized profile to be about 10 Å, or 3 silicon layer spacings.

In-plane structural information is contained in the relative intensities of the different 1/7 order reflections. We measured 30 positions including some symmetry equivalents; the reproducibility was only 30%, and it was clear that a better understanding of the relation between resolution function and sample alignment would be necessary to obtain 1% accuracy. There is, however, excellent *qualitative* agreement between these data and published transmission electron diffraction photographs of Si(111)7 × 7 [10.16].

Finally, we followed the departure of long-range order through the 7 × 7 to "1 × 1" order-disorder transition at 1100 K, as shown in Fig.10.4. We found the transition to be very sharp, with no evidence of residual scattering beyond. We saw no evidence of hysteresis or irreversible behavior.

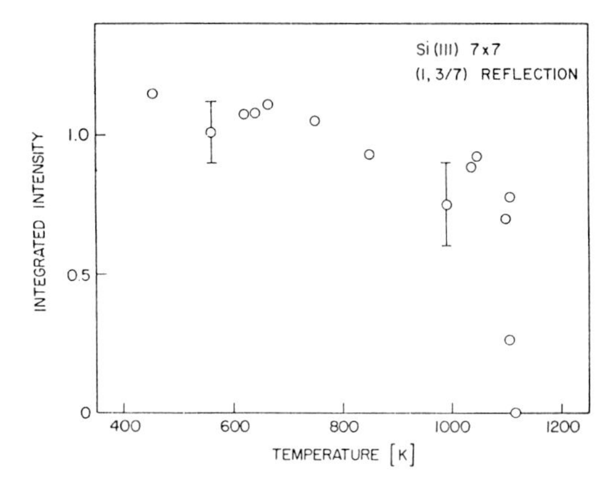

Fig.10.4. Temperature dependence of the integrated (1, 3/7) intensity. Low temperatures were measured from the silicon lattice constant; high temperatures (over 1000 K) were measured by optical pyrometry. In both cases the precision is about ±5 K and the accuracy ±20 K. The error bars on intensity values arise from misalignment

References

10.1 B.E. Warren: *X-ray Diffraction* (Addison Wesley, Reading 1969)
10.2 D.E. Moncton, G.S. Brown: Nucl. Inst. Meth. **208**, 579 (1983)
10.3 W.C. Marra, P. Eisenberger, A.Y. Cho: J. Appl. Phys. **50**, 6927 (1979)
10.4 P.H. Fuoss, I.K. Robinson: Nucl. Inst. Meth. **222**, 171 (1984)
10.5 P. Eisenberger, W.C. Marra: Phys. Rev. Lett. **46**, 1081 (1981)
10.6 I.K. Robinson: Phys. Rev. Lett. **50**, 1145 (1983)
10.7 W.C. Marra, P.H. Fuoss, P. Eisenberger: Phys. Rev. Lett. **49**, 1169 (1982)
10.8 K. D'Amico, D.E. Moncton: To be published
10.9 I.K. Robinson, Y. Kuk, L.C. Feldman: Phys. Rev. B**29**, 4762 (1984)
10.10 Y. Kuk, L.C. Feldman, I.K. Robinson: Surf. Sci. **138**, L168 (1984)
10.11 G. Binning, H. Rohrer, C. Gerber, E. Weibel: Surf. Sci. **131**, L379 (1983)
10.12 A. Guinier: *X-ray Diffraction* (Freeman, San Francisco 1963)
10.13 R.E. Schlier, H.E. Farnsworth: J. Chem. Phys. **30**, 917 (1959)
10.14 M. Henzler: Surf. Sci. **132**, 82 (1983)
10.15 Y. Tanishiro, K. Takayanagi, K. Yagi: Ultramicroscopy **11**, 95 (1983)
10.16 K. Takayanagi, Y. Tanishiro, M. Takahashi, H. Motoyoshi, K. Yagi: Proc. 10th Int. Cong. on Elec. Micros. (Frankfurt 1982)

11. Optical Transitions and Surface Structure

P. Chiaradia and A. Cricenti

Istituto di Struttura della Materia of the CNR, Frascati, Italy

G. Chiarotti, F. Ciccacci, and S. Selci

Dipartimento di Fisica, II Universita' di Roma, Italy and
Istituto di Struttura della Materia of the CNR, Frascati, Italy

To obtain information on the structure of a solid surface probes should be employed having a wavelength as short as the interatomic distance, which is typically of the order of an angstrom. In the case of electromagnetic radiation, this corresponds to X-rays. On the other hand, optical transitions involving electronic surface states are excited by near IR, visible and near UV light, whose wavelength is too large to probe the atomic structure directly. In some cases, however, valuable structural information can be obtained from optical spectroscopy indirectly, that is via the electronic structure. This is accomplished by exploiting symmetry arguments and making use of models. The present article exemplifies the above concepts by showing their application to the cleavage surface of the covalent semiconductors Si and Ge.

11.1 The Si(111)2 × 1 and Ge(111)2 × 1 Surface Structures

It is well known that after the buckling model failed to explain the angle-resolved photoemission spectroscopy (ARPES) data for the Si(111)2 × 1 surface [11.1,2] *Pandey* put forward an entirely new model, in which surface atoms are disposed along zigzag chains running in the [$\bar{1}$10] direction [11.3]. In spite of its great geometrical complexity, Pandey's model is successful for total energy minimization [11.3] and agrees with an increasing number of experimental results [11.1,2,4-6]. Moreover, this chain-like structure can be derived from the ideal (111) crystal termination by following a path with a very low energy barrier [11.7]. Still, some controversy remains in settling the Si(111)2 × 1 structure definitively, due to the fact that Pandey's model (as well as all other current models) disagrees with dynamical LEED data [11.8].

Although the chain model was originally proposed only for silicon, it is widely assumed to be valid for Ge(111)2 × 1 as well. This stems from the similarity between the two surfaces, as indicated by ARPES [11.9] and confirmed by theoretical calculations [11.10] and more recently by IR optical spectroscopy with polarized light [11.11].

We wish to underline that strangely enough at the present time all experimental evidence supporting Pandey's model, with the exception of the ion channeling-blocking experiment by *Tromp* et al. [11.4], comes from spectroscopies of the electronic structure (ARPES and light absorption or reflection) rather than genuine structural techniques. No tunneling microscopy, SEXAFS or atom scattering experiments on either Si(111)2 × 1 or Ge(111)2 × 1 have been reported so far.

11.2 Optical Experiments and Surface Structure

Let us examine in more detail how the results of optical experiments on Si and Ge cleavage surfaces yield structural information and test structural models. Symmetry properties of the surface unit cell engender selection rules for the dipole matrix element in optical transitions [11.12]. For example, in Si(111)2 × 1 the existence of a $(\bar{1}10)$ mirror plane [11.13] causes electron wave functions at high symmetry points of the Surface Brillouin Zone (SBZ), where important optical transitions usually take place, to have definite parity. This in turn sets a condition for the direction of the electric field $\vec{E}$ of the light in order to have allowed transitions.

The experiment can be performed either in reflection or absorption mode: in both cases the product $\varepsilon_2 d$ is obtained, where ε_2 is the imaginary part of the surface dielectric function and d is a phenomenological surface thickness [11.14]. Strictly speaking, the above statement is true only at photon energies lower than the fundamental gap, though under special conditions the same result is obtained also when the substrate is itself absorbing [11.15].

Recently the polarization dependence of the surface optical properties below the gap has been measured in single-domain surfaces by differential reflection spectroscopy [11.5,16] as well as by a unconventional new technique, named Photothermal Displacement Spectroscopy (PTDS) [11.6,11,17].

11.3 Results and Discussion

11.3.1 Si(111)2 × 1

The experimental results for the Si(111)2 × 1 surface are shown in Fig.11.1, where the differential reflectivity spectrum in the range 0.3 to 3.5 eV is plotted. The LEED pattern is schematically shown in the inset. At photon energies below the gap a strongly anisotropic peak is present for an electric vector in the $[\bar{1}10]$ direction. Above the gap we observe a broad band with opposite polarization dependence and a much reduced anisotropy [11.16]. As already mentioned, PTDS below the gap yields essentially the same results [11.6].

The interpretation of these experimental findings is straightforward in the framework of the chain model, which is essentially a 1 D structure [11.5,6,18]. The reduced anisotropy of the higher energy peak is most probably related to a mixing of the dangling-bond (DB) wave functions with degenerate bulk bands [11.16].

On the other hand, the buckling model, being much more isotropic, cannot account for these experimental results, not even qualitatively [11.5,6,18].

Theoretical results of the dangling-bond (DB) bands (filled and empty) in Si(111)2 × 1 obtained with a self-consistent pseudopotential calculation [11.7] are shown in Fig.11.2. The relevant part of the SBZ is sketched in the inset. Strong optical transitions are expected to occur near Γ,J' and along the JK line. It can be demonstrated [11.19] that initial and final states along ΓJ' and along JK have the same and opposite parity, respectively, due to the existence of the $(\bar{1}10)$ mirror plane. Indeed the SBZ of the ideal chain model contains also a glide plane, which should cause the two DB bands of Fig.11.2 to merge along JK [11.20]. However, this degeneracy is lifted by either a departure from the ideal configuration or by interaction with the back bonds or both. Then independently of the model one can predict that op-

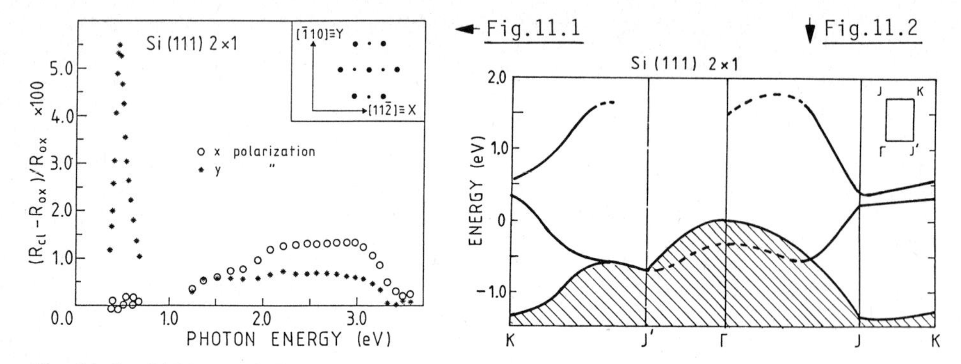

Fig.11.1. Differential reflectivity spectrum of a Si(111)2 × 1 single-domain surface in the visible and near IR range. The inset shows a sketch of the LEED pattern, with extra spots in the [11$\bar{2}$] direction. Light was polarized along the x = [11$\bar{2}$] (*open circles*) and y = [$\bar{1}$10] (*asterisks*) directions

Fig.11.2. Calculated surface dangling-bond bands along the main symmetry directions of the surface Brillouin zone (shown in the inset) for the chain model of Si(111)2 × 1 [11.7]. The hatched area indicates the projected bulk bands. *Dashed lines* point out surface resonances as opposed to true surface states (*solid lines*)

tical transitions taking place along ΓJ' and JK are allowed for electric fields in [11$\bar{2}$] and [$\bar{1}$10] directions, respectively. Thus the absorption peak at 0.45 eV originates from transitions along JK, while the broad band above the gap is essentially due to transitions near the ΓJ' line. This picture is consistent with calculations based on the chain model [11.18]. In the buckling model, on the contrary, the DB bands are much less dispersive throughout the SBZ [11.21] and we can assume that transitions at J' and along JK occur at the same energy [11.22]. In this case, the anisotropy ratio should be 3:1 with a maximum for polarization along [11$\bar{2}$] [11.18], contrary to the experimental results of Fig.11.1.

11.3.2 Ge(111)2 × 1

The surface absorption spectrum of Ge(111)2 × 1 as measured below the gap by means of PTDS [11.11] is reported in Fig.11.3a. As in the case of Si($\bar{1}$11)2 × 1, optical transitions occur for light polarized along the [$\bar{1}$10] direction.

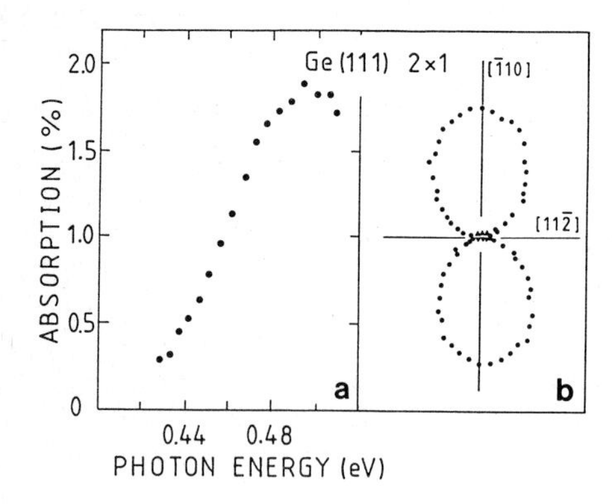

Fig.11.3. (**a**) Ge(111)2 × 1 surface state absorption spectrum obtained by PTDS [11.11]. Incident light is polarized to [$\bar{1}$10]. (**b**) Polar plot of the polarization dependence at 0.496 eV (dots). Triangular data points were taken after an exposure to 10^4L of O_2

The similar behavior of Si(111)2 × 1 and Ge(111)2 × 1 clearly demonstrates that the chain model is valid for both surfaces.

11.4 Conclusions

Optical spectroscopies can yield unambiguous information about surface reconstruction models for the cleavage faces of Si and Ge. The reason why optical techniques are so powerful lies in the selection rules associated with symmetry properties of the surface unit cell. In the case of Si(111)2 × 1 and Ge(111)2 × 1, by investigating the electronic structure with polarized light one can test the various models proposed for these reconstructions. As a result, only Pandey's chain model has been found to agree with the experimental data.

References

11.1 F.J. Himpsel, P. Heimann, D.E. Eastman: Phys. Rev. B**24**, 2003 (1981)
11.2 R.I.G. Uhrberg, G.V. Hansson, J.M. Nicholls, S.A. Flodström: Phys. Rev. Lett. **48**, 1032 (1982)
11.3 K.C. Pandey: Phys. Rev. Lett. **47**, 1913 (1981); Phys. Rev. Lett. **49**, 223 (1982)
11.4 R.M. Tromp, L. Smit, J.F. Van der Veen: Phys. Rev. Lett. **51**, 1672 (1983)
11.5 P. Chiaradia, A. Cricenti, S. Selci, G. Chiarotti: Phys. Rev. Lett. **52**, 1145 (1984)
11.6 M.A. Olmstead, N.M. Amer: Phys. Rev. Lett. **52**, 1148 (1984)
11.7 J.E. Northrup, M.L. Cohen: Phys. Rev. Lett. **49**, 1349 (1982)
11.8 H. Liu, M.R. Cook, F. Jona, P.M. Marcus: Phys. Rev. B**28**, 6137 (1983)
11.9 J.M. Nicholls, G.V. Hansson, R.I.G. Uhrberg, S.A. Flodström: Phys. Rev. B**27**, 2594 (1983);
J.M. Nicholls, G.V. Hansson, U.O. Karlson, R.I.G. Uhrberg, R. Engelhardt, K. Seki, S.A. Flodström, E.E. Koch: To be published
11.10 J.E. Northrup, M.L. Cohen: Phys. Rev. B**27**, 6553 (1983)
11.11 M.A. Olmstead, N.M. Amer: Phys. Rev. B**29**, 7048 (1984)
11.12 F. Bassani: Proc. of the Int. School of Physics, Varenna, course 34, p.36;
F. Bassani, Pastori-Parravicini: *Electronic States and Optical Transitions in Solids* (Pergamon, New York 1975) p.95
11.13 R. Feder, W. Mönch, P.P. Auer: J. Phys. C**12**, L179 (1979)
11.14 P. Chiaradia, G. Chiarotti, S. Nannarone, P. Sassaroli: Solid State Commun. **26**, 813 (1978)
11.15 P. Chiaradia, G. Chiarotti, I. Davoli, S. Nannarone, P. Sassaroli: Inst. Phys. Conf. Ser. No.43, 1979, Chap.6;
S. Nannarone, P. Chiaradia, F. Ciccacci, R. Memeo, P. Sassaroli, S. Selci, G. Chiarotti: Solid State Commun. **33**, 593 (1980);
P. Chiaradia, G. Chiarotti, F. Ciccacci, R. Memeo, S. Nannarone, P. Sassaroli, S. Selci: Surf. Sci. **99**, 70 (1980)
11.16 S. Selci, F. Ciccacci, A. Cricenti, P. Chiaradia, G. Chiarotti: To be published
11.17 N.M. Amer, M.A. Olmstead: Surf. Sci. **132**, 68 (1983)
11.18 R. Del Sole, A. Selloni: Solid State Commun. **50**, 825 (1984)
11.19 A. Selloni: Private communication
11.20 V. Heine: Group Theory in Quantum Mechanics (Pergamon, New York 1960) p.287
11.21 K.C. Pandey, J.C. Phillips: Phys. Rev. Lett. **34**, 1450 (1975);
M. Schlüter, J.R. Chelikowski, S.G. Louie, M.L. Cohen: Phys. Rev. B**13**, 4200 (1975)
11.22 D.J. Chadi, R. Del Sole: J. Vac. Sci. Technol. **21**, 319 (1981)

12. High-Resolution Infrared Spectroscopy and Surface Structure

Y.J. Chabal

AT&T Bell Laboratories, Murray Hill, NJ 07974, USA

The potential of high-resolution surface infrared spectroscopy as a structural tool is assessed by combining results on the Si(100)-(2 × 1)H system with ab initio cluster calculations. It is shown that a complete determination of the geometry and vibrational parameters can be obtained.

12.1 Introduction

Since the pioneering work of *Francis* and *Allison* [12.1] on metal substrates and of *Becker* and *Gobeli* [12.2] on semiconductor substrates, infrared spectroscopy has been used increasingly to probe vibrational modes of adsorbates on single-crystal surfaces [12.3]. Like all vibrational spectroscopies, Surface Infrared Spectroscopy (SIS) gives direct information on the *chemical* nature of adsorbates [12.4]. Further, by monitoring the vibrational frequency of strong internal modes such as CO on metal surfaces, SIS has proven useful in distinguishing unambiguously [12.5] between different adsorption sites (on-top, bridge, etc.). However, it is the high resolution of SIS which sets it apart from other surface vibrational techniques such as Electron Energy Loss Spectroscopy (EELS). The high resolution makes it possible to monitor small frequency shifts resulting either from isotopic substitutions or from coverage variations and thus to separate dipole interactions [12.6] from chemical interactions [12.7]. Line shapes can also be studied to uncover the formation of islands [12.8,9] reflected in the inhomogeneous broadening or to probe the vibrational deexciation mechanisms [12.10] manifested in the homogeneous broadening. Clearly, the study of internal modes yields important information on the surface arrangement and the vibrational parameters of adsorbates.

With the development of sensitive techniques, the observation of substrate-adsorbate modes has been possible [12.11,12] and opened a new and potentially rich area. Indeed, these modes are a direct result of the chemisorption process and reflect more sensitively the adsorption geometry. They tend to be weak so that dipole interactions do not overshadow fine frequency shifts arising from chemical interactions. Finally, the study of such modes yields important information on the gas-solid interaction potential and potentially on the vibrational energy decay channels into the substrate [12.13]. The purpose of this paper is to assess the potential of studies of substrate-adsorbate vibrational modes with high-resolution infrared spectroscopy as a *structural* tool. The system chosen for this study, H on the Si(100)-(2 × 1) surface, presents several advantages. First, the stretching vibration of SiH (~ 2100 cm^{-1}) is far removed from the substrate phonon frequencies ($\leqslant 500$ cm^{-1}) and from the electronic continuum ($\geqslant 8800$ cm^{-1}) so that the lifetime broadening by either phonon decay [12.13] or electron-hole pair coupling [12.14] will be negligible. Second, the dynamic dipole moment associated with the SiH vibration is extremely small [12.15] so that the dipole interactions are expected to be small ($\leq 3\ cm^{-1}$). Finally, the use of a substrate transparent in the frequency region of the SiD and SiH stretch makes possible the study of parallel as well as perpendicular components of the vibrational modes by means of polarized radiation. We find that by combining high-resolution SIS measurements on single domain Si(100)-(2 × 1) H surfaces with ab initio cluster calculations, a complete structural and dynamic description of the most stable hydride phase on Si(100)-(2 × 1) can be obtained.

12.2 Experimental

The experimental apparatus, comprising a Nicolet interferometer coupled to a UHV chamber by means of CsI windows, has been previously described [12.15, 16]. A multiple internal reflection geometry shown in Fig.12.1 increases the sensitivity sufficiently so that 1% of a monolayer of the weak SiH mode can be probed with a nominal resolution of 1 cm^{-1} [12.17]. More importantly for the structural analysis, the infrared beam can be polarized so as to probe components parallel (s-pol.) and perpendicular to the surface as shown in Fig.12.1.

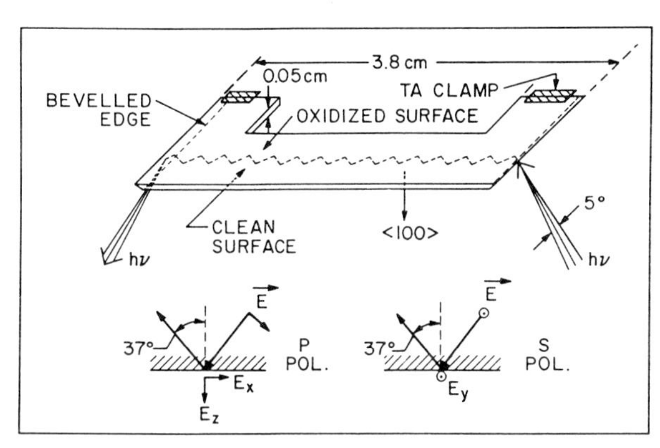

Fig.12.1. Sketch of the sample cut in a U-shape to allow easy clamping without interfering with the bevelled edges and to give a uniform resistive heating of the portion probed by the IR beam. Radiation is incident on the input bevelled edge and is totally internally reflected 50 times on the surface under study with an internal angle of incidence of 37°. The other surface is kept oxidized so that no hydrogen or deuterium chemisorption can take place. The adsorbate-induced spectrum is obtained by subtracting from the spectrum obtained after a given exposure that of the clean surface. The electric field components at the surface for the two polarizations used are shown at the bottom

The flat Si(100) surfaces exhibit a 2 × 1 reconstruction when clean. This reconstruction is believed to be due to the dimerization of two neighboring surface Si atoms [12.18]. For a given surface plane, the dimers must all be aligned in the same direction. However, the LEED patterns indicate that domains are present oriented perpendicular to each other. The flat surface must therefore be made up of several terrace planes, some of which are an odd number of layers deep relative to others. The separation between these domains is presumably due to steps or defects. Upon H chemisorption on a substrate at ~300°C, the LEED patterns display a very sharp 2 × 1, indicating that the Si dimers are preserved and that there is one H bound to each surface Si atom [12.19] in the plane defined by the dimer direction and the normal to the surface. However, the presence of two domains on the flat Si(100) surface precludes the use of polarization to characterize the contribution of step atoms fully (which must be present to account for the two domains) and that of dimerized terrace atoms. Single domains can be obtained, however, following the observations by *Kaplan* [12.20] that surfaces cut along the <011> direction at some angle, i.e., surfaces vicinal to the (100) plane are characterized by a regular array of terraces separated by *double-layer* steps. For this reason, we have performed measurements on a stepped surface cut at 9° to the Si(100) surface along the <011> direction. On this surface, the dimers are all aligned to the step edge as shown in Fig.12.2. Moreover, the step atoms all have dangling bonds directed in the plane defined by the <011> and <100> vectors, i.e., perpendicular to the plane of the terrace dangling bonds defined by the <100> and <011> vectors. The LEED patterns show no change upon H chemisorption at 250°C except for small intensity variations, indicating that the dimers all remain parallel to the step edge. Figure 12.2 further shows how modes of H on the terraces (terrace modes) can be unambiguously distinguished from modes of H adsorbed on step atoms (step modes) by polarization studies. Using configurations I and II, the modes associated with H on step atoms (i.e., polarized along the step dangling bond direction) can be determined even if their frequencies are identical with those associated with H on terrace atoms.

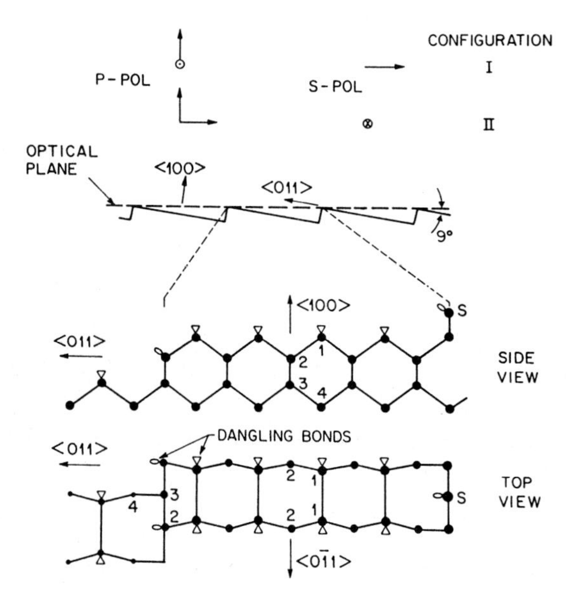

Fig.12.2. (*top*) Geometry of the two configurations used with electric field directions indicated. In configuration I, P-pol. radiation probes modes parallel to the step edge and perpendicular to the terrace plane (i.e., terrace modes) while S-pol. probes only modes perpendicular to the step edge (i.e., step modes). In configuration II, P-pol. probes modes normal to the terrace and the step edge (i.e., both step and terrace modes) while S-pol. probes only modes parallel to the step edge (i.e., terrace modes). (*bottom*) Atomic representation of a clean terrace with adjacent steps for a 9° cut as deduced by LEED. The side view shows the polarization of the unique dangling bond associated with the step atom S. The top view shows the positions of the four dimers on the terrace and the relative position of dimers on adjacent terraces

12.3 Results

Selected results for D and H adsorption are summarized in Fig.12.3 (*A*, *B* and *B*'). At 80 K, the main terrace modes occur at 2087.5 and 2098.8 cm^{-1} for H and 1519 and 1528 cm^{-1} for D. The step modes (dashed curves), probed by S-pol. (conf. I) and P-pol. (conf. II), occur in between the terrace modes at low coverages (e.g., at 2094 cm^{-1} for 0.3 monolayer H in Fig.12.3*B*). They shift to lower frequency as the coverage increases (e.g., see the two peaks at 1524 and 1519 cm^{-1} for 0.5 monolayer D in Fig.12.3*A*), as confirmed by

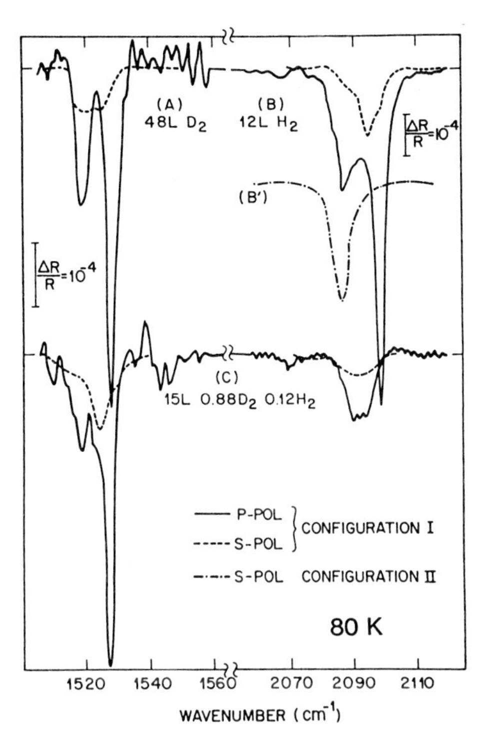

Fig.12.3. Infrared spectra ($\Delta R/R$) induced by (*A*) 0.5 monolayer of deuterium; (*B*) and (*B'*) 0.3 monolayer of hydrogen; and (*C*) 0.4 monolayer of 84% deuterium and 16% hydrogen on a Si(100) 9° surface. All exposures were done with a substrate temperature T = 250°C and the data was recorded at T = 80 K. The dashed lines (S-pol.) are smoothed for clarity. The apodized resolution is 1 cm^{-1}

P-pol. data (conf. II), not shown for clarity. At saturation coverage, the step modes have a single peak at 1519 cm^{-1} and 2087.5 cm^{-1} for D and H, respectively, thus clearly interfering with intensity measurements on flat surfaces where two domains are present and step modes cannot be distinguished with polarization studies. Hence, the step modes are now identified, and we can establish that the terraces are characterized by two modes, one polarized parallel to the surface ($M_{\parallel}$) along the dimer direction and the other polarized perpendicular to the surface ($M_{\perp}$). The strength of these modes yield a constant ratio $(I_{\perp}/I_{\perp})=0.67/\varepsilon^2_{\infty\perp}$ after correction for field amplitudes and the 9° inclination. Here, $I_{\perp}$ and $I_{\parallel}$ are the unscreened intensities and $\varepsilon_{\infty\perp}$ is the electronic screening [12.15,21], which can be measured experimentally once the dipole interaction is determined. To establish on the one hand that $M_{\parallel}$ and $M_{\perp}$ are associated with one local structure, i.e., not due to two different chemisorption sites with equal sticking coefficients, and to determine on the other hand the origin of the frequency splitting, isotopic mixture experiments were performed. For instance, a surface exposed to a dilute amount of H in a D-rich mixture exhibits only one mode at 2092 cm^{-1} associated with H (Fig.12.3*C*). Spectra obtained with configuration II show that this mode is a terrace mode parallel to the step edge with a screened ratio $\varepsilon^2_{\infty\perp}(I_{\parallel}/I_{\perp})$ ~ 0.7. Hence, the two-site hypothesis can be ruled out since a parallel mode would have no dipole interaction with a neighboring perpendicular mode [12.22]. Having established that the two modes $M_{\parallel}$ and $M_{\perp}$ are two normal modes of the same local structure, we must now understand the origin of the frequency splitting of the doublet $M_{\parallel}$ and $M_{\perp}$. In general, there will be a contribution from the dipole interaction between neighboring SiH units and a contribution due to "chemical" interactions which include dynamical effects. We can determine the chemical interaction by measurements over a wide range of coverages. Since at very low coverages ($\theta < 0.1$) we observe a broad mode due partly to isolated H on a Si-Si dimer and partly to isolated monohydride, we must use data from $\theta = 0.1$ to $\theta = 1$ monolayer in which the two modes can be seen clearly. We find $\Delta_{chem}(H) = 9\ cm^{-1}$ and $\Delta_{chem}(D) = 7.5\ cm^{-1}$. These values are confirmed by carrying out constant coverage ($\theta = 1$ monolayer) isotopic experiments (H/D = 0 to ∞) which also establish that the largest dipole/dipole interaction (at $\theta = 1$ monolayer) is 3 cm^{-1} for H and 2 cm^{-1} for D. Finally, we use the measured absolute and relative absorption strengths of each mode $M_{\parallel}$ and $M_{\perp}$ along with the knowledge of the dipole-dipole splitting to determine the screening $\varepsilon_{\infty} = 1.4 \pm 0.05$. We are now able to extract the unscreened ratio of the dynamic dipole moments associated with $M_{\parallel}$ and $M_{\perp}$, $(\mu_{\parallel}/\mu_{\perp}) \simeq (I_{\parallel}/I_{\perp})^{1/2} = 0.6 \pm 0.1$. Now that the vibrational parameters (chemical and dipole interactions) have been determined, we turn to the structure determination.

12.4 Discussion

The two possible bonding configurations for hydrogen which would give rise to two normal modes in the SiH stretching region are the monohydride and dihydride [12.19] shown in Fig.12.4. Although it is tempting to assign $M_{\parallel}$ and $M_{\perp}$ to the antisymmetric and symmetric stretching modes of the dihydride phase [12.15], for which the dimer is broken and two H are bound to each surface Si atom [12.19], several experimental observations do not support this assignment. The modes persist upon annealing to 350°C at which point the LEED pattern is sharp 2×1. More importantly, recent EELS data [12.22] show that the scissor mode associated with the dihydride species vanishes upon annealing to 350°C. Further, the dihydride stretching modes have recently been observed by SIS at 2106 and 2116 cm^{-1} [12.21]. While these results indicate that the observed modes $M_{\parallel}$ and $M_{\perp}$ are not associated with the dihydride, calculations must be carried out to assign them unambiguously to the monohydride structure.

Using an ab initio cluster approach well suited for the study of semiconductor surfaces for which delocalization effects are usually not significant, *Raghavachari* and *Chabal* [12.23] calculated the optimized geometry by energy minimization procedures. The monohydride structure, with a dimer length of 2.51 Å, a Si-H bond length of 1.48 Å and an angle θ(Si-Si-H) = 110°, was found

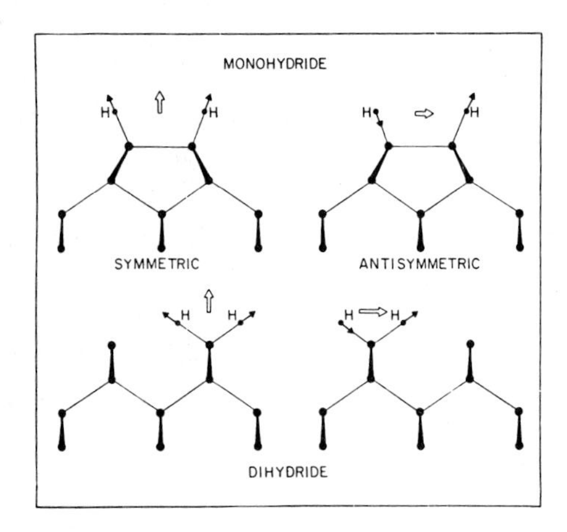

Fig.12.4. Sketches (to scale) of the monohydride (*top*) and dihydride (*bottom*) with motion of the hydrogens indicated by an arrow (not to scale). The polarization associated with the symmetric and antisymmetric stretching modes for both hydride configurations is marked by a double arrow (not to scale)

to be the most stable. Further, the frequency splittings, due to chemical interactions only, were found to be +11 cm^{-1} for H and +9 cm^{-1} for D, where the + sign indicates that the symmetric stretch is at higher frequency than the antisymmetric stretch. These values agree excellently with the experimental findings: +9 cm^{-1} for H and +7.5 cm^{-1} for D. In sharp contrast, the dihydride structure gave +5 cm^{-1} for H and -14 cm^{-1} for D, in clear disagreement with the data.

Finally, the dipole moments associated with each normal mode were evaluated by taking finite steps along the mode direction, yielding 1.36 D/Å and 0.80 D/Å for $M_{\|}$ and $M_{\perp}$, respectively. The ratio $(\mu_{\|}/\mu_{\perp})$ was thus calculated to be 0.59 in remarkable agreement with the experimentally determined unscreened ratio. Note that the back projection of $\mu_{\|}$ and $\mu_{\perp}$ along the Si-H bond does not give the same value for the dynamic dipole moment. In other words, if we had assumed that the dynamic dipole moments associated with $M_{\|}$ and $M_{\perp}$ were the same, then the Si-H direction with respect to the surface normal would have been $\alpha = \tan^{-1}(\mu_{\|}/\mu_{\perp}) \simeq 30^{\circ}$ instead of 20°. Hence, the geometric angle of an adsorbate cannot, in general, be deduced by polarized infrared data alone without some knowledge of the relative strengths of the dynamic dipole moments associated with the normal modes. For example, in the study of the Si(100): H_2O system [12.21] where only the effective dipole moment directions were measured, the actual geometric bonds will probably be oriented differently.

12.3 Conclusions

This work demonstrates that high-resolution infrared spectroscopy of substrate-adsorbate modes can be used to arrive at an accurate structure determination for semiconductor substrates. For the system considered, H on Si(100)-(2 × 1), ab initio cluster calculations are well suited to calculate the geometry and the associated vibrational frequencies because the modes are well decoupled from the substrate phonon vibrations. This is not the case for heavier adsorbates such as Cl or O for which other approaches may be preferred [12.24,25] to take into account the influence of the substrate surface phonons.

For metallic substrates selection rules dictate that only components perpendicular to the surface can be probed by SIS, which in general reduces the amount of information available from experiments. However, motion parallel to the surface can, in some cases, be detected by SIS [12.26]. Along with the increasing successes of SIS for metallic substrates, for which different experimental approaches are used [12.27], the level of theory is now good enough [12.28] to indicate that accurate structure determinations can also be expected from SIS for adsorbates on metals.

Acknowledgements. The author is grateful to K. Raghavachari, D.R. Hamann, E.G. McRae and M.J. Cardillo for stimulating discussions, to S.B. Christman, E.E. Chaban and M.E. Sims for critical technical assistance, and to A.M. Bradshaw for a critical reading of the manuscript.

References

12.1 S.A. Francis, A.H. Allison: J. Opt. Soc. Am. **49**, 131 (1959)
12.2 G.E. Becker, G.W. Gobeli: J. Chem. Phys. **38**, 2942 (1963)
12.3 See, for example, the excellent review of F.M. Hoffmann: Surf. Sci. Reports 3, 107 (1983) and references therein
12.4 M.K. Debe: Appl. Surf. Sci. **14**, 1 (1982-1983)

12.5 E.G.A. Ortega, F.M. Hoffmann, A.M. Bradshaw: Surf. Sci. **119**, 79 (1982)
12.6 R. Shigeishi, D.A. King: Surf. Sci. **58**, 379 (1976)
12.7 P. Hollins, J. Pritchard: Surf. Sci. **89**, 489 (1979)
12.8 A. Crossley, D.A. King: Surf. Sci. **95**, 131 (1980); D.A. King: In *Vibrational Spectroscopy of Adsorbates*, ed. by R.F. Willis (Springer, Berlin, Heidelberg, New York 1980) p.179
12.9 H. Pfnür, D. Menzel, F.M. Hoffmann, A. Ortêga, A.M. Bradshaw: Surf. Sci. **93**, 431 (1980)
12.10 B.N.J. Persson, R. Ryberg: Phys. Rev. Lett. **48**, 549 (1982)
12.11 Y.J. Chabal, A.J. Sievers: Phys. Rev. B**24**, 2921 (1981)
12.12 S. Chiang, R.G. Tobin, P.L. Richards, P.A. Thiel: Phys. Rev. Lett. **52**, 648 (1984)
12.13 J.C. Ariyasu, D.L. Mills, K.G. Lloyd, J.C. Hemminger: Phys. Rev. B**28**, 6123 (1983)
12.14 M. Persson, B. Hellsing: Phys. Rev. Lett. **49**, 662 (1982)
12.15 Y.J. Chabal, E.E. Chaban, S.B. Christman: J. Electr. Spectrosc. Related Phenom. **29**, 35 (1983)
12.16 Y.J. Chabal, G.S. Higashi, S.B. Christman: Phys. Rev. B**28**, 4472 (1983)
12.17 Y.J. Chabal: Phys. Rev. Lett. **50**, 1850 (1983)
12.18 J.A. Appelbaum, D.R. Hamann: Surf. Sci. **74**, 21 (1978)
12.19 T. Sakurai, H.D. Hagstrum: Phys. Rev. B**14**, 1593 (1976)
12.20 R. Kaplan: Surf. Sci. **93**, 145 (1980)
12.21 Y.J. Chabal: Phys. Rev. B**29**, 3677 (1984) and Ref.12 therein
12.22 B.N.J. Persson, R. Ryberg: Phys. Rev. B**24**, 6954 (1981)
12.23 Y.J. Chabal, K. Raghavachari: Phys. Rev. Lett. **53**, 282 (1984)
12.24 G.B. Bachelet, M. Schlüter: Phys. Rev. B**28**, 2302 (1983)
12.25 I.P. Batra, P.S. Bagus, K. Hermann: Phys. Rev. Lett. **52**, 384 (1984)
12.26 B.E. Hayden, K. Prince, D.P. Woodruff, A.M. Bradshaw: Phys. Rev. Lett. **51**, 475 (1983)
12.27 A discussion of the different techniques used for metal substrates, such as polarization and wavelength modulation, is beyond the scope of this article and the reader is referred to Ref.12.3. To date, there is no IR work which uses substrate-adsorbate vibrations to determine the structure of the adsorbates on metallic substrates. For both results presented in Ref.12.11,12, the bonding configuration was known from EELS work and no new structural details were obtained from the the IR data
12.28 K.-M. Ho, C.L. Fu, B.N. Harmon: Phys. Rev. B**29**, 1575 (1984)

13. Optical Second Harmonic Generation for Surface Studies

Y.R. Shen

Department of Physics, University of California and
Materials and Molecular Research Division, Lawrence Berkeley Laboratory
Berkeley, CA 94720, USA

The progress of surface science relies heavily on our ability to probe surfaces and interfaces. For this purpose, many techniques have been developed in the past [13.1]. Recently, laser methods for material studies have advanced to a highly sophisticated level; one therefore wonders if they can also be applied to surface studies. Indeed, there have been a number of very interesting recent discoveries in this area. It is found that laser-induced fluorescence or resonant ionization can be used to probe angular, velocity, and internal energy distributions of molecules scattered or desorbed from a surface [13.2]. Coherent Raman spectroscopy [13.3], laser-induced desorption [13.4], photoacoustic spectroscopy [13.5], and photothermal deflection spectroscopy [13.6] can be used to study surface states and molecular vibrations of adsorbates. Lately, we have shown that surface second harmonic generation (SHG) is also an effective tool for surface studies [13.7]. We describe here some of our recent work on this topic.

Among the existing surface techniques, the optical ones are generally attractive for the following reasons. They have inherently a high spectral and time resolution; they are nondetrimental and suitable for remote probing; they can be applied to surfaces under high gas pressure or interfaces between two condensed media. The disadvantage is that the detection sensitivity is often low. Surface SHG obviously has all the virtues of the optical methods. As we shall see below, it also has enough sensitivity to detect a submonolayer of adsorbed atoms or molecules and enough flexibility to be useful for investigation of many different types of surface problems.

The basic idea of using SHG for surface studies is quite simple [13.7]. The SHG process is forbidden in a centrosymmetric medium, but allowed at a surface or interface. As a result, the SH signal generated from the interface between two centrosymmetric bulk media can dominate the SH signal from the bulk. In this sense, SHG is surface specific. It is known that the typical value of a second-order nonlinear susceptibility is around 10^{-30} esu/molecule. From the solution of Maxwell's equations [13.8] we can then show that the SH signal from a molecular monolayer with 10^{15} molecules/cm^2 can yield 10^4 photons/pulse if a 1.06 μm laser pulse with a pulse energy of 40 mJ, a pulsewidth of 10 nsec, and a beam cross section of 1 cm^2 is used. Since a detection system can easily detect an average signal of less than 1 photon/pulse, we readily see that surface SHG has the sensitivity to detect a fraction of a molecular monolayer.

Experimentally, the submonolayer sensitivity of surface SHG has been confirmed at various interfaces: air/solid [13.7,9], liquid/solid [13.9], and air/liquid [13.10]. That the technique can be used for spectroscopic study of molecular adsorbates [13.7], for probing the structural symmetry of a surface layer [13.7,11], for determining the orientation of molecular adsorbates

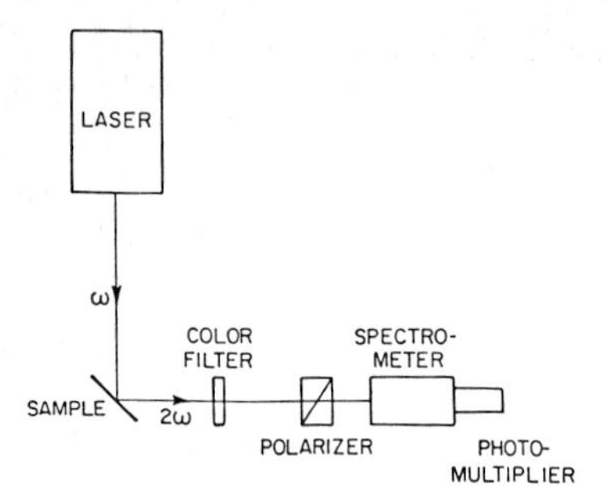

Fig.13.1. Experimental arrangement for surface second harmonic generation

[13.9], and for measuring adsorption isotherms [13.9] has also been demonstrated. More recently, to show that surface SHG is also an effective method for studying well-characterized surfaces, we have carried out a series of SHG measurements on both semiconductor and transition metal surfaces in an ultrahigh vacuum (UHV) chamber supported by LEED, Auger, and mass spectrometers [13.12,13]. We discuss briefly some of the results here.

The experimental arrangement for surface SHG, schematically shown in Fig.13.1, is fairly simple. The SH output, which is coherent and highly directional, can be selectively detected through a filtering system. The samples used in our experiment were crystalline Si and Rh. Their oriented surfaces were cleaned in the UHV chamber by the usual procedure. Cleanliness of the surface and adsorbates on the surface were monitored by LEED, Auger, and thermal desorption spectroscopies. To avoid laser-induced damage or desorption of molecules, we kept the input laser fluence sufficiently low.

Transition metal surfaces are important for catalytic reactions [13.1]. We were interested in adsorption of O_2, CO, and alkali atoms on Rh(111) since they are related to catalysis for hydrocarbon formation and oxidation reactions. In all cases, we found that surface SHG was sensitive enough to detect the presence of submonolayers of adsorbates on the Rh(111) surface. A Nd:YAG laser at 1.06 μm or 0.532 μm with a ~7 nsec pulsewidth was used for the SHG measurements. The laser energy was kept at ~6 mJ and focused to $\sim 10^{-2}$ cm^{-2}. The SH signal from a clean Rh(111) surface was as large as 10^3 photons/pulse.

When the clean Rh surface was exposed to O_2, the SH signal decreased immediately in response to the adsorption of O_2 on Rh(111), Fig.13.2. The signal

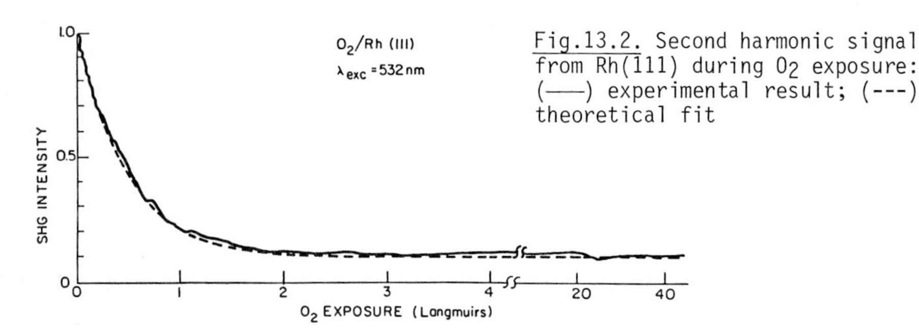

Fig.13.2. Second harmonic signal from Rh(111) during O_2 exposure: (——) experimental result; (---) theoretical fit

drops to 12% of the bare metal value at ~1.8 Langmuirs of O_2; the sharp 2 × 2 LEED pattern for a saturated oxygen layer was observed only when the exposure exceeded ~20 L. Oxygen appears on Rh(111) in the atomic form. Thus, the result of Fig.13.2 is a clear demonstration of the capability of SHG to monitor time-resolved adsorption of submonolayers of atomic species. A model assuming that all adsorption sites are equivalent and noninteracting can explain the result. Following the model, the surface nonlinear susceptibility can be written as

$$\chi^{(2)} = A + B\theta/\theta_s \quad , \qquad (13.1)$$

where A and B are constants, θ is the fractional surface coverage of oxygen with respect to Rh surface atoms, and θ_s is the saturation value of θ. In addition, the adsorption process should obey the Langmuir kinetics [13.1], which yields

$$\theta(t) = \theta_s[1 - \exp(-K\, pt/\theta_s)] \qquad (13.2)$$

if the desorption rate is negligible, as is in the present case. Here, K is a constant accounting for the sticking coefficient, and p is the oxygen pressure. Knowing that the SH signal is proportional to $|\chi^{(2)}|^2$, and taking B/A and K/θ_s as adjustable parameters, we can actually use (13.1,2) to fit the experimental result in Fig.13.2 very well. That the Langmuir kinetics govern the oxygen adsorption on Rh has also been confirmed in the experiment by *Yates* et al. [13.14]. The value 0.93/L deduced for K/θ_s from our result agrees well with 0.78/L obtained by *Yates* et al., considering the limited accuracy of the pressure gauge.

The SH signal from Rh(111) also decreased rapidly upon adsorption of CO. We plot in Fig.13.3 the measured SH signal as a function of CO surface coverage calibrated by LEED. The data exhibit a rather sudden change in slope at $\theta = 1/3$. This suggests that for $\theta \geqslant 1/3$, a new site may have appeared for CO adsorption on Rh. Indeed, previous studies [13.15] have shown that CO adsorbs to Rh on the top sites if $\theta \leqslant 1/3$, and on both the top sites and the bridge sites if $\theta \geqslant 1/3$. Using the model of noninteracting adsorption sites, we can again write the surface nonlinear susceptibility as

$$\begin{aligned} \chi^{(2)} &= A + B\theta/\theta_s && \text{for} \quad \theta \leqslant 1/3 \\ &= A + B/3\theta_s + C(\theta - 1/3)/\theta_s && \text{for} \quad 1/3 \leqslant \theta \leqslant 3/4 \quad . \end{aligned} \qquad (13.3)$$

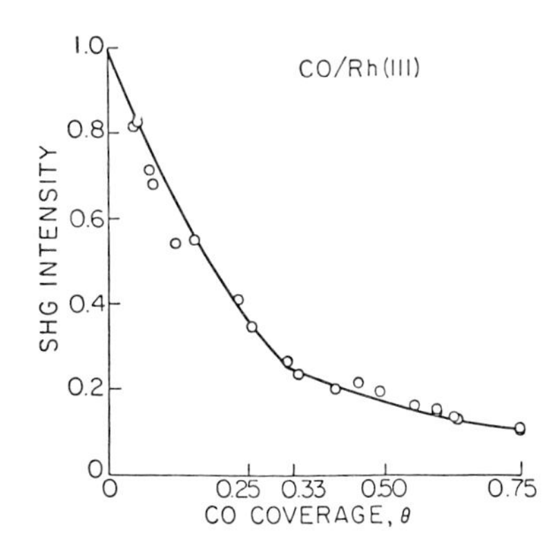

Fig.13.3. Second harmonic signal from Rh(111) vs CO coverage. Solid curve is the theoretical fit

With B/A and C/A taken as adjustable parameters, the expression of $|\chi^{(2)}|^2$ versus θ following (13.3) actually describes the experimental data very well, as shown in Fig.13.3. The result here indicates that surface SHG is also site specific for CO on Rh(111).

One may ask why the SH signal from Rh should decrease upon adsorption of either O_2 or CO. This can be understood as follows. For metals, the surface nonlinearity is generally dominated by delocalized electrons. It is known that oxygen and CO are electron acceptors. Their adsorption on metals would localize part of the delocalized electrons. Atomic or molecular adsorption on metals that results in electron localization would then reduce the effective surface nonlinearity if the adsorbed species is not as nonlinear as the metal surface. This being the case for O_2 and CO absorption on Rh, we expect that adsorption of electron donors on Rh should increase the surface nonlinearity, and hence enhance the SHG. Alkali atoms are known to be effective electron donors. Figure 13.4 shows the SH signal as a function of surface coverage of Na, K, and Cs on Rh(111) at 210 K with 1.06 μm laser excitation. Indeed, as expected, the signal increases monotonically with surface coverage for $\theta/\theta_s < 0.3$ in all three cases. For $0.3 < \theta/\theta_s < 2$, surface plasmon studies have suggested that interaction between adsorbed alkali atoms is important and the plasmon frequency increases from one associated with an electron-enhanced metal surface to that of a pure alkali metal surface [13.16]. This change of surface electronic properties is presumably responsible for the complex variation of the SH signal in this range. For $\theta/\theta_s > 2.0$, the SH signal is nearly a constant, indicating that it arises solely from the top two layers of alkali atoms. This, therefore, is another clear demonstration of the surface specificity of SHG.

The presence of alkali atoms on Rh(111) should affect the adsorption of O_2 and CO. In Fig.13.5a, we show how SHG from Rh(111) with preadsorption of 0.6 monolayer of Na responds to exposure to CO. Compared with the case of adsorption of CO on bare Rh(111), the curve in Fig.13.5a exhibits two, instead of one, sudden changes in slope, one at ~0.5 L and the other at ~2 L, suggest-

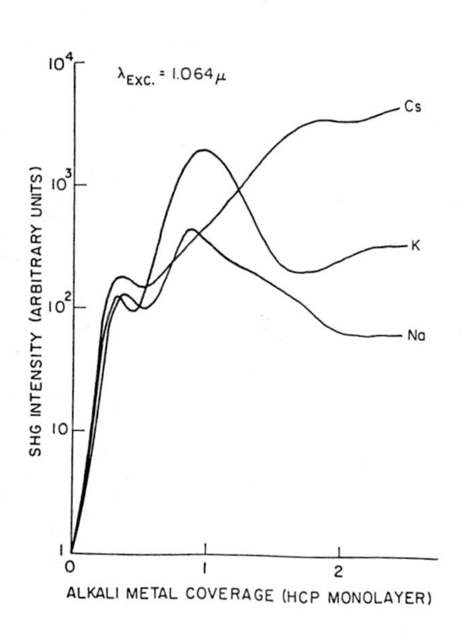

Fig.13.4. Second harmonic signal from Rh(111) vs alkali coverage

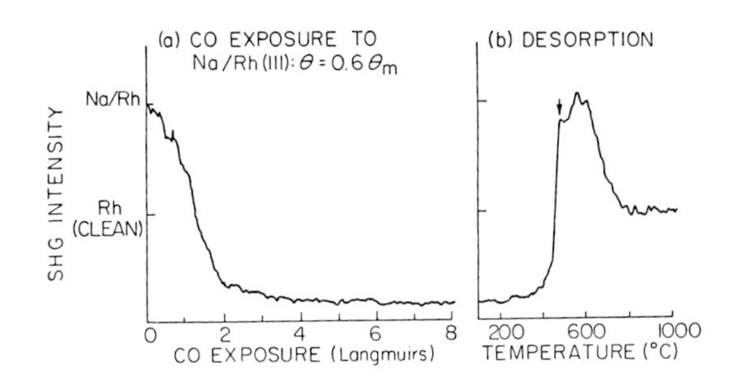

Fig.13.5a,b. Second harmonic signal during (**a**) CO exposure from Rh(111) predeposited by 0.6 monolayer of Na, and (**b**) thermal desorption from Rh(111) predeposited by 0.6 monolayer of Na and saturated by CO adsorption

ing that there are now three different adsorption sites for CO. Aside from the usual top and bridge sites, a new site, presumably close to Na, has appeared with a stronger binding energy. This conclusion is supported by the thermal desorption curve, monitored by SHG, in Fig.13.5b. The Na/Rh(111) surface was first saturated by CO coverage and then heated up at a rate of 15°C/sec. The desorption of CO began at ~200°C. In Fig.13.5b, the SH signal increases first slowly from 200 to 400°C and then rapidly from 400 to 450°C, indicating the desorption of CO from two sites with two different binding energies. After a narrow plateau, the signal rises again at ~520°C, showing the desorption of CO at the third site. It soon reaches the level of Rh with Na only, and starts to decrease at ~600°C. Finally, at ~800°C, the clean Rh(111) surface is obtained.

Previous studies [13.17] showed that CO adsorption on Pt(111) or Rh(111) can be greatly reduced by preadsorption of K. Using SHG to monitor, we have found the same for CO adsorption on Na/Rh(111). With a monolayer of Na on Rh, the signal dropped only 20% after 1200 L exposure. This is in sharp contrast to the result in Fig.13.5a, where the signal dropped to 25% after only 2 L exposure.

The SHG technique can also be used to study adsorbates on semiconductors, e.g., Si [13.13]. It is known that clean Si surfaces are generally characterized by dangling bonds. Being highly asymmetric, these bonds are strongly nonlinear and contribute significantly to the surface nonlinearity of Si. Upon adsorption of atoms or molecules, such as oxygen, that quench the dangling bonds, the surface nonlinearity should decrease accordingly. We have used SHG to monitor the oxidation of Si surfaces. Figure 13.6 displays the variation of the SH signal from a Si(111) surface exposed to 10^{-6}Torr of O_2 at room temperature and 800 C; the surface was initially cleaned and annealed, as denoted by the appearance of the sharp 7×7 LEED pattern. The room-temperature curve shows the decrease of SHG saturating at ~100 L. This agrees with the result of previous studies [13.18] that at room temperature, O_2 chemisorbs to Si(111) and forms an atomic monolayer at ~100 L exposure. At 800°C, the SH signal indicates that the oxygen monolayer is formed at ~30 L. With longer O_2 exposure, the signal decreases much less significantly, and remains constant after ~80 L. It is known that at high temperatures, multilayers of oxide can result on Si, but the SHG is apparently not very sensitive to the growth of more than one oxide layer because it no longer involves quenching of the dangling bonds. The oxygen atoms in these additional layers have lower binding energy than the chemisorbed oxygen, and can be more easily thermally desorbed than the latter. Indeed, the SH probe shows that the desorption rate of such oxygen is 5 times faster than the chemisorbed one.

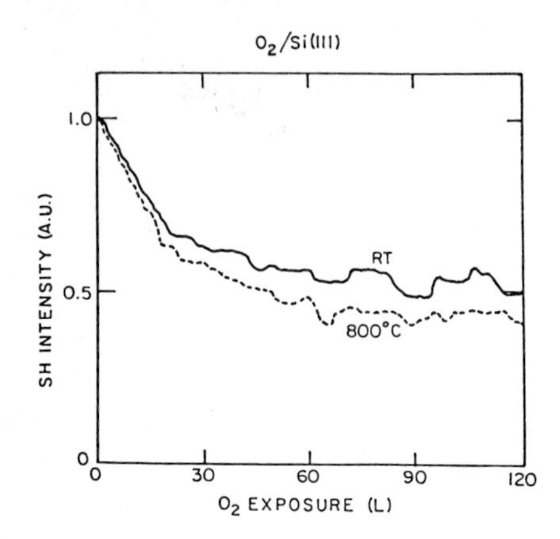

Fig.13.6. Second harmonic signal during oxidation of Si(111) at room temperature (RT) and at 800°C

The SHG is also useful for investigation of bare Si surfaces. From simple symmetry arguments, it is easily seen that the SH signal as a function of sample rotation about the surface normal should reflect the structural symmetry of the surface layer [13.7,11]. *Shank* et al. [13.19] have actually used the technique to study laser melting. More recently, *Heinz* and *Loy* [13.20] have succeeded in using SHG to monitor the transformation of the freshly cleaved 2×1 Si(111) surface to the annealed 7×7 surface.

What we have discussed are just a few examples of surface studies by SHG. They help in establishing SHG as a potentially useful surface tool. Numerous other possible applications of SHG to surfaces and interfaces are yet to be explored. In particular, surface dynamic studies with possible subpicosecond resolution, and studies of surface reactions under high gas pressure or in solution, are most interesting. A variation of the technique is the sum-frequency generation. With a tunable infrared laser, infrared spectroscopic measurements on surface states and adsorbates should become possible.

The UHV work described here was the result of a joint effort between Prof. Somorjai's group and our group. The following individuals deserve full credit: H. Tom, X.D. Zhu, C.M. Mate, and T.F. Heinz. J.E. Crowell also made a significant contribution. This work was supported by the Director. Office of Energy Research, Office of Basic Energy Sciences, Materials Sciences Division of the U.S. Department of Energy under Contract Number DE-AC03-76SF00098.

References

13.1 G.A. Somorjai: *Chemistry in Two Dimensions: Surfaces* (Cornell Univ. Press, Ithaca, NY 1981)
13.2 F. Frenkel et al.: Phys. Rev. Lett. **46**, 831 (1981); J.S. Hayden, G.J. Diebold: J. Chem. Phys. **77**, 4767 (1982)
13.3 C.K. Chen, A.R.B. de Castro, Y.R. Shen, F. DeMartini: Phys. Rev. Lett. **43**, 946 (1979); J.P. Heritage, D.L. Allara: Chem. Phys. Lett. **74**, 507 (1980)
13.4 T.J. Chuang, H. Seki: Phys. Rev. Lett. **49**, 382 (1982)
13.5 F. Trager, H. Coufal, T.J. Chuang: Phys. Rev. Lett. **49**, 1720 (1982)
13.6 A.C. Boccara, D. Fournier, J. Badoz: Appl. Phys. Lett. **36**, 130 (1980); M.A. Olmstead, N.M. Amer: Phys. Rev. Lett. **52**, 1148 (1984)

13.7 T.F. Heinz, C.K. Chen, D. Ricard, Y.R. Shen: Phys. Rev. Lett. **48**, 478 (1982)
13.8 N. Bloembergen, P.S. Pershan: Phys. Rev. **128**, 606 (1962)
13.9 T.F. Heinz, H.W.K. Tom, Y.R. Shen: Phys. Rev. A**28**, 1883 (1983)
13.10 T. Rasing, M.W. Kim, Y.R. Shen: To be published
13.11 H.W.K. Tom, T.F. Heinz, Y.R. Shen: Phys. Rev. Lett. **51**, 1983 (1983)
13.12 H.W.K. Tom et al.: Phys. Rev. Lett. **52**, 348 (1984)
13.13 H.W.K. Tom, X.D. Zhu, Y.R. Shen, G.A. Somorjai: Proc. XVII Int. Conf. on the Phys. of Semiconductors (San Francisco 1984)
13.14 J.T. Yates, P.A. Thiel, W.H. Weinberg: Surf. Sci. **71**, 519 (1978)
13.15 M.A. Van Hove, R.J. Koestner, G.A. Somorjai: Phys. Rev. Lett. **50**, 903 (1982)
13.16 S.A. Lindgren, L. Wallden: Phys. Rev. B**22**, 5969 (1980); U. Jostell: Surf. Sci. **81**, 333 (1979)
13.17 J.E. Crowell, E.L. Garfunkel, G.A. Somorjai: Surf. Sci. **121**, 303 (1982)
13.18 J. Onsgaard, W. Heiland, E. Taglauer: Surf. Sci. **99**, 112 (1980) and references therein
13.19 C.V. Shank, R. Yen, C. Hirlimann: Phys. Rev. Lett. **51**, 900 (1983)
13.20 T.F. Heinz, M.M.T. Loy: Conference on Lasers and Electro-Optics (Anaheim, CA, 1984), postdeadline paper PD-1

14. NMR and Surface Structure

C.P. Slichter

University of Illinois at Urbana-Champaign
1110 W. Green Street, Urbana, IL 61801, USA

14.1 Introduction

My students and postdoctoral students [14.1] have been using nuclear magnetic resonance (NMR) to investigate a number of aspects of surfaces in collaboration with Dr. John Sinfelt of the Exxon Research and Development Laboratory. Although NMR has played an enormously important role in solid-state physics and in chemistry, it has been little used to study surfaces since the number of nuclei needed to observe a signal is large, but the number of atoms on a surface is typically small. Nevertheless, NMR is such a powerful spectroscopic technique that we have pursued its use for the study of surfaces.

We have focused on the surfaces of metals. To obtain an adequate number of surface nuclei, we are thus forced to go to small particles (typically 10 to 100 Å in diameter). Such particles have a large percentage of their atoms on the surface. For our samples, the percentage ranges from about 80% on the smallest particles to 5% on the largest. The percentage of atoms on the surface is called the dispersion. The values we quote are measured by hydrogen chemisorption [14.2]. One can conceive of experiments in which one studies the NMR of the metal, and others in which one studies adsorbed atoms or molecules. We have performed both [14.2-6]. We have chosen to emphasize the Group VIII metals (Fe, Co, Ni, Ru, Rh, Pd, Os, Ir, Pt) which play such an important role in catalysis. While all of these elements possess isotopes with nuclear moments, all but Fe, Rh, and Pt possess electric quadrupole moments, which complicates the interpretation of their NMR. Of these three, only ^{195}Pt possesses both adequate abundance (33.7%) and large enough gyromagnetic ratio (the resonance occurs at 9.153 MHz in 10 kG) to give a good NMR signal. Accordingly we commenced our studies with Pt particles, supported on alumina. We have investigated both the ^{195}Pt NMR, and also the ^{13}C and ^{1}H NMR of adsorbed ^{1}H, ^{13}CO, and $^{13}C_2H_2$ or $^{13}C_2D_2$. We are extending our studies to Rh, Os, Ir, and mixed Rh-Pt particles. In these cases, however, we ordinarily observe either adsorbed molecules or the ^{195}Pt.

Many applications of NMR in chemistry or physics are to liquids where the NMR spectral lines are exceedingly narrow owing to the rapid molecular reorientation. High-resolution NMR apparatus is employed. Our spectra are all much broader. At the temperatures we employ the Pt atoms do not diffuse: molecules on the surface are in fixed positions at 77 K or 4.2 K. At modestly high temperatures what motion exists can be thought of as two dimensional (hops occur on a particular crystal face of the metal particle). Only at the highest temperatures of our apparatus (above 420 K for ^{13}CO on Pt) did motional narrowing occur, resulting from diffusion of the adsorbed molecule over the entire surface of the particle. Thus over most of the range of temperatures we employ we are in the so-called broad-line regime of NMR. We observe our signals using the technique of spin echoes [14.2] in which we apply two radio frequency pulses spaced in time by t_d and record the NMR signal, the so-called echo, which occurs at a time t_d after the second pulse. From the NMR results, one can deduce information about wave functions and bonding, molecular structure of adsorbates, atomic motion of adsorbates, and about rates of reaction. Examples of the sort of information one can obtain are illustrated below.

14.2 Studies of the Metal Particles – Information About the Electrons

In bulk Pt metal, the NMR signal occurs at a magnetic field some 3.4% higher than it does in H_2PtI_6, the standard reference compound which is typical of diamagnetic molecules. The large displacement of the metal compared to diamagnetic compounds arises from polarization of the spins of the conduction electrons in the metal. It is called the Knight shift. Prior to observing a Pt resonance in small particles, we had reasoned that the Knight shift at a clean surface should be at most 40% of that in the bulk metal [14.2]. The hybridization of the 6s orbitals into the 5d orbitals could easily reduce this further (even changing the sign) since the 6s orbitals produce a Knight shift of opposite sign to that from the d band. This expectation is borne out by the data of Fig.14.1 taken from [14.2]. The strong line at H_0/ν_0 equal to 1.138 kG/MHz in Fig.14.1a (a sample of 4% dispersion with a coating designated as "R") is at the frequency of bulk Pt. The ^{195}Pt resonance in diamagnetic compounds extends from H_0/ν_0 of about 1.085 kG/MHz for $[PtF_6]^{2-}$ to 1.095 in H_2PtI_6. These differences are manifestations of the well-known chemical shift originating in the induced orbital magnetism of the Pt d electrons. It is immediately evident that as one increases the dispersion (decreases the Pt particle radius) the strong absorption at the bulk metal position gets weaker and the NMR intensity is confined more and more to the region of diamagnetic compounds (i.e., zero Knight shift). *Weinert* and *Freeman* have carried out a relativistic band-theory calculation of the Pt Knight shift for a clean Pt (001) film 5 atom layers thick [14.7]. They found that the Knight shift at the clean surface is reduced sevenfold from its value in

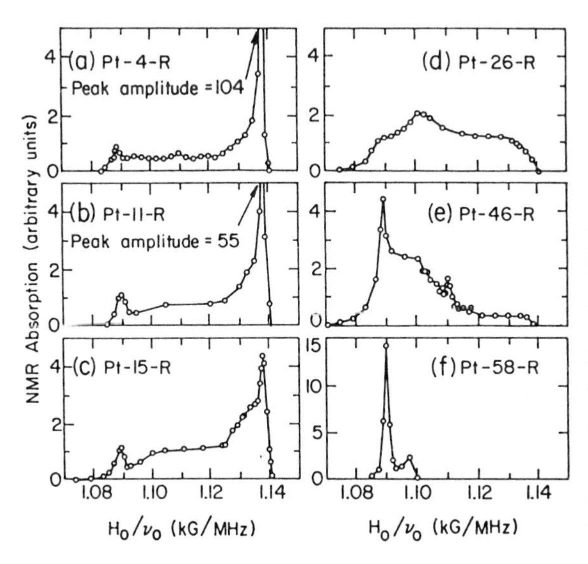

Fig.14.1a-f. The dependence of Pt NMR line shape on particle size: ^{195}Pt NMR absorption versus H_0/ν_0, where H is the applied static magnetic field, and ν_0 the frequency of the NMR apparatus [14.2]. The dispersions of the samples are (a) 4%, (b) 11%, (c) 15%, (d) 26%, (e) 46%, (f) 58%. (The notation such as Pt-4-R is explained in [14.2]

the bulk metal. The result therefore agrees qualitatively with the data. Of course, the absorption in Fig.14.1 arises not only from the surface layer, but also from the next layer deeper, the layer below that, and so on. Each layer should in principle have a different resonance position, but evidently the resonances overlap sufficiently that we do not see an isolated absorption from the surface layer. Fortunately, NMR provides a method, so-called double resonance, for resolving the absorption of the surface layer if one coats the surface with a molecule possessing a nuclear magnetic moment.

14.3 Resolving the NMR Line of the Surface Pt Atoms

To isolate the NMR signal of the surface Pt, *Makowka* et al. [14.5] prepared a sample in which the Pt surface was covered with a layer of adsorbed CO molecules enriched to 90% with ^{13}C. The general principle of the method is based on using the proximity of the surface Pt to the adsorbed layer as a signature to identify Pt nuclei at the surface. The particular technique we employ is called spin-echo double resonance (SEDOR). It involves two sets of rf pulses, one tuned to the ^{195}Pt resonance as in the usual spin echo, the other tuned to the ^{13}C resonance to enable us to flip the ^{13}C spins. The presence of the ^{13}C nuclei on the surface produces an extra magnetic field on the nearby ^{195}Pt nuclei, the direction of which reverses if a ^{13}C spin is flipped from up to down (or down to up). Since this field falls off as $1/r^3$, where r is the distance between the C and Pt nuclei, it is only felt by the Pt close to the ^{13}C. Application of the ^{13}C pulse flips the ^{13}C causing the neighboring ^{195}Pt nuclei to precess at a different frequency. It can thereby be used to modify the ^{195}Pt echo for those Pt close to the ^{13}C, while leaving unaffected the echo of the ^{195}Pt nuclei farther away (deeper in the sample). To collect data, we first record the ^{195}Pt echo *without* flipping ^{13}C spins, then record the echo produced *with* a ^{13}C spin flip. When we substract the second echo from the first, the echoes from the "distant" Pt's cancel, but the echo from the surface Pt's remain. In practice, only the first layer of Pt contributes. Figure 14.2 shows the ^{195}Pt normal absorption in open circles, and the ^{195}Pt signal from such a subtraction, the so-called SEDOR signal, in solid circles for a 26% dispersion Pt sample coated with ^{13}CO (the amplitude of the solid circle signal has been renormalized for purposes of the plot). The solid circles thus give the NMR absorption line of the surface layer of Pt atoms.

The SEDOR signal agrees qualitatively with the result of Weinert and Freeman as to position. It would be of great interest if their calculation could be extended to include Pt covered with a layer of CO so that one could also make a quantitative comparison.

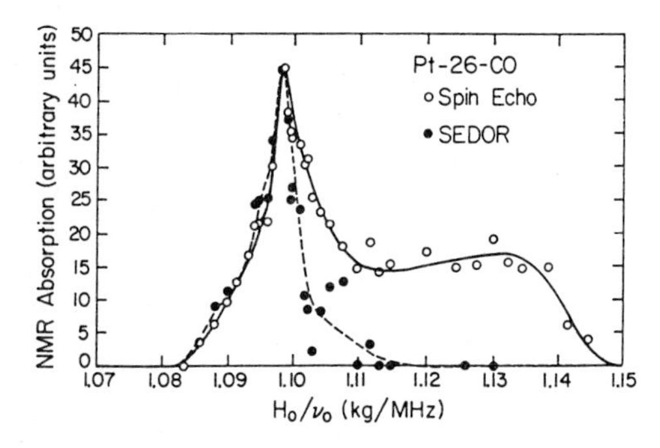

Fig.14.2. ^{13}CO line shape for a 76% dispersion sample: spin each amplitude versus σ (ppm), where σ is the fractional shift in applied magnetic field relative to the value to produce resonance in TMS. The ^{13}C NMR positions of various compounds, including platinum carbonyls, are shown. Also shown (----) is the NMR line shape of a powder of solid CO

14.4 Adsorbed Molecules

14.4.1 CO Line Position — The Mixing of Molecular Wave Functions with the Conduction Band

Serge Rudaz and Jean-Philippe Ansermet have been studying the ^{13}C resonance of ^{13}CO adsorbed on Pt. The most striking feature of the resonance is that it is strongly displaced (to lower applied magnetic fields) from the ^{13}C resonance in diamagnetic compounds (Fig.14.3). The latter range from about +50 ppm to -200 ppm relative to the conventional reference material TMS, but the ^{13}CO on Pt occurs at -320 ppm. (The negative sign signifies that the ^{13}C resonance occurs at a lower magnetic field than in TMS.) It seemed unlikely that such a large difference could arise from the usual mechanism of chemical shifts (induced orbital magnetization of the C valence electrons) since the shift was so far outside the normal range. Rather, a shift arising from electron spin polarization (a Knight shift) seemed likely. Such a shift might arise if the CO valence electrons were to mix with the Pt conduction electrons when the CO is bonded to the metal. It is well known in metals that the conduction electrons not only can shift the NMR frequency but also produce a spin-lattice relaxation, a relationship established many years ago by *Korringa* [14.8,9]. He showed that if one knows the Knight shift, one can predict the nuclear relaxation time T_1 at any temperature, as well as its dependence on the frequency of the NMR apparatus. Thus one can test whether or not the shift arises from electron spin polarization by seeing whether or not there is an accompanying relaxation which obeys the Korringa relationship. Ansermet and Rudaz have used measurements of the ^{13}C spin-lattice relaxation at several frequencies and temperatures to demonstrate that the Korringa relationship holds for the ^{13}CO line on Pt, thereby proving that the ^{13}CO has a Knight shift when adsorbed on Pt. Their result is a direct and quantitative demonstration of the mixing of the CO orbitals with the Pt conduction band states. It would be of great interest to have a theoretical calculation of the shift to compare with experiment.

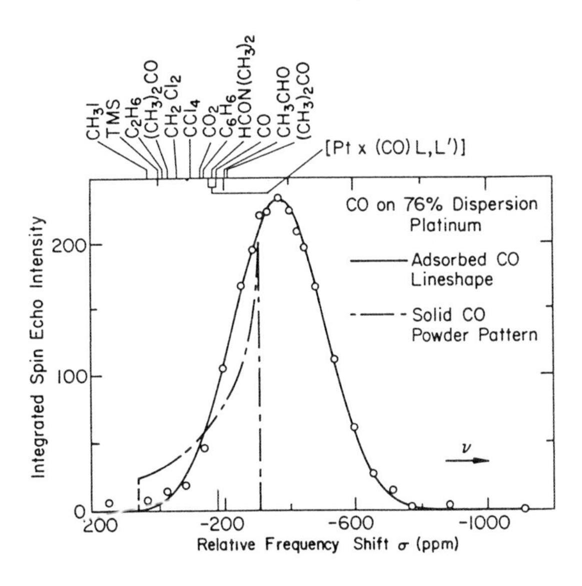

Fig.14.3. ^{195}Pt NMR (o) and SEDOR (•) signals versus H_0/ν_0 for a 26% dispersion sample coated with ^{13}CO [14.5]. The solid and dashed curves are a guide to the eye. The SEDOR line gives the position and shape of the NMR absorption of the surface layer of Pt atoms

14.4.2 CO Relaxation Times and Line Narrowing — Evidence for Diffusion on the Surface

In his studies of ^{13}CO spin lattice relaxation T_1, Rudaz found evidence that at room temperature the CO molecules could sample the entire surface within the measured T_1. This suggests a diffusion process which turns out quantitatively to be consistent with the results of *Lewis* and *Gomer* [14.10]. Ansermet developed an NMR probe for our apparatus which enabled him to go to temperatures of 500 K. He found that by 485 K the NMR line is much narrower than at low temperatures. For the narrowing to occur, there must be bodily motion of the CO molecules over the surface of the particle at a rate fast compared to the low temperature NMR linewidth.

The ratio of the rates found by Rudaz and Ansermet are consistent with an activation energy of diffusion approximately 11 kcal/mole.

14.4.3 C_2H_2 on Pt — Determination of the Molecular Structure

Since the earliest days of NMR, one of its most useful applications has been to determine structures of molecules [14.11]. The manner in which this is done differs between the cases in which the NMR lines are strongly motionally narrowed (so-called high-resolution NMR), in which case one utilizes information about chemical shifts [14.12] and indirect spin-spin couplings [14.13, 14], and the case of rigid lattices [14.11]. While in the former case the direct dipole-dipole coupling between neighboring nuclear moments is averaged to zero, in the latter case it is not. We have been investigating the structure of adsorbed molecules, beginning with acetylene (C_2H_2).

Figure 14.4 shows spin-echo data of *Wang* obtained from ^{13}C NMR of acetylene adsorbed on Pt for a sample enriched in ^{13}C to 90% [14.6]. Shown is the amplitude of the ^{13}C spin echo versus the time t_d between the applied pulses. The data may be described as the sum of a pure exponential decay and a decaying oscillation. The oscillatory signal arises from the dipolar coupling of the two ^{13}C atoms in a single acetylene molecule. From the period of the oscillation *Wang* et al. concluded that the C-C bond length is 1.44 Å, midway between a single and a double bond. Predictions for a single bond (1.54 Å) and a double bond (1.34 Å) are also shown. One source of the pure exponential decay is the 10% of the ^{13}C nuclei which are bonded to a ^{12}C neighbor. Another source is molecules in which the C-C bond is broken. *Wang* et al. concluded that only about 10% of the C-C bonds are broken.

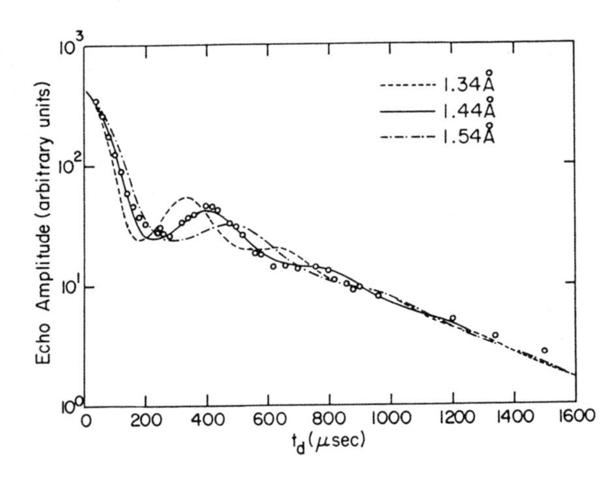

Fig.14.4. ^{13}C spin echo versus time t_d between applied pulses for C_2H_2 adsorbed on Pt [14.6]. The oscillating component of the decay arises from ^{13}C-^{13}C nuclear dipole coupling within the molecule. The theoretical curves correspond to several C-C bond lengths

From the work of various other workers (see [14.6] for references), it is believed that at room temperature the acetylene molecule on a (111) Pt face attaches by one carbon, stands upright on the surface, and has either 2 or 3 hydrogens attached to the other carbon atom. On a (100) surface, the acetylene is thought to be an HCCH molecule lying parallel to the surface. By means of ^{13}C-^{1}H SEDOR, Wang et al. showed that their acetylene is either a) 1/4 HCCH, 3/4 CCH_2, or b) 100% CCH, or c) 1/2 CCH, 1/2 CCH_3. They then employed a technique known as multiple quantum coherence to the proton NMR to determine how many hydrogens are attached to the carbons. They found there are CH_2 groups, thereby concluding that alternative (a) is correct.

Wang et al. are now extending their studies of acetylene on Pt to other metals (Os and Ir), including investigations at elevated temperatures. It is clear that any metals will do if one is willing to confine one's study to adsorbed molecules.

This research was supported in part by the Department of Energy, Division of Materials Research, Contract NO. DE-AC02-76ER01198.

References

14.1 J.-P. Ansermet, C.D. Makowka, H.E. Rhodes, S.L. Rudaz, S. Shore, H.T. Stokes, P.-K. Wang, Z. Wang: To be published
14.2 H.E. Rhodes, P.-K. Wang, H.T. Stokes, C.P. Slichter, J.H. Sinfelt: Phys. Rev. B**26**, 3559-3568 (1982)
14.3 H.E. Rhodes, P.-K. Wang, C.D. Makowka, S.L. Rudaz, H.T. Stokes, C.P. Slichter, J.H. Sinfelt: Phys. Rev. B**26**, 3569-3574 (1982)
14.4 H.T. Stokes, H.E. Rhodes, P.-K. Wang, C.P. Slichter, J.H. Sinfelt: Phys. Rev. B**26**, 3575-3581 (1982)
14.5 C.D. Makowka, C.P. Slichter, J.H. Sinfelt: Phys. Rev. Lett. **49**, 379-382 (1982)
14.6 P.-K. Wang, C.P. Slichter, J.H. Sinfelt: Phys. Rev. Lett. **35**, 82-85 (1984)
14.7 M. Weinert, A.J. Freeman: Phys. Rev. B**28**, 6262 (1983)
14.8 J. Korringa: Physica **16**, 601 (1950)
14.9 C.P. Slichter: *Principles of Magnetic Resonance*, Second revised and expanded ed., corrected second printing (Springer, Berlin, Heidelberg, New York 1980) p.397
14.10 R. Lewis, R. Gomer: Nuovo Cimento **5**, Suppl.2, 506 (1967)
14.11 H.S. Gutowsky, G.B. Kistiakowsky, G.E. Pake, E.M. Purcell: J. Chem. Phys. **17**, 972 (1949)
14.12 L.H. Meyer, A. Saika, H.S. Gutowsky: J. Am. Chem. Soc. **75**, 4567 (1953)
14.13 H.S. Gutowsky, D.W. McCall, C.P. Slichter: J. Chem. Phys. **21**, 229 (1953)
14.14 E.L. Hahn, D.E. Maxwell: Phys. Rev. **84**, 1246 (1951)

Part III

Developments in Existing Techniques

III.1 LEED and Electron Propagation

15. Determination of Surface Structure by LEED

F. Jona[1]
Department of Materials Science, State University of New York
Stony Brook, NY 11794, USA
J.A. Strozier, Jr.
Empire State College, State University of New York, Stony Brook, NY 11794, USA
P.M. Marcus
IBM Research Center, P.O. Box 218, Yorktown Heights, NY 10598, USA

After a brief review of some noteworthy recent achievements of LEED crystallography we describe two important advances in methodology. One is experimental and concerns the rapid acquisition of LEED intensity data from display-type equipment with a computer-assisted television camera. The other is theoretical and concerns the calculation of LEED intensities with a new cluster approach that offers notable advantages over the schemes presently used, especially for structures involving large numbers of atoms and for high electron energies.

15.1 Introduction

Among the many techniques that have been developed for determining the atomic structure of solid crystal surfaces, low-energy electron diffraction (LEED) is the oldest and most successful. We refer here in particular to the technique based on the match of theoretical to experimental LEED intensities, commonly called LEED intensity analysis. The main advantages of this technique are that it can determine atomic positions within the first several layers (4 to 8, depending on the scattering power of the surface atoms) of an exposed crystal face with precision of about 0.03 Å, that it can deal with clean as well as with ordered adsorbate-covered surfaces, and that, when properly executed, it has a high probability of being right. To be sure, a few of the structures reported as solved by LEED intensity analysis were later found to be incorrect, but the evidence shows that these failures could be ascribed either to insufficient or unreliable experimental data, or to inaccurate theoretical parameters and calculations, or to poor standards in the evaluation of the fit between theory and experiment. The main disadvantage of the technique is that, owing to the multiple-scattering nature of the interaction between low-energy electrons and atoms in condensed matter, the experimental data cannot be inverted directly to produce the atomic structure, and one must rely on trial-and-error procedures that can be lengthy, tedious and costly. Nevertheless, more surface structures have been determined by LEED than by all other surface-sensitive techniques combined. In this brief review we plan first to give a concise description of some recent noteworthy achievements of LEED structure analysis and then to discuss the advances that have been or are being made in experimental procedures and theoretical approaches.

---

1 Sponsored in part by the National Science Foundation and the Office of Naval Research

15.2 Recent Achievements

LEED intensity analysis has been applied extensively to clean surfaces of metals, alloys and semiconductors, and to ordered adsorbate structures on the same surfaces. Clean crystal surfaces may have so-called 1×1 structures (meaning that they have the same periodicities in the surface plane as in corresponding parallel planes in the bulk) or be reconstructed (i.e., with surface periodicities which are multiples of the corresponding bulk periodicities). Both types of structure have been analyzed, the former being, of course, easier to study and therefore more frequently reported. We summarize below some of the most significant among the recent results in LEED crystallography.

An important discovery in the study of 1×1 structures with fundamental implications for metal theory is the phenomenon of multilayer relaxation of metal surfaces. Relaxation means here rigid translation of the outer atomic layers of a crystal from their bulk positions without change of the two-dimensional unit mesh. On low-index surfaces, e.g., Cu(110) [15.1,2], Re($10\bar{1}0$) [15.3], Al(110) [15.4], Ag(110) [15.5], Cu(001) [15.6], and V(001) [15.7], the relaxation is oscillatory and damped: it consists of contraction, expansion and contraction of the first, second and third interlayer spacings, respectively, with decreasing magnitude. Higher-index surfaces, e.g., Fe(211) [15.8], Fe(310) [15.9] and Fe(210) [15.10], exhibit both perpendicular and parallel relaxations, as atom rows are shifted away from bulk positions in directions not only perpendicular but also parallel to the surface plane, and the relaxations are not always oscillatory. The contractions of the first interlayer spacings are found to increase monotonically from less to more open surfaces, and the overall relaxation extends to a depth of approximately 2 Å, at least for the high-index Fe surfaces that have been investigated [15.11]. These findings constitute one of the most interesting aspects of surface science and have led to new developments in the theory of metal surfaces [15.12,13].

Much less is known about the atomic structure of alloy surfaces, although extensive LEED observations [15.14-16] and low-energy ion scattering (LEIS) investigations [15.17] have been made on low-index surfaces of Cu_3Au. These surfaces, and the corresponding ones of the isomorphous ordered Ni_3Al alloy, have only recently been selected for quantitative structure determination and are presently under scrutiny by LEED intensity analysis. The most important result obtained so far concerns the (001) surfaces of these A_3B-type alloys. In the bulk, (001) planes alternate between 50-50% AB and 100% A, but a (001) surface always has the structure of the 50-50% AB planes [15.18]. This result has been corroborated by first-principle calculations of the cohesive energies of crystal slabs which show greater stability (larger cohesive energy) for the mixed surface layer [15.18].

The study of reconstructed semiconductor surfaces has seen slow but encouraging progress. For Si(001)2×1, where the existence of buckled dimers seems to be universally accepted, LEED analysis has produced an acceptable model [15.19], which has recently been further improved [15.20] and produces a much better fit with experiment than other models, including one recently proposed [15.21]. For Si(111)2×1 (the cleaved surface of silicon) a structural model based on buckled chains and derived from the π-bonded model of *Pandey* [15.22] is the only one among the several models which were tested that produces encouraging fit with experiment, although it cannot yet be labelled as the final solution of this structural problem [15.23]. The only other reconstructed semiconductor surface for which a LEED solution was claimed is the GaAs(111)2×2 surface, where a vacancy-buckling model was shown to produce acceptable fit with experiment [15.24] and to be consistent with total-energy calculation [15.25].

Among the 1 × 1, i.e., relaxed (as opposed to reconstructed), semiconductor surfaces, the impurity- and laser-stabilized Si(111)1 × 1 surface is an object of unsettled controversy (LEED analysis supports a relaxed bulk-like model [15.26,27], whereas photoemission [15.28,29] and ion-scattering [15.30] experiments favor a disordered reconstruction), but the (110) surfaces of III-V and II-VI compounds have well-understood and noncontroversial distorted bulk-like structures [15.31].

The field of ordered adsorbate structures on solid surfaces is too large to be reviewed here, even from the viewpoint of LEED crystallography only. Atomic adsorption has been successfully studied in many systems [15.32], and molecular adsorption is currently stimulating the development of novel and powerful methods for LEED structure analysis [15.33].

15.3 Recent Advances

Two major advances stand out in the continuing effort toward improvement of LEED crystallographic techniques. One is experimental and has been successfully in use for a few years – it concerns the rapid acquisition of intensity data with a computer-controlled television camera system. The other is theoretical and is in the development stage – it concerns the calculation of diffracted intensities with a cluster approach that offers considerable advantages over the conventional layer-by-layer schemes. Other theoretical approaches, aimed in particular at reducing the computer time for complex calculations [15.33], will not be discussed here.

15.3.1 Rapid Acquisition of LEED Data

We describe here a data-acquisition system (called VLA for Video LEED Analyzer) that we have used in our laboratory during the past few years for the rapid collection of large sets of intensity data from the fluorescent screen of display-type LEED equipment. A system based on similar principles but with different execution was developed independently at Erlangen, Germany, and is described in several literature reports [15.34,35]. The principle consists in using a television (TV) camera associated with a computer to measure, digitize and process the intensity of LEED spots on the fluorescent screen of conventional display-type equipment. Accordingly, the VLA consists of five basic components: a 64 K microcomputer (Motorola 6800) with dual 5.25 inch disc drives, a TV camera, a TV monitor, a computer terminal and suitable interfaces to camera, monitor, LEED gun and oscilloscope or/and x-y recorder. The TV camera is a model 65 DAGE MTI, Inc. camera with a "Newicon" tube capable of producing usable pictures with face-plate illumination as low as 0.0008 foot-candles. Both the sensitivity and the dynamic range of this tube are adequate for reliable measurements of LEED intensities (with incident beam currents of the order of microamperes) to roughly one part in 256. The noise level of the camera is typically between 5% and 10%.

The heart of the VLA system is a video interface, which allows the computer to access a given pixel "randomly" (a TV image consists of a frame of 256 × 256 pixels) by passing one-byte values for both the x and y coordinates of the pixel to the video interface. When the TV raster scan reaches the addressed pixel, a sample-and-hold circuit is opened for 200 ns. The accumulated charge is then digitized to 8 bits (one part in 256 resolution) in approximately 10 μs, and the one-byte value of the intensity for that pixel is passed back to the computer for further processing. A raster scan for one interlace is called a "frame" and is completed in 1/60 of a second. Thus, the access time for acquiring one pixel is always less than or equal to 16.7 ms (1/60 s). The time to access a group of pixels depends on their relative arrangement. It takes roughly 40 μs (or 2/3 of the scan time for one horizontal line) to access the video interface and store the digitized one-byte intensity value in memory. At the beginning of the frame the raster scan starts at the top left corner of the image, proceeds to the right for one line then drops down to the next (lower) line, starting again at the extreme left-hand side and moving to the right. If the group of pixels to be accessed consists of a line of N pixels in the vertical direction, one pixel can be accessed on each horizontal raster line, so that the entire group of pixels can be acquired in 1/60 s. However, if the group of pixels consists of a line of N pixels in the horizontal direction (i.e., along a horizontal raster-scan line) then it would take N/60 s to acquire the entire group as only one pixel could be picked up during each frame. For example, a square window of 10×10 pixels would require $10/60 = 167$ ms to allow for acquisition and storage of all 100 intensity values.

The VLA system can operate in either of two modes: the *beam mode*, designed for measuring the integrated LEED spot intensities as functions of incident electron energy (I-V curves or spectra), and the *pixel mode*, in which individual pixel intensities are available for beam profile measurements. We give below a brief description of the *modus operandi* for both modes.

In the beam mode, the software establishes a square window of 10×10 pixels over each LEED spot to be measured. We have noted above that the time necessary to measure, digitize and store all 100 pixels in such a window is 167 ms. For the purposes of the present discussion we will call this time the Single Beam Time (SBT). The system described in [15.34,35] can access up to 20 pixels in a horizontal line in one scan, so that the SBT is correspondingly shorter. The selection of the beams to be measured is done by the operator in a sequence of interactive steps on the computer. The operator enters the two basic vectors that define the net under consideration. The net thus calculated by the computer is displayed on the TV monitor and brought into coincidence with the actual LEED pattern also visible on the TV monitor by the operator entering on the "live" terminal keyboard commands which can expand (contract), translate and/or rotate the calculated net. This matching operation usually requires less than 30 s and is independent of the position of the TV camera in front of the observation window. The operator can then select the beams to be measured, up to 15 beams out of 49 being possible in any given run. The software automatically arranges the order of access of the selected beams in such a way that the beam intensities are acquired in the shortest possible time consistent with the raster nature of the video scan. Thus, the time required for acquisition of the integrated intensities of N beams for one energy point will on the average be proportional to $\mathrm{SBT} \times \sqrt{N}$ rather than grow linearly with N. For example, the intensity data for all beams in a 3×3 square array (9 beams) for 200 energy points can be measured in $200 \times 3 \times 0.167$ or 100 s, corresponding to about 11 s per beam. The change of incident energy is accomplished by the software instructing a D/A converter to output a control voltage to the LEED power supply. As the incident electron energy changes, the positions of all beams except the specular beam change on the fluorescent screen. It is therefore necessary to "track" the LEED spots so that the system knows where on the screen the next series of measurements is to be done. There are two tracking modes available which are software selectable. In the first, or "calculated", mode the computer calculates from a linearized version of the equations governing the positions of the LEED spots the coordinates of the spots and adjusts the windows accordingly. In the second, or "dynamic", mode the window positions are selected in such a way that the coordinates of the most intense pixel in the preceding frame establish the center of the window for the current measurement. With both tracking modes in operation it is practically impossible to "lose" LEED spots in the background when their intensities reach a minimum.

In the beam mode, the intensity values of the 100 pixels over each LEED spot are summed (we call this the HSUM) and the resulting value is assigned as the beam intensity of the corresponding LEED spot. Another sum is also calculated (we call this the NSUM) in which only those pixels which have intensities larger than a user-selected fraction of the maximum pixel intensity are included. The availability of both the HSUM and the NSUM makes it possible to define and calculate a "background" B of the LEED pattern. A universally accepted definition of such a quantity does not exist but a possible definition of B is: $B = (HSUM-NSUM)/(100-N)$, where N is here the number of pixels included in the NSUM. Such background may be subtracted from the integrated intensities off line, if desired. Both the NSUM and the HSUM are available for further data processing.

The VLA in the beam mode has proven to be useful not only for the acquisition of intensity data for purposes of structure analysis but also for establishing normal-incidence conditions, for testing the lifetime of surface structures in the vacuum chamber and, generally, for rapid and reliable characterization of surface structures. Published reports involving data collected with the VLA in the beam mode include the studies of the Si(001)2×1 [15.19], $NiSi_2$(111) [15.36], and Si(111)2×1 [15.23] structures.

The pixel mode was designed to measure the intensity distribution within, i.e., the cross-section profile of, a LEED beam in k space. A variable-size window is introduced which can be positioned and dimensioned by the user via a "live" terminal keyboard. The intensities of all individual pixels within such a window are stored and are available for plotting or general data processing. The resolution in the vertical direction can be increased by a factor of 2 under software control by acquiring both interlaced frames. The signal-to-noise ratio can be enhanced by signal averaging also under software control. Window dimensions of $1\times N$ have proven useful for studying the intensity distribution along lines in k space. This feature was exploited by *Yang* and *Jona* [15.37] in establishing the 2×1 nature of the Si(111)7×7 surface.

15.3.2 A Cluster Approach to the Calculation of LEED Intensities

The extension of LEED intensity analysis to more complicated structures or to higher electron energies is limited by the growth of intensity computation time. For the methods currently in use this time grows as the cube of the product of the number N of translationally inequivalent atoms by the number P of spherical-wave components in the expansion of the wave function of the scattered electron around any atom, since such methods depend either on eigenvalue determination or on inversion of a matrix of dimension $(N\times P)$. The calculation of LEED intensities can be made much simpler and faster by exploiting the strong attenuation of the electrons, which limits the path-length and the number of scattering atoms which contribute to the backscattered wave from the crystal. Two features should be recognized: 1) after the incident plane wave scatters at an initial atom the neighbor atoms which contribute significantly to the reflected electron amplitude form a small cluster around the initial atom; 2) the calculation of the scattering from a cluster does not have to be repeated when the first stage of scattering is around an atom that is translationally equivalent to the initial atom. Rather we can readily superpose the scattering from the initial cluster when repeated at all points of a two-dimensional net by using a simple kinematic formula. This superposition generates a set of (diffracted) beams with wave numbers obtained by adding the reciprocal net vectors to the parallel component of the reflected wave vector.

The procedure suggested above has been quantitatively implemented by finding the cluster around each nonequivalent atom in a given surface structure such that it includes all paths shorter than a critical value LC. For a number of LC values the total outgoing spherical-wave amplitudes at each atom of the cluster have been found by summing the scatterings coherently over all paths smaller than or equal to LC. The convergence of the amplitudes of the beams can then be studied as a function of LC. The mathematical basis of the calculation rests on certain theorems satisfied by spherical waves: 1) a translation theorem which expands a spherical wave around one center in spherical waves around a second center, hence provides a propagation matrix that converts the spherical-wave expansion coefficients at one atom into the expansion coefficients at a second atom by a matrix-vector multiplication; 2) a combination or kinematic theorem which converts the sum of outgoing spherical waves centered at the lattice points of a two-dimensional net, with phases at each atom of the net fixed by the phase of an incident plane wave

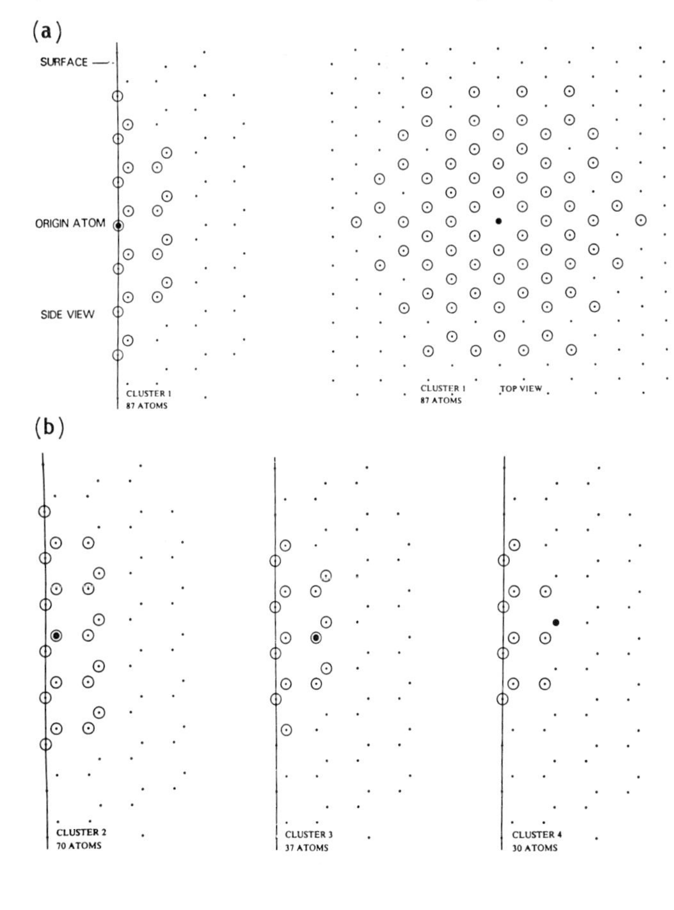

Fig.15.1a,b. Clusters of atoms (open circles) on Si(111)1 × 1 (not relaxed) for LC = 14 Å. (**a**) Side and top view of first cluster (87 atoms). The solid line indicates the plane of surface atoms; the origin atom of the cluster is a solid circle; underlying dots show the complete lattice. (**b**) Side views of clusters with origins in layers 2, 3 and 4 (70, 37 and 30 atoms, respectively). The solid line indicates the plane of surface atoms; the origin atom of the cluster is a solid circle; underlying dots show the complete lattice

at that atom, into a sum of outgoing plane waves (beams); hence, as noted above, this theorem converts the cluster of outgoing spherical waves into amplitudes of diffracted beams.

The great gain in efficiency of the cluster procedure arises from the fact that the work of calculating scattered intensities increases linearly with the number of nonequivalent atoms (equal to the number of clusters) and quadratically with the number of spherical-wave components, since a matrix-vector multiplication of N components has N^2 operations.

The procedure is illustrated by a double-scattering calculation on Si(111) 1 × 1. The clusters (of 87, 70, 37 and 30 atoms) for LC = 14 Å extending over 4 layers are shown in side view in Fig.15.1a, and the first cluster of 87 atoms is shown in top view in Fig.15.1b. The resulting LEED intensity spectra are compared in Fig.15.2 with full-dynamical calculations made with the CHANGE program [15.26] based on the layer-KKR procedure. We note that much of the multiple-scattering contributions have emerged, although quantitative agreement with the full-dynamical result is not expected until higher stages of scattering are added.

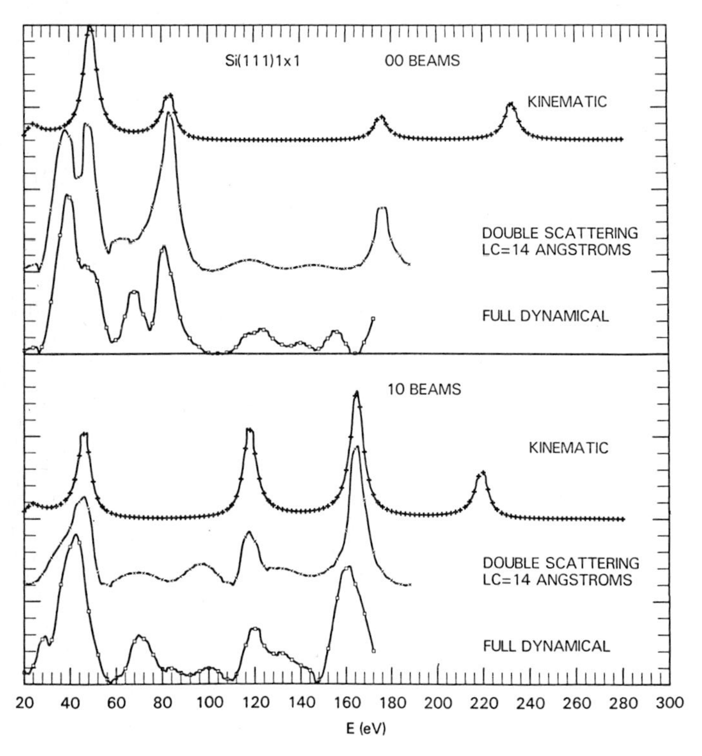

Fig.15.2. Si(111)1 × 1: double-scattering for LC = 14 Å (224 atoms) compared with kinematic and full-dynamical spectra for the 00 and 10 beams

References

15.1 H.L. Davis, J.R. Noonan, L.H. Jenkins: Surf. Sci. **83**, 559 (1979)
15.2 D.L. Adams, H.B. Nielsen, J.N. Andersen, I. Stengaard, R., Feidenhans'l, J.E. Sorensen: Phys. Rev. Lett. **49**, 669 (1982)
15.3 H.L. Davis, D.M. Zehner: J. Vac. Sci. Technol. **17**, 190 (1980)
15.4 H.B. Nielsen, J.N. Andersen, L. Petersen, D.L. Adams: J. Phys. C: Solid State Phys. **15**, L1113 (1982)
15.5 J.R. Noonan, H.L. Davis: Bull. Am. Phys. Soc. **26**, 224 (1981)
15.6 H.L. Davis, J.R. Noonan: Surf. Sci. **126**, 245 (1983)
15.7 V. Jensen, J.N. Andersen, H.B. Nielsen, D.L. Adams: Surf. Sci. **116**, 66 (1982)
15.8 J. Sokolov, H.D. Shih, U. Bardi, F. Jona, P.M. Marcus: J. Phys. C: Solid State Phys. **17**, 371 (1984)
15.9 J. Sokolov, F. Jona, P.M. Marcus: Phys. Rev. B**29**, 5402 (1984)
15.10 J. Sokolov, F. Jona, P.M. Marcus: Submitted to Phys. Rev. B
15.11 J. Sokolov, F. Jona, P.M. Marcus: Solid State Commun. **49**, 307 (1984)
15.12 U. Landman, R.N. Hill, M. Mostoller: Phys. Rev. B**21**, 448 (1980)
15.13 R.N. Barnett, U. Landman, C.L. Cleveland: Phys. Rev. B**27**, 6534 (1983)
15.14 V.S. Sundaram, R.S. Alben, W.D. Robertson: Surf. Sci. **46**, 653 (1974)
15.15 H.C. Potter, J.M. Blakely: J. Vac. Sci. Technol. **12**, 635 (1975)
15.16 E.G. MacRae, R.A. Malic: Preprint (1984)
15.17 T.M. Buck, G.H. Wheatley, L. Marchut: Phys. Rev. Lett. **51**, 43 (1983)
15.18 D. Sondericker, J. Jona, V.L. Moruzzi, P.M. Marcus: To be published
15.19 W.S. Yang, F. Jona, P.M. Marcus: Phys. Rev. B**28**, 2049 (1983)
15.20 W.S. Yang, F. Jona, P.M. Marcus: Phys. Rev., and this conference
15.21 B.W. Holland, C.B. Duke, A. Paton: To be published (1984)
15.22 K.C. Pandey: Phys. Rev. Lett. **47**, 1913 (1981)
15.23 F.J. Himpsel, P.M. Marcus, R. Tromp, I.P. Batra, M.R. Cook, F. Jona, H. Liu: Phys. Rev. B**30**, XXX (1984)
15.24 S.Y. Tong, G. Xu, W.N. Mei: Phys. Rev. Lett. **52**, 1693 (1984)
15.25 D.J. Chadi: Phys. Rev. Lett. **52**, 1911 (1984)
15.26 D.W. Jepsen, H.D. Shih, F. Jona, P.M. Marcus: Phys. Rev. B**22**, 814 (1980)
15.27 D.M. Zehner, J.R. Noonan, H.L. Davis, C.W. White: J. Vac. Sci. Technol. **18**, 852 (1981)
15.28 D.E. Eastman, F.J. Himpsel, J.F. Van der Veen: Solid State Commun. **35**, 345 (1980)
15.29 D.M. Zehner, C.W. White, F. Heimann, B. Reihl, F.J. Himpsel, D.E. Eastman: Phys. Rev. B**24**, 4875 (1981)
15.30 R.M. Tromp, E.J. van Loenen, M. Iwami, F.W. Saris: Solid State Commun. **44**, 971 (1982)
15.31 C.B. Duke, R.J. Meyer, P. Mark: J. Vac. Sci. Technol. **17**, 971 (1980)
15.32 G.A. Somorjai, M.A. Van Hove: *Adsorbed Monolayers on Solid Surfaces* (Springer, Berlin, Heidelberg, New York 1979)
15.33 M.A. Van Hove: This conference
15.34 P. Heilmann, E. Lang, K. Heinz, K. Müller: *Proceeding of the Conference on Determination of Surface Structures by LEED*, IBM Research Center 1980 (Plenum, New York 1984)
15.35 K. Heinz, K. Müller: *Experimental Progress and New Possibilities of Surface Structure Determination*, Springer Tracts in Modern Physics, Vol.91 (Springer, Berlin, Heidelberg, New York 1983); K. Müller, K. Heinz: This Volume, p.105
15.36 W.S. Yang, F. Jona, P.M. Marcus: Phys. Rev. B**28**, 7377 (1983)
15.37 W.S. Yang, F. Jona: Solid State Commun. **48**, 377 (1983)

16. Structure Determination of Molecular Adsorbates with Dynamical LEED and HREELS

M.A. Van Hove

Materials and Molecular Research Division, Lawrence Berkeley Laboratory
and
Department of Chemistry, University of California, Berkeley, CA 94720, USA

Molecular adsorbate structures can be determined by combining Low-Energy Electron Diffraction (LEED) with High-Resolution Electron Energy Loss Spectroscopy (HREELS). The HREEL vibrational spectroscopy identifies the molecular species, whose bond lengths and angles are obtained by LEED. Appropriate calculational methods are required in LEED to solve complex molecular structures: such methods will be discussed for large molecules, for large unit cells and for disordered adsorbates. Results have been obtained for the following molecular species adsorbed on several low-Miller-index metal surfaces: CO, C_2H_2,→ CCH_3, C_3H_4,→ CCH_2CH_3 and C_6H_6. The presently available methods should be capable of solving a multitude of molecular adsorbate structures, including large molecules, coadsorbates and disordered species.

16.1 Introduction

The combination of High-Resolution Electron Energy Loss Spectroscopy (HREELS) and Low-Energy Electron Diffraction (LEED) has been found to be powerful for determining the crystallography of molecular species adsorbed on surfaces. HREELS is particularly convenient for determining the identity of molecular species, by comparing their vibrational frequencies with those of known molecules, including organometallic clusters [16.1]. Often, orientational information can also be obtained with HREELS [16.1]. Knowledge of identiy and orientation markedly simplifies the subsequent LEED analysis, which compares measured with calculated intensities for a large number of still plausible adsorption geometries [16.2,3]. This allows bond lengths and bond angles to be determined.

The multiple-scattering (i.e., dynamical) nature of LEED complicates structural determinations by strongly increasing the computational cost relative to that of crystallographic techniques based on "kinematic" electron scattering. Therefore, appropriate calculational methods are required to handle the large unit cells and large molecules that can be found at surfaces. In addition, since not all molecules adsorb in an ordered fashion, it is desirable also to perform LEED analyses in the absence of long-range order.

16.2 Calculational Methods in LEED

LEED theory was designed in the years 1968-1971 for dense, strongly scattering materials such as metals, in which multiple scattering is dominant [16.2,3]. Molecular adsorbates are less dense and scatter less strongly, but do not give a kinematic behavior. Therefore, computational savings are to be sought in selective neglect of certain multiple-scattering terms [16.4].

Since the substrates used today for molecular adsorption are largely the same metals considered in 1968-1971, the full multiple-scattering formalism must be maintained in that part of the surface. Through the "combined-space" [16.3] method, one may add molecular overlayers treated with suitably approximated formalisms, to be described next.

The simplest case is that of hydrogen atoms, which are such weak scatterers that they can be ignored (unless they are themselves responsible for a superlattice).

The approximation called "near-neighbor muliple scattering" [16.4] recognizes that any atom in a molecule is surrounded by few neighboring atoms, most of which are relatively weak scatterers compared with the situation in the bulk metal. As a result, only a few multiple-scattering paths need to be taken into account. In particular, one may choose to let each segment in such a scattering path link only neighboring, bonding atoms. Accuracy is gradually lost as fewer scattering paths are included. Nevertheless, the ultimate approximation, which treats the molecular layer kinematically, has been used to eliminate many structural models which might appear plausible at first sight [16.5,6]. Successively more severe approximations do of course reduce the accuracy attainable for the correct structure: the best calculation that can be afforded should be used to refine the best structure which has been identified with more approximate calculations.

One of the computational difficulties with most conventional LEED formalisms is the large number of beams that must be considered when the surface unit cell is large [16.3]. The computational effort is proportional to the square or the cube of the unit cell area. Molecules are found frequently to order in lattices with intractably large unit cells. Therefore, the "beam-set-neglect" approximation has been designed [16.7] to avoid this unfavorable scaling law. The resulting computational effort is at worst linear in the unit cell area, and even constant if one so desires.

The beam-set-neglect method can be easily extended to disordered overlayers. The resulting diffuse intensity can be shown to contain the local bonding information (short-range ordering) that one is most interested in with molecular adsorption [16.8]. Thus, *local* structure determination is possible with LEED in the absence of long-range order, as it is with SEXAFS, ARPES and other surface-sensitive techniques. This method has been successfully compared to a different theory of diffraction by disordered adsorbates [16.8,9].

16.3 Molecular Adsorbate Structures

Only *organic* molecules have been subjected so far to surface structure determination by LEED, coupled with HREELS. Carbon monoxide (CO) was found to be a favorable candidate for such work and has served well as a test case for experimental and theoretical procedures. The adsorption sites of CO on metal surfaces were successfully predicted on the basis of the CO-stretch vibration frequencies, using an empirical site assignment derived for metal-carbonyl clusters [16.10]. Indeed, either the top site or the bridge site occurs in the six CO adsorption structures determined so far: CO on Ni(100) [16.11], CO on Cu(100) [16.11a], CO on Pd(100) [16.12], CO on Rh(111) at two coverages [16.3,5,6] and CO on Ru(0001) [16.14].

The other molecular structures determined with LEED concern hydrocarbons having two to six carbon atoms: C_2H_2 (acetylene) on Pt(111) [16.15], Ni(111) [16.16] and Ni(100) [16.17], $\rightarrow CCH_3$ (ethylidyne) on Pt(111) [16.18] and Rh(111) [16.19], C_3H_4 (methylacetylene) on Rh(111) [16.20], $\rightarrow CCH_2CH_3$ (propylidyne) on Rh(111) [16.21] and C_6H_6 (benzene) on Rh(111) [16.7]. The complete molecules in this list (C_2H_2, C_3H_4 and C_6H_6) are found primarily to π-bond to the metal surface. For C_2H_2 on Pt(111) and Ni(111), C_3H_4 ($HC \equiv C - CH_3$) and C_6H_6 on Rh(111), the unsaturated C-C bonds are parallel to the metal surface. The adsorption site is somewhat uncertain for C_2H_2 on Pt(111) and Ni(111), top and μ bridge being preferred, respectively. Centering on the hollow site is obtained for C_2H_2 on Ni(100), C_3H_4 on Rh(111) and C_6H_6 on Rh(111). The C_3H_4 molecule is found to be considerably bent, with its methyl ($-CH_3$) group pointing up at an angle of about 35° from the surface plane. This reflects rehybridization due to adsorption and was predicted already for the terminal hydrogens of adsorbed C_2H_2, based on HREELS data [16.22] and theoretical calculations [16.23]. Such upward bending of hydrogen is also predicted for adsorbed benzene [16.24,25].

The alkylidynes (ethylidyne and propylidyne) have been observed an Rh(111) and Pt(111) as products of acetylene or ethylene (C_2H_4) adsorption (for ethylidyne), or of methylacetylene or propylene (C_3H_6) adsorption (for propylidyne). The multiple C-C bond of the parent molecule becomes a single bond during hydrogenation or dehydrogenation, oriented perpendicularly to the surface. The adsorbed species are bonded through a carbon atom to three metal atoms surrounding a hollow site. The upper part of the species is a saturated hydrocarbon terminating in a methyl group. In the case of propylidyne, the methyl group is tilted with respect to the surface normal.

16.4 Prospects

Work is underway at Berkeley on some molecular adsorbate structures whose complexities illustrate the present capabilities of LEED calculations coupled with HREELS. One such structure is naphthalene ($C_{10}H_8$) on Rh(111), which produces both a (3×3) and a $(3\sqrt{3} \times 3\sqrt{3})$ R30° unit cell: these have areas 9 and 27 times that of the (1×1) unit cell and presumably contain 10 and 30 carbon atoms per unit cell, respectively. Another such structure concerns benzene on Pt(111), which can form both a $(2\sqrt{3} \times 4)$rect and a $(2\sqrt{3} \times 5)$rect unit cell, with areas 16 and 20 times the (1×1) area, respectively. If, as HREELS indicates, these cells contain 4 and 6 CO molecules, respectively, in addition to the presumed two benzene molecules per cell, there are 20 and 24 atomic scatterers per unit cell, respectively (hydrogen being ignored).

In view of the many structural parameters that these structures contain, it is important to gather complementary information restricting the number of possible structures. Besides HREELS data, one source of information is to consider Van der Waals radii that identify forbidden molecular overlaps [16.26]. A more accurate approach to this question comes from force-field calculations, which evaluate Van der Waals interactions in more detail [16.27]. Also, geometries of adsorbed molecular species may be studied by energy-minimization techniques: for large molecules on extended surfaces, extended Hückel cluster calculations [16.23,28] or corresponding two-dimensional tight-binding band-structure calculations [16.25,28] can serve as useful guides.

Another important aspect of the work concerns experimental procedures. Apart from the standard need for well-defined surface conditions, adsorbed molecules present a special difficulty: great susceptibility to electron-beam-induced damage. New approaches to reduce the LEED beam exposure have been implemented, principally the use of video cameras [16.29], which shorten the time needed to take the data. A further development, which is about to come into use, replaces the LEED display screen by a position-sensitive detector coupled with microchannel plates [16.30]: this allows the intensity of the incident LEED beam to be reduced by a factor of 10^6, essentially eliminating surface damage.

Acknowledgments. Most of the work discussed here was performed with Professor G.A. Somorjai and collaborators, as well as with Professor M. Simonetta and collaborators, and with Dr. C. Minot. This work was supported in part by the Director, Office of Energy Research, Office of Basic Energy Sciences, Materials Sciences Division of the U.S. Department of Energy under Contract No. DE-AC03-76SF00098. Other funding was received from an NSF-Italy Cooperative Science Grant and a NATO Exchange Grant.

References

16.1 H. Ibach, D.L. Mills: *Electron Energy Loss Spectroscopy and Surface Vibrations* (Academic, New York 1982)
16.2 J.B. Pendry: *Low-Energy Electron Diffraction* (Academic, London 1974)
16.3 M.A. Van Hove, S.Y. Tong: *Surface Crystallography by LEED* (Springer, Berlin, Heidelberg 1979)
16.4 M.A. Van Hove, G.A. Somorjai: Surf. Sci. **114**, 171 (1982)
16.5 M.A. Van Hove, R.J. Koestner, G.A. Somorjai: Phys. Rev. Lett. **50**, 903 (1983)
16.6 M.A. Van Hove, R.J. Koestner, J.C. Frost, G.A. Somorjai: Surf. Sci. **129**, 482 (1983)
16.7 M.A. Van Hove, R.F. Lin, G.A. Somorjai: Phys. Rev. Lett. **51**, 778 (1983)
16.8 D.K. Saldin, J.B. Pendry, M.A. Van Hove, G.A. Somorjai: To be published
16.9 J.B. Pendry: This Volume, p.124;
D.K. Saldin, D.D. Vredensky, J.B. Pendry: This Volume, p.131
16.10 R.P. Eischens, W.A. Pliskin: Adv. Catal. **10**, 1 (1958)
16.11a S. Andersson, J.B. Pendry: Phys. Rev. Lett. **43**, 363 (1979); and J. Phys. C**13**, 3547 (1980);
16.11b M.A. Passler, A. Ignatiev, F. Jona, D.W. Jepsen, P.M. Marcus: Phys. Rev. Lett. **43**, 360 (1979);
16.11c K. Heinz, E. Lang, K. Müller: Surf. Sci. **87**, 595 (1979);
16.11d S.Y. Tong, A.L. Maldonado, C.H. Li, M.A. Van Hove: Surf. Sci. **94**, 73 (1980)
16.12 R.J. Behm, K. Christmann, G. Ertl, M.A. Van Hove, P.A. Thiel, W.H. Weinberg: Surf. Sci. **88**, L59 (1979);
R.J. Behm, K. Christmann, G. Ertl, M.A. Van Hove: J. Chem. Phys. **73**, 2984 (1980)
16.13 R.J. Koestner, M.A. Van Hove, G.A. Somorjai: Surf. Sci. **107**, 439 (1981)
16.14 G. Michalk, W. Moritz, H. Pfnür, D. Menzel: Surf. Sci. **129**, 92 (1983)
16.15 L.L. Kesmodel, R.C. Baetzold, G.A. Somorjai: Surf. Sci. **66**, 299 (1977)
16.16 G. Casalone, M.G. Cattania, F. Merati, M. Simonetta: Surf. Sci. **120**, 171 (1982)
16.17 G. Casalone, M.G. Cattania, M. Simonetta: Surf. Sci. **103**, L121 (1981)
16.18 L.L. Kesmodel, L.H. Dubois, G.A. Somorjai: J. Chem. Phys. **70**, 2180 (1979)
16.19 R.J. Koestner, M.A. Van Hove, G.A. Somorjai: Surf. Sci. **121**, 321 (1982)

16.20 G. Casalone, M. Simonetta, R.J. Koestner, D.F. Ogletree, M.A. Van Hove, G.A. Somorjai: To be published
16.21 M.A. Van Hove, R.J. Koestner, G.A. Somorjai: J. Vac. Sci. Technol. **20**, 886 (1982);
R.J. Koestner, M.A. Van Hove, G.A. Somorjai: J. Chem. Phys. **87**, 203 (1983)
16.22 H. Ibach, S. Lehwald: J. Vac. Sci. Technol. **15**, 407 (1978)
16.23 A. Gavezzotti, M. Simonetta: Surf. Sci. **99**, 453 (1980)
16.24 B.E. Koel, J.E. Crowell, C.M. Mate, G.A. Somorjai: J. Phys. Chem. (in press)
16.25 E.L. Garfunkel, C. Minot, A. Gavezzotti, M. Simonetta: To be published
16.26 A. Gavezzotti, M. Simonetta, M.A. Van Hove, G.A. Somorjai: To be published
16.27 A. Gavezzotti, M. Simonetta, M.A. Van Hove, G.A. Somorjai: Surf. Sci. **122**, 292 (1982);
A. Gavezzotti, M. Simonetta: Surf. Sci. **134**, 601 (1983)
16.28 C. Minot, M.A. Van Hove, G.A. Somorjai: Surf. Sci. **127**, 441 (1983)
16.29 P. Heilmann, E. Lang, K. Heinz, K. Müller: Appl. Phys. **9**, 247 (1976);
E. Lang, P. Heilmann, G. Hanke, K. Heinz, K. Müller: Appl. Phys. **19**, 287 (1979);
K. Müller, K. Heinz: This Volume, p.105
16.30 P.C. Stair: Rev. Sci. Instrum. **51**, 132 (1980);
D.F. Ogletree, J.E. Katz, J.C. Frost, G.A. Somorjai: To be published

17. Computer Controlled LEED Intensity and Spot Profile Determination

K. Müller and K. Heinz

Lehrstuhl für Festkörperphysik, Universität Erlangen-Nürnberg
Erwin-Rommel-Straße 1, D-8520 Erlangen, FRG

Surface-structure analysis, investigations of the degree of order and studies of surface phase changes by means of LEED require collecting as well as handling a large amount of data. Rapid data acquisition was achieved through the computer-controlled system DATALEED designed several years ago, which in the meantime has shown its capabilities in various examples. Applications with respect to surface structure transitions [K/Ir(100), Ir(100)$1\times1\rightarrow1\times5$] are given.

The time required to determine the integral intensity of a LEED beam at a specific energy amounts to 20 ms including background subtraction. A spectrum of LEED intensities is completed in about 10 s depending on the energy steps and the energy range covered. Several spots can be addressed quasisimultaneously. Spot profiles of a selected number of beams are displayed on an oscilloscope. Again it requires only 20 ms to determine their halfwidths, which may be affected by a phase transition or any other dynamic surface process.

17.1 Introduction

Reconstructed surfaces and adsorbates often display complicated LEED patterns. A large number of nonequivalent beams, poor stability of the surface order or the observation of phase transitions require fast data acquisition and handling. One possible answer to this demand was the development of a computer-controlled system for LEED intensity and spot profile determination which was introduced several years ago [17.1-3]. A brief description of the systems's most recent version and its performance is given before we discuss some of the current applications.

17.2 Method

The video signal generated by a TV camera viewing a LEED pattern is proportional over three orders of magnitude to the flux of diffracted electrons exciting the luminescent screen. Instead of the direct beam current measurement by a Faraday collector, the computer-controlled system generates an electronic window of variable size and shape around a certain spot or group of spots and stores up to 10^3 pixels of digital information contained in the window in a fast memory. Its access time of 20 ns is short enough to be used on line with the video signal and to be readdressed with each TV half-frame generation. Just like a Faraday collector the electronic window is made to follow the spot as it moves over the screen with varying beam energy. However, since the data stored in the fast memory can be processed (to calculate integral intensity, FWHM of profiles, etc.) within the period of time required

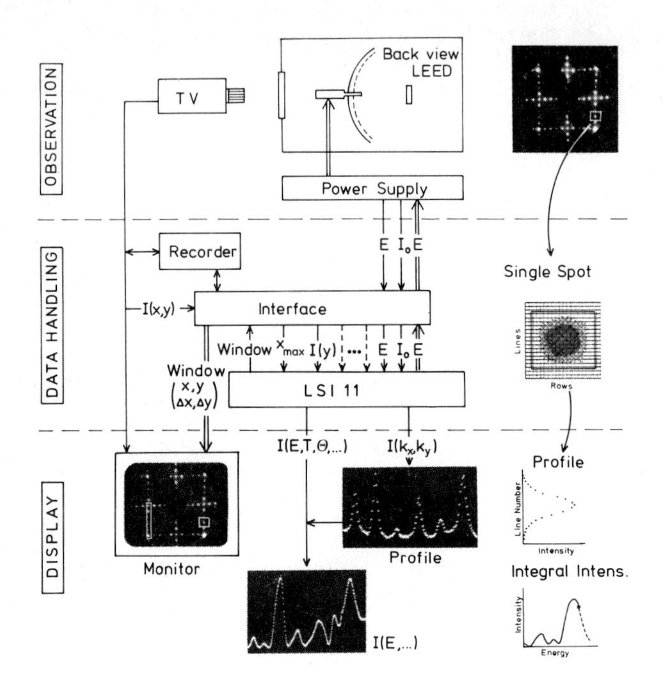

Fig.17.1. Information flow diagram from observation through data handling to display of intensities by the computer-controlled TV method DATALEED

by the TV electronics to generate one half-frame (20 ms), the energy can be stepped forward every video half-frame. Thus a spectrum of 300 data points is taken within 6 s, considerably faster than using a manually operated Faraday collector.

The procedure is illustrated in Fig.17.1, where one can follow the flow of information from the state of observation by a back view LEED optics through data handling to display from top to bottom. Single lines carry information, double lines represent control channels. Two kinds of data processing at every energy step are shown: a) the development of about 20 parallel profiles through the spot and their subsequent integration to give the spot intensity, b) the generation and display of spot profiles along a slit-like window.

17.3 Interface

Rather than the computer itself (16 bit/64 kByte LSI 11/2) the interface must be considered the heart of the system. It is designed to generate the electronic window, to digitize the incoming video signal by an 8 bit/13 MHz ADC, to store temporarily the window information in the 1 k × 9 bit fast memory and to slow down the data rate passed from the camera to the computer. For the latter purpose the interface integrates the video signal I(x,y), in the horizontal x direction within the edges of the window for every TV line.

The resulting vertical profile $I(y) = \int I(x,y)\,dx$ is passed on to the computer together with the x-coordinate value of the intensity maximum determined simultaneously. The latter is used to center the window when the energy is stepped forward. Thus the window position is always one step behind, which, however, does not matter as long as the energy steps are limited to below 2 eV. The interface also transmits the energy ramp control via opto-couplers and handles additional data transfer from and to the computer, such as the actually measured energy E, the primary beam current I_0 and other parameters for data reduction, through an 8 channel multiplexer.

17.4 Performance

Due to a fast assembler program the software-controlled data processing can be carried out within the period of time required for one half-frame generation, i.e., 20 ms for European TV standard (16.67 ms US standard). Within that time the computer subtracts the background determined at the window edges and integrates I(y) to the final integral intensity. The window position is adjusted as well. So a one-beam spectrum consisting of 500 points develops within 10 s, and the total measuring time increases linearly with the number of beams to be considered. The use of a high quality recorder stores, of course, the information of all beams simultaneously in the first run. Further characteristics of the system are listed in Table 17.1.

Table 17.1. Specifications of the DATALEED system

SPEED	20 ms per energy and beam (20 ms per energy for all beams if intermediate tape storage is used)
SENSITIVITY	$5 \cdot 10^{-12}$A for diffracted beam for a CdSe Camera (Siemens K2B) $\lesssim 10^{-13}$A for diffracted beam for an image intensifier camera (Bosch TYC 9A)
LINEARITY	3 orders of magnitude (tested with Faraday collector)
ACCURACY	mainly limited by the homogeneity of the luminescent screen (typical variations: $<15\%$), much improved by averaging of equivalent beams

As an example of intensity measurements, Fig.17.2 shows three (10) beam spectra taken from an Ir(100) surface ($T \approx 100$ K). The upper curve results from the reconstructed 1×5 phase, the middle curve represents the clean 1×1 metastable phase, while the third one was taken from a slightly contaminated surface. These are experimental input data for surface structure determinations [17.4], but it should also be pointed out that I(E) spectra are sensitive fingerprints for specific surface conditions or states, much more reliable than any pattern can be.

It has occasionally been argued that the high speed of measurement causes distortions of the true spectra [17.5,6]. To disprove this point, Fig.17.3 shows a series of I(E) spectra taken from the same Cu(100) sample with different speeds. They are all identical. Due to the specific time constant of the power suppley there is, however, an energy shift proportional to the speed of measurement, i.e., a constant shift value for any selected speed. The curves in Fig.17.3 have been shifted according to these values. This correction, taken into account or not, has no effect on structure determinations: it just changes the constant inner potential input in the calculations.

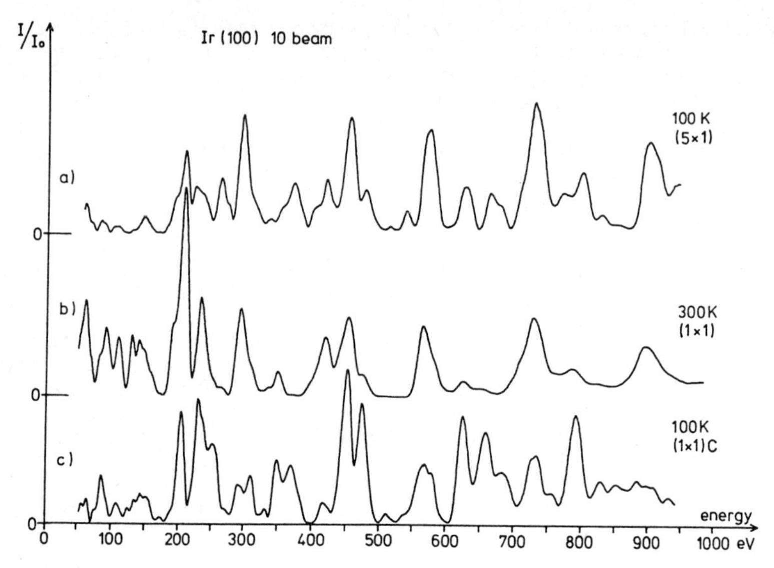

Fig.17.2. Spectra of the (10) beam taken from three phases of the Ir(100) surface

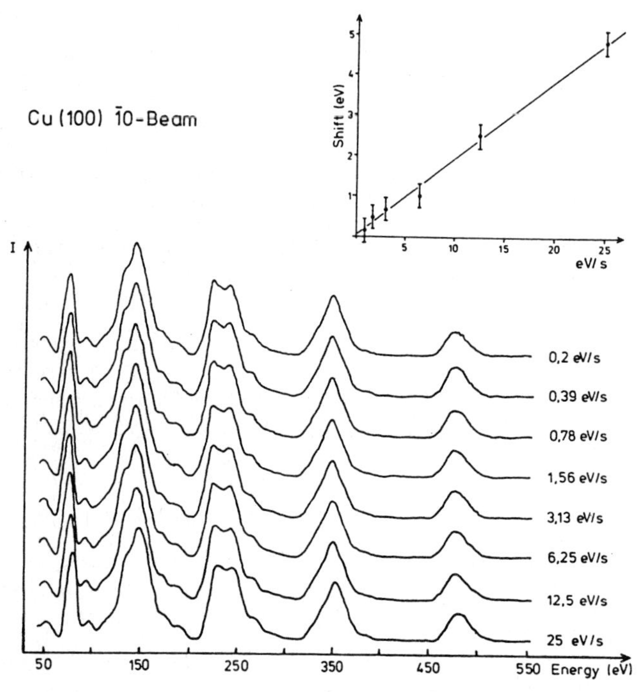

Fig.17.3. Intensity spectra of the ($\bar{1}0$) beam of Cu(100) at normal incidence and $T \approx 100$ K for various speeds of measurement. Curves are shifted on the energy scale according to the speed by values given in the insert

Another point of concern is the proper primary-beam alignment. For normal incidence it has been demonstrated several times [17.2-4,7] how sensitive spectra of symmetrically equivalent beams respond to slightly off normal ($<1^{\circ}$) incidence. Therefore the computer-controlled system provides up to four windows quasisimultaneously, i.e., within four subsequent TV half-frames. Thus the adjustment can be improved by comparing the spectra, remaining differences can be leveled out by automatic averaging of the spectra. These averaged data have been shown to be insensitive to up to 3° misalignment [17.3].

Since I(E) spectra are sensitive to the primary-beam angle it is generally desirable for surface structure determinations to include data taken with deliberately chosen off-normal incidence. This is more difficult but still quite accurate. Starting with normal incidence, we chose two different beams $(hk)_1$ and $(hk)_2$. Their relation to each other is such that $(hk)_1$ appears with off-normal incidence (θ) at the same spot of the screen as $(hk)_2$ does at normal incidence using the same energy $E((hk)_{1,2},\theta)$. This location and consequently the rotation of the sample by the angle θ can be determined ($\pm 20'$) by the video LEED system. To demonstrate the accuracy of the angular adjustment, two off-normal spectra have been compared. They were measured after independent rotation of the sample into the same position. The spectra differ by a Zanazzi-Jona r factor of $r = 0.023$.

17.5 Application to Surface Structure Transitions

Structure transitions at surfaces are one of the most interesting subjects of current activities in surface science. Such transitions are observed in adsorption systems as a function of surface coverage or of temperature with fixed coverage. Clean surfaces show structural transitions as well, e.g., reversible as a function of temperature in the case of W(100) or irreversible as activated processes, e.g., for the (100) surfaces of Ir, Pt and Au. To observe such transitions by LEED requires measuring intensity profiles and/or integral intensities as a function of coverage, temperature or time. Consequently, a large amount of data must be measured and handled, and for time-dependent processes even with sufficient speed. As demonstrated above, the video-LEED system meets these requirements. Two examples are given in the following.

In the first example the adsorption of potassium on the nonreconstructed surface of Ir(100) at low temperature ($T \approx 100$ K) is treated. As demonstrated in [17.8] in more detail, several well-resolved coincidence superstructures appear with varying coverage. They are repeated schematically at the vertical margins of Fig.17.4. In addition to the treatment in [17.8], intensity profiles are presented for the coincidence structures as well as for intermediate coverages in the figure. The directions of the profiles in k space are indicated by the hatched areas in the schematic diffraction patterns. The profiles, each of which was taken within a TV half-frame, demonstrate that continuous spot splitting and shifting appears with varying coverage. This could be interpreted by continuous contraction of the real space unit mesh. However, it would require adsorption sites of the K atoms out of registry with the substrate. Therefore, in [17.8] the continuous shift of spots was attributed to a continuous mixing of microscopic adsorption domains with high-symmetry adsorption sites only.

The second example to demonstrate the capability of the TV LEED system is the structure transition $Ir(100) 1\times 1 \rightarrow 1\times 5$. The metastable 1×1 structure can be prepared by thermochemical treatment of the stable reconstructed 1×5

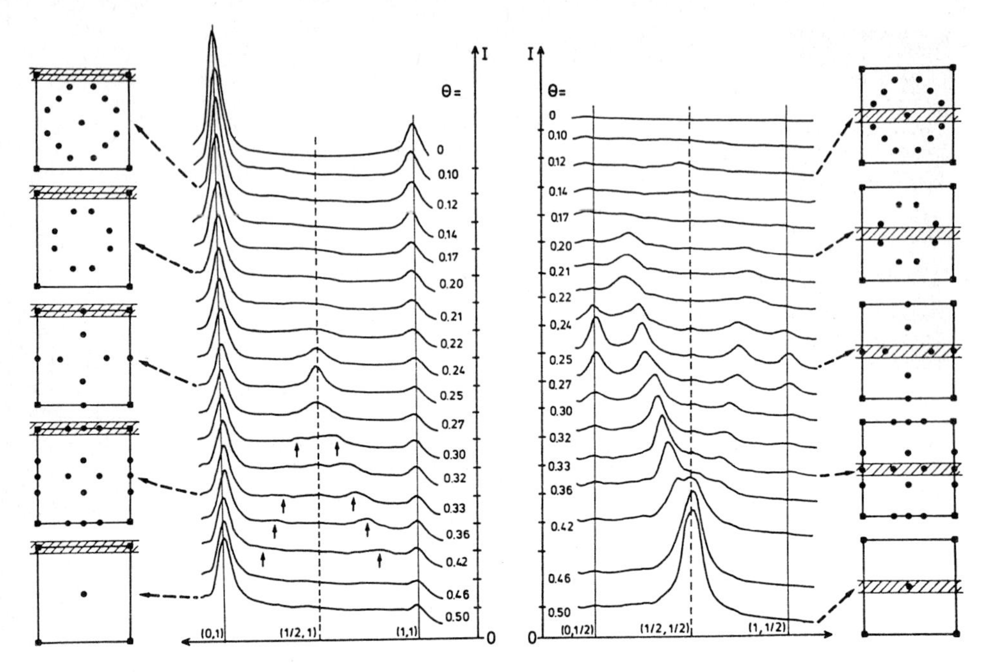

Fig.17.4. Intensity profiles for various coverages Θ of K on Ir(100)1 × 1 at T ≈ 100 K. Schematic LEED patterns of coincidence superstructures are given at the left- and right-hand sides for Θ = 1/8, 1/5, 1/4, 1/3, 1/2 together with the direction of the profiles (*hatched area*)

phase [17.9,10]. Then the structure transition is initiated by thermal activation (T ≧ 800 K) and develops irreversibly with time. So, for observation by LEED, intensity profiles and integral intensities have to be recorded on a time scale much shorter than that of the transition. This was done using the TV method. Figure 17.5 presents intensity profiles along an array of appearing superstructure spots as indicated in the schematic final LEED pattern. It is clear that the superstructure spots narrow and move with time. Other data, which cannot be shown here because of lack of space, demonstrate that they develop from streaks of increased background between integer order spots (in Fig.17.5 the background level is subtracted as described above). This is interpreted by the assumption that the 1 × 1 phase disorders via shifting atomic rows. The hcp 1 × 5 structure develops by nucleation of more or less ordered domains which grow and finally coalesce [17.10]. The activation energy for the whole process can be taken from the developement of intensities with time. Figure 17.6 shows that the superstructure spots increase at the expense of the integer order spot intensities. The characteristic time in the first period of the transition is strongly temperature dependent as demonstrated for the (2/5,0) spot. Of course this is expected for an activated process and from the Arrhenius plot an activation energy of 0.88 ± 0.03 eV results (Fig.17.6). The observed growth of the superstructure spot after the initial increase is similar to that observed by *Wang* and *Lu* for the time-dependent ordering of oxygen on W(112) [17.11]. However, in both cases it is not yet well understood.

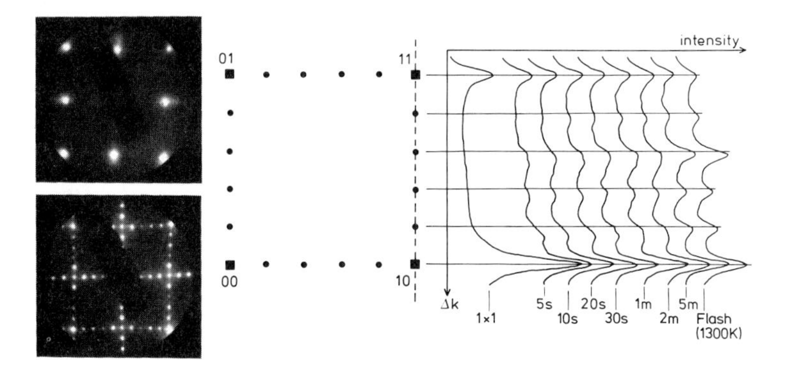

Fig.17.5. *Left*: LEED patterns of the metastable 1 × 1 phase and the reconstructed 1 × 5 phase of Ir(100); *center*: schematic 1 × 5 LEED pattern; *right*: intensity profiles for various annealing times at T = 990 K (data taken after quenching to ~100 K)

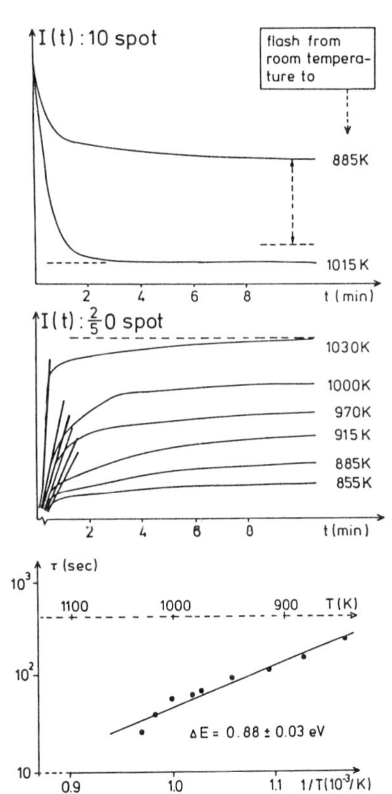

Fig.17.6. Integal intensities of the integer order (10) spot and the (2/5,0) superstructure spot varying with time for different temperatures. *Below*: Arrhenius plot for the initial characteristic times of the (2/5,0) spot

17.6 Conclusion

The TV computer system described in this paper provides a fast method for measuring LEED intensity profiles and integral intensities. The more or less automatic procedure is easy to handle, quite sensitive and accurate and therefore encourages also convenient control measurements for sample adjustment or surface characterization by fingerprints. The speed of the method allows surfaces to be investigated with rapidly varying structures. Typical examples are surface phase transitions.

Acknowledgments. The able cooperation of L. Hammer, H. Hertrich, H. Lindner and G. Schmidt is gratefully acknowledged.

References

17.1 P. Heilmann, E. Lang, K. Heinz, K. Müller: Appl. Phys. **9**, 247 (1976)
17.2 P. Heilmann, E. Lang, K. Heinz, K. Müller: *The Necessity for fast LEED Intensity Measurements*, in Proc. Conf. Determination of Surface Structure by LEED, Yorktown Heights (1980)
17.3 K. Heinz, K. Müller: *LEED Intensities – Experimental Progress and New Possibilities in Surface Structure Determination*, in Springer Tracts in Mod. Phys., Vol.91, (Springer, Berlin, Heidelberg, New York 1982) pp.1-53
17.4 E. Lang, K. Müller, K. Heinz, M.A. Van Hove, R.J. Koestner, G.A. Somorjai: Surf. Sci. **127**, 347 (1983)
17.5 M.A. Stevens, G.J. Russell: Surf. Sci. **104**, 354 (1981)
17.6 M. Maglietta: J. Phys. C: Solid State Phys. **17**, 363 (1984)
17.7 H.L. Davis, J.R. Noonan: Surf. Sci. **115**, L75 (1982)
17.8 K. Heinz, H. Hertrich, L. Hammer, K. Müller: Submitted to Surf. Sci., Proceedings of the 6th European Conference on Surface Science (1984)
17.9 J. Küppers, H. Michel: Appl. Surf. Sci. **3**, 179 (1979)
17.10 K. Heinz, K. Müller: Abstract, American Crystallographic Association, Annual Meeting, Lexington, USA (1984)
17.11 G.C. Wang, T.M. Lu: Phys. Rev. Lett. **50**, 2014 (1983)

18. On the Role of Space Inhomogeneity of Electron Damping in LEED

I. Bartoš*
Department of Applied Mathematics, University of Waterloo
Waterloo, Ontario N2L 3G1, Canada
*Permanent address: Institute of Physics, Czech. Acad. Sci., Praha

J. Koukal
Institute of Physics, Czech. Acad. Sci., 18040 Praha 8, Czechoslovakia

18.1 Introduction

Surface sensitivity of Low-Energy Electron Diffraction (LEED) results from incident electron damping, theoretically described by a complex optical potential. The imaginary part of this effective potential is then responsible for electron losses from coherent channels in LEED; it reflects the consequences of such processes like excitation of plasmons, creation of electron-hole pairs and interactions of an electron with lattice vibrations. In LEED calculations, an approximate form with uniform magnitude of the imaginary component of the crystal potential is used. Though this description appears to suffice for surface crystallography, the nonuniform nature of several processes involved indicates that the inhomogeneity of the damping should be considered [18.1], especially if subtler features in the intensity profiles are analyzed.

18.2 The Nearly Free Electron Model

18.2.1 Damping and Electron Structure

In this model, the crystal potential is described by its single Fourier component V_g, and the semiinfinite system with the step potential at the surface has

$$V(x) = (V_0 + V_g \cdot \cos gx) \cdot \theta(x) \qquad \theta(x) = \begin{Bmatrix} 1 & \text{for} & x \geq 0 \\ 0 & & x < 0 \end{Bmatrix} , \tag{18.1}$$

where g is the reciprocal lattice vector, $g = 2\pi/a$, a being a lattice constant.

If both V_0 and V_g are real, there is no damping in the crystal, and if only V_0 has a nonzero imaginary component V_{0i} ($V_0 = V_{0r} + iV_{0i}$, $V_{0i} \leq 0$) the damping is homogeneous. The inhomogeneous contribution to the crystal damping is taken into account by $V_{gi} \neq 0$ ($V_g = V_{gr} + iV_{gi}$); this optical potential satisfies general requirements [18.2].

For the quantum-mechanical description of the system, the formalism of surface Green's functions has been adopted: the surface Green's function is expressed in terms of surface Green's functions of the infinite crystal G_c and vacuum G_v. Surface density of electron states is then $-(1/\pi)$ Im G and the reflectivity R is $R = (G - G_v)G_v^{-1}$, according to *Velický* and *Bartoš* [18.3]. For simplicity, only two symmetrical terminations of the crystal potential are considered, corresponding to cutting the real potential in its minimum or maximum at $x=0$ ($V_{gr}<0$, $V_{gr}>0$, respectively).

Before discussing electron reflectivity, and because of the close relationship between the two (as manifested by the surface Green's functions), local densities of electron states are mentioned. Surface density of states (local in r space) is simply reduced or enhanced if inhomogeneous damping is introduced. The effect is determined by the phase of the oscillating imaginary part of the potential: if effective damping is increased at the surface, the surface density of states is suppressed and vice versa. The behavior of the density of states for a given k (local in the reciprocal k space) is not as simple: for k vectors close to the Brillouin zone boundary the inhomogeneous damping gives rise to increased density of states at one of the two gap edges, while at the other one, it decreases. In this case, the relation between the phase of the oscillatory damping and the localization of electrons (bonding and antibonding states) determines the effect [18.4].

18.2.2 Damping and Electron Reflectivity

The electron reflectivity of the one-dimensional nearly free electron semi-infinite crystal is well known, both for the undamped and uniformly damped models. In the former case, reflectivity is unity for gap energies with decaying tails; this bell-shaped curve erodes into a Gaussian-like curve if uniform damping is included [18.3,5]. What happens if, in addition, oscilating inhomogeneous damping takes place?

Classically, the following effects might be expected in the two limiting situations: if electrons penetrated only infinitesimally into the crystal, the reflectivity, governed by the effective damping right at the surface, would be enhanced or decreased in the whole range of energies. If, on the other hand, electrons penetrated deep into the crystal, then contributions from alternating regions with increased and decreased damping should average out and no effect should be observed.

The evaluated profiles confirm neither of the two predictions. In comparison with a uniformly damped model, reflectivities from the inhomogeneously damped crystals exhibit both energy regions where reflectivity is enhanced as well as those where it is suppressed. Symmetrical profile (I/V curve) becomes asymmetrical: as one part of the symmetrical peak is suppressed while the other is enhanced, the position of the maximum is shifted from the middle of the gap and the peak width is reduced.

The classical picture fails to interpret the changes in reflectivities, because it ignores inhomogeneities of electron distribution within a crystal. Bonding states (with energies close to the lower gap edge) are localized predominantly in regions around minima of the real part of the periodical potential, whereas antibonding ones (with energies at the upper gap edge) are concentrated mainly around the maxima. When the imaginary part of the crystal potential is in phase with its real part ($V_{gr} \cdot V_{gi} > 0$), the effective damping will be increased around minima of the real potential, thus leading to suppressed reflectivity at energies corresponding to the lower part of the band gap. The opposite effect consisting in enhancing the reflectivity due to decreased effective damping around the potential maxima then occurs in the upper part of the gap. For the out-of-phase case ($V_{gr} \cdot V_{gi} < 0$) the effect should be just reversed. Both predictions agree with the results presented in Fig.18.1.

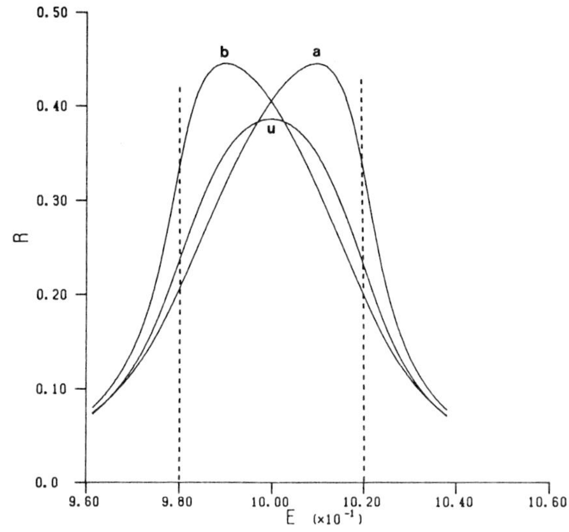

Fig.18.1. Energy dependence of electron reflectivity from the semiinfinite one-dimensional nearly-free-electron model with damping. The potential is zero for $x < 0$ and is periodic inside the crystal: $V(x) = V_0 + V_g \cdot \cos gx$ for $x \geq 0$; ($g = 2$, $V_{gr} = 0.02$, so that the gap is between 0.98 and 1.02). Damping is described by imaginary components of the potential: $V_0 = V_{0r} + iV_{0i}$, $V_g = V_{gr} + iV_{gi}$. Reflectivity R from a crystal with homogeneous damping ($V_{0r} = 0$, $V_{0i} = -0.01$, $V_{gi} = 0$, *curve u*) is compared with reflectivities from the inhomogeneously damped crystal for the (*a*) in-phase case ($V_{gi} = 0.005$) and (*b*) out-of-phase case ($V_{gi} = -0.005$). Atomic units are used, also for the energy scale

18.3 Intensity Profiles for Cu(100)

The copper crystal potential, composed of atomic muffin-tin potentials, has been taken to demonstrate the role of inhomogeneous damping in a three-dimensional case. For the evaluation of diffracted intensity of the (00) beam (at the normal incidence), the renormalized forward scattering program by *Van Hove* and *Tong* [18.6] has been adopted. All parameters were kept except the position of the vacuum level which was artificially lowered to reveal the whole gap at the Brillouin zone boundary (X). The two gap edges are then: ($E(X_4') \cong 1.4$ eV and $E(X_1^c) \cong 6.7$ eV above the vacuum level.

The uniformly damped crystal was in this low-energy region represented by $V_{0i} = -2$ eV: the corresponding intensity profile is shown by the curve u in Fig.18.2. The inhomogeneous damping was introduced by bringing additional damping into the muffin-tin spheres, and modifying the complex phase shifts δ_l accordingly (first-order Taylor expansion has been used). From the intensity profiles for inhomogeneously increased (a) and decreased (b) damping, again an enhancement of the reflectivity in the upper part of the gap is seen at an increased damping accompanied by the suppression in the lower part of the gap. The opposite effect takes place for inhomogeneous decrease of damping.

Again, as in the nearly-free electron model, the role of inhomogeneous damping on the intensity profiles is neither zero nor simply uniform.

18.4 Conclusion

The examples demonstrated indicate the importance of damping inhomogeneities for the shape of LEED intensity profiles. In comparison with the results for uniformly damped models, both enhancements and suppressions of reflectivities take place under additional inhomogeneous damping. The close connection be-

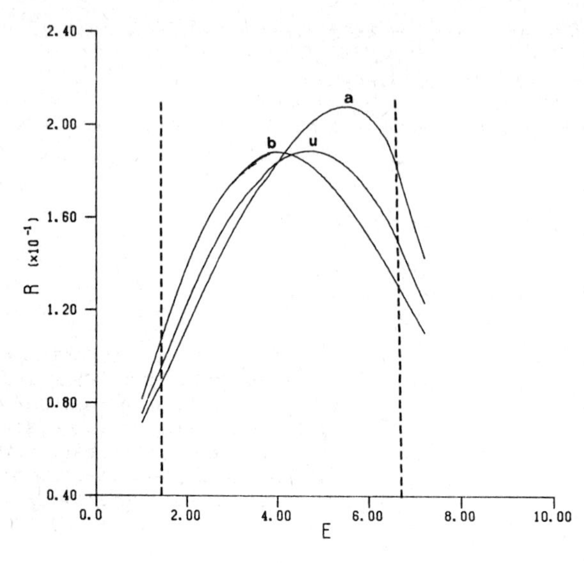

Fig.18.2. The energy dependence of the (00) beam, diffracted from Cu(100), for homogeneous and inhomogeneous damping within the crystal (program and parameters according to *Van Hove* and *Tong* [18.6]; the work function has been artificially lowered to reveal the whole gap at X). Homogeneous damping is represented by V_{0i} = -2 ev uniformly within the whole crystal, strongly inhomogeneous damping is introduced by additional contributions ΔV_{0i} into the imaginary part of the potential inside the muffin-tin spheres: (*a*) increased damping ΔV_{0i} = -2 eV, (*b*) decreased damping ΔV_{0i} = -2 eV. The energy scale is in eV

tween the electron structure of crystals and corresponding intensity profiles stresses the importance of band-structure effects in the low-energy region of LEED. Detailed analyses of the profiles should bring information about the damping of excited electrons in crystals complementary to that provided by angular resolved photoelectron spectroscopy [18.7-9].

Though the effect of finite temperatures is also treated by means of complex phase shifts δ_1, it can be separated from the electronic contribution. The latter has to change significantly with energy of incident electrons, where, e.g., below the plasmon threshold the dominant source of delocalized damping is turned off.

Recently, a simplified inhomogeneity of electron damping in crystals has been used to interpret narrow peaks, observed in "subthreshold" regions in LEED from Cu(111) by *Lindgren* et al. [18.10].

References

18.1 A. Howie: In *Electron Diffraction 1927-1977*, ed. by P.J. Dobson, J.B. Pendry, C.J. Humphreys (The Institute of Physics, Bristol 1978) pp.1-12
18.2 P.H. Dederichs: In *Solid State Physics 27*, ed. by H. Ehrenreich, F. Seitz, D. Turnbull (Academic, New York 1972)
18.3 B. Velický, I. Bartoš: In *LEED-Surface Structures of Solids*, ed. by M. Lãznička (JČMF, Praha 1982) p.423
18.4 I. Bartoš: Phys. Stat. Sol.(b) **122**, K159 (1984)
18.5 J.C. Slater: Phys. Rev. **51**, 840 (1937)
18.6 M.A. Van Hove, S.Y. Tong: *Surface Crystallography by LEED* (Springer, Berlin, Heidelberg, New York 1979)
18.7 J.A. Knapp, F.J. Himpsel, D.E. Eastman: Phys. Rev. B**19**, 4952 (1979)
18.8 J.K. Grepstad, B.J. Slagsvold, I. Bartoš: J. Phys. F**12**, 1679 (1982)
18.9 I. Bartoš, J. Koukal: Surf. Sci. **138**, L151 (1984)
18.10 S.A. Lindgren, L. Waldén, J. Rundgren, P. Westrin: Phys. Rev. B**29**, 576 (1984)

19. Attenuation of Isotropically Emitted Electron Beams

R. Mayol, F. Salvat, and J. Parellada-Sabata

Facultat de Fisica, Universitat de Barcelona
Diagonal 645. E-08028 Barcelona, Spain

The transport features of isotropically emitted electrons inside amorphous samples are analyzed using a Monte Carlo method. This method is used in several electron spectroscopies, e.g., Auger, ESCA, CEMS, which differ only in the excitation mechanism. The scattering model includes realistic cross sections and it uses the continuous slowing down approximation corrected for energy straggling. The energy and angular distributions of the electrons emerging from the surface are computed for several sample compositions and electron initial conditions. Simple scaling rules, which may be useful for a quantitative approach, are found.

19.1 Introduction

In recent years, several groups [19.1-3] have developed theoretical methods to use Conversion Electron Mössbauer Spectroscopy (CEMS) for quantitative analysis of surface regions of solids containing Mössbauer nuclei (mainly ^{57}Fe and ^{119}Sn). The central problem in these methods regards the flux attenuation and slowing down of the electrons emitted as a consequence of nuclear deexcitations in the sample. These electrons are generated mainly by internal conversion (IC) and Auger effect (which takes place after the IC process). Both IC and Auger electrons are emitted isotropically from the sample atoms with discrete initial energies. Isotropically emitted electron beams are also present in other electron spectroscopies in which the external bombardment causes inner shell ionizations and subsequent Auger processes. Quantitative analysis in these spectroscopies requires knowing the so-called weight functions [19.3] which give the probability for an electron generated (isotropically) at a given depth inside the sample with known initial energy to reach the surface of the sample with the appropriate energy and direction to be detected.

A detailed scattering model, based on the continuous slowing down and the Born approximation, is used in this work to compute accurate weight functions in conjunction with the Monte Carlo method. The adopted scattering model has proved to give results in good agreement with experiments of transmission and backscattering of parallel monoenergetic electron beams through thin films for energies in the range between a few keV and several hundreds of keV (depending on the atomic number of the diffusing material) [19.4]. Some scaling properties of the weight functions are derived from the simulated results for different materials and electron energies.

19.2 Monte Carlo Procedure

The natural tool to solve transport problems is the Monte Carlo method, i.e., computer simulation of random particle tracks. The path of an electron is described as a series of connected straight segments or "free flights". The angular deflections at the end of each free flight are due to elastic and inelastic collisions with the atoms of the medium. The length of each free flight, the energy loss and the angular deflections are independent random variables following probability distributions which are fixed by the adopted cross sections. Electron transport properties are inferred from a large series of electron trajectories simulated by random sampling. The random paths are simulated step by step, and at the end of each free flight, boundary conditions are checked. In this work, the electrons are assumed to be effectively absorbed when their energy becomes lower than 0.4 keV.

The scattering model adopted here has been described in detail in [19.4]. It is based on the continuous slowing down approximation, i.e, the average rate of energy loss is computed from the Bethe stopping power formula. The continuous slowing down approximation is corrected for energy straggling by using the distribution of energy losses found in the classical theory of binary collisions. The total inelastic cross section is derived from the Born approximation following the method proposed by *Lenz* [19.5]. Elastic collisions are described by using the Mott cross section, i.e., the Born approximation for the screened Coulomb potential. In this way, both elastic and inelastic cross sections are determined by the atomic form factor.

To compute accurate atomic densities and form factors, a Dirac-Hartree-Fock-Slater code [19.6] has been used. The self-consistent calculations have been performed under Wigner-Seitz boundary conditions to take solid-state effects into account. It should be pointed out that the differential elastic cross section for small scattering angles computed from the Wigner-Seitz density is about two times lower than that derived from the free atom density. Clearly, this reduction of the cross section for atoms in solids must be taken into account in the Monte Carlo procedure.

The FORTRAN IV code MCSDA applies this scattering model, allowing the use of variable boundary conditions with laminar geometry. Electrons are assumed to be generated isotropically at a fixed depth x in the sample (or according to a uniform distribution over the sample volume) with fixed initial energy E. The MCSDA code provides the probability $P(E,x;e,\mathbf{\Omega})$ for an electron to reach the surface with energy e and moving in the direction $\mathbf{\Omega}$. As the emerging flux has axial symmetry around the normal to the surface, the escaping probability $P(E,x;e,\mathbf{\Omega})$ depends explicitly only on e and on the polar angle θ relative to the surface normal. It is printed on the output of the program as a two-dimensional histogram.

Some care must be taken when Monte Carlo results are compared with experimental data because of its dependence on the spectrometer characteristics. The aspects to be considered are the acceptance angles and energy resolution of the spectrometer. For a given spectrometer setting, $W(e,\mathbf{\Omega})$ denotes the detection efficiency for electrons entering the spectrometer with direction $\mathbf{\Omega}$ and energy e. Following the usual practice in CEMS, we shall define the weight function $T(E,x)$ as the probability for an electron generated at a depth x with initial energy E to be detected. Clearly

$$T(E,x) = \int_{2\pi} \int_0^E P(E,x;e,\mathbf{\Omega}) W(e,\mathbf{\Omega}) de \, d\mathbf{\Omega} \quad . \qquad (19.1)$$

This function can be easily calculated from the escaping probability $P(E,x;e,\Omega)$ when the spectrometer characteristics are known.

The numerical simulation was carried out using an IBM 4341/M2 computer. It covers several materials and electron energies which are usual in CEMS with ^{57}Fe and ^{119}Sn Mössbauer isotopes. For each energy and each material, the escaping probability $P(E,x;e,\Omega)$ was computed for several fixed values of the reduced depth $r = x/R(E)$, where $R(E)$ is the Bethe range of electrons of energy E in the considered material. Each simulation involves computing about 5000 random tracks to ensure that the statistical uncertainty in the transmitted fraction is smaller than 1%. Although a more extensive calculation would be desirable for general purposes, these results are enough to point out the general properties of the energy-direction distribution of the emerging electrons.

19.3 Simulation Results

For a given material and a given energy, the mean and the most probable energy loss of the emerging electrons in a given exit direction θ are found to be an increasing function of θ. Then, the recorded energy distribution is shifted to lower energies when the spectrometer entrance covers larger emergence angles. As a consequence, for a given spectrometer setting, the detected electrons originate closer to the surface if the entrance diaphragm is moved towards larger emergence angles. This fact has been recently exploited to obtain a sharp surface selectivity in CEMS by detecting conversion electrons emerging at glancing angles [19.7,8].

Unfortunately, the statistical uncertainty of the partial scores in the Monte Carlo results practically invalidates the method proposed here when the spectrometer covers only narrow angular and energy intervals. In this case, the calculation method must be improved by means of suitable variance reduction techniques. This drawback is avoided when global energy distributions (irrespective of the exit direction) or angular distributions (irrespective of the exit energy) are considered.

In integral CEMS the emerging electrons are usually detected by means of a gas flow proportional counter covering the whole 2π hemisphere. In this case the detection efficiency is practically one for all the emerging electrons, and from (19.1)

$$T(E,x) = \int_{2\pi} \int_{0}^{E} P(E,x;e,\Omega)de\, d\Omega \quad . \tag{19.2}$$

For all the simulated cases the integral weight function (19.2) becomes independent of the initial electron energy when depth is measured in units of the Bethe range for this energy (energy scaling) [19.9]. It has been found that energy scaling holds with high accuracy for a wide range of materials and electron energies, Fig.19.1. As a consequence, if the integral weight function $T(E,x)$ for a given material and a given energy is known, we can easily obtain the weight function for other energies. It should be noticed that $T(E,0)$, Table 19.1, strongly depends on the sample composition, in the same way as the backscattering coefficient of parallel beams depends on the atomic number of the target material.

Table 19.2 shows the mean and the most probable energies of the emerging electrons as functions of reduced depth r (i.e., depth in units of the Bethe range) for metallic iron. It is evident that both quantities practically obey

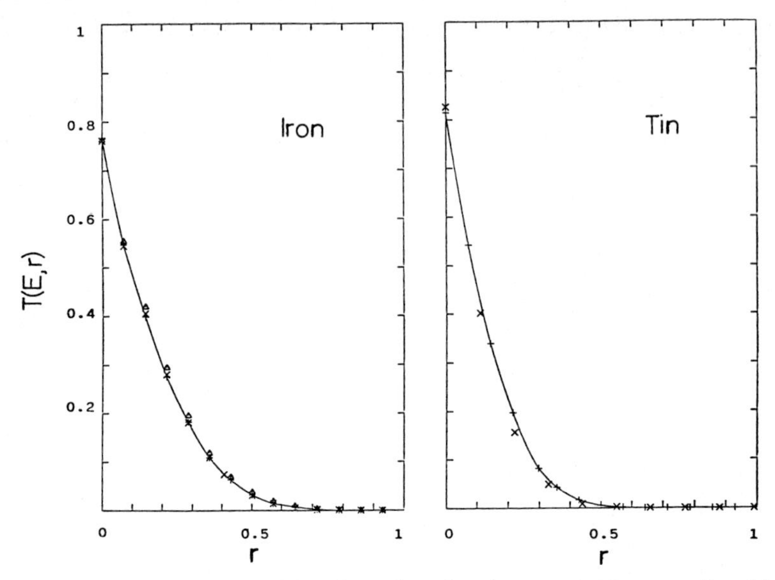

Fig.19.1. Integral weight function for iron and tin as a function of the reduced depth. Special symbols refer to different initial electron energies. *Iron*: (+), 14.4 keV; (x), 7.3 keV; (Δ), 5.4 keV. *Tin*: (+), 23.8 keV; (x), 3.6 keV

Table 19.1. Integral CEMS weight function (19.2) at the surface of homogeneous samples for different compositions and electron energies. The statistical uncertainties of the Monte Carlo results are indicated between parentheses

Material	Energy [keV]	T(E,0)
Al_2O_3	14.4	0.694 (0.006)
	5.4	0.702 (0.005)
Al	5.4	0.714 (0.006)
Fe_2O_3	14.4	0.754 (0.005)
	5.4	0.740 (0.006)
Fe	14.4	0.763 (0.006
	7.3	0.761 (0.005)
	5.4	0.766 (0.004)
Cu	5.4	0.779 (0.007)
Ag	5.4	0.826 (0.006)
Sn	23.8	0.814 (0.004)
	3.6	0.826 (o.004)
Au	5.4	0.851 (0.005)
Pb	5.4	0.854 (0.005)

Table 19.2. Mean E_m and most probable E_p energies (in keV) of the emerging electrons for iron. The last two columns show that the mean and most probable energies depend only on the reduced depth r if they are expressed in units of the initial energy E_0

E_0 = 5.4 keV

r	E_p	E_m	$(E_p/E_0)^2$	$(E_m/E_0)^2$
0	5.4	5.1	1.00	0.89
1/14	5.2	4.5	0.93	0.69
2/14	5.1	4.2	0.89	0.60
3/14	4.7	3.9	0.76	0.52
4/14	4.2	3.5	0.60	0.42
5/14	3.9	3.4	0.52	0.40

E_0 = 7.3 keV

r	E_p	E_m	$(E_p/E_0)^2$	$(E_m/E_0)^2$
0	7.3	6.9	1.00	0.89
1/14	7.1	6.1	0.95	0.70
2/14	6.8	5.6	0.87	0.59
3/14	6.5	5.2	0.79	0.51
4/14	-	4.8	-	0.43
5/14	-	4.5	-	0.38

E_0 = 14.4 keV

r	E_p	E_m	$(E_p/E_0)^2$	$(E_m/E_0)^2$
0	14.4	13.6	1.00	0.89
1/14	14.0	12.0	0.95	0.69
2/14	13.4	11.1	0.87	0.59
3/14	12.5	10.3	0.75	0.51
4/14	-	9.6	-	0.44
5/14	-	8.9	-	0.38

energy scaling, i.e., they depend only on the reduced depth at least up to $r = 5/14$. Then, the energy distributions, which are practically determined by the mean and the most probable energies, also obey energy scaling. For larger depths there are considerable uncertainties in our simulation results, due to the loss of statistics, which can invalidate further speculations. Similar results have been found for other materials (Sn, Al, Fe_2O_3 and Al_2O_3). A Thompson-Widhington law [19.10] does not hold for isotropically emitted electrons, neither for the most probable energy nor for the mean one.

Angular distributions of emerging electrons for iron are shown in Fig. 19.2. There are no appreciable differences between the (normalized) angular distributions of electrons of different energies coming from the same reduced depth. The same result holds for the other materials studied. As a consequence, the energy scaling can also be used to obtain the angular distribution for a given reduced depth and any electron energy from a unique Monte Carlo calculation.

Energy scaling of the transmitted fraction and the energy and angular distributions of isotropically emitted electron beams can be useful for a quantitative approach, at least in the keV range analyzed here. The most important drawback of the Monte Carlo simulations is the loss of statistics. This drawback can be partially avoided by using energy scaling, because then just a single calculation for a fixed energy is needed, and scaling may provide the desired results for other energies.

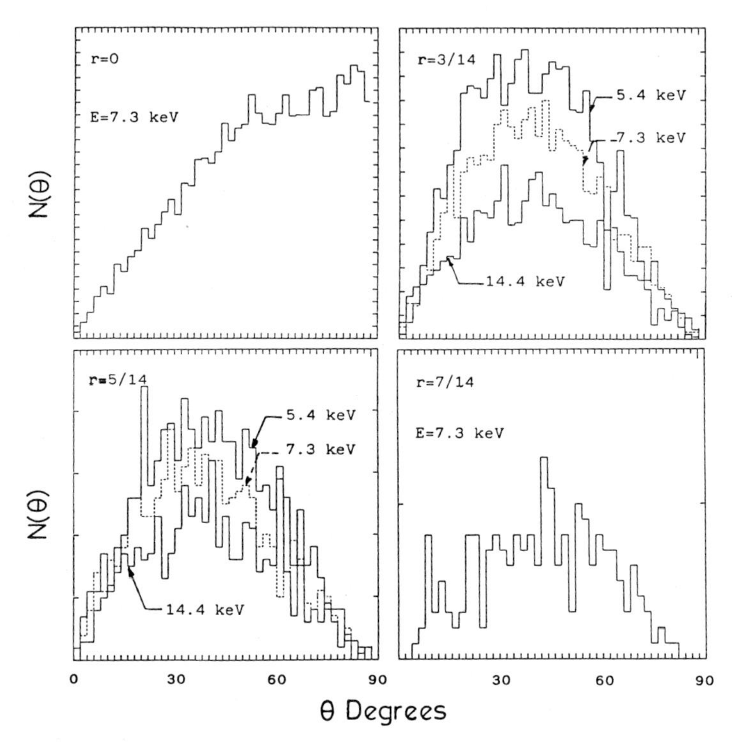

Fig.19.2. Angular distributions of emerging electrons in iron for several initial energies and reduced depths. Each division of the vertical scale corresponds to ten counts. The histograms are not normalized

References

19.1 F. Salvat, J. Parellada: Nucl. Instr. Method B**1**, 70 (1984)
19.2 D. Liljequist, T. Ekdahl, U. Bäverstam: Nucl. Instr. Method **155**, 529 (1978)
19.3 D. Liljequist: USIP Report 80-07 (1980)
19.4 F. Salvat, J. Parellada: J. Phys. D: Appl. Phys. **17**, 171 (1984); **17**, 1545 (1984)
19.5 F. Lenz: Z. Naturf. **3**A, 78 (1954)
19.6 R. Mayol, J.D. Martinez, F. Salvat, J. Parellada: An. Fisica A**80**, 130 (1984)
19.7 K. Saneyoshi, K. Debusmann, W. Keune, R.A. Brand, D. Liljequist: 25th Meeting of the Mössbauer Spectroscopy Discussion Group, Oxford (July 1984)

19.8 D. Liljequist, K. Saneyoshi, D. Debusmann, W. Keune, M. Ismail, R.A. Brand, W. Kiauka: 25th Meeting of the Mössbauer Spectroscopy Discussion Group, Oxford (July 1984)
19.9 F. Salvat, R. Mayol, J.D. Martinez, J. Parellada: To be published
19.10 V.E. Cosslett, R.N. Thomas: Brit. J. Appl. Phys. **15**, 1283 (1964)

III.2 Diffuse LEED, NEXAFS/XANES and SEXAFS

20. LEED, XANES and the Structure of Disordered Surfaces

J.B. Pendry

The Blackett Laboratory, Imperial College, London SW7 2BZ, Great Britain

Surface crystallography is now a well-established discipline, but, like its bulk counterpart, high quality "surface crystals" tend to be required for a complete structural analysis. This has been particularly true where low-energy electron diffraction has been concerned. However, there are techniques for studying surfaces which are insensitive to long-range order. I want to talk about two of them: X-ray absorption near edge structure (XANES) and a new technique only recently proposed: diffuse low-energy electron diffraction (DLEED).

X-ray absorption by an atom adsorbed on a surface results in an ejected electron which diffracts from the neighboring atoms, modifying the absorption cross section in a way that enables bond lengths to be found [20.1-3]. Near the absorption edge the ejected electron will have low kinetic energy and if the energy is less than about 50 eV multiple-scattering effects are likely to be pronounced. This gives the XANES region more information about the environment of the absorbing atom because of the more complete exploration of the neighborhood by the multiply scattered electron. Figure 20.1 taken from [20.4] shows how by comparing calculated with theoretical XANES spectra we can arrive at the geometry of an oxygen atom adsorbed on a nickel (100) surface.

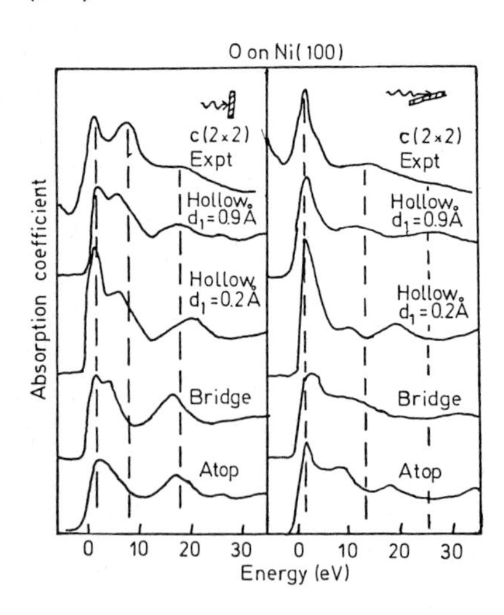

Fig.20.1. Experimental and calculated XANES spectra for a C(2 × 2)O on Ni(100) overlayer. The calculated spectra assumed different chemisorption sites as indicated

The great virtue of surface XANES is that it is insensitive to the long-range order of the surface. The ejected electron is a local probe of the geometry surrounding the absorbing atom. In contrast, conventional LEED is concerned with measuring the sharp diffraction spots resulting from a nearly perfectly ordered surface and this order is built into the theoretical interpretation at a basic level. So the techniques have so far been largely complementary: LEED confined to well-ordered surfaces but within this restriction allowing a wide range of different mixes of atoms to be considered that would confuse a XANES signal. XANES is available for structures without long-range order but requires that the absorbing atom have a more or less unique local environment.

Recently it has been realized that the virtues of each approach could be united by measuring the diffuse scattering that a disordered structure produces between the sharp spots in a LEED experiment.

The theory of this diffuse scattering contains at its core the same calculations involved in the theory of XANES.

The idea amounts to observing that instead of using X-rays to inject an electron in the vicinity of the adsorbed atom we can use an external electron gun. The external beam can be diffracted by the clean surface before it reaches the adsorbate, but this is easily calculated. Once the electron reaches the adsorbate the same processes that characterize XANES take place and provide the same sort of information. Finally the electron is detected after escaping from the adsorbate, and again we can correct for the interaction with the clean surface. Provided that the surface is clean and well ordered except for the adsorbate, we know that any diffusely scattered electron must have interacted with an adsorbate molecule. The pre- and post-XANES stages, far from being a nuisance in the interpretation, provide additional valuable information about the orientation of the molecule relative to the clean surface.

We can in fact do it better than XANES, partly because of the additional information present in the diffuse scattering, but much more importantly for the following reasons. XANES suffers from the severe limitation that there is one and only one data set available for the system, or three if a polarized source is available. Thus in XANES information about the surrounding atoms falls rapidly with the distance of the atom from the absorber. Also, there are just not enough data to resolve the structure in a really complex coordinating shell.

The new technique is not without its pitfalls. The measurement of diffuse as opposed to discrete beams imposes much stronger constraints on the system. The diffuse intensities are much weaker per unit solid angle and require efficient detectors. They also are susceptible to interference from other mechanisms which produce diffuse scattering such as defects, steps and thermal vibrations, all of which must either be eliminated or accurately subtracted. Fast low-intensity data taking techniques [20.5] will be essential, especially in systems susceptible to e-beam damage.

We need to calculate the intensities of electrons emerging in diffuse (i.e., non-Bragg) directions from a surface containing an adsorbed molecule. In order that an electron emerge in a non-Bragg direction, it needs to have followed a path which, at some stage, involved scattering off the nonperiodic part of the surface—in this case the adsorbed molecule. If we examine all possible paths for an electron which scatters at least once off the molecule, each path can be split into three parts. In the first part are all scattering

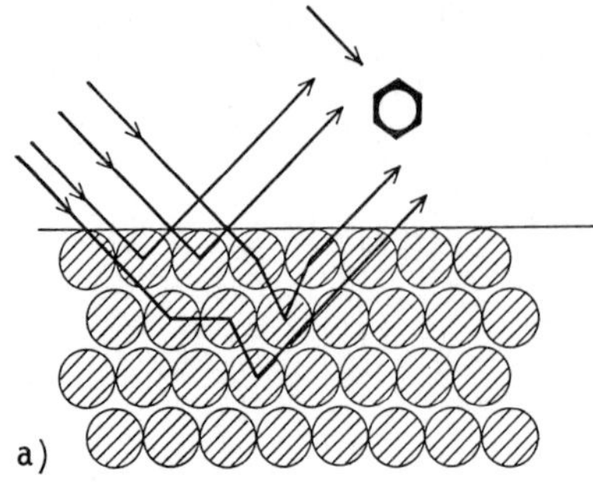

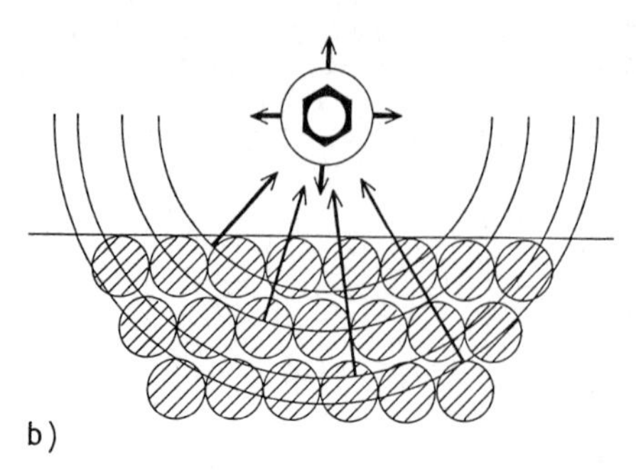

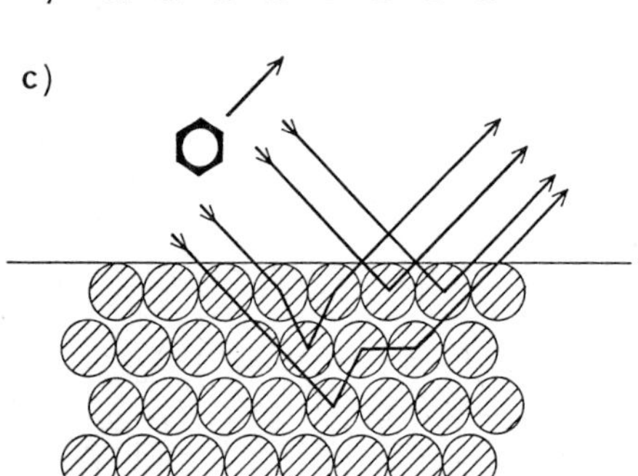

Fig.20.2. (**a**) *Step 1*: All scattering events prior to hitting the molecule replace the K-shell excitation process in a conventional SEXAFS experiment. (**b**) *Step 2*: All scatterings between the first and last encounter with the molecule are treated on the same footing as in a SEXAFS calculation. The symmetry of the wave function is, however, different in general. (**c**) *Step 3*: All scatterings after the final encounter with the molecule

events preceding the first encounter with the molecule (Fig.20.2a). These generate the total incident wave field which can be calculated by simple LEED theory for perfect, clean surfaces. In the second part are all paths which start with the first encounter with the molecule and end with the last (Fig.20.2b). In this step we cannot use conventional LEED theory because the presence of the molecule breaks the translational symmetry of the surface. Instead we use a cluster calculation as for EXAFS [20.6,7] or XANES [20.8]. The cluster need only contain a finite number of atoms because inelastic events prevent the electron from escaping from the molecule by more than a few Angstroms. Finally, in the third part are all paths which begin after the last encounter with the molecule and terminate in the detector (Fig. 20.2c). These paths sum to a "time-reversed-LEED wave function" familiar from theories of photoemission [20.9] and electron energy loss spectroscopy [20.10]. Now let us clothe these generalities with some formulae.

Step 1. Waves Incident on the Molecule. The first task is to evaluate the amplitude of the wave initially incident on the molecule but corrected for all interactions with the clean surface. Consider an incident plane wave

$$\sum_{\mathbf{g}} A^{+}_{\mathbf{g}} \exp[i\mathbf{K}^{+}_{\mathbf{g}} \cdot (\mathbf{r} - \mathbf{r}_0)] \quad , \tag{20.1}$$

where $A^{+}_{\mathbf{g}} = \delta_{\mathbf{g}0}$ and the incident wave vectors may be written $\mathbf{K}^{\pm}_{\mathbf{g}} = (\mathbf{k} + \mathbf{g},\ K^{\pm}_{\mathbf{g}z})$ in terms of its components parallel and perpendicular to the surface. Then

$$K^{\pm}_{\mathbf{g}z} = \pm(K^2 - |\mathbf{k} + \mathbf{g}|^2)^{1/2} \quad , \tag{20.2}$$

where K is the wave vector appropriate to the constant potential between the atomic potentials (we assume the muffin-tin approximation) and is given in terms of the energy E of the electron in vacuo, and the height of the step in potential at the surface V_0:

$$\frac{1}{2}K^2 = E - V_0 \quad . \tag{20.3}$$

Here V_0 has both a real part V_{0r} and imaginary part V_{0i}, which is responsible for the finite escape length of the electron. We have assumed the so-called no-reflection boundary condition at the surface whereby all incident and reflected beams are transmitted across the step in potential at the surface without reflection. Hence all calculations of the reflected intensities can be made inside the surface barrier. The latter appears in subsequent calculations when relating the vacuum energy to that inside the surface barrier. After multiple scattering with the substrate above, the electron wave field may be written:

$$\sum_{\mathbf{g}} A_{\mathbf{g}}^{+} \exp[i\mathbf{K}_{\mathbf{g}}^{+} \cdot (\mathbf{r} - \mathbf{r}_0)] + A_{\mathbf{g}}^{-} \exp[i\mathbf{K}_{\mathbf{g}}^{-} \cdot (\mathbf{r} - \mathbf{r}_0)] \quad , \tag{20.4}$$

where

$$A_{\mathbf{g}}^{-} = \sum_{\mathbf{g}'} R_{\mathbf{gg}'}(\mathbf{k}_{\parallel}) A_{\mathbf{g}'}^{+} \tag{20.5}$$

and $\mathbf{r}$ is a position vector.

The reflection matrix $\underline{\mathbf{R}}$, whose elements are used in (20.5), are defined with respect to an origin $\mathbf{r}_0$ and are the quantities calculated by conventional layer-type LEED computer programs [20.11-13].

Suppose the molecule is centered about $\mathbf{r}_M$. The total incident wave field (20.4) may be expanded as a sum of spherical waves incident on $\mathbf{r}_M$:

$$\psi_i = \sum_{\ell=0}^{\infty} \sum_{m=-\ell}^{+\ell} 4\pi i^{\ell} j_{\ell}(K|\mathbf{r}-\mathbf{r}_M|)(-)^m Y_{\ell m}(\widehat{\mathbf{r}-\mathbf{r}_M})$$

$$\times \sum_{\mathbf{g}} \left\{A_{\mathbf{g}}^{+} \exp[i\mathbf{K}_{\mathbf{g}}^{+} \cdot (\mathbf{r}_M - \mathbf{r}_0)] Y_{\ell -m}(\hat{\mathbf{K}}_{\mathbf{g}}^{+}) + A_{\mathbf{g}}^{-} \exp[i\mathbf{K}_{\mathbf{g}}^{-} \cdot (\mathbf{r}_M - \mathbf{r}_0)] Y_{\ell -m}(\hat{\mathbf{K}}_{\mathbf{g}}^{-})\right\}$$

$$= \sum_{\ell m} B_{\ell m}^{-} j_{\ell}(K|\mathbf{r} - \mathbf{r}_M|) Y_{\ell m}(\widehat{\mathbf{r} - \mathbf{r}_M}) \quad , \tag{20.6}$$

where

$$B_{\ell m}^{-} = 4\pi i^{\ell}(-)^m \sum_{\mathbf{g}} \left\{A_{\mathbf{g}}^{+} \exp[i\mathbf{K}_{\mathbf{g}}^{+} \cdot (\mathbf{r}_M - \mathbf{r}_0)] Y_{\ell -m}(\hat{\mathbf{K}}_{\mathbf{g}}^{+})\right.$$

$$\left. + A_{\mathbf{g}}^{-} \exp[i\mathbf{K}_{\mathbf{g}}^{-} \cdot (\mathbf{r}_M - \mathbf{r}_0)] Y_{\ell -m}(\hat{\mathbf{K}}_{\mathbf{g}}^{-})\right\} \quad . \tag{20.7}$$

Step 2: The Molecular Scattering. Instead of tracing through the multiplicity of paths in detail we can make a simple summary of the scattering by defining a local "reflection coefficient" as the response of the surface to a set of outgoing spherical waves centered on the molecule. Our task is then to sum all multiple scattering between the molecule and the surface. From (20.6) we have calculated the wave field incident on the molecule, now we want the total outgoing wave field

$$\psi_s = \sum_{\ell m} B^+_{\ell m} h^{(1)}_\ell (\kappa|\mathbf{r} - \mathbf{r}_M|) Y_{\ell m}(\hat{\mathbf{r} - \mathbf{r}_M}) \quad , \tag{20.8}$$

where $h^{(1)}_\ell$ is the first-order Hankel function of the first kind. We can write B^+ in terms of $\underline{S}$ and the molecular scattering matrix $\underline{T}_M$:

$$\begin{aligned} \mathbf{B}^+ &= \mathbf{B}^-(\underline{T}_M + \underline{T}_M\underline{S}\underline{T}_M + \underline{T}_M\underline{S}\underline{T}_M\underline{S}\underline{T}_M + \dots) \\ &= \mathbf{B}^-\underline{T}_M(1 - \underline{S}\underline{T}_M)^{-1} \quad . \end{aligned} \tag{20.9}$$

Expression (20.8) may be rewritten in a plane-wave representation as [20.14]

$$\psi_s = |C^{-1}N^{-2}| \sum_{\mathbf{k}'_\parallel \mathbf{g}} \Big\{ D^+_{M\mathbf{g}}(\mathbf{k}'_\parallel) \exp[i\mathbf{K}'^+_{\mathbf{g}} \cdot (\mathbf{r} - \mathbf{r}_0)] + D^-_{M\mathbf{g}}(\mathbf{k}'_\parallel) \exp[i\mathbf{K}'^-_{\mathbf{g}} (\mathbf{r} - \mathbf{r}_0)] \Big\} \quad , \tag{20.10}$$

where $K'^\pm_{\mathbf{g}}$ is defined in terms of $\mathbf{k}'_\parallel$

$$K'^\pm_{\mathbf{g}} = (\mathbf{k}'_\parallel + \mathbf{g}, \mathbf{K}'^\pm_{gz}) \quad ,$$

$$\mathbf{K}'^\pm_{\mathbf{gz}} = \pm(\kappa^2 - |\mathbf{k}'_\parallel + \mathbf{g}|^2)^{\frac{1}{2}} \quad ,$$

$$D^\pm_{M\mathbf{g}} = \sum_{\ell m} (\kappa \mathbf{K}'^+_{\mathbf{gz}})^{-1} 2\pi^2 B^+_{\ell m} i^{-\ell} Y_{\ell m}(\hat{\mathbf{K}}'^\pm_{\mathbf{g}}) \exp[i\mathbf{K}'^\pm_{\mathbf{g}} (\mathbf{r}_0 - \mathbf{r}_M)] \tag{20.11}$$

and CN^2 is the area of the surface illuminated by the beam.

Step 3: Final Scattering by Crystal Alone. Consider now a detector so placed as to receive only diffuse electrons whose wave vector parallel to the surface lies in a small range around $\mathbf{k}'$. The amplitude of these electrons could arise from those electrons scattered there directly after the molecular-cluster scattering of step 2 and also from those electrons which suffer subsequent scattering from the substrate alone. The total detected amplitude is

$$D^-_{\mathbf{g}}(\mathbf{k}'_\parallel) = D^-_{M\mathbf{g}}(\mathbf{k}'_\parallel) + \sum_{\mathbf{g}'} R'_{\mathbf{gg}'}(\mathbf{k}'_\parallel) D^+_{M\mathbf{g}'} \quad , \tag{20.12}$$

where $\underline{R}'$ is the reflection matrix for electrons with parallel wave vector $\mathbf{k}'_\parallel$. Finally we may write the intensities of electrons emerging with parallel wave vectors $\mathbf{k}'_\parallel + \mathbf{g}$ as

$$\frac{dI}{d\Omega}(\mathbf{k}'_\parallel + \mathbf{g}) = |D^-_{\mathbf{g}}(\mathbf{k}'_\parallel)|^2 \left| \frac{K'^+_{gz}}{K^+_{0z}} \right| \frac{\cos\theta}{(2\pi)^2} n_M \tag{20.13}$$

in units of electrons per incident electron per steradian, where n_M is the density of the molecules per unit surface area and θ is the polar angle of emergence from the surface.

As in the case of conventional LEED, we feel that rather than the bare intensities, the quantity most suitable for comparison with experiment is the Y function [20.15], defined in terms of the logarithmic derivative L(E) of the intensity I(E) with respect to energy:

$$Y(E) = \frac{L^{-1}(E)}{L^{-2}(E) + V_{0i}^2} \quad , \quad \text{where} \tag{20.14}$$

$$L(E) = I^{-1} dI/dE \tag{20.15}$$

and V_{0i} is the imaginary part of the electron self-energy. Then L(E) is evaluated by calculating I at three closely spaced energies around E. This necessitates the calculation and transfer of three different matrices from the cluster program.

Phase shifts for O and Ni were calculated from the MUFPOT program using *Clementi* and *Roetti* [20.16] for atomic wave functions and a NaCl structure with a O-Ni spacing of 1.948 Å. This corresponds to the O-Ni spacing as determined by LEED [20.17-19] for the O/Ni(100) system. The exchange parameter was adjusted so as to give an O p-resonance around 0.05 Hartrees. The value $x = 1$ was found to be appropriate.

LEED studies [20.17-19] have suggested that O is adsorbed in the hollow sites of the Ni(100) surface and that the distance (d) of the O atom from the uppermost layer of Ni atoms is 0.9 Å. Figure 20.3 depicts the variation of Y as a function of $(k_x, k_y) = \mathbf{k}_\parallel$ as calculated by our scheme for this value of d. The subscripts x and y above refer to the [100] and [010] directions of the bulk nickel. The electron beam was normally incident on the surface with energy 2 Hartrees and V_{0i} was taken to be -0.147 Hartrees. Only one quadrant is shown for reasons of symmetry.

It is possible to use these diffuse intensities to determine surface geometry with approximately the same sensitivity and accuracy as conventional diffraction patterns, but now with the restriction of ordered layers removed. The technique has yet to be proved, of course, but holds out the possibility not only of an extension of the powers of surface crystallography to new systems, but also offers computationally efficient procedures for the determination of complex structures.

I thank the BP Venture Research Unit for their support.

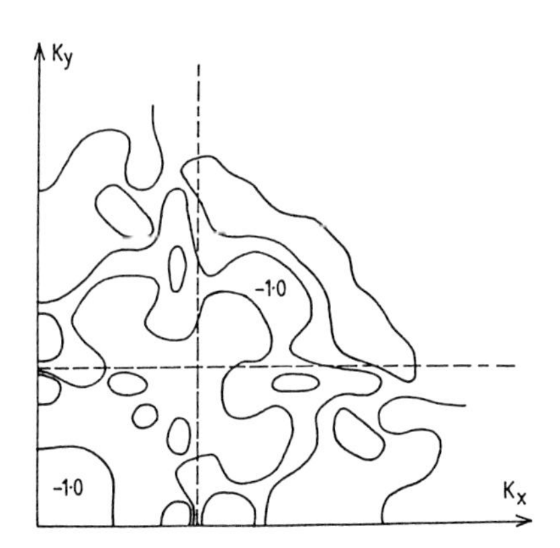

Fig.20.3. Diffuse intensity distribution in one quadrant for electrons normally incident on O/Ni(100). The variation of Y is plotted as a function of $\mathbf{k}_\parallel = (k_x, k_y)$. Note that $|k_\parallel| < 2E$. Electron energy (E) = 2.0 Hartrees. Normal spacing (d) of O from uppermost layer of Ni atoms = 0.9 Å. Contours are plotted for only those polar angles of emergence (with respect to the outward surface normal) less than 60°

References

20.1 P.A. Lee: Phys. Rev. B**13**, 5261 (1976)
20.2 P.H. Citrin, P. Eisenberger, R.C. Hewitt: Surf. Sci. **89**, 28 (1979)
20.3 J. Stöhr, D. Denley, P. Perfetti: Phys. Rev. B**18**, 4132 (1978)
20.4 D. Norman, J. Stöhr, R. Jaeger, P.J. Durham, J.B. Pendry: Phys. Rev. Lett. **51**, 2052 (1983)
20.5 K. Heinz, K. Müller: In *Structural Studies of Surfaces*, Vol.91, Springer Tracts in Modern Physics, ed. by G. Hohler (Springer, Berlin, Heidelberg, New York 1980) and this Volume, p.105
20.6 P.A. Lee, J.B. Pendry: Phys. Rev. B**11**, 2795 (1975)
20.7 C.A. Ashley, S. Doniach: Phys. Rev. B**11**, 1279 (1975)
20.8 P.J. Durham, J.B. Pendry, C.H. Hodges: Comput. Phys. Commun. **25**, 193 (1982)
20.9 J.B. Pendry: Surf. Sci. **57**, 679 (1976)
20.10 G. Aers, J.B. Pendry: Comput. Phys. Commun. **25**, 389 (1982)
20.11 J.B. Pendry: *Low Energy Electron Diffraction* (Academic, London 1974)
20.12 P.M. Marcus, D.W. Jepsen: Phys. Rev. Lett. **20**, 925 (1968)
20.13 M.A. Van Hove, S.Y. Tong: *Surface Crystallography by LEED* (Springer, Berlin, Heidelberg, New York 1979)
20.14 J.B. Pendry: J. Phys. C**8**, 2413 (1975)
20.15 J.B. Pendry: J. Phys. C**13**, 937 (1980)
20.16 E. Clementi, C. Roetti: In *Atomic Data and Nuclear Data Tables* (Academic, London 1974)
20.17 J.E. Demuth, P.M. Marcus, D.W. Jepsen: Phys. Rev. Lett. **46**, 1469 (1981)
20.18 S. Andersson, B. Kasemo, J.B. Pendry, M.A. Van Hove: Phys. Rev. Lett. **31**, 595 (1973)
20.19 S. Andersson, J.B. Pendry: Solid State Commun. **16**, 563 (1975)

21. The Structure of Organic Adsorbates from Elastic Diffuse LEED

D.K. Saldin, D.D. Vvedensky, and J.B. Pendry
The Blackett Laboratory, Imperial College, London SW7 2BZ, Great Britain

The novel method of interpreting the diffuse intensity distributions of elastically scattered electrons generated by disordered adsorbates on surfaces [21.1] is shown to be capable of being applied even to the case of adsorbates consisting of large organic molecules. Use is made of a matrix formulation of renormalized forward scattering (RFS) perturbation theory in the cluster calculations. The prospects for accurately determining the adsorption sites and orientations of isolated or disordered complex molecules by LEED are discussed.

21.1 Introduction

The restriction of traditional methods of surface-structure determination by LEED to the case of ordered surfaces is artificial, arising from the form of the theories used to interpret the intensity distributions. Many important surface structures are inevitably disordered, for example, those arising in the catalysis of hydrocarbons on transition metal surfaces. Any order present is generally of the lattice-gas type and not long range in character. The diffraction pattern due to even elastically scattered electrons thus ceases to consist of sharp Bragg spots but takes the form of a diffuse distribution of intensities. There is nevertheless important structural information in these intensities, namely that of the adsorption sites of the molecules, which undoubtedly has a crucial bearing on the nature of, for example, any catalytic action.

The first attempt to decode this information was by *Van Hove* et al. [21.2]. They employed a perturbation scheme applied to standard LEED methods based on the scattering properties of layers. This scheme is known as the beam-set-neglect method and has been successfully employed to determine the ordered structure of benzene adsorbed on Rh(111). This substantially extended the scope of LEED due to the complexity of the surface unit cell, consisting not only of the atoms of the adsorbate but also of the substrate atoms of the large areas between the adsorbates. It also marks a transition towards the limiting case of isolated or disordered adsorbates that is the subject of this paper, for although in their case the adsorbate is still ordered the unit cell is so large that multiple scattering amongst the adsorbate molecules becomes small. In reciprocal space, the density of Bragg spots is large due to the size of the adsorbate superlattice and this may be regarded as a discrete prototype of a continuous diffuse distribution. Indeed the formal equivalence of the two approaches in the limiting case has been demonstrated in the paper by *Saldin* et al. [21.3].

Pendry and *Saldin* [21.1] have concentrated on the limiting case of isolated (or disordered) adsorbates and shown how the continuous intensity distribution resulting can be calculated by a combination of traditional layer-type LEED methods and a local cluster calculation involving the atoms of the adsorbate and its immediate environs. They showed the results of calculations for disordered atoms on surfaces, viz., O/Ni(100). In this paper we demonstrate how the method may be extended to more complicated molecular adsorbates of practical interest using a matrix formulation of renormalized forward scattering perturbation theory for the cluster calculations.

21.2 Evaluation of the Renormalized Molecular Scattering Matrix

Pendry and *Saldin* [21.1] have outlined a method to compute the diffuse intensities. When the adsorbate is a molecule the part of the calculation concerned with the diffraction of the electrons before their first encounter and after their last encounter with the molecule, which is of standard layer type, is unaltered. The only increase in complication comes about in the calculation of the renormalized molecular scattering matrix $\underline{\tau}$. This must now be written as

$$\underline{\tau} = \underline{t}_M(1 - \underline{S}\underline{t}_M)^{-1} \quad , \tag{21.1}$$

where $\underline{S}$ is "out-in" reflection matrix of the cluster of atoms immediately surrounding the adsorbate and $\underline{t}_M$ is the molecular scattering matrix. If the molecule is sufficiently small to be represented by a single shell, the latter quantity is merely the "in-out" scattering matrix of this shell and can be calculated by standard cluster methods from the inverse of the real-space KKR matrix [21.4].

An assumption of the RFS perturbation scheme is that backscattering is small compared to the forward scattering of electrons in the LEED energy range. Since both matrices $\underline{t}_M$ and $\underline{S}$ are of backscattering type, it would be expected that (21.1) may be adequately represented by the expansion

$$\underline{\tau} = \underline{t}_M + \underline{t}_M\underline{S}\underline{t}_M + \dots \quad . \tag{21.2}$$

Although $\underline{t}_M$ may be evaluated from the scattering matrix of a single shell, $\underline{S}$ may need to be constructed from those of several concentric shells. The multiple scattering between shells may then be calculated by the methods of *Durham* et al. [21.4]. These can be very time consuming, however, since they involve finding inverses of matrices. At LEED energies above 20 eV the RFS perturbation scheme, [21.5,6,10] allows a much swifter calculation of $\underline{S}$ with very little loss of accuracy. In fact the first-order approximation to the perturbation series for $\underline{S}$, one which allows for just one backscattering by any of the shells of atoms, turns out to be quite adequate for our purposes:

$$\underline{S} = \underline{T}_{OI}^{(1)} + \underline{T}_{OO}^{(1)}\underline{T}_{OI}^{(2)}\underline{T}_{II}^{(1)} + \dots + \underline{T}_{OO}^{(1)}\underline{T}_{OO}^{(2)}\dots\underline{T}_{OI}^{(N)}\dots\underline{T}_{II}^{(2)}\underline{T}_{II}^{(1)} \quad , \tag{21.3}$$

where $\underline{T}$'s are the scattering matrices of the shells, the subscripts O and I standing for "out" and "in" respectively, and the shells are numbered by the superscripts. Thus $\underline{T}_{OO}$ and $\underline{T}_{II}$ are forward-scattering matrices and therefore large in magnitude, while $\underline{T}_{OI}$ is a reflection matrix and much smaller. Substituting (21.3) in (21.1) yields an expression for τ involving only those terms in the perturbation series which involve 3 or less shell backscatterings. It should be noted also that in the spherical-wave approximation used, $\underline{\tau}$ will have the same dimensionality as $\underline{t}_M$ which will generally be much smaller than that of $\underline{S}$.

21.3 Diffuse LEED from Benzene on Ni(111)

The adsorption of benzene on Ni(111) has been studied using electron energy loss spectroscopy (EELS) and LEED by *Bertolini* et al. [21.7] and *Lehwald* et al. [21.8]. Both groups concluded that benzene molecules retain their aromatic character and are π-bonded with their rings parallel to the surface. They also concluded that the adsorption sites must have at least threefold rotation symmetry about the normal to the surface. This narrows down the absorption sites to be either "on-top" or "hollow".

The results reported below explore the possible uses of our technique for the structural determination of molecular adsorbates, and in particular we have addressed the question of whether it would be possible to distinguish these benzene sites by measuring the diffuse LEED (DLEED) patterns from isolated adsorbates (assuming the molecules adsorb on one type of site only). The DLEED patterns expected from "top" site adsorption are shown in Fig.21.1 and from adsorption in the hcp "hollow" site in Fig.21.2. The benzene molecule was modeled by its carbon ring only (due to the small scattering power of its hydrogen atoms). The maximum angular momentum quantum number ℓ_{out}

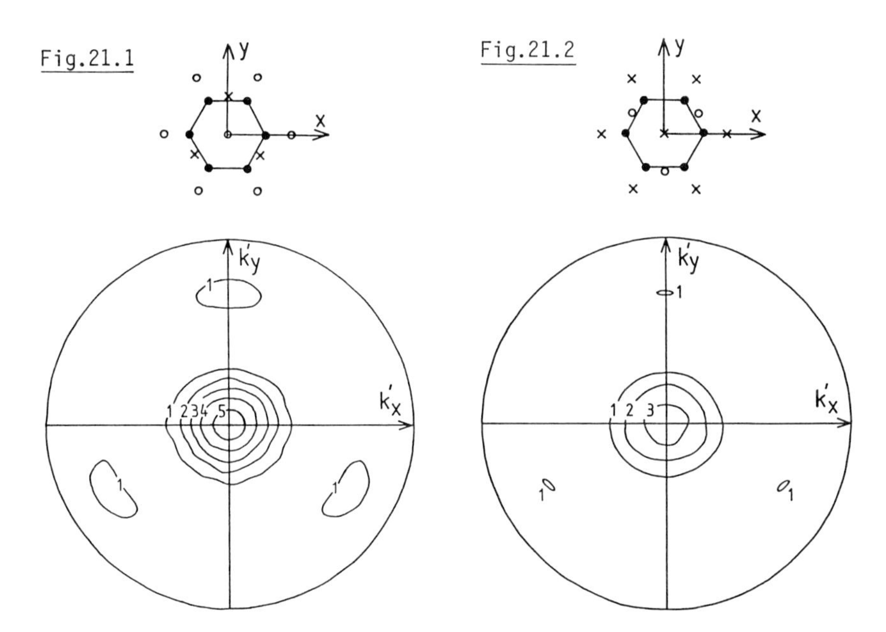

Fig.21.1. Elastic diffuse LEED pattern from benzene adsorbed on a "top" site on Ni(111). The plane of the benzene ring is parallel to the surface and 1.48 Å from the top surface layer. The energy of the electrons normally incident on the surface is 2 Hartrees (=54.4 eV). The whole of the projected backscattering hemisphere is shown. The contour values shown are 0.1 times the diffracted flux/incident flux unit/steradian/adsorbed molecule in unit area. The upper diagram shows the corresponding plan view of the adsorption site of the benzene with the dark circles representing carbon atoms, the open circles the top layer Ni atoms and the crosses the Ni atoms on the second layer

Fig.21.2. Same as Fig.21.1 but with the benzene molecule adsorbed on the hcp "hollow" site

used in the evaluation of $\underline{t}_M$ was 6. These two patterns may be distinguishable by careful experiments. As pointed out by *Pendry* and *Saldin* [21.1], if the adsorbates are so close that they cannot be considered isolated, a more suitable quantity to calculate would be the Y function (21.9), since this would filter out any information about the long-range order. This quantity is easily computed from three DLEED patterns closely spaced in energy.

Both Figs.21.1,2 display a threefold symmetry emphasizing the important contribution of the (ordered) substrate to the DLEED patterns, thus yielding information about the relative orientation and bonding geometry of the adsorbate, as is also clear in DLEED from $Ni(CO)_4$ adsorbed on Ni(111) [21.6]. The three mirror planes of the structure in real space give rise to corresponding mirror planes in the DLEED patterns.

21.4 Conclusions

The method proposed by *Pendry* and *Saldin* [21.1] for calculating the elastic diffuse LEED patterns of adsorbates has been shown to be easily extended to the case where the adsorbates are substantial molecules with the use of the cluster formulation of renormalized-forward-scattering perturbation theory. This holds out considerable hope for the use of this technique in determining the structures of complex organic adsorbates (e.g., those partaking in important industrial catalytic processes) even when the adsorbates do not form an ordered overlayer, while avoiding the disadvantages of the SEXAFS technique [21.1].

References

21.1 J.B. Pendry, D.K. Saldin: Surf. Sci. **145**, 33 (1984); see also J.B. Pendry, Paper No.20, this Volume, p.124
21.2 M.A. Van Hove, R. Lin, G.A. Somorjai: Phys. Rev. Lett. **51**, 778 (1983)
21.3 D.K. Saldin, J.B. Pendry, M.A. Van Hove, G.A. Somorjai: Phys. Rev.B, in press (1985)
21.4 P.J. Durham, J.B. Pendry, C.H. Hodges: Comput. Phys. Commun. **25**, 193 (1983)
21.5 J.B. Pendry: In *EXAFS and Near Edge Structure*, ed. by A. Bianchoni, L. Incoccia, S. Stipcich (Springer, Berlin, Heidelberg 1983)
21.6 D.D. Vvedensky, D.K. Saldin, J.B. Pendry: Presented at the 3rd. Int. Symposium on Small Particles and Inorganic Clusters, Berlin, FRG (1984), and to be published in Surf. Sci.
21.7 J.C. Bertolini, G. Dalmai-Imelik, J. Rousseau: Surf. Sci. **67**, 478 (1977)
21.8 S. Lehwald, H. Ibach, J.E. Demuth: Surf. Sci. **78**, 577 (1978)
21.9 J.B. Pendry: J. Phys. C**13**, 937 (1980)
21.10 D.D. Vvedensky, D.K. Saldin, J.B. Pendry: This Volume, p.135

22. Multiple Scattering Effects in Near-Edge X-Ray Absorption Spectra

D.D. Vvedensky, D.K. Saldin, and J.B. Pendry

The Blackett Laboratory, Imperial College, London SW7 2BZ, Breat Britain

We investigate the importance of various multiple-scattering corrections to electron propagation in near-edge X-ray absorption spectra using the O/Ni(100) system as an example. We find that within approximately 10 eV of the absorption edge multiple scattering contributions cannot be neglected, though for greater energies the convergence of all perturbation schemes is quite rapid. We also outline our recently developed geometric series representation of renormalized forward scattering perturbation theory.

X-ray absorption spectroscopy has in recent years become an established family of probes of local structure for both surface and bulk geometries. The interpretation of the absorption spectra, and indeed the type of information that may be inferred from the data, is determined by the nature of the photoelectron scattering processes. In the EXAFS regime ($\gtrsim$ 50 eV above the absorption edge) a single scattering theory suitably interprets the data [22.1,2] and structural parameters may thereby be extracted by Fourier transform methods. As is typical of single-scattering probes, EXAFS provides information only in the form of a two-point correlation function, giving the radial distribution of atoms from the absorbing atom. On the other hand, within approximately 50 eV of the absorption edge (the XANES regime) the strong photoelectron scattering and relatively long mean-free path lead to multiple-scattering corrections to electron propagation and thus to a sensitivity to multiatom correlations. Though the extraction of structural parameters from XANES data using multiple-scattering computation schemes [22.3] is not as straightforward as for EXAFS, the XANES spectra may provide information that is inaccessible to EXAFS, e.g., bond angles. We discuss below the sensitivity of XANES spectra to different types of multiple-scattering events, within the computational scheme of *Durham* et al. [22.3] and using the O/Ni(100) system as an example [22.4]. The importance of multiple scattering in XANES has also recently been discussed by *Bunker* and *Stern* [22.5].

The XANES code [22.3] is based upon a cluster method and includes full multiple-scattering contributions. The cluster is divided into concentric shells of atoms centered around the absorbing atom. The multiple-scattering equations are solved first within each shell and then between the shells themselves and the central atom. Since the solution of the multiple-scattering equations at each step inevitably involves a matrix inversion, this procedures becomes very time consuming for large clusters and for geometries with low symmetry. We have been implementing approximations to multiple scattering that involve representing the matrix inversions as geometric series. The advantages of such approximation schemes are 1) the convergence of the geometric-series representations is determined by the spectral radius of a matrix and thus 2) where the series converges the truncation error may be readily determined.

For intrashell multiple scattering the convergence of the multiple scattering series is determined by the maximum modulus $|\lambda|_{max}$ of the eigenvalues of the shell X matrix, which describes electron propagation between every distinct pair of atoms in the shell. (The series converges if $|\lambda|_{max} < 1$ and diverges otherwise.) The importance of assessing intrashell multiple scattering lies not only in the economy of computation time achieved in bypassing the inversion of $\underline{\mathbf{X}}$, but also in identifying the extent to which intrashell multiatom correlations can be obtained from the near-edge data. If $|\lambda|_{max} \ll 1$ intrashell multiple scattering is unimportant and multiatom correlations for the given shell cannot be realistically extracted from the XANES spectrum. Alternatively, as $|\lambda|_{max} \to 1^-$ successively more intrashell scattering events must be included in the photoelectron propagator, indicating a correspondingly greater sensitivity to multiatom correlation functions. If $|\lambda|_{max} \geqq 1$, the geometric series representation of intrashell multiple scattering does not exist and the matrix inversion must be performed explicitly.

The approximation we have employed for intershell multiple scattering is based upon renormalized forward scattering (RFS) perturbation theory [22.6-8]. We first give a very brief review of the original recursive form of RFS following *Pendry* [22.8], and then outline our geometric series representation (see *Adams* [22.9] for a complementary discussion). A detailed discussion will be published elsewhere. In RFS perturbation theory we begin with a zeroth-order wave field that takes account of all forward-scattering events and corrections are made in orders of backscattering events. We denote by a_j and b_j the outgoing and incoming wave fields between the $(j-1)^{th}$ and j^{th} shells. Out-out, out-in, in-out, and in-in shell scattering matrices of the jth shell are denoted by T_j^{OO}, T_j^{OI}, T_j^{IO}, T_j^{II}, respectively and we suppose that there are N shells in the cluster. The wave emitted from the central atom is

$$\sum_{\ell m} a_{1,\ell m}^{(0)} h_\ell^{(1)}(kr) Y_{\ell m}(\theta,\phi) \quad , \tag{22.1}$$

where $k = \sqrt{2E}$ and $h^{(1)}$ is the spherical Hankel function of the first kind. The zeroth-order wave field between the $(j-1)^{th}$ and j^{th} shell is

$$a_j^{(0)} = a_{j-1} T_{j-1}^{OO} \quad (j = 2,\ldots,N) \quad , \tag{22.2}$$

where we have suppressed the partial wave indices. In general, the $2n^{th}$-order correction is given by

$$\begin{aligned} a_1^{(2n)} &= b_1^{(2n-1)} t \\ a_j^{(2n)} &= b_j^{(2n-1)} T_{j-1}^{IO} + a_{j-1}^{(2n)} T_{j-1}^{OO} \quad (j = 2,\ldots,N) \quad , \end{aligned} \tag{22.3}$$

where t is the scattering matrix of the central atom and the $(2n+1)^{th}$ correction is given by

$$\begin{aligned} b_j^{(2n+1)} &= a_j^{(2n)} T_j^{OI} + b_j^{(2n+1)} T_j^{II} \quad (j = 1,\ldots,N-1) \\ b_N^{(2n+1)} &= a_N^{(2n)} T_N^{OI} \quad . \end{aligned} \tag{22.4}$$

The XANES transition rate σ is thereby given as

$$\sigma = \sum_{\ell m} k^{-1}(|a_{1,\ell m}|^2 - |b_{1,\ell m}|^2) \quad , \tag{22.5}$$

where

$$a_1 = \sum_{n=0}^{\infty} a_1^{(2n)} \quad , \quad b_1 = \sum_{n=0}^{\infty} b_1^{(2n+1)} \quad . \tag{22.6}$$

To cast the RFS recursion relations into the form of a geometric series, we define row vectors $\mathbf{a}^{(n)} = (a_1^{(n)},\ldots,a_N^{(n)})$ and $\mathbf{b}^{(n)} = (b_1^{(n)},\ldots,b_N^{(n)})$, an upper triangular matrix $\mathbf{U}$ with entries $U_{ij} = \delta_{i+1,j}T_i^{OO}$, a lower triangular matrix $\mathbf{L}$ with entries $L_{ij} = \delta_{i-1,j}T_i^{II}$, and block diagonal matrices $\mathbf{D}$ and $\mathbf{D}'$ with entries $D_{ij} = \delta_{ij}T_i^{OI}$ and $D'_{ij} = \delta_{ij}T_i^{IO}$, respectively. The partial wave indices have been suppressed in these definitions. Then denoting by $\mathbf{a}^{(0)} = (a_1^{(0)},0,\ldots,0)$ the initial vector (22.1), the recursion equations (22.2-4) may be written in matrix form as

$$\begin{aligned} \mathbf{a}^{(0)} &= \mathbf{a}^{(0)}\underline{U} + a_1^{(0)} \\ \mathbf{a}^{(2n)} &= \mathbf{b}^{(2n-1)}\underline{D}' + \mathbf{a}^{(2n)}\underline{U} \\ \mathbf{b}^{(2n+1)} &= \mathbf{a}^{(2n)}\underline{D} + \mathbf{b}^{(2n+1)}\underline{L} \quad . \end{aligned} \tag{22.7}$$

The matrix equations (22.7) may be solved for a_1 and b_1 in the form

$$a_1 = a_1^{(0)} \sum_{n=0}^{\infty} \mathbf{B}^n \quad , \quad b_1 t = a_1^{(0)} \sum_{n=1}^{\infty} \mathbf{B}^n \quad , \tag{22.8}$$

where the matrix $\underline{B}$, which we call the backscattering matrix, is given by

$$\underline{B} = (1 - \underline{U})^{-1}\underline{D}(1 - \underline{L})^{-1}\underline{D}' \quad . \tag{22.9}$$

The matrix inversions can be easily performed and we find that the matrix elements B_{ij} of $\underline{B}$ consist of all paths that transform an outgoing wave amplitude between the $(j-1)^{th}$ and j^{th} shells to an outgoing wave amplitude between the $(i-1)^{th}$ and i^{th} shells with two backscattering and all forward-scattering events. For example, B_{11} is given by

$$B_{11} = T_1^{OI}t + T_1^{OO}T_2^{OI}T_1^{II}t + \ldots + T_1^{OO}T_2^{OO} \ldots T_N^{OI} \ldots T_1^{II}t \quad . \tag{22.10}$$

Note that the lowest-order RFS expression for σ is given entirely in terms of $a_1^{(0)}$ and B_{11}. As we stated earlier, the convergence of RFS perturbation theory may be assessed directly in terms of the eigenvalues of $\mathbf{B}$. Our development is not restricted to cluster geometry and may equally well be applied to LEED calculations [22.9].

Figure 22.1 shows the calculated O K-edge XANES spectrum for a $c(2\times2)$ overlayer with fourfold hollow adsorption sites. [22.4]. We have also calculated approximate spectra by including 1) full intershell multiple scatter-

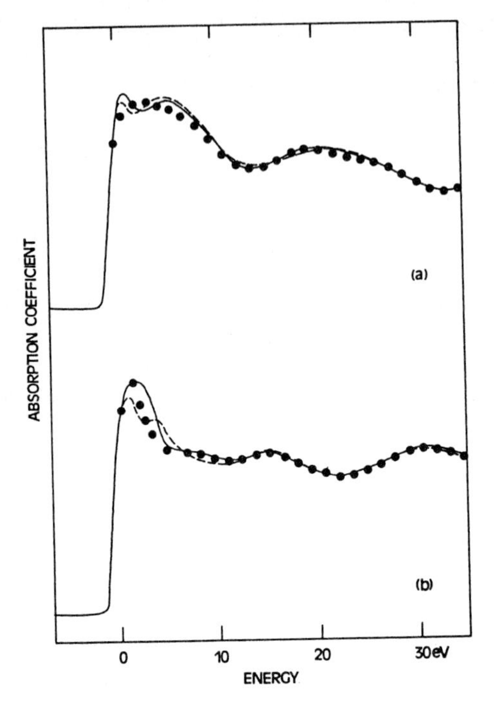

Fig.22.1. Calculated O K-edge spectra for a c(2 × 2) overlayer with fourfold hollow adsorption sites on Ni(100) for polarizations of the electric-field vector (*a*) in the surface plane and (*b*) along the surface normal. Calculations shown include full multiple scattering (——), full intershell multiple scattering but no intrashell multiple scattering (----) and full intrashell multiple scattering with intrashell multiple scattering given by the lowest order B-matrix expression from (22.5,6,8,10) (····). The spectra have been folded with a room-temperature Fermi function

ing but no intrashell multiple scattering, and 2) full intrashell multiple scattering but with intershell multiple scattering given by the lowest-order B-matrix expression in terms of B_{11}. For energies $\gtrsim$ 10 eV above the absorption edge both approximations yield quite good results, though marked deviations arise within 10 eV of the edge. In fact, $|\lambda|_{max} < 1$ for both $\underline{X}$ and $\underline{B}$ over the entire energy range though near the edge $|\lambda|_{max} \to 1$ due to the influence of the nickel 3d levels, which leads to slow convergence of the respective scattering series. We see in particular that because intrashell multiple scattering *cannot* be neglected in the near-edge region for this system, the XANES spectrum does contain local bond-angle information. Moreover, the rapid convergence of these approximations in the region >10 eV above the edge makes them particularly suitable for use in the SEXAFS step of the elastic diffuse LEED theory recently developed by *Pendry* and *Saldin* [22.10,11]. We shall report elsewhere on the effectiveness of these approximation schemes for the near-edge spectra of molecular adsorbates.

Acknowledgment. The support of the British Petroleum Venture Research Unit is gratefully acknowledged.

References

22.1 P.A Lee, J.B. Pendry: Phys. Rev. B**11**, 2795 (1975)
22.2 E.A. Stern, D.E. Sayers, F.W. Lytle: Phys. Rev. B**11**, 4836 (1975)
22.3 P.J. Durham, J.B. Pendry, C.H. Hodges: Compt. Phys. Commun. **25**, 193 (1982)

22.4 D. Norman, J. Stohr, R. Jaeger, P.J. Durham, J.B. Pendry: Phys. Rev. Lett. **51**, 2052 (1983)
22.5 G. Bunker, E.A. Stern: Phys. Rev. Lett. **52**, 1990 (1984)
22.6 J.B. Pendry: J. Phys. C**4**, 3095 (1971)
22.7 J.J. Rehr, E.A. Stern: Phys. Rev. B**14**, 4413 (1976)
22.8 J.B. Pendry: In *EXAFS and Near Edge Structure*, ed. by A. Bianconi, L. Incoccia, S. Stipcich (Springer, Berlin, Heidelberg 1983)
22.9 D.L. Adams: J. Phys. C**14**, 789 (1981)
22.10 J.B. Pendry, D.K. Saldin: Surf. Sci. **145**, 33 (1984)
22.11 D.K. Saldin, D.D. Vvedensky, J.B. Pendry: This Volume, p.131

23. NEXAFS and SEXAFS Studies of Chemisorbed Molecules: Bonding, Structure and Chemical Transformations

J. Stöhr

Corporate Research Science Laboratories, Exxon Research and Engineering Company, Clinton Township, Route 22 East Annandale, NJ 08801, USA

We discuss the application of the surface extended X-ray absorption fine structure (SEXAFS) and near edge X-ray absorption fine structure (NEXAFS) techniques to study the intramolecular and chemisorption bonds of carbon-, nitrogen-, oxygen-, and sulfur-containing molecules and their reaction intermediates on metal surfaces. NEXAFS is dominated by *intramolecular* multiple scattering resonances and is, therefore, preferentially sensitive to the intramolecular bonding and structure. It provides the intramolecular bond length, the hybridization of the molecular bond and the molecular orientation on the surface. SEXAFS mainly arises from single scattering processes off the atomic cores of *substrate* neighbor atoms and determines the chemisorption site and the chemisorption bond lengths. An example is given of a complete structure determination for formate (HCO_2) on Cu(100). The power of NEXAFS is pointed out to monitor surface-induced molecular bond length variations, e.g., C_2H_2 on Pt(111), CO/Na/Pt(111) as well as chemical transformations of complex molecules, e.g., C_4H_4S/Pt(111).

23.1 Introduction

Surface scientists have had to learn that accurate determination of the geometric arrangement of molecules on surfaces is all but easy [23.1]. The problem is complex and involves the identification of the molecular species (fragment), the determination of the internal structure of the species and of its orientation and distances relative to the surface substrate atoms. Finally, it is important to understand a particular chemisorption geometry in terms of the molecular and surface valence charge distribution (orbitals). The molecular chemisorption geometry is to first order a local problem, determined by the immediate environment of the molecule. Therefore, the problem can be well addressed by "local" techniques which tune into a specific atom and probe its environment. Two such techniques, the near edge X-ray absorption fine structure (NEXAFS) [23.2] and surface extended X-ray absorption fine structure (SEXAFS) [23.3] are the subject of this article.

In both techniques a surface atom is selected by one of its characteristic absorption edges (typically K edge). The edge corresponds to the threshold of a core electron excitation by means of X-rays. Above threshold the excitation probability, i.e., the absorption coefficient, is modulated by the scattering processes of the created photoelectron. These intensity modulations are called NEXAFS near the edge (within ~30 eV) and SEXAFS well above the edge. The two regimes are characterized by different scattering processes of the excited photoelectron owing to its different kinetic energy. At small excitation (kinetic) energies the effective scattering potential is created by the atomic cores of the excited atom and its neighbors as well as by the detailed charge distribution of the valence electrons. Therefore, the NEXAFS region is best described by a multiple scattering (MS) theory. At higher excitation energies the scattering processes of the photoelectron are dominated by the atomic cores of the absorbing atom and its neighbors. Single-scattering (SS) theory can be used, and it is this simple theoretical framework which allows us to analyze SEXAFS data by convenient and accurate Fourier transform techniques [23.4,5]. SEXAFS is governed by the same physical processes, theory and data analysis procedures as its well-established bulk analog, EXAFS [23.5].

In the following, we specifically discuss NEXAFS and SEXAFS measurements above the K edge of C, N and O atoms in molecules chemisorbed on single-crystal metal surfaces. The two techniques are shown to provide complementary information. NEXAFS is dominated by *intramolecular* multiple-scattering resonances and is, therefore, preferentially sensitive to intramolecular bonding and structure. It provides the intramolecular bond lengths, the hybridization of the molecular bond and the molecular orientation relative to the surface. SEXAFS arises mainly from single-scattering processes off the atomic cores of *substrate* neighbor atoms and is best used to determine the molecule-substrate distances and geometry. A combination of the techniques allows us to obtain a complete picture of the local molecular chemisorption geometry and bonding.

23.2 SEXAFS of Low-Z Molecules

SEXAFS and its bulk analogue, EXAFS, have been extensively reviewed such that only a few general comments are necessary. For K edges the EXAFS signal as a function of the wave vector k of the photoelectron is given by [23.4,5]

$$\chi(k) = \sum_i N_i^* A_i(k) \sin[2kR_i + \phi_i(k)] \quad . \tag{23.1}$$

The sum extends over shells which contain identical atoms at a distance R_i from the central (excited) atom. The amplitude function contains several factors, most notably a backscattering amplitude F(k) which depends on the atomic number Z of the neighbors. The total scattering phase shift characteristic for the central and the backscattering atoms is $\phi(k)$. Further, N^* is a geometry factor which arises from the fact that for s initial states the photoelectrons are emitted into p-like lobes along the electric field vector **E** of the X-rays. If α is the angle between **E** and a given neighbor direction we have [23.4b]

$$N^* = 3 \cos^2\alpha \quad . \tag{23.2}$$

Thus, the **E** vector acts as a search light for nearest-neighbor atoms. In order to determine the structure of a molecule on a surface, one records the SEXAFS signal above the K edge of one of the atoms (e.g., O) in the molecule. This is illustrated in Fig.23.1 for a diatomic molecule consisting of a C and an O atom chemisorbed in a fourfold hollow site of Cu atoms. We assume for simplicity that the O-C axis is perpendicular to the plane of the four Cu atoms and that the O atom sits in that plane. First we orient our hypothetical sample such that the **E** vector lies along the O-C bond. The SEXAFS spectrum above the O K edge is then completely determined by the backscattering of the C atom at a distance R_1. This is a result of the "search light" effect (23.2) which prevents the Cu atoms from being seen. If we rotate the sample by 90° such that **E** lies in the plane of the O and Cu atoms, the SEXAFS signal will reflect only the O-Cu distance R_2. Model calculations for the two cases using tabulated backscattering amplitudes

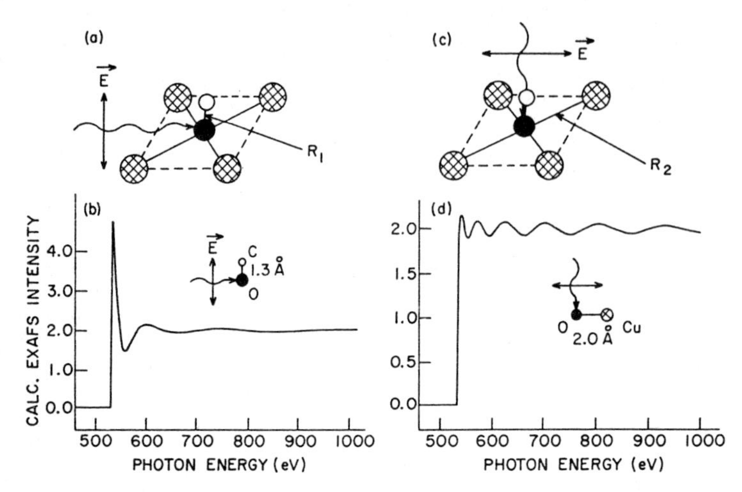

Fig.23.1. (**a**) Hypothetical diatomic molecule (filled and open circles) in a fourfold site of four metal atoms (*dashed circles*). The molecular axis is perpendicular to the plane containing the black and the four hatched atoms. The X-rays excite the black atom and are incident with the electric field vector **E** along the molecular axis. (**b**) Model calculation of the SEXAFS spectrum (23.1) for an O central and C backscattering atom at a distance of 1.3 Å with **E** along the O-C internuclear axis. (**c**) Same model sample as in (**a**) for **E** parallel to the plane containing the black and the four hatched atoms. (**d**) Calculated SEXAFS spectrum for an O central and Cu backscattering atom separated by 2.0 Å with **E** along the O-Cu axis. Note the SEXAFS signal is normalized to the same edge jump as in (**b**)

F(k) for C and Cu [23.6] and appropriate O-C and O-Cu scattering phase shifts [23.4c] are also shown in Fig.23.1. We assumed typical O-C ($R_1 = 1.3$ Å) and O-Cu ($R_2 = 2.0$ Å) distances. The O-C SEXAFS signal shows the largest structure near threshold since for C, F(k) peaks at low energy (wave vector) [23.6] and falls off rapidly with increasing energy. Also, the short O-C distance results in widely spaced SEXAFS oscillations. Substituting the C neighbor by Cu gives rise to a sizeable SEXAFS signal over a wide energy range due to the fact that F(k) for Cu has a broad maximum at $k \sim 7.5$ Å^{-1} (~ 200 eV above the edge) [23.6]. The larger O-Cu distance also increases the SEXAFS oscillation frequency.

Figure 23.1 illustrates how, in principle, the complete geometry of the molecule on the surface can be obtained. In practice, however, it is difficult to measure the O-C SEXAFS oscillations because of their small size. Furthermore, the long wavelength of the oscillations is difficult to distinguish from slow variations in the background of the measured signal due to data normalization problems. It is clear from Fig.23.1 that if it can be analyzed, the near-edge NEXAFS region would be better suited to yielding information on the short bonds between low-Z atoms. At this point, it should be pointed out that an extrapolation of the results in Fig.23.1 clearly shows that SEXAFS (and EXAFS) cannot measure the distance to a hydrogen neighbor atom because F(k) is too small. On the other hand, SEXAFS is well suited to measuring the distances and, from the angular **E** vector dependence, the site of a low-Z atom relative to the substrate [23.4]. This has been demonstrated previously for a variety of atomic chemisorption systems [23.4c].

23.3 NEXAFS of Low-Z Molecules

The question arises whether the intramolecular distances which are not easily obtainable from SEXAFS can be determined by NEXAFS. Fortunately, the NEXAFS spectra of the chemisorbed molecules can be understood quite easily since they are dominated by two types of structures, so-called σ and π resonances [23.2,7].

The σ *resonance* is observed if **E** has a finite projection along the internuclear axis between two atoms (e.g., C-O) in the molecule, similar to SEXAFS (23.2). It is an MS resonance where the photoelectron remains quasi-trapped by the intramolecular field [23.8]. In an MS picture the photoelectron can be envisioned as scattering back and forth between the central atom and a neighbor. In a molecular orbital (MO) picture the σ resonance corresponds to a transition to a σ^* antibonding orbital of the molecule [23.9]. The σ resonance is observed for all molecules with bonds between two low-Z atoms like C, N or O. The σ resonances associated with bonds to H atoms are usually too weak to be seen [23.10,11].

The π *resonance*, if present, is the lowest energy structure in the NEXAFS spectrum. In fact, its excitation energy is always less than the 1s binding energy relative to the vacuum level (1s ionization potential) and it is therefore often called a bound-state transition. Like the σ resonance, it can be viewed as an MS resonance with the photoelectron trapped by the intramolecular potential [23.8]. It is most easily understood as a transition to a π^* antibonding orbital of the molecule. Therefore, it is observed only if the **E** vector has a projection along a π orbital, i.e., perpendicular to the σ bond axis between atoms in a molecule. Thus, opposite to the σ resonance intensity which follows a $I_\sigma \sim \cos^2\alpha$ law, the π intensity varies as $I_\pi \sim \sin^2\alpha$ for an oriented molecule. The π resonance is observed only if the bond between two low-Z atoms has a π contribution (i.e., double or triple bonds). The π resonances or related Rydberg structures associated with bonds to H atoms [23.12] are weak [23.10,11,13].

The above concepts are illustrated by the NEXAFS spectra in Fig.23.2 for gas-phase [23.14] and chemisorbed [23.15] CO and methanol (CH_3OH). The fact that the corresponding spectra for gas-phase and chemisorbed molecules are dominated by the same π and σ resonances demonstrates the strong localization of the excitations on the molecule [23.8]. The relative intensity of the intramolecular to extramolecular (i.e., molecule-substrate) scattering processes is seen from Fig.23.2b by comparing the O-C σ resonance intensity to the O-Cu scattering resonance at ~555 eV. The strong difference in the O-C to O-Cu scattering intensities near the O K edge is also revealed by the single-scattering calculations in Fig.23.1, although the SS theory cannot quantitatively account for the details of the NEXAFS region [23.16]. Figure 23.2 also demonstrates that the π resonance is only present for triple-bonded CO, but not for single-bonded CH_3OH and for methoxy (CH_3O) on Cu(100). The dependence of the spectra on **E** vector orientation [23.7] clearly allows us to determine the molecular orientation on the surface, as indicated. Finally, the location of the σ resonance shifts from 552.5 eV for CO to 539 eV for CH_3O on Cu(100). This demonstrates the sensitivity of the σ shape resonance to the intramolecular bond length R which changes by 0.3 Å between the two molecules. The correlation between R and the σ resonance position has been discussed in detail elsewhere for gas phase [23.10,11] and chemisorbed [23.15,17] molecules.

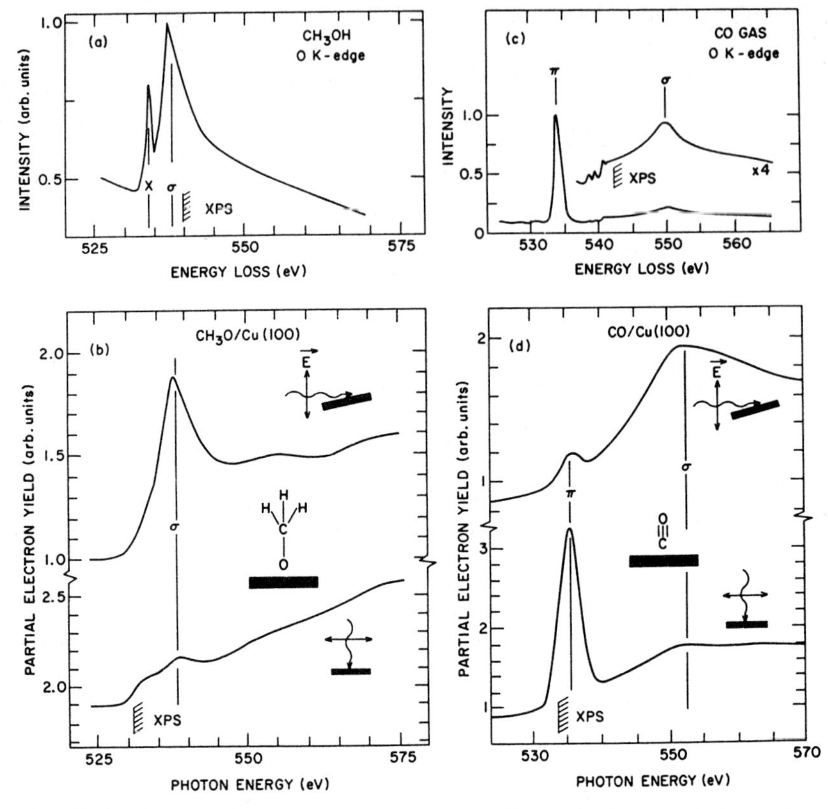

Fig.23.2a-d. Gas-phase near-edge spectra (O K edge) recorded by electron energy loss spectroscopy [23.14] for (**a**) methanol (CH_3OH) and (**c**) carbon monoxide (CO) in comparison to the NEXAFS spectrum [23.15] of the corresponding molecules (**b**) methoxy (CH_3O) and (**d**) CO chemisorbed on Cu(100). Resonances are labelled π and σ according to the final state symmetry. Peak X in (**a**) corresponds to a transition to a Rydberg final state. The label "XPS" marks the 1s binding energies relative to the vacuum level (gas phase) and Fermi level (chemisorbed phase), respectively

23.4 Applications

In the following we shall discuss an example of a complete structure determination of a chemisorbed molecule by means of NEXAFS and SEXAFS and point out the power of NEXAFS to monitor changes in the molecular composition and intramolecular bonding of a molecule caused by its interaction with the surface.

23.4.1 Structure Determination Using NEXAFS and SEXAFS

a) Formate (HCO_2) on Cu(100)

At room temperature the interaction of formic acid (HCOOH) and a Cu(100) surface results in the formation of a formate (HCO_2) complex as revealed by thermal desorption (TDS) [23.18] and electron energy loss (EELS) spectroscopy

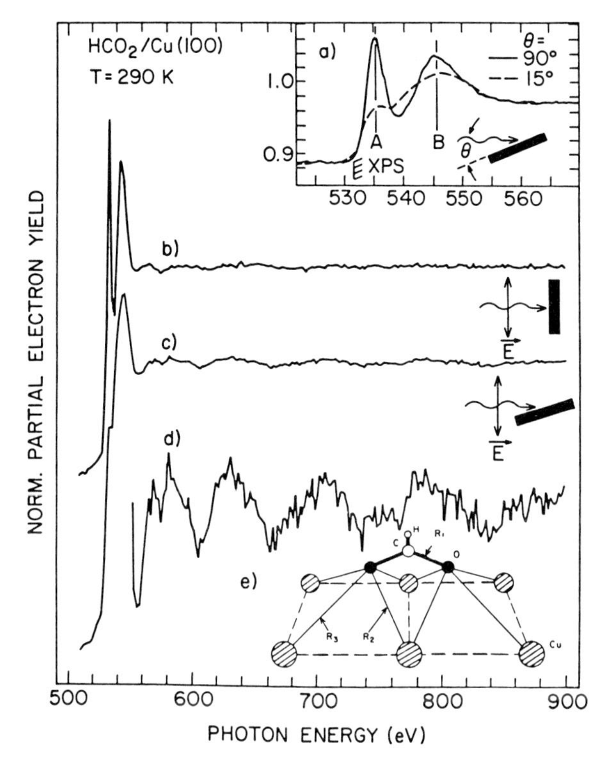

Fig.23.3. NEXAFS (*a*) and SEXAFS (*b* and *c*) spectra for formate (HCO_2) chemisorbed on Cu(100) at room temperature, for **E** parallel to the surface ($\Theta = 90^\circ$) and 15° from the sample normal. The background subtracted SEXAFS signal of (*c*) is shown enlarged in (*d*). The model in (*e*) represents the derived structure

[23.19]. Figure 23.3 shows the O K edge NEXAFS and SEXAFS spectra recorded for this molecular species [23.20]. The NEXAFS spectra in Fig.23.3a exhibit a π (peak A) and σ (peak B) resonance, both resonances being more pronounced when the **E** vector lies in the surface plane. Quantitative analysis [23.7] of the angular dependence of peaks A and B shows that the π^* orbitals of the molecule are parallel ($\pm 10^\circ$) to the surface plane. The position of peak B, the σ resonance, can be used to derive the O-C bond length, $R_1 = 1.27 \pm 0.05$ Å [23.20]. From the SEXAFS oscillation frequencies in both spectra in Fig.23.3 we determine the average O-Cu nearest-neighbor distance, $\bar{R} = 2.38 \pm 0.03$ Å. The SEXAFS amplitude is significantly larger at grazing ($\Theta = 15^\circ$) than normal ($\Theta = 90^\circ$) incidence (factor ~ 1.7), which is only compatible with a fourfold hollow site occupation of the O atoms.

These results can be used to construct the formate chemisorption geometry on Cu(100) shown in Fig.23.3e. The molecule is symmetrically bonded via the two oxygen atoms with the O-C-O plane perpendicular to the surface, a O-C-O bond angle of $\sim 125^\circ$ and bond lengths $R_1 = 1.27 \pm 0.05$ Å and 2.30 Å $\leq R_2 \leq R_3 \leq$ 2.45 Å. The surprising result of this structure is the unusually long O-Cu bond length which is more than 0.35 Å longer than for all known O-Cu bulk and surface nearest-neighbor bond lengths such as for $(2\sqrt{2} \times 2\sqrt{2})R45^\circ$ O on Cu(100): $R = 1.84 \pm 0.03$ Å [23.21], $c(2 \times 2)$ O on Cu(100): $R = 1.94 \pm 0.03$ Å [23.21], or methoxy on Cu(100): $R = 1.97 \pm 0.04$ Å [23.20]. The formate bonding is the first observation of a long sought-after phenomenon, namely that a surface chemical bond may be uniquely different from what is expected from conventional rules [23.22].

23.4.2 Surface Reactions Monitored by NEXAFS

a) C_2H_2 and C_2H_4 on Pt(111)

NEXAFS studies of ethylene [23.17,23] and acetylene [23.17] on Pt(111) reveal substantial changes of the intramolecular C-C bond relative to the gas phase. At 90 K C_2H_4 is found to lie on the surface. The position of the σ resonance reveals a C-C bond length of 1.49 ± 0.03 Å [23.17], an increase of 0.15 Å over the gas phase value [23.24]. The NEXAFS results indicate significant rehybridization of ethylene on the surface in good accord with EELS results [23.25]. At 300 K the C-C bond is found to rotate by 90° and is perpendicular to the surface. The C-C distance in the created ethylidyne ($\equiv C-CH_3$) species is slightly shorter (1.47 ± 0.03 Å) than in ethylene at 90 K. The experimental NEXAFS spectra of ethylene and ethylidyne are well accounted for by SCF Xα calculations using a cluster consisting of the molecules and the nearest-neighbor Pt atoms [23.23]. The most pronounced changes of the C-C bond are observed for C_2H_2 on Pt(111) at 90 K. Acetylene is found to lie on the surface with a bond length of 1.45 ± 0.03 Å, a remarkable 0.25 Å increase relative to the gas phase. This is attributed to a rehybridization to a species with a π orbital parallel to the surface and an effective bond order around 1.4.

b) Cyclic and Heterocyclic Hydrocarbons on Pt(111)

NEXAFS is also well suited to determining the orientation of molecular ring structures on the surface, as has been demonstrated for benzene (C_6H_6) and pyridine (C_5H_5N) on Pt(111) [23.26]. When chemisorbed at room temperature, C_6H_6 lies down and C_5H_5N stands up on the surface, as is evident from the angular dependence of the π and σ resonances. In addition, the π resonance was found to be significantly broadened for benzene relative to pyridine, which is direct evidence for the interaction of the π^* system with the surface in parallel bonded benzene. Temperature-dependent studies in the range 90-420 K revealed strong orientational changes of the ring plane for pyridine in contrast to benzene which always remained parallel bonded. Pronounced temperature-dependent orientation changes were also found for toluene ($C_6H_5CH_3$) [23.26]. The power of NEXAFS to follow chemical transformations of molecules on surfaces has been demonstrated by the study of thiophene (C_4H_4S) on Pt(111) [23.27]. At monolayer coverage (~250 K) thiophene was found to lie on the surface. Both the C K edge and S $L_{2,3}$ edge revealed a pronounced resonance arising from the C-S bond. This resonance diminished in intensity at temperatures above 270 K and had vanished at 470 K while all other resonances remained. This is direct evidence for the desulfurization of thiophene with increasing temperature. The S in the hydrocarbon ring is substituted by a Pt surface atom, resulting in a metallocycle, while S remains on the surface in atomic form.

c) CO on Na Promoted Pt(111)

A variety of spectroscopic techniques indicate that the coadsorption of alkali atoms on metal surfaces significantly weakens the intramolecular bond in CO. NEXAFS studies for the CO/Na/Pt(111) system [23.28] demonstrate convincingly that the C-O bond weakening is a direct consequence of charge backdonation from the metal to the CO $2\pi^*$ antibonding orbital. For both CO/Pt(111) and CO/Na/Pt(111), at a variety of CO and Na coverages, CO is found to stand up. The increased interaction of the metal with the $2\pi^*$ orbital in the presence of Na is directly revealed by a pronounced broadening and intensity reduction of the π resonance which corresponds to a $1s \rightarrow 2\pi^*$ excitation. At the same time, a 4 eV shift of the σ resonance reveals a 0.12 ± 0.03 Å stretch of the C-O distance, induced by Na.

23.5 Conclusions

We have discussed and demonstrated the complementary nature of two surface structural tools, NEXAFS and SEXAFS, for the determination of molecular chemisorption geometries. Combination of these techniques allows us not only to determine internuclear distances and geometries, but through the sensitivity of the NEXAFS π resonance to the valence charge distribution, a more complete picture of the bonding can be obtained. It appears that NEXAFS is particularly powerful in monitoring chemical transformations of molecules on surfaces.

Acknowledgments. The studies reported here were carried out in numerous collaborations as is evident from the references listed. I have particularly benefitted from discussions with F. Sette concerning the principles of NEXAFS and I would like to thank him and my other collaborators for their significant contributions. The experimental work was performed at SSRL which is supported by the office of Basic Energy Sciences of the DOE and the Division of Materials Research of NSF.

References

23.1 G.A. Somorjai: *Chemistry in Two Dimensions: Surfaces* (Cornell University Press, Ithaca 1981)
23.2 J. Stöhr, K. Baberschke, R. Jaeger, R. Treichler, S. Brennan: Phys. Rev. Lett. **47**, 381 (1981)
23.3 The first SEXAFS measurements were published in 1978: P.H. Citrin, P. Eisenberg, R.C. Hewitt: Phys. Rev. Lett. **41**, 309 (1978); J. Stöhr, D. Denley, P. Perfetti: Phys. Rev. B**18**, 4132 (1978); J. Stöhr: Jpn. J. Appl. Phys. **17**, Suppl. 17-2, 217 (1978)
23.4 For a review of SEXAFS spectroscopy see: a) J. Stöhr: In *Emission and Scattering Techniques*, ed. by P. Day (Reidel, Dordrecht 1981); b) J. Stöhr: In *Chemistry and Physics of Solid Surfaces V*, ed. by R. Vanselow and R. Howe, Springer Series in Chemical Physics, Vol.35 (Springer, Berlin, Heidelberg 1984) p.231; c) J. Stöhr: In *X-ray Absorption: Principles, Applications, Techniques of EXAFS, SEXAFS and XANES*, ed. by R. Prins and D. Koningsberger (Wiley, New York 1985)
23.5 For reviews of EXAFS spectroscopy see: a) E.A. Stern: Contemp. Phys. **19**, 289 (1978); b) P. Eisenberger, B.M. Kincaid: Science **200**, 1441 (1978); c) D.R. Sandstrom, F.W. Lytle: Ann. Rev. Phys. Chem. **30**, 215 (1979); d) P. Rabe, R. Haensel: In *Festkörperprobleme*, Vol.20 (Pergamon-Vieweg, Stuttgart 1980) p.43; e) P.A. Lee, P.H. Citrin, P. Eisenberger, B.M. Kincaid: Rev. Mod. Phys. **53**, 769 (1981); f) B.K. Teo, D.C. Joy (eds.): *EXAFS Spectroscopy, Techniques and Applications* (Plenum, New York 1981)
23.6 Boon-Keng Teo, P.A. Lee, A.L. Simons, P. Eisenberger, B.M. Kincaid: J. Am. Chem. Soc. **99**, 3854 (1977); Boon-Keng Teo, P.A. Lee: J. Am. Chem. Soc. **101**, 2815 (1979)
23.7 J. Stöhr, R. Jaeger: Phys. Rev. B**26**, 4111 (1982)
23.8 J.L. Dehmer, D. Dill: J. Chem. Phys. **65**, 5327 (1976); J.L. Dehmer, D. Dill, A.C. Parr: In *Photophysics and Photochemistry in the Vacuum Ultraviolet*, ed. by S.P. McGlynn, G. Findley, R. Huebner (Reidel, Dordrecht 1983)
23.9 N. Padial, G. Csanak, B.V. McKoy, P.W. Langhoff: J. Chem. Phys. **69**, 2992 (1978);

W. Thiel: J. Electron. Spectrosc. **31**, 151 (1983)
23.10 A.P. Hitchcock, S. Beaulieu, T. Steel, J. Stöhr, F. Sette: J. Chem. Phys. **80**, 3927 (1984)
23.11 F. Sette, J. Stöhr, A.P. Hitchcock: Chem. Phys. Lett. **110**, 517 (1984)
23.12 W.L. Jorgensen, L. Salem: *The Organic Chemist's Book of Orbitals* (Academic, New York 1973)
23.13 R. Jaeger, J. Stöhr, T. Kendelewicz: Surf. Sci. **134**, 547 (1983)
23.14 G.R. Wright, C.E. Brion: J. Electron. Spectrosc. **4**, 25 (1974); A.P. Hitchcock, C.E. Brion: J. Electron. Spectrosc. **18**, 1 (1980)
23.15 J. Stöhr, J.L. Gland, W. Eberhardt, D. Outka, R.J. Madix, F. Sette, R.J. Koestner, U. Döbler: Phys. Rev. Lett. **51**, 2414 (1983)
23.16 D. Norman, J. Stöhr, R. Jaeger, P.J. Durham, J.B. Pendry: Phys. Rev. Lett. **51**, 2052 (1983)
23.17 J. Stöhr, F. Sette, A.L. Johnson: Phys. Rev. Lett. **53**, 1684 (1984)
23.18 I.E. Wachs, R.J. Madix: J. Catalysis **53**, 208 (1978); M. Bowker, R.J. Madix: Surf. Sci. **102**, 542 (1981)
23.19 B.E. Sexton: Surf. Sci. **88**, 319 (1979)
23.20 J. Stöhr, D. Outka, R.J. Madix, U. Döbler: Phys. Rev. Lett. (to be published)
23.21 U. Döbler, K. Baberschke, J. Stöhr, D. Outka: Phys. Rev. B**31**, 2532 (1985)
23.22 L. Pauling: *The Nature of the Chemical Bond* (Cornell University, Ithaca 1960)
23.23 R.J. Koestner, J. Stöhr, J.L. Gland, J.A. Horsley: Chem. Phys. Lett. **105**, 332 (1984); J.A. Horsley, J. Stöhr, R.J. Koestner: To be published
23.24 Landolt-Börnstein (Eds.): *Structure Data of Free Polyatomic Molecules*, New Series Group II, Vol.7 (Springer, Berlin, Heidelberg 1976)
23.25 H. Ibach, S. Lehwald: J. Vac. Sci. Technol. **15**, 407 (1978); H. Steininger, H. Ibach, S. Lehwald: Surf. Sci. **117**, 685 (1982)
23.26 A.L. Johnson, E.L. Muetterties, J. Stöhr: J. Am. Chem. Soc. **105**, 7183 (1983); A.L. Johnson, E.L. Muetterties, J. Stöhr, F. Sette: To be published
23.27 J. Stöhr, J.L. Gland, E.B. Kollin, R.J. Koestner, A.L. Johnson, E.L. Muetterties, F. Sette: Phys. Rev. Lett. **53**, 2161 (1984)
23.28 F. Sette, J. Stöhr, E.B. Kollin, D.J. Dwyer, J.L. Gland, J.L. Robbins, A.L. Johnson: Phys. Rev. Lett. **54**, 935 (1985)

24. Current Status and New Applications of SEXAFS: Reactive Chemisorption and Clean Surfaces

P.H. Citrin

AT&T Bell Laboratories, Murray Hill, NJ 07974, USA

The development of surface extended X-ray absorption fine structure (SEXAFS) as a technique for determining local geometry of adsorbates on ordered single crystals has been extended to the study of surfaces lacking either order or adsorbates. Two examples briefly described here are the reaction of Ni on Si(111) to form a disordered silicide overlayer and the surface structure of clean amorphized Si.

24.1 Introduction

The technique of surface extended X-ray absorption fine structure (SEXAFS) has made significant progress since its feasibility demonstration seven years ago [24.1]. The determination of adsorbate-substrate bond lengths from sub- and monolayer coverages of high-Z [24.2] and low-Z [24.3] adatoms, the identification of adsorption sites using absolute and polarization-dependent SEXAFS amplitudes [24.4-6] coupled with the use of higher neighbor distances [24.7-10], and the extension from adsorbate studies on simple close-packed metal surfaces [24.4-7,9] to those on reconstructed semiconductors [24.8,10, 11] and more open metal faces [24.12] all provide evidence for the ongoing development of the technique. The primary factors responsible for this growth derive from the straightforward analysis and information content of the data. The analysis relies on the same single-scattering and theory-independent procedures used in bulk EXAFS [24.13] and thus the same short-range information with the same degree of accuracy is obtained. Other features of the method, as it has been applied so far, are also important factors. It is inherently surface sensitive because only adsorbate-specific photoabsorption is monitored, it is polarization dependent because absorption of synchrotron radiation from a 2D overlayer is anisotropic, and it uses essentially the same equipment (sans synchrotron) as that found in typical UHV systems for preparing and characterizing surfaces.

Along with the above features, however, must come some qualifications. First, the virtues of using intense polarized synchrotron radiation weigh against the fact that synchrotron facilities are sparsely distributed and available for the general scientific community on only a limited basis. At least part of this problem should be lifted with the full commissioning of current and planned dedicated facilities. Second, the desirable short-range and atom-specific properties of SEXAFS necessarily rule out its ability to address questions of long-range surface order or bonding to hydrogen [24.14] studies of this kind obviously require other methods. Finally, the advantages of the polarization dependence and surface sensitivity from ordered adlayers become academic in systems with disordered adatoms or no adatoms at all, i.e., clean surfaces. Here, however, these limitations are less problematic than they might appear. The remainder of this paper briefly reviews two new applications of SEXAFS which directly address these apparently limiting cases.

24.2 Reactive Chemisorption: Ni on Si(111)

The Ni-Si system has particular importance not only for its implications in silicon technology, but also as a prototype for studying solid-state reactions, epitaxial growth, and Schottky barrier formation. Extensive efforts using a broad range of experimental techniques have been applied to this system, but no satisfactory model for the room-temperature nucleation of nickel silicide had been proposed. The reason for this is that the lack of long-range order, even in the first stages of Ni deposition, precludes most structural probes from following the silicide formation process *as it occurs*, i.e., *without* subsequent annealing of the system. The short-range nature of SEXAFS, however, is ideally suited to this study [24.15].

In Fig.24.1 Ni KLL Auger yield data from the model compound $NiSi_2$ are compared with those obtained from increasing coverages of Ni on Si(111). Standard analysis procedures confirm the similarity between $NiSi_2$ and the 0.5 ML Ni/Si(111) system, with the first coordination shell around Ni being composed of 6-7 Si atoms at a distance of 2.37 ± 0.03 Å (Ni in $NiSi_2$ is surrounded by 8 Si atoms at a distance of 2.34 Å). These results directly establish the Ni chemisorption site to be the sixfold hollow *between* the first and second Si layers. To accommodate the observed Ni-Si bond length, the three surface silicon atoms must expand outward 0.8 Å and break the Si-Si interlayer bond. The resulting configuration of the as-deposited silicide is structurally similar to cubic $NiSi_2$(111) with partially unsaturated bonds lying along the [111] direction. Other proposed nucleation sites, unreacted Ni clusters, or compounds with NiSi or Ni_2Si structures in this coverage regime have all been ruled out by extensive computer simulations and comparisons with other experimental EXAFS results. The determined chemisorption site naturally accounts for a variety of other measurements [24.16] and forms a basis for understand-

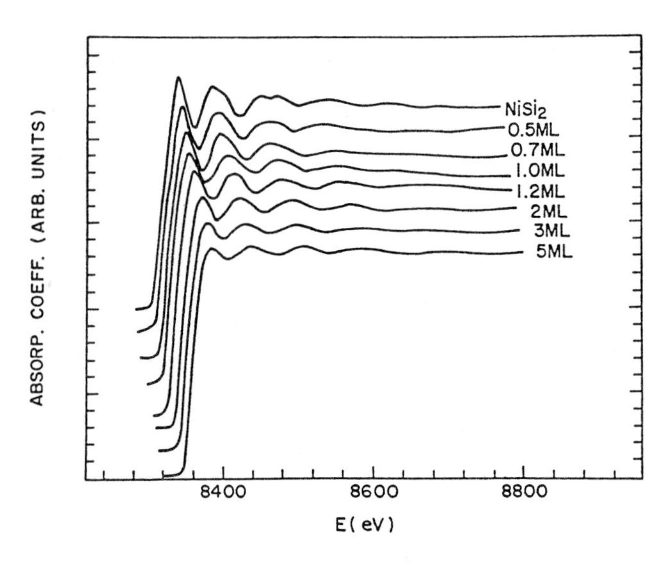

Fig.24.1. Ni K-edge absorption data from bulk $NiSi_2$ and from Ni deposited on Si(111) for different coverages (ML = monolayer)

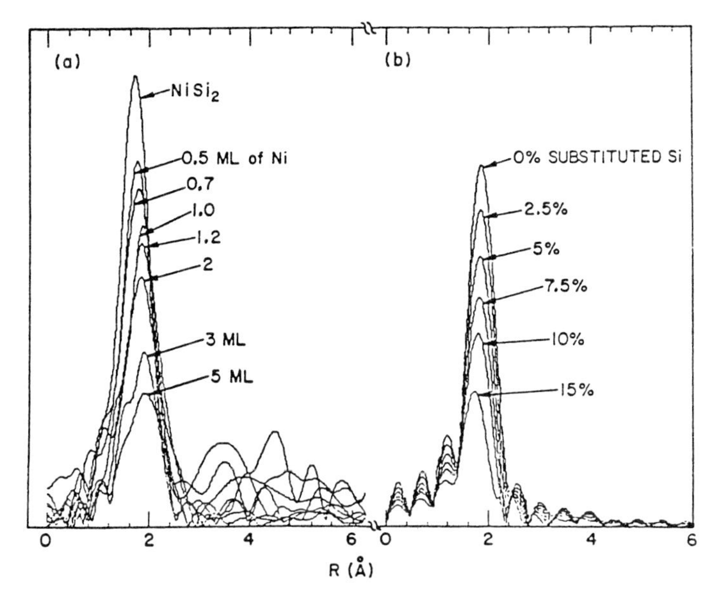

Fig.24.2. (**a**) Fourier transforms of data shown in Fig.24.1 after background subtraction and k^2 multiplication. Peaks at $R > 3$ Å are due to noise. (**b**) Transforms of simulated data assuming the $NiSi_2$ structure with varying amounts of Ni substituting Si sites. The 0% data have been normalized to the 0.5 ML data in (**a**). Side peaks are due to truncation errors

ing the observed 180° rotation of the epitaxial layer which results from light annealing of low coverages of Ni [24.17].

At coverages up to 5 ML, the SEXAFS data exhibit a monotonic decrease of intensity, see Figs.24.1 and 2a, and a modification of the envelope function, with the first shell bond length remaining nearly unchanged. This behavior is compatible with progressive Ni substitution of the Si atoms within the $NiSi_2$ overlayer (antisite defects). Because the Ni-Ni and Ni-Si phase shifts differ by approximately π, each substituted atom contributes antiphase and leads to an even greater decrease of the coordination number [24.18]. Computer simulations have been used to estimate the concentration of antisite defects, Fig. 24.2b. This behavior, combined with other measurements, lead to other interesting implications of the silicide growth mechanism in this coverage regime [24.5].

The sum of these findings clarifies a number of questions concerning the Ni-Si system. Equally important is the demonstration that SEXAFS can be effectively applied to the study of reactive interface formation. The lack of polarization dependence is not a limiting factor, and the experiments can be carried out in systems where most other structural probes are inapplicable.

24.3 SEXAFS From Clean Surfaces: Amorphized Silicon

Dealing with clean surfaces removes the surface sensitivity offered by the atom selectivity of SEXAFS from an adsorbate. Total or partial yield detection modes are not effective because they probe far below the surface (even secondaries with low escape depth originate well within the bulk). However, *elastically* scattered Auger electrons with low escape depth do probe a large fraction of the surface region. Linear combinations of absorption data obtained with these electrons and analogous data obtained with electrons of high escape depth can therefore be used to isolate the surface layer from the bulk (such procedures are common in photoemission [24.19]). Almost every element has both low and high escape depth Auger transitions; for Si these are the 90 eV LVV and 1615 eV KLL Auger electrons with escape depths of ~5 and ~25 Å, respectively [24.20]. Additional surface sensitivity can also be obtained using shallow electron take-off angles, *s* polarization, and optimized energy resolution.

Clean single-crystal Si samples with different crystal orientations were prepared following usual procedures [24.21]. The samples were amorphized by bombardment (sputtering) with saturation doses of 2 kV rare gas ions. Amorphous Si was prepared by in situ electron beam evaporation. Absorption data using Si KLL and Si LVV Auger electrons were collected from the crystalline, amorphous, and amorphized samples under conditions which essentially left the escape depths of these transitions unaltered [24.20]. The Si K-edge absorption data were analyzed in the range of 10-90 eV following standard EXAFS procedures. This range is not formally applicable to single-scattering analysis procedures, although theoretical work [24.22] suggests this possibility. It is not the aim of the present work to demonstrate the validity of such an approach. Rather, since a single atomic species is involved here, questions of phase and amplitude transferability [24.13] become unimportant in using the low-k regime to monitor relative degrees of order. A noteworthy advantage of using this low-k region, in fact, is that the Debye-Waller-like damping of the EXAFS from the 2nd and 3rd coordination shells is less than it would be in data of more extended range. This allows the degree of medium-range order (MRO) to be easily detected by following the ratio between the 1st and the 2nd and 3rd Fourier peak intensities [24.23,24].

In Fig.24.3 the raw and background subtracted Si K-edge absorption data from single-crystal and amorphous Si are shown, and in Fig.24.4a their

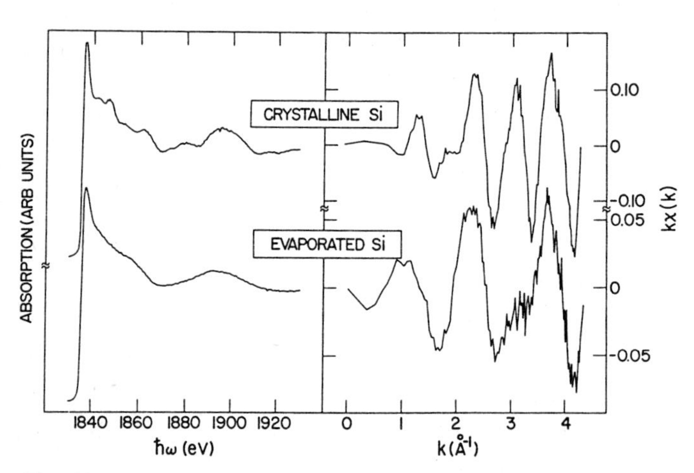

Fig.24.3. Si K-edge absorption data from single-crystal and evaporated Si measured with Si KLL Auger electrons before (*left*) and after (*right*) background substraction

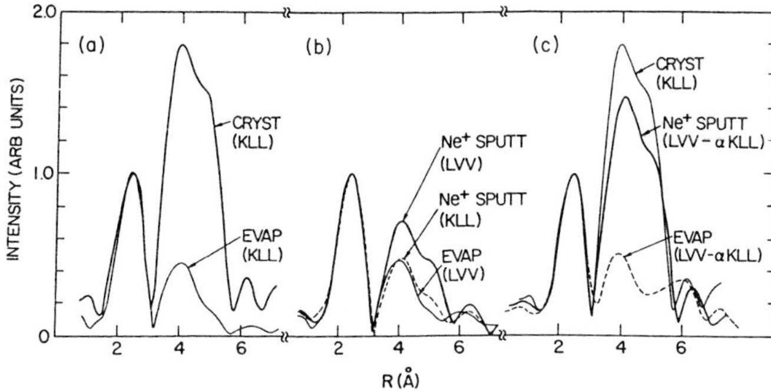

Fig.24.4. Fourier transforms of Si K-edge data as a function of sample preparation and measurement detection mode. Intensities of 1st coordination shell peak at ~2 Å are normalized for all data. Intensity variations of unresolved 2nd and 3rd coordination shell structure at ~4-5 Å reflect changes in medium-range order. (**a**) Single-crystal vs evaporated Si measured with KLL electrons (raw data shown in Fig.24.3). (**b**) Ne^+ sputtered (amorphized) vs evaporated (amorphous) Si measured with KLL or LVV electrons (raw amorphized data shown in upper half of Fig.24.5). (**c**) (——): Fourier transform of raw difference spectrum of amorphized Si, shown in lower half of Fig.24.5 as bold curve. (----): Same procedure applied to evaporated Si. (——): Single-crystal Si data reproduced from panel (**a**) for comparison

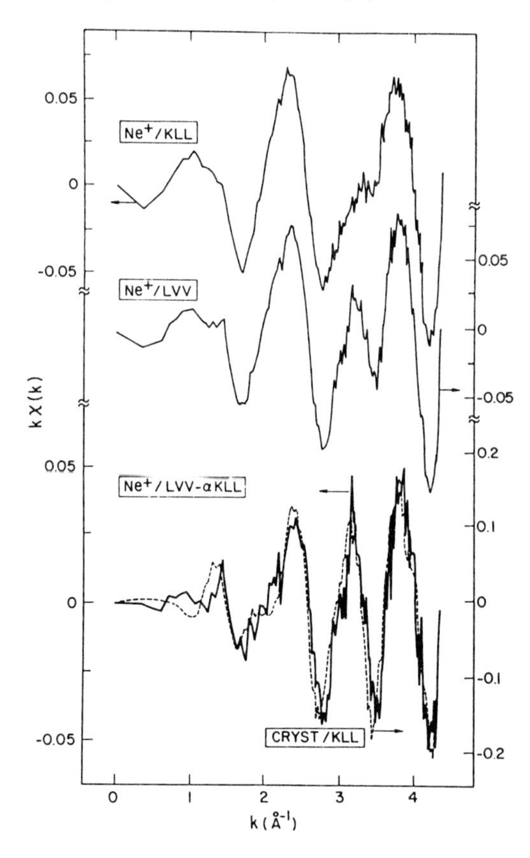

Fig.24.5. *Upper*: Ne^+ sputtered Si measured with KLL and LVV electrons. *Lower*: difference between LVV data and an amount α of KLL data. Here α is the ratio of fractional surface contributions in the different data and in these experiments is 0.8. (----) single-crystal data reproduced from Fig.24.1

Fourier transforms are compared to illustrate the greater expected MRO in the crystalline data. The same comparison is made in Fig.24.4b between the amorphized Si data taken with Si KLL and Si LVV Auger electrons. The increased MRO of the more surface sensitive LVV amorphized data, as well as the lack of such an increase in the KLL amorphized or LVV evaporated Si data, is apparent and surprising. This is a clear signature of *reordering* of the topmost layer of amorphized Si. The crystallinity of this layer is most clearly seen by comparing the difference spectrum, shown in the bottom half of Fig.24.5, with the corresponding data from single-crystal Si. The analogous Fourier transform comparison is given in Fig.24.4c.

These unexpected results have direct implications on the characterization of amorphous systems and on the processes of amorphization and crystal regrowth [24.21]. More generally, however, is the demonstration that a proper choice of detection mode allows SEXAFS to be meaningfully interpreted from *substrate surfaces* either in the presence or absence of a chemisorbed species. A complete SEXAFS characterization of adsorbate and substrate surface atoms before and after chemisorption should therefore be possible. Furthermore, the lack of requiring long-range order and the wide availability of elements with appropriate Auger transitions allow these techniques to be applied to almost any system. Future experiments will include the complete EXAFS range, extending beyond that portion of the data where substrate photoelectrons enter the low-energy Auger window.

24.4 Conclusions

It has been shown that new applications are open to the SEXAFS technique. Clean and disordered surfaces can be studied, with the possibility of addressing such problems as surface reconstruction and reactive chemisorption pathways.

This work was performed in close collaboration F. Comin, along with L. Incoccia, P. Lagarde, G. Rossi, and J.E. Rowe. Experiments at the Stanford Synchrotron Radiation Laboratory were supported by the NSF Division of Material Research and the U.S. DOE Office of Basic Energy Sciences.

References

24.1 P.H. Citrin: Bull. Am. Phys. Soc. **22**, 359 (1977)
24.2 P.H. Citrin, P. Eisenberg, R.C. Hewitt: Phys. Rev. Lett. **41**, 309 (1978)
24.3 L.I. Johansson, J. Stöhr: Phys. Rev. Lett. **43**, 1882 (1979); J. Stöhr, L.I. Johansson, I. Lindau, P. Pianetta: Phys. Rev. B**20**, 664 (1979)
24.4 P.H. Citrin, P. Eisenberger, R.C. Hewitt: Phys. Rev. Lett. **45**, 1948 (1980)
24.5 F. Comin, P.H. Citrin, P. Eisenberger, J.E. Rowe: Phys. Rev. B**26**, 7060 (1982)
24.6 J. Stöhr, R. Jaeger, T. Kendelewicz: Phys. Rev. Lett. **49**, 142 (1982)
24.7 S. Brennan, J. Stöhr, R. Jaeger: Phys. Rev. B**24**, 4871 (1981)
24.8 P.H. Citrin, P. Eisenberger, J.E. Rowe: Phys. Rev. Lett. **48**, 802 (1982)
24.9 P.H. Citrin, D.R. Hamann, L.F. Mattheiss, J.E. Rowe: Phys. Rev. Lett. **49**, 1712 (1982)
24.10 P.H. Citrin, J.E. Rowe, P. Eisenberger: Phys. Rev. B**28**, 2299 (1983)
24.11 J. Stöhr, R. Jaeger, G. Rossi, T. Kendelewicz, I. Lindau: Surf. Sci. **134**, 813 (1983)

24.12 U. Döbler, K. Baberschke, J. Haase, A. Puschmann: Phys. Rev. Lett. **52**, 1437 (1984)
24.13 P.A. Lee, P.H. Citrin, P. Eisenberger, B.M. Kincaid: Rev. Mod. Phys. **53**, 769 (1981), and references therein
24.14 In special cases, hydrogen can be indirectly detected, cf. B. Lengeler: Phys. Rev. Lett. **53**, 74 (1984)
24.15 F. Comin, J.E. Rowe, P.H. Citrin: Phys. Rev. Lett. **51**, 2402 (1983)
24.16 See references 1-11 in Ref.24.15 and others cited in R.T. Tung: J. Vac. Sci. Technol. B**2**, 465 (1984)
24.17 R.T. Tung, J.M. Gibson, J.M. Poate: Phys. Rev. Lett. **50**, 429 (1983)
24.18 This effect is also sensitive to bond distances, cf., J. Mimault, A. Fontaine, P. Lagarde, D. Raux, A. Sadoc, D. Spanjaard: J. Phys. F**11**, 1311 (1981)
24.19 P.H. Citrin, G.K. Wertheim, Y. Baer: Phys. Rev. Lett. **41**, 1425 (1978); Phys. Rev. B**29**, 3160 (1983)
24.20 M.P. Seah, W.A. Dench: Surf. and Interface Anal. **1**, 2 (1979), and references therein
24.21 F. Comin, L. Incoccia, P. Lagarde, G. Rossi, P.H. Citrin: To be published
24.22 R. Natoli: In *Proceedings of 1st Int. Conf. on EXAFS and XANES, Fracati, Italy, 1982*, Springer Series on Chem. Phys., Vol.27 (Springer, Berlin, Heidelberg 1983) p.43
24.23 F. Evangelisti, M.G. Proietti, A. Balzarotti, F. Comin, L. Incoccia, S. Mobilio: Solid State Commun. **37**, 413 (1981)
24.24 The distances of the 1st, 2nd, and 3rd coordination shells in Si are 2.35, 3.84, and 4.50 Å, respectively

III.3 High-Resolution Electron Energy Loss Spectroscopy (HREELS)

25. Electron-Phonon Scattering and Structure Analysis

M. Rocca[1], H. Ibach, and S. Lehwald
Institut für Grenzflächenforschung und Vakuumphysik, Kernforschungsanlage Jülich, Postfach 1913, D-5170 Jülich, FRG

M.-L. Xu, B.M. Hall[2], and S.Y. Tong
Department of Physics and The Laboratory for Surface Studies, University of Wisconsin-Milwaukee, Milwaukee, WI 53201, USA

We have calculated the inelastic cross section of the Ni(001) surface phonons S_4 and S_6 for an extended energy range of 50-250 eV. The calculated results indicate regions of energy where the S_6 mode has a cross section comparable to that of S_4. The theoretical results are confirmed by measurements which observed the S_6 mode at the predicted energies. The simultaneous measurement of S_4 and S_6 modes allows a discrimination between different structural models for the Ni(001) surface. From the S_4/S_6 intensity ratio, we determined a 1.7-3.3% contraction in the surface interlayer spacing of Ni(001).

25.1 Introduction

Electron energy loss spectroscopy is widely used as a tool for studying vibrational properties of clean and adsorbate covered surfaces. Recently the experiments have been extended into the energy range up to 400 eV. By simultaneous measurements of energy loss and momentum exchange, the dispersion of surface phonons of clean surfaces and adsorbate layers thereon have been determined [25.1-3]. Analysis of the intensities requires a fully dynamical treatment of the scattering process. However, at such high energies, the electrons scatter mainly from the near-core region of the atoms. A theory for the phonon intensities has been developed [25.4,5] which incorporates the muffin-tin approximation for the scattering potential and the additional assumption that the ions move rigidly when the nuclei are displaced through the excitation of a phonon. In this paper we compare experimental data on the intensities of inelastic electron scattering to this rigid-ion multiple scattering (RIMS) method and explore the possibilities of using the intensity data to extract structural information. We also report the observation of a surface mode of the Ni(001) surface which is polarized parallel to the surface. While this mode would contribute only very weakly to the spectrum when the Born-approximation (kinematic scattering) is used, multiple scattering effects can cause this mode to become strong at particular energies and for particular diffraction conditions.

---

1 Permanent address: Dipartimento di Fisica, Università degli Studi, Via Dodecaneso 33, I-16100 Genova, Italy

2 Permanent address: Department of Physics, University of California, Irvine, CA 92717, USA

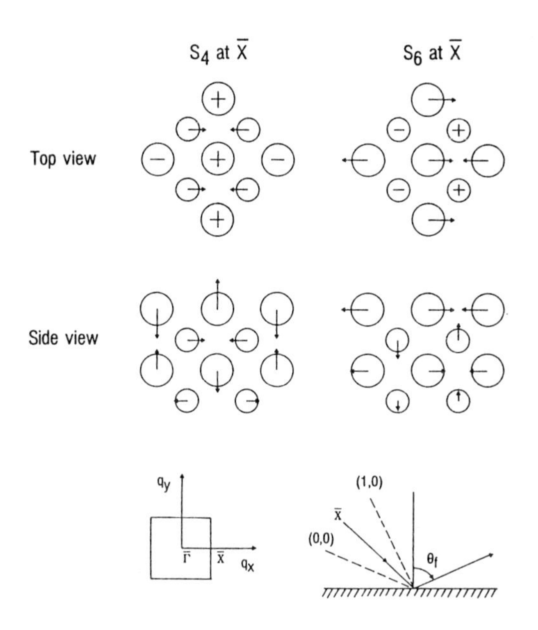

Fig.25.1. Direction of atomic displacements of S_4 and S_6 modes at $\bar{X}$, (+) up, (-) down. In top view, the smaller circles denote second layer atoms. Lower right figure shows the scattering plane

Intensities are compared with two modes at the $\bar{X}$ point of the surface Brillouin zone for the clean Ni(001) surface. The polarization of the two modes S_4 and S_6 and the scattering geometry are depicted in Fig.25.1. A third surface model, labeled S_1, is a shear horizontal wave. The polarization is odd with respect to the ($\bar{1}$10) plane which is also the scattering plane in the experiments. Selection rules applicable to the scattering process considered here require that this mode is not observable [25.5].

25.2 Measurement of Electron-Phonon Intensities

The Ni(001) single crystal was prepared following conventional procedures. Sulfur and carbon were removed from the sample by cycles of neon bombardment and annealing. The carbon level in the sample was reduced to the extent that even after flashing to 1350 K and cooling to room temperature the concentration of carbon on the surface was below the detection limit of the CMA-Auger spectrometer (carbon peak less than 1/40 of the Ni 102 eV peak). The low carbon level is of importance since the frequency of the S_4 mode at $\bar{X}$ was found to be rather sensitive to carbon. Phonon loss intensities were measured with the double pass spectrometer used in earlier experiments [25.1]. The azimuthal and polar orientation of the sample with respect to the scattering plane was controlled in situ by LEED and by the diffraction peaks measured with the spectrometer. The error in the alignment and the determination of the scattering angle is less than 1°. To have reasonably high count rates, intensity profiles were measured with a moderate resolution (60 cm^{-1}), which corresponds to a pass energy of 0.85 eV in the energy dispersive parts. Measuring Intensity-Energy (I-V) profiles for phonon scattering is significantly more demanding than for diffracted beams. Between the monochromator and the analyzer electrons are accelerated and decelerated by a factor of about 100. Since the lens potentials as seen by the electrons are subject to local variation of the surface potential on the electrodes, the long-term reproducibility of the intensities is only within a factor of

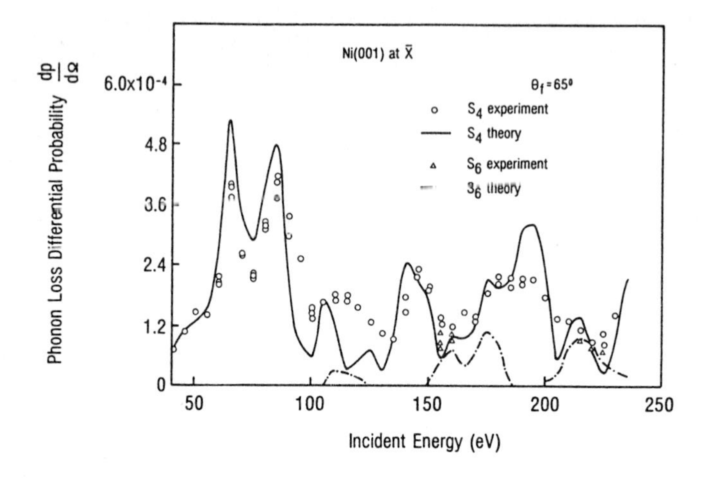

Fig.25.2. Comparison between experimental and theoretical phonon loss I-V profiles of S_4 and S_6 surface modes, at $\bar{X}$ and $\theta_f = 65°$

two. The I-V profiles generated in different runs therefore had to be scaled by a constant factor in order to match to previous results. With this overall scaling, however, the features in the I-V profiles were fairly reproducible, as can be seen from the relatively small scattering of the data in Fig.25.2. We have measured I-V profiles for the S_4 and S_6 modes (when possible) at the $\bar{X}$ point for a constant polar angle of the emerging beam (taken at two values: $\theta_f = 65°$ in Fig.25.2 and $\theta_f = 60°$ in Fig.25.3, respectively).

As the solid angle of detection can be energy dependent, I-V profiles do not directly represent the energy dependence of the cross section. We have determined that the image size of the beam on the sample stays constant with impact energy E_0. Because of the symmetric construction of the spectrometer the same is to be expected for the area viewed by the analyzer. Phase space conservation then requires the solid angle of detection $\Delta\Omega$ to be proportional to E_0^{-1}. Therefore experimental data are represented as intensities multiplied by E_0 which should make them proportional to the differential cross section. The procedure described above cannot, however, correct for energy-dependent image aberrations of the lenses. In comparing with the theoretical result over a range of energies, more emphasis should therefore be placed on the position of maxima and minima than on the peak heights when at substantially different energy.

25.3 Comparison with Theory

The measured phonon loss I-V profiles are shown in Figs.25.2,3 for the S_4 and, when observed, for the S_6 modes together with the calculated results. The angle of the incident beam is fixed by the condition that the parallel component of the difference between initial and final wave vectors of the electron should be $\mathbf{k}_\parallel(I) - \mathbf{k}_\parallel(F) = \mathbf{q}_\parallel = 1.26$ Å^{-1} in order to match the phonon wave vector at $\bar{X}$. There is a close similarity between theory and experiment in the energy dependence of the intensities. Part of the remaining mismatch is probably due to contributions of bulk phonons to the observed peaks. In the energy loss spectra we occasionally notice a shift in the apparent peak position of the loss assigned to S_4 from 130 cm^{-1} to 145 cm^{-1}. As the value for the S_4 frequency is 130 cm^{-1}, the shift indicates a contribution to the loss peak from bulk modes. The spectral densities at $\bar{X}$ show a broad maximum

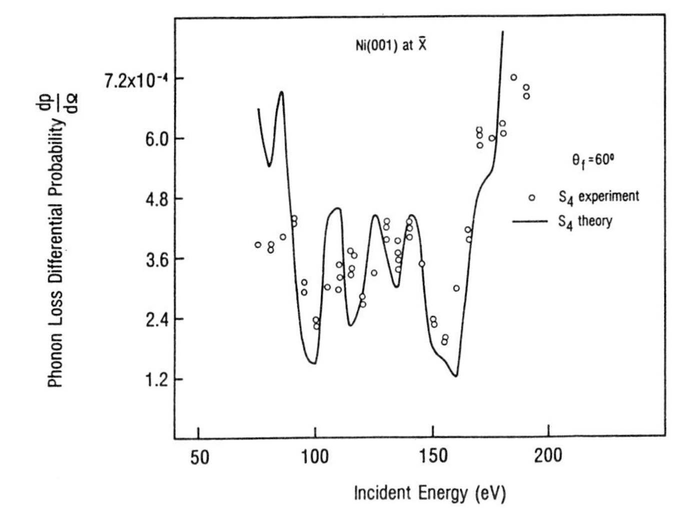

Fig.25.3. Same as Fig.25.2, for $\theta_f = 60°$

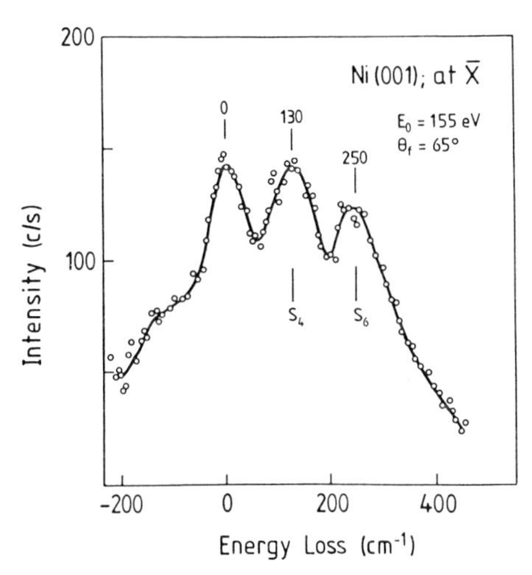

Fig.25.4. Experimental loss spectrum at $\bar{X}$ ($q_{\|} = 1.25$ Å^{-1}) and $E_i = 155$ eV, $\theta_f = 65°$ showing both S_4 and S_6 modes

of bulk phonons centered around ~168 cm^{-1} [25.6]. The deviation between theory and experiment is therefore partly attributed to the pick up of bulk modes. The same consideration holds for the "S_6 peaks" where bulk phonons can also contribute and shift the observed peak to high frequencies.

The comparison between theory and experiment could be used to determine the interlayer spacings between the surface and deeper Ni layers. As the experimental measurement of the I-V profiles are subject to technical and methodical errors which are difficult to assess in magnitude (see discussion above) a different procedure is suggested which has no equivalent in LEED. Spectra which display the S_4 and S_6 modes simultaneously as in Fig. 25.4 allow a sensitive determination of the ratio of the intensity of the two modes. This ratio is independent of the particular focusing conditions

Table 25.1. Phonon loss intensity ratio S_4/S_6 vs surface interlayer spacing at 155 eV and $\theta_f = 65^\circ$

	S_4/S_6
Experiment	1.15
Theory	
no contraction (d = 1.765 Å)	0.57
1% contraction (d = 1.747 Å)	0.70
2% contraction (d = 1.730 Å)	0.87
3% contraction (d = 1.712 Å)	1.16
4% contraction (d = 1.694 Å)	1.65
5% contraction (d = 1.677 Å)	2.60

of the lens elements. The remaining sources of error are then the limited accuracy of the energy and angle calibration for which we have estimated the error to be ±0.5 eV and ±0.5°, respectively. This estimate is based on the measurement of the position of diffracted beams. The other source of error is the multiphonon background and bulk phonon contribution. We consider the bulk phonon contribution to be small when the observed peak positions agree with the frequencies of S_4 and S_6 (known from data in other energy ranges). By assuming either that the background is zero or is as high as compatible with the intensity in the minimum between S_4 and S_6 (Fig.25.4), we determine a lower and an upper bound to the intensity ratio S_4/S_6, which for Fig.25.4 are 1.00 and 1.30, respectively. This intensity ratio can be very sensitive to the spacing between Ni layers, especially the spacing between the surface and the layer immediately below it. For example, Table 25.1 shows the calculated S_4/S_6 ratio for different % contractions of this interlayer spacing for the case corresponding to Fig.25.4. The calculated ratio varies, on the average, by 40% for every 1% change in the interlayer spacing. Figure 25.5 compares the calculated S_4/S_6 ratio with data.

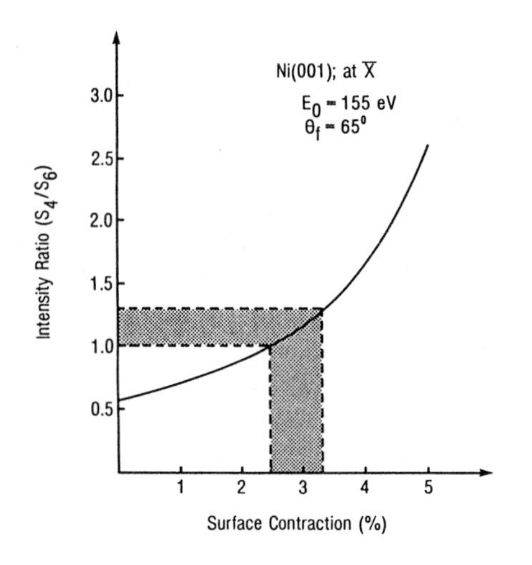

Fig.25.5. Variation of calculated S_4/S_6 intensity ratios (——) with surface contraction. The dotted area represents the measured result and its uncertainty

Table 25.2. Experimental phonon loss intensity ratio at different energies and scattering angles

E_0	θ_f	S_4/S_6	Remarks
160	60	$0.5^{+0.5}_{-0.25}$	contribution of bulk resonance to "S_4" peak
220	65	1.4 ± 0.2	
155	65	1.15 ± 0.15	
160	65	1.2 ± 0.5	contribution of bulk resonances to both peaks

Data which included other energies and θ_f where the S_6 mode is observed simultaneously with S_4 are summarized in Table 25.2. Using these data and the corresponding calculated intensity ratios, we evaluated a reliability factor defined as

$$R = \frac{\sum_i |S_i(\text{exp}) - S_i(\text{theory})|^2}{\sum_i |S_i(\text{exp})|^2} \quad , \tag{25.1}$$

where $S_i(\text{exp})$ or $S_i(\text{theory})$ are the measured or calculated intensity ratios respectively for the ith scattering condition. A plot of R vs interlayer spacing using the upper bounds of the measured data yields a surface contraction of $\leqslant 3.3\%$. A similar plot using the lower bounds yields a surface contraction of $\geqslant 1.7\%$ (see Fig.25.6). Thus, the surface interlayer spacing contraction is 1.7-3.3%.

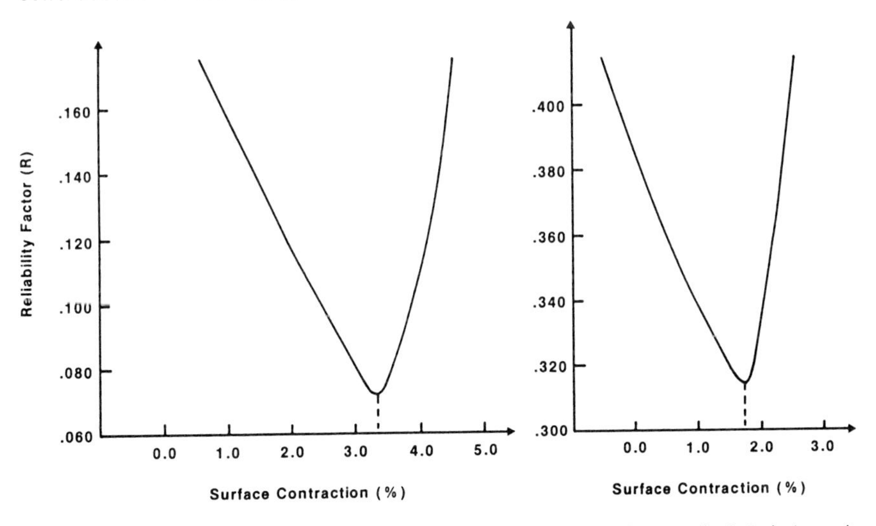

Fig.25.6. R-factor analysis using upper (*left*) and lower (*right*) bounds of measured S_4/S_6 intensity ratios and the calculated ratios corresponding to scattering conditions listed in Table 25.2

25.4 Conclusion

The method proposed here can be extended to structure determination of reconstructed surfaces and of ordered overlayers of atoms or molecules. The method could be particularly suitable for large molecules weakly bound to the surface. There, the normal coordinates of intramolecular vibrations do not significantly depend on the bonding to the surface. Thus, one may keep the same normal mode displacements in the theory and use the phonon loss intensities (or their ratios) to discriminate between various bonding geometries of the molecule on the surface.

The work at Wisconsin is supported in part by the U.S. Department of Energy under Grant No. DE-FG02-84ER45076. M.R. wishes to thank the Centro Nazionale delle Ricerche of Italy for fellowship support. We also acknowledge helpful discussions with D.L. Mills, T.S. Rahman, and J.M. Szeftel.

References

25.1 S. Lehwald, J.M. Szeftel, H. Ibach, T.S. Rahman, D.L. Mills: Phys. Rev. Lett. **50**, 518 (1983)
25.2 J.M. Szeftel, S. Lehwald, H. Ibach, T.S. Rahman, J.E. Black, D.L. Mills: Phys. Rev. Lett. **51**, 268 (1983)
25.3 M. Rocca, S. Lehwald, T.S. Rahman: Surf. Sci. **138**, L123 (1984)
25.4 S.Y. Tong, C.H. Li, D.L. Mills: Phys. Rev. Lett. **44**, 407 (1980)
25.5 C.H. Li, S.Y. Tong, D.L. Mills: Phys. Rev. **21**, 3057 (1980); S.Y. Tong, C.H. Li, D.L. Mills: Phys. Rev. **24**, 806 (1981)
25.6 M. Rocca, S. Lehwald, H. Ibach, T.S. Rahman: To be published

26. Shape Resonances in OH Groups Chemisorbed on the (100)Surface of Ge-Si Alloys

H.H. Farrell
Bell Communications Research, Murray Hill, NJ 07974, USA

J.A. Schaefer
Center for Research in Surface Science, Department of Physics
Montana State University, Bozeman, MA 59717, USA

J.Q. Broughton
Department of Material Science, State University of New York
Stony Brook, NY 11794, USA

J.C. Bean
AT&T Bell Laboratories, Murray Hill, NJ 07974, USA

26.1 Introduction

Andersson and *Davenport* [26.1] have previously used high-resolution electron-energy-loss spectroscopy (HREELS) to study hydroxyl groups on NiO(111). They found that the dipole interaction mechanism completely fails to describe this system as the relative loss intensity was a decreasing function of primary energy for both the O-H stretching fundamental and overtone as opposed to the monotonically increasing function predicted by the dipole interaction mechanism. In addition, they found that the overtone intensity was more than an order of magnitude larger than a reasonable dipole estimate. Therefore, they concluded that a negative ion resonance was responsible for the observed anomalies.

More recently, *Stucki* et al. [26.2] used HREELS to study the dissociative chemisorption of H_2O on Si(100)(2 × 1) surfaces. They also observed anomalous resonance behavior in the intensities of the O-H stretching mode overtone. In fact the enhancement of the overtone intensities is so strong in this case that it allowed the first observation of a second vibrational overtone in a chemisorbed species (though such observations have been made on physisorbed systems). The similarity in behavior observed for O-H groups chemisorbed on rather different substrates suggests that these phenomena are properties of the hydroxyl group per se.

26.2 Experimental

The HREELS spectra reported here were obtained at the Center for Research in Surface Science (CRISS) at Montana State University, Bozeman, Montana. The samples used in this study were (100) surfaces of Ge_xSi_{1-x} alloys prepared by molecular beam epitaxy on Si(100) substrates at AT&T Bell Laboratories, Murray Hill, NJ [26.3,4]. For the HREELS experiment, the samples were ion-bombarded and annealed until a clean, well-ordered (2 × 1) surface was obtained. This procedure produced Ge enrichment of the surface. Direct exposure of the surface to 1 Langmuir of H_2O at room temperature resulted in dissociative chemisorption that saturated the Si dangling bonds to produce Si-H and Si-OH units on the surface. Cleanliness and order were monitored using ESCA and LEED. A more detailed discussion of the experimental prccedure is given elsewhere [26.5-7]. Typical elastic intensities were on the order of 10^5 cps and those for the first overtone of the O-H stretching mode were on the order of 30 cps. When scanning over wide ranges of primary energy, all spectrometer settings were kept constant and the ratio of inelastic to elastic intensities was determined to minimize instrumental factors.

26.3 Results

A detailed presentation of our experimental results will also be published elsewhere [26.6], but a summary of the significant features is given below. Aside from the elastic peak, the strongest feature in the HREELS spectrum is the O-H bending mode at 98 meV which tends to obscure the Si-H and Ge-H bending modes at somewhat lower loss energies. Both the Ge-H and the Si-H stretching modes are observed just below and above 250 meV respectively. The OH stretching mode fundamental and first overtone are seen at approximately 460 meV and 890 meV and the second overtone was occasionally observed just above 1300 meV. The O-H bending mode is sufficiently strong for it to produce combination losses at about 100 meV above both fundamentals and overtones. All of these features are similar to those seen for H_2O chemisorbed on Si(100)(2×1) [26.2,8,9].

When the primary energy and final angle are held constant, and the angle of incidence is varied over a range of about 30°, the elastic peak, O-H bending mode and Ge-H stretching mode decrease by two to three orders of magnitude as expected for a dipole interaction mechanism. Both the O-H stretching mode fundamental and first overtone, however, display anomalous behavior and decrease only by about a factor of six and three, respectively.

The relative intensity (i.e., the intensity normalized to that of the elastic peak) of both the O-H stretching fundamental and overtone is observed to decrease as a function of increasing primary energy as was observed for O-H on NiO(111) and in contradistinction to normal dipole interaction behavior [26.1]. None of the above observations was totally unexpected. What was unexpected, however, was the observation of modulations on the I/I_0 vs E_0 plots that produced maxima that were dependent upon the experimental geometry. For example, at specular reflection for an angle of incidence of 50°, no maximum is seen for the fundamental, only a monotonic decrease in E_0. However, for an angle of incidence of 38° and a final angle of 50°, a maximum is observed near 4.5 eV for the fundamental and 3.5 eV for the first overtone. These and similar observations led us to explore the quantitative aspects of the resonance interaction in some detail as is outlined in the discussion section.

26.4 Discussion

As *Andersson* and *Davenport* [26.1] suggest, the resonant state is probably the OH 4σ level coupled to the continuum. This state is unoccupied in the neutral OH species and is briefly occupied (e.g., $\tau\sim10^{-15}$s) by the HREELS electron during the resonance interaction. As this state has a node between the H and O nuclei, it is antibonding in character and the O-H bond will begin to elongate during the resonance. This motion is essentially identical with that experienced by the O-H group during the expansion phase of the stretching mode. Therefore, this mode has the proper symmetry to couple with the resonance interaction in contrast to the O-H bending mode, for example. Furthermore, in both the harmonic and the anharmonic oscillator approximations, when the hydroxyl group is returned to its ground electronic state, on the average it will have a bond length larger than the ground-state bond length. This facilitates transitions into the $n=1$ and higher vibrational states.

To understand the more qualitative aspects of the experimental data, it is necessary to investigate the scattering process in some detail. Following *Davenport* et al. [26.10], the matrix element describing the resonance interaction is given by

$$\mu(\mathbf{k}^0,\mathbf{k}) = (-m/2\pi\hbar^2)\langle\Phi^0\psi^0|V(\mathbf{r},\mathbf{R})|\Phi'\psi'\rangle \quad , \tag{26.1}$$

where $V(\mathbf{r},\mathbf{R})$ is the interaction potential between the HREELS electron and the hydroxyl group. Here, $\Phi^0=\Phi^0(\mathbf{r})=\exp(i\mathbf{k}^0\cdot\mathbf{r})$ is the plane wave incident electron eigenfunction and $\Phi'=\Phi'(\mathbf{r})=\exp(i\mathbf{k}\cdot\mathbf{r})$ describes the scattered electron to within the first Born approximation. As the only significant change in state of the neutral O-H group before and after the resonance interaction involves the vibrational state, we may take $\psi=\psi(R)=\chi_n(R)$, where χ_n is the vibrational eigenfunction of the neutral species. Within the simple harmonic oscillator (SHO) approximation

$$\chi_n(R) = \chi_n(Z) = C_n[\exp(-\xi^2/2)]H_n(\xi) \quad , \tag{26.2}$$

where C_n is a constant, $H_n(\xi)$ is the nth Hermite polynomial and $\xi=\alpha Z=\sqrt{\mu\omega/\hbar}Z$. The internuclear separation is given by

$$\mathbf{R} = \mathbf{R}_e + \mathbf{Z} \tag{26.3}$$

where R_e is the mean ground-state interatomic spacing and $\mathbf{Z}$ is the vibrational displacement around $\mathbf{R}_e$. At room temperature essentially all of the hydroxyl groups will initially be in the ground vibrational state.

As we now simply wish to explore the general nature of the resonance interaction, we will approximate $V(\mathbf{r},\mathbf{R})$ with shielded Coulombic potentials centered on the H and O nuclei as

$$V(\mathbf{r},\mathbf{R}) = e^2Z_0'[\exp(-\lambda_0 r)/r] + e^2Z_H'[\exp(-\lambda_H|\mathbf{r}-\mathbf{R}|)]/|\mathbf{r}-\mathbf{R}| \tag{26.4}$$

where the coordinate system has arbitrarily been centered on the O nuclei. The matrix element may now be expressed as a sum of contributions from the O and H atoms as

$$\mu(\mathbf{k}^0,\mathbf{k}) = \mu_0(\mathbf{k}^0,\mathbf{k}) + \mu_H(\mathbf{k}^0,\mathbf{k}) \quad , \tag{26.5}$$

where

$$\mu_0(\mathbf{k}^0,\mathbf{k}) = a_0 \langle\chi_0 \exp(i\mathbf{k}^0\cdot\mathbf{r})|r^{-1}\exp(-\lambda_0 r)|\chi_n \exp(i\mathbf{k}\cdot\mathbf{r})\rangle \tag{26.6}$$

and $a_0=(-me^2Z_0'/2\pi\hbar^2)$. Noting that the operator is only a function of r and not of R or Z,

$$\mu_0(\mathbf{k}^0,\mathbf{k}) = a_0\langle\exp(i\mathbf{k}^0\cdot\mathbf{r})|r^{-1}\exp(-\lambda_0 r)|\exp(i\mathbf{k}\cdot\mathbf{r})\rangle\cdot\langle\chi_0|\chi_n\rangle \tag{26.7}$$

from which it may be seen that $\mu_0(\mathbf{k}^0,\mathbf{k})$ will contribute only to the elastic $\Delta n=0$ intensity. The integral over r is simply the Fourier transform of the shielded Coulombic potential and produces

$$\mu_0(\mathbf{k}^0,\mathbf{k}) = 4\pi a_0/(\lambda_0^2+Q^2); \quad \Delta n = 0 \tag{26.8}$$

where Q is the scattering vector

$$\mathbf{Q} = \mathbf{k}^0 - \mathbf{k} \quad . \tag{26.9}$$

Evaluation of the hydrogenic contribution to the scattering matrix element

$$\mu_H(\mathbf{k}^0,\mathbf{k}) = a_H \langle \chi_0 \exp(i\mathbf{k}^0 \cdot \mathbf{r}) | [\exp(-\lambda_H|\mathbf{r} - \mathbf{R}|)]/|\mathbf{r} - \mathbf{R}|| \chi_n \exp(i\mathbf{k} \cdot \mathbf{r}) \rangle \tag{26.10}$$

is somewhat more elaborate because of the inclusion of **R** as well as of **r** in the operator. Consider first the $\Delta n = 1$ transition that produces the fundamental OH stretching mode. It can be shown (Appendix 26.A) that

$$|\mu_H|^2 = |C(Q)|^2 \Big\{ \cos^2\theta_0 [Q \cos QR_e - \lambda_H \sin QR_e]^2 + [2\lambda_H \cos QR_e + Q(1 + \lambda_H^2/Q^2)\sin QR_e]^2 \Big\} \quad , \tag{26.11}$$

where

$$|C(Q)|^2 = [\sqrt{2}me^2\alpha Z_H'/\hbar^2(\lambda_H^2 + Q^2]^2 \exp(-2\lambda_H R_e) \tag{26.12}$$

and where θ_0 is the angle between **Q** and **R** and is therefore dependent upon E_0 as well as θ_i and θ_f. Note that while $\mu_0 = 0$ for $\Delta n = 1$, μ_H has a finite value. This is an artifact that results from the choice of coordinate system that does not affect the fundamental physics at the level of the shielded Coulombic potential approximation. Here again we have an inverse proportionality in $(\lambda_H^2 + Q^2)^2$ for the intensity. At a fixed geometry, this will lead to a decreasing dependence of the intensity on the primary energy in strong contrast to the monotomic increase expected for dipole interaction mechanisms [26.1]. In addition this decrease in intensity with increasing primary energy is modulated by the oscillatory terms sin QR_e and cos QR_e. The term $\cos^2\theta_0$ is sensitive to the alignment of the scattering vector relative to the O-H bond direction. The aggregate effect of these terms is to produce structure on the I/I_0 versus E_0 curves that is dependent upon the geometry of the experiment.

In deriving the expression for $|\mu|^2$, it has been assumed that the mean vibrational displacements are much smaller than Q^{-1} or λ^{-1} such that $Q^2/\alpha^2 \ll 1$ and $Q\lambda/\alpha^2 \ll 1$. Within this approximation, if we consider the limiting case of $\lambda_H \ll Q$, then

$$|\mu_H|^2 \simeq |C(Q)|^2 Q^2 (\cos^2\theta_0 \cos^2 QR_e + \sin^2 QR_e)$$

$$= |C(Q)|^2 Q^2 (1 - \sin^2\theta_0 \cos^2 QR_e) \quad . \tag{26.13}$$

In this limit, it may be seen that when **Q** is parallel to $\mathbf{R}_e$ (or, more properly, to **Z**) there is no modulation of the monotomic decrease in intensity with primary energy. In this limit, keeping in mind that Q varies with E_0, it would be possible experimentally to determine the direction of $\mathbf{R}_e$ without resorting to calculations. If, however, $\theta_0 \neq 0$, then $|\mu|^2$ is expected to have a maximum at energies somewhat below $QR_e = (2n+1)\pi/2$. Care should be taken to include the appropriate inner potential when determining Q or E_0.

Conversely, when $\lambda_H \gg Q$, then $|\mu_H|^2 \propto Q^{-2} \sin^2 QR_e$, and again, normalized intensity maxima are expected at primary energies somewhat below $QR_e = (2n+1)\pi/2$. Intermediate values of λ_H/Q require a more detailed determination of $|\mu|^2$. Within the approximations used in this model (e.g., shielded Coulombic potentials, first Born approximation, etc.) a good qualitative fit to the experimental data is not expected. However, the observed modulations in the normalized intensity vs primary energy curves as a function of both the initial and final angles are predicted in a general fashion. Detailed calculations using this simple model will be presented in a subsequent paper [26.6]. Finally, it is to be hoped that more quantitative calculations will allow this behavior to be used to determine molecular geometries for those adsorbates showing resonance interactions.

26.A Appendix

From (26.10)

$$\mu_H = a_H \langle \chi_0(Z)|L_H(Z)|\chi_n(Z)\rangle \quad , \tag{26.A1}$$

where

$$L_H(Z) = \langle \exp(i\mathbf{k}^0 \cdot \mathbf{r}) \Big| [\exp(-\lambda_H|\mathbf{r}-\mathbf{R}|)]/|\mathbf{r}-\mathbf{R}| \Big| \exp(i\mathbf{k}\cdot\mathbf{r})\rangle \ . \tag{26.A2}$$

Let $\rho = |\mathbf{r}-\mathbf{R}|$, then

$$L_H(Z) = \exp(i\mathbf{Q}\cdot\mathbf{R}) \left\{ \int_R^0 d^3\rho\rho^{-1} \exp(-i\mathbf{Q}\cdot\boldsymbol{\rho} - \lambda_H\rho) + \int_0^\infty d^3\rho\rho^{-1} \exp(+i\mathbf{Q}\cdot\boldsymbol{\rho} - \lambda_H\rho) \right\} \ . \tag{26.A3}$$

Integrating over θ and ϕ and rearranging, then

$$\begin{aligned} L_H(Z) &= (2\pi/iQ)\exp(i\mathbf{Q}\cdot\mathbf{R}) \int_R^\infty d\rho[\exp(-i\mathbf{Q}\cdot\boldsymbol{\rho}) - \exp(+i\mathbf{Q}\cdot\boldsymbol{\rho})]\exp(-\lambda_H\rho) \\ &= [4\pi/(\lambda^2+Q^2)][\cos QR - (\lambda_H/Q)\sin QR]\exp[i\mathbf{Q}\cdot\mathbf{R} - \lambda_H R] \ . \end{aligned} \tag{26.A4}$$

Noting that $\mathbf{R} = \mathbf{R}_e + \mathbf{Z}$ and defining

$$F(Q) = \cos QR_e - (\lambda_H/Q)\sin QR_e \tag{26.A5}$$

and

$$G(Q) = -\sin QR_e - (\lambda_H/Q)\cos QR_e \quad , \tag{26.A6}$$

then

$$\mu_H = b_H\langle \chi_0|(F\cos QZ + G\sin QZ)\exp[i\mathbf{Q}\cdot\mathbf{Z} - \lambda_H Z]|\chi_n\rangle \quad , \tag{26.A7}$$

where

$$b_H = [-2me^2Z_H'/\hbar^2(\lambda_H^2+Q^2)]\exp[i\mathbf{Q}\cdot\mathbf{R}_e - \lambda_H R_e] \ . \tag{26.A8}$$

Defining

$$M_\pm = \langle \chi_0|\exp[-\lambda_H Z + i(\mathbf{Q}\cdot\mathbf{Z} \pm QZ)]|\chi_n\rangle \ , \tag{26.A9}$$

$$\mu_H = b_H[(F(Q)(M_+ + M_-)/2 + G(Q)(M_+ - M_-)/2i] \ . \tag{26.A10}$$

Consider the $\Delta n = 1$ vibrational transition,

$$\chi_0(Z) = \sqrt{\alpha\pi}\ \exp(-\alpha^2 Z^2/2) \qquad (26.A11)$$

$$\chi_1(Z) = \sqrt{\alpha/2\pi}\ \alpha Z \exp(-\alpha^2 Z^2/2) \quad , \qquad (26.A12)$$

therefore

$$M_\pm = (\alpha/\sqrt{2})\langle\exp(-\alpha^2 Z^2/2)|\exp[-\lambda_H Z + iQZ(\cos\theta_0 \pm 1)]|$$
$$\times\ \alpha Z \exp(-\alpha^2 Z^2/2)\rangle \quad , \qquad (26.A13)$$

where θ_0 is defined as the angle between the scattering vector **Q** and the vibrational displacement vector Z. Integrating over Z

$$M_\pm = M_0[\lambda_H - iQ(\cos\theta_0 \pm 1)]\exp\{\pm 2[(Q/2\alpha)^2 + i\lambda Q/4\alpha^2]\} \quad , \qquad (26.A14)$$

where

$$M_0 = -\sqrt{\pi/2}\ \exp\left\{\left[(\lambda_H/2\alpha)^2 - (Q/2\alpha)^2(1 + \cos^2\theta_0)\right] - i(\lambda_H Q/2\alpha^2)\cos\theta_0\right\} \ . \qquad (26.A15)$$

For $Q^2/4\alpha^2 \ll 1$ and $\lambda Q/4\alpha^2 \ll 1$

$$M_\pm = -\sqrt{\pi/2}[\lambda_H - iQ(\cos\theta_0 \pm 1)] \qquad (26.A16)$$

and

$$\mu_H = C(Q)[F(\lambda_H - iQ\cos\theta_0) - GQ] \ . \qquad (26.A17)$$

Therefore

$$|\mu_H|^2 = |C(Q)|^2[F^2(\lambda_H^2 + Q^2\cos^2\theta_0) + G^2Q^2 - 2FGQ\lambda]$$
$$= |C(Q)|^2\left\{\cos^2\theta_0[Q\cos QR_e - \lambda_H \sin QR_e]^2\right.$$
$$\left. + [2\lambda\cos QR_e + Q(1 + \lambda_H^2/Q^2)\sin QR_e]^2\right\} \quad , \qquad (26.A18)$$

where

$$|C(Q)|^2 = \left[\sqrt{2}me^2\alpha Z_H'/\ ^2(\lambda^2 + Q^2)\right]^2 \exp(-2\lambda_H R_e) \ . \qquad (26.A19)$$

Acknowledgements. We should like to acknowledge the many valuable discussions with J. Anderson, J.W. Davenport, D.J. Frankel and G.J. Lapeyre. Experiments done at the CRISS facility which is supported by NSF Grant DMS-8309460, were supported in part by NSF grant DMR-8205581.

References

26.1 S. Andersson, J.W. Davenport: Solid State Commun. **28**, 677 (1977)
26.2 F. Stucki, J. Anderson, D.J. Frankel, G.J. Lapeyre, H.H. Farrell: Surf. Sci. (1984)
26.3 J.C. Bean, T.T. Sheno, L.C. Feldman, A.T. Fiory, R.J. Lynch: Appl. Phys. Lett. **44**, 102 (1984)
26.4 J.C. Bean, L.C. Feldman, A.T. Fiory, S. Nakahara, I.K. Robinson: J. Vac. Sci. Technol. A**2**, 436 (1984)

26.5 J.A. Schaefer, J.Q. Broughton, J.C. Bean, H.H. Farrell: To be published
26.6 J.Q. Broughton, J.A. Schaefer, J.C. Bean, H.H. Farrell: To be published
26.7 H.H. Farrell, J.A. Schaefer, J.Q. Broughton, J.C. Bean: To be published
26.8 H. Ibach, H.D. Bruchmann, H. Wagner: Solid State Commun. **42**, 457 (1982)
26.9 Y.J. Chabal, S.B. Christman: Phys. Rev. B**29**, 6974 (1984)
26.10 J.W. Davenport, W. Ho, J.R. Schrieffer: Phys. Rev. B**17**, 3115 (1978)

27. Structure and Temperature-Dependent Polaron Shifts on Si(111) (2 × 1)

C.D. Chen, A. Selloni*, and E. Tosatti

International School for Advanced Studies, Trieste, Italy
*Dipartimento di Fisica, Università La Sapienza, Roma, Italy

27.1 Introduction

Recently *Demuth* et al. [27.1] studied the temperature dependence of ultraviolet photoemission (UPS) and electron-energy-loss (EELS) spectra of Si(111)(7 × 7). They found that significant temperature-dependent changes occur in occupied surface states and their transitions, thus suggesting important electron-phonon coupling at the surface.

We have independently carried out a first attempt at understanding theoretically some effects produced by coupling of the surface state electrons (and holes) to the vibrating surface state lattice [27.2]. To be specific, and also because of its high current interest, we haven chosen the Si(111)(2× 1) reconstructed surface as our working example. Since at least two widely discussed models – the buckling model [27.3] and the π-bonded chain model [27.4] – have been proposed for this surface, we consider both of them. As it turns out, these two cases are different enough to give also a broad idea of what would happen in a more general case.

For both the buckling and chain models we start by constructing a model Hamiltonian and fix its parameters by requiring a reasonable comparison of the resulting band structure with known experimental and/or theoretical results. Both of our model Hamiltonians are strictly one-electron plus coupling to a surface lattice. While of course electron-electron interactions may often be relevant in a real situation, they are not an essential ingredient of the physical effects we want to describe and have thus been dropped. Two provisions must, however, be made in this respect. One is that the one-electron or hole states to be considered must always be energetically close enough to the gap – or the Fermi energy in a metal – that their lifetimes, due to electron-electron interactions, are long enough for any lattice relaxation to play a role. The second provision is that we do in fact reintroduce some effects of electron-electron interactions when dealing with electron-hole pairs bound to form an exciton.

What we study is the energy shift and the lattice deformation when one extra electron or extra hole is injected into a surface state. This state, i.e., a surface state electron (hole) plus its accompanying surface lattice deformation, is what we call a surface state polaron. A surface state polaron will also build up around a bound electron-hole pair, i.e., a surface state exciton. This situation, typically for optical absorption, is of course not just the linear superposition of the polarons of a free electron and a free hole and requires a separate calculation. Finally, the line shape of the optical absorption between surface states is calculated. A study of this quantity and of its temperature dependence is probably the most direct way to detect surface polaron effects.

27.2 Surface State Polarons in the Buckling Model

A buckled (nonmagnetic) model of Si(111)(2×1) can be described by the single-particle Hamiltonian

$$H = \sum_{n,i} \varepsilon_{ni} |ni\rangle\langle ni| + t \sum_{ni,mj}{}' [|ni\rangle\langle mj| + h.c.] \quad , \qquad (27.1)$$

where $|n,i\rangle$ is the dangling-bond (DB) orbital carried by the ith atom ($i = 1,2$) in the nth unit cell, ε_{ni} is the corresponding energy, and t is the nearest-neighbor hopping integral (the prime in the 2nd summation indicates restriction to nearest-neighbor pairs). This model has two bands separated by a gap which increases with the buckling magnitude. Only the lowest band is filled at T = 0 K. The electron-lattice coupling is present through the dependence of the on-site DB energy ε_{ni} upon the atomic coordinate (neglecting additional Fröhlich-like couplings). The simplest picture of a DB orbital is a combination of s and p_z wave functions, with coefficients which depend on the distance H_{ni} of the atom from the 2nd atomic plane. Correspondingly,

$$\varepsilon_{ni} = \varepsilon_p - \frac{C}{2}\left(\frac{H_{ni}}{a}\right)^2 \quad , \qquad (27.2)$$

where $C = 12\,(\varepsilon_p - \varepsilon_s)$, ε_s and ε_p are the s and p atomic energies, and a is the surface lattice constant (a = 3.85 Å).

We first of all find the T = 0 K equilibrium positions of the surface atoms which minimize the total electronic plus elastic energy. The first is obtained from (27.1), while the latter is parameterized in terms of a single force constant, whose value can be roughly determined using available experimental and/or theoretical results [27.2]. We find $H_1 = 0.99$ Å and $H_2 = 0.66$ Å, where 1 (2) is the raised (lowered) atom, as in Fig.27.1. Correspondingly, the electronic charge is almost completely localized on type 1 atoms.

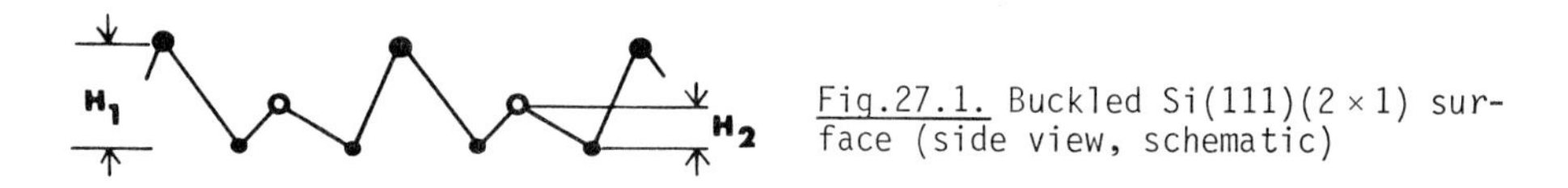

Fig.27.1. Buckled Si(111)(2 × 1) surface (side view, schematic)

Consider next an extra electron added to the system, otherwise at equilibrium. In the absence of coupling to the lattice the extra electron is in a Bloch state of the empty upper band. When the coupling to the lattice is turned on, the electron is subjected to two competing tendencies: one towards delocalization, so as to minimize the kinetic energy; another towards localization via a lattice distortion, which reduces the electron on-site energy. An efficient characterization of the strength of the electron-phonon coupling is provided by the value of the Huang-Rhys factor [27.5,6] ($S \lesssim 1$ for weak coupling and $S \gg 1$ for strong coupling). For one extra electron on the buckled surface $S_e = 3.6$, which corresponds to a mass enhancement factor $\exp(S_e) \sim 40$ and thus to strong coupling. Even stronger coupling is predicted for an extra hole, for which the calculated Huang-Rhys factor is $S_h = 8.3$. The mass enhancement factor is then $\exp(S_h) \sim 4000$, and the hole is self-trapped.

Similarly we calculate the localized lattice distortion surrounding an electron-hole pair excitation, taking care to include the electron-hole interaction as part of the total energy to be minimized. The electron and hole are found to be strongly localized on nearest-neighbor sites, with the electron (hole) on an atom of type 2 (1). Correspondingly there is a strong relaxation of these atoms: from $H_2 = 0.66$ Å to 0.73 Å for the electron, and from $H_1 = 0.99$ Å to 0.86 Å for the hole. With the lattice frozen in its ground-state configuration, the local single-particle gap is found to be 1.02 eV and the exciton binding energy 0.55 eV, resulting in a "vertical" transition energy of 0.47 eV. Our parameters have been chosen precisely so that this energy, which gives the position of the absorption peak according to the Franck-Condon principle, is close to the absorption peak energy of *Chiarotti* et al. [27.7]. For the corresponding emission process we find an energy of almost zero, implying an extremely large Stokes shift.

In the adiabatic and Condon approximation the absorption line shape for transitions from the electronic ground state (0) to the exciton state (1) is given by [27.5]

$$I(E) = I_e \sum_k P_k^0 \sum_\ell |\langle\chi_k^0|\chi_\ell^1\rangle|^2 \delta(\varepsilon_{1,\ell} - \varepsilon_{0,k} - E) \quad , \tag{27.3}$$

where χ_k^0 and χ_ℓ^1 are the vibrational wave functions for the electronic ground and excited states, with total quantum numbers k and ℓ respectively, P_k^0 is the probability of the state χ_k^0 at thermal equilibrium, and I_e is the electronic squared matrix element. The calculated line shapes for various temperatures are displayed in Fig.27.2 in the form of histograms, the vertical lines being approximately the zeros of the δ-function argument in (27.3). The main lines are accompanied by satellites forming a fine structure which becomes increasingly richer with increasing temperature. These fine-structure oscillations, due to coexisting ground-state and excited state vibrations of different frequencies, may in reality be washed out by finite phonon lifetimes. The line-shape envelope is asymmetrical and Poisson-like at

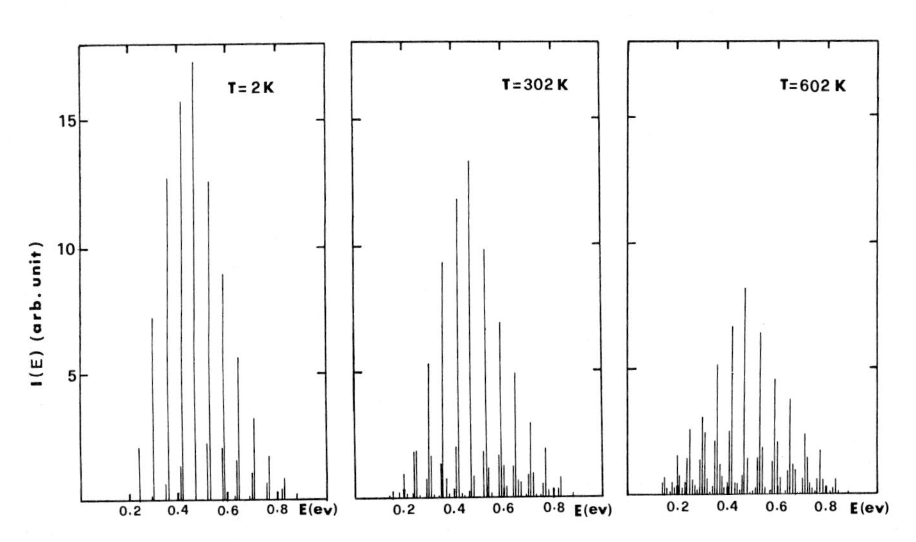

Fig.27.2. Calculated optical absorption for the buckled surface

low temperature, and evolves slowly towards a Gaussian shape with increasing temperature, as in usual strong coupling situations [27.5]. There is no detectable shift of the peak position within the accuracy of our calculations.

27.3 Surface State Polarons in the π-Bonded Chain Model

In the π-bonded chain model the surface atoms are each bonded to two other surface atoms and form zigzag chains along the $[1\bar{1}0]$ direction, similar to those occurring on the Si(110) surface [27.4]. We assume that the ground-state configuration of the surface is characterized by uniformly dimerized chains, with alternating short (contracted) and long (stretched) bonds [27.8], as indicated in Fig.27.3. The situation is thus very similar to that of a Peierls-distorted quasi-one-dimensional system, particularly polyacetylene [27.9]. Contrary to the buckling model, no charge transfer occurs between DB's so that the surface ground state is in this case purely covalent. To describe the electronic structure of this model we assume the one-electron Hamiltonian

$$H = \varepsilon \sum_{ni} |ni\rangle\langle ni| + \sum_{n} t_{1n}[|n1\rangle\langle n2| + \text{h.c.}]$$

$$+ \sum_{n} t_{2n}[|n1\rangle\langle n-1,2| + |n2\rangle\langle n+1,1|] \quad . \qquad (27.4)$$

Here ε is the DB on-site energy (the same for all DB's, t_{1n} is the hopping integral between the two DB's connected by a short bond in the cell n, t_{2n} is the hopping integral between the two DB's connected by a long bond in neighboring cells along a given chain. Energy minimization, again performed using reasonable surface elastic constants [27.2], yields $\Delta d_1 = -0.194$ Å and $\Delta d_2 = 0.065$ Å for the contractions and expansions with respect to the ideal (bulk) nearest-neighbor distance $d_0 = 2.35$ Å.

Injection of an extra electron or an extra hole (they are symmetric in this model) gives rise to a local reduction of the dimerization amplitude, resulting in an intermediate-coupling polaron (the Huang-Rhys factor is $\sim$2). An accurate evaluation of polaron binding energies and radii is quite difficult for this case. Approximate calculations yield $\Delta p = 0.05$-0.06 eV for the binding energy and $r_p \sim 2$-3 unit cells for the polaron radius. These

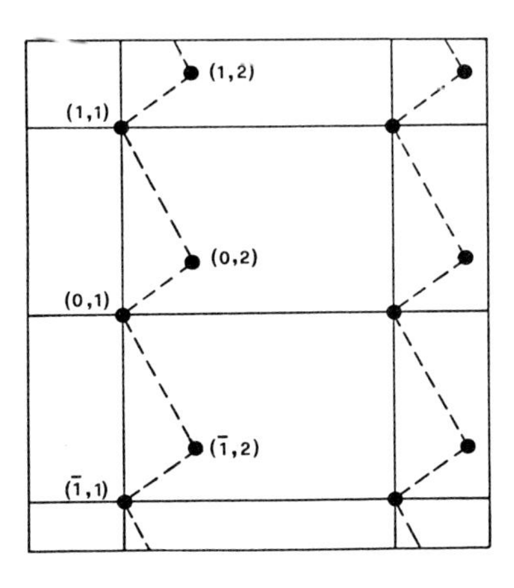

Fig.27.3. Dimerized chains on Si(111)(2 × 1) (top view, schematic)

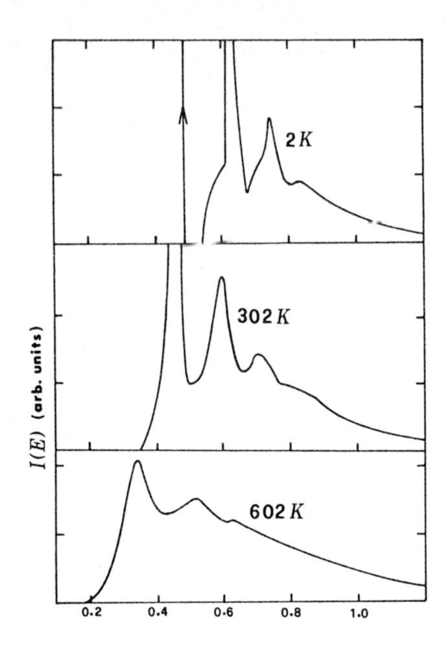

Fig.27.4. Calculated optical absorption for the dimerized chain model of Si(111)(2 × 1). The lowest peak corresponds to the zero-phonon line

values of r_p are also of the same order as the calculated exciton radius when the lattice is frozen in the ground-state configuration. In such cases a strong interference between the electron-hole and electron–lattice interactions can occur, so that the two terms should be treated simultaneously and on the same footing. We again use an approximate scheme, which yields $\Delta_{ex} \sim 0.2$ eV for the exciton-polaron binding energy.

We calculate the absorption line shape for this intermediate coupling model using [27.6]

$$I(E) = \frac{I_e}{\pi} \frac{\Gamma_{ex}(E)}{[E - \varepsilon_0 - \Delta_{ex}(E)]^2 + \Gamma_{ex}^2(E)} \quad , \tag{27.5}$$

where ε_0 is the exciton energy in the frozen lattice, while $\Delta_{ex}(E)$ and $\Gamma_{ex}(E)$ are the real and imaginary parts of the exciton self-energy caused by coupling to the surface lattice. Our calculated absorption spectra at various temperatures are shown in Fig.27.4. The overall shape of the spectra retains the typical one-dimensional character which is appropriate for the present model. The various peaks are multiple phonon structures which can be interpreted as indirect transitions involving phonon emission and/or absorption. The position of the first peak, which we identify with the peak observed experimentally in [27.7], shifts from 0.49 eV at $T = 0$ to 0.45 eV at $T = 300$ K and 0.35 eV at $T = 600$ K, while the corresponding halfwidth varies from 1.5×10^{-5} eV to 4.4×10^{-2} eV and 0.13 eV.

27.4 Discussion and Conclusions

The results in Figs.27.2,4 clearly illustrate the very different temperature-dependent changes of the optical absorption spectra which ought to be expected for the buckling and π-bonded chain models of Si(111)(2 $\times$ 1) (note that an additional T-dependent shift, due to gradual change of the reconstruction magnitude, should be added before direct comparison with experiment [27.2]). For the buckling model (strong coupling) the absorption is similar to that in color centers, which implies a weak to negligible blue shift with increasing T, associated with a linewidth increasing asymptotically as $\sqrt{T}$ [27.5]. In the alternative case of intermediate coupling to π-bonded chains, the absorption mechanism is more akin to that of bulk Si: transitions occur between a ground state and a fully relaxed excited state. The peak position shifts towards the red with increasing T, as in bulk Si, due chiefly to the usual Fan mechanism of increasing self-energy with T [27.10]. Line-shape effects are mostly due to the electronic band dispersion, which is large in this case, but also to phonons which have been assumed to be – but need not be – Einstein-like.

Summarizing, the main purpose of this note has been to discuss the relevance of polaron effects on electrons that belong in surface states. To this end, the 2 $\times$ 1 reconstructed surface of Si(111) has been selected; in particular the two most popular models for this reconstruction, i.e., the buckling and the π-bonded chain models have been considered. Polaron effects are found to be very important in either model, and quantitatively more than one order of magnitude larger than in bulk silicon. Since surface-state transport will probably never be measurable, the main impact of surface state polarons should be on the spectroscopy of surface states. We have here discussed optical absorption in some detail. Effects of surface state polarons on other surface state spectroscopies, such as luminescence or UPS or voltage-dependent scanning tunneling microscopy, have been outlined in [27.2], where a more detailed account of our work can also be found.

References

27.1 J.E. Demuth, B.N.J. Persson, A.J. Schell-Sorokin: Phys. Rev. Lett. **51**, 2214 (1983)
27.2 C.D. Chen, A. Selloni, E. Tosatti: To be published
27.3 D. Haneman: Phys. Rev. **121**, 1093 (1961)
27.4 K.C. Pandey: Phys. Rev. Lett. **47**, 1913 (1981); **49**, 223 (1982)
27.5 G. Chiarotti: In *Theory of Imperfect Crystalline Solids*, Trieste Lectures 1970 (IAEA, Vienna 1971)
27.6 Y. Toyozawa: In *The Physics of Elementary Excitations*, ed. by S. Nakajima, Y. Toyozawa, and R. Abe, Solid State Sci. Vol.12 (Springer, Berlin, Heidelberg, New York 1980)
27.7 G. Chiarotti, S. Nannarone, R. Pastore, P. Chiaradia: Phys. Rev. B**4**, 3398 (1971)
27.8 K.C. Pandey: Phys. Rev. B**25**, 4338 (1982)
27.9 W.P. Su, J.R. Schrieffer, A.J. Heeger: Phys. Rev. B**22**, 2099 (1980)
27.10 H.Y. Fan: Phys. Rev. **82**, 900 (1951)

III.4. Atom and Ion Scattering

28. Surface Structure Analysis by Atomic Beam Diffraction

J. Lapujoulade, B. Salanon, and D. Gorse

Service de Physique des Atomes et des Surfaces, Centre d'Etudes Nucléaires de Saclay, 91191 Gif-sur-Yvette Cedex, France

The main advantages of atomic beam diffraction for surface structure analysis are:

- high surface selectivity (no penetration),
- no surface damage (even for physisorbed layers),
- high sensitivity to structural disorder.

However, in order to get quantitative information a good knowledge of the atom surface potential and its relation to the surface atomic structure is needed. We review recent advances in the field and show how this goal is now almost achieved.

Then we review some recent applications with special emphasis on two systems studied in our laboratory by helium beam diffraction.

1) The thermal stability of stepped surfaces of copper. Evidence is given for a kind of roughening transition on such surfaces.
2) The reconstruction of W(100). We discuss how atomic beam diffraction casts some light on this controversial system.

28.1 Introduction

Atomic beam diffraction (ABD) has been known since the early thirties but it was only during the sixties that nozzle beam techniques have allowed its actual expansion. At this time it was already realized that ABD would be a very promising tool for surface structure analysis. The expected advantages were:

- high surface selectivity since low-energy atoms (< 0.2 eV) do not penetrate at all into the solid;
- a very high sensitivity to structural disorder in the first layer as a consequence of the high surface selectivity;
- no surface damage even for weakly bound adsorbed species like physisorbed adatoms;
- ability to probe hydrogen adatoms directly.

However, much had to be done before achieving this goal. More precisely, it was necessary to show that diffraction patterns could be unambiguously related to the surface atomic structure using an easily tractable model. Important progress has been made in this direction in the last few years, making it now possible to use ABD for a quantitative analysis of surface structure.

In this paper we shall review the present situation and expose two recent applications: one is the study of the thermal stability of copper stepped surfaces, the other is an attempt to cast some new light on an old but controversial problem, the reconstruction of W(100).

28.2 Models for Atomic Beam Diffraction

We cannot cover the whole subject of ABD within the limits of this paper so we shall restrict ourselves to the diffraction of helium from metal surfaces. Note that Ne diffraction behaves in a similar way.

A beam of monoenergetic parallel helium atoms is equivalent to a plane wave whose wave vector **k** is given by the De Broglie relation

$$\frac{\hbar^2 |k|^2}{2m} = E_i$$

where E_i is the kinetic energy of the atom, and m, the mass of the atom.

For thermal energies ($20 < E < 200$ meV) the corresponding wavelength ranges between 0.1 and 1 Å, which is very well suited for crystal surface analysis. This plane wave experiences a surface potential $V(\mathbf{R},z)$ (**R** parallel coordinate, z normal coordinate).

For a perfect surface the potential is periodic in **R** and the Floquet (or Bloch) theorem states that the diffracted waves are given by

$$\psi_r(\mathbf{R}) = \sum_{\mathbf{G}} A_{\mathbf{G}} \exp(i\mathbf{G}\cdot\mathbf{R}) \quad , \tag{28.1}$$

where **G** is the reciprocal surface lattice vector.

Then the diffraction pattern consists of δ-function diffraction peaks with intensities

$$I_{\mathbf{G}} = |A_{\mathbf{G}}|^2 \quad .$$

If the surface has defects which break the periodicity then the scattering pattern becomes more or less diffuse according to the shape of the correlation function of these defects. The diffracted intensity is now a continuous function of $\mathbf{Q} = \mathbf{K}_r - \mathbf{K}$, i.e. $I(\mathbf{Q})$.

There are two distinct steps in the theoretical treatment of the diffraction problem:

1) to construct a potential function $V(\mathbf{R},z)$ from an assumed surface atomic structure;
2) to solve the scattering equation for this potential in order to get the scattering intensities.

Historically the second step was the first to be developed, first using an empirical form for $V(R,z)$. It is only the availability of reliable experimental diffraction data which have stimulated recent work in order to solve the first step. However, for clarity, we shall use in the following the logical rather than the historical order.

28.2.1 Scattering Potential

The potential $V(\mathbf{R},z)$ which is schematically shown in Fig.28.1 is repulsive for small z and becomes attractive at large z. The attractive part is due to Van der Waals dispersion forces which are long range. Its asymptotic value for large z is known to behave like z^{-3}. Due to its long range this part of the potential is not very much affected by the crystal atomic structure and as a first approximation may be assumed to be independent of $\mathbf{R}$. On the contrary, the repulsive part is due to Pauli repulsion between the electron clouds of the atom and the metal, which are short-range forces. It is thus modulated by the atomic structure and this part is primarily responsible for the diffraction pattern. A decisive advance in this field was achieved in 1980 when *Esbjerg* and *Nørskov* [28.1] showed, using the effective medium approach, that the potential in the repulsive region is proportional to the local electron density $\rho(\mathbf{R},z)$ of the unperturbed solid

$$V(\mathbf{R},z) = \bar{\alpha}_{eff}\rho(\mathbf{R},z) \quad , \tag{28.2}$$

where the constant is $\bar{\alpha}_{eff} = 255$ eV a_0^3 [28.2].

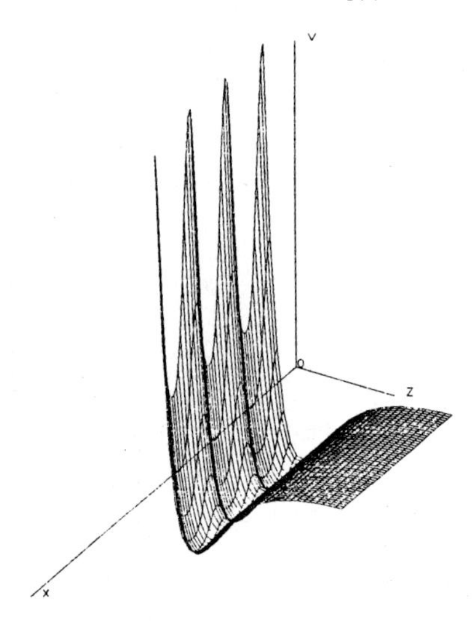

Fig.28.1. Helium-metal interaction potential

With this approximation the problem is now reduced to determining this electron density. Only few self-consistent calculations are available [28.3, 4], but a very useful heuristic approximation has been suggested by *Haneman* and *Haydock* [28.5]. They assume that ρ can be obtained from a simple superposition of free metal atom densities

$$\rho(\mathbf{R},z) = \sum_n \rho_a(\mathbf{R} - \mathbf{R}_n, z - z_n) \quad , \tag{28.3}$$

where $(\mathbf{R}_n, z_n)$ are the position coordinates of the metal atoms.

Besides its obvious tractability, this method presents the advantage that very refined calculations are available for the atomic function including nonlocal exchange correlation terms which are expected to be important at the distance considered here (~ 6-8 a.u.). Nonlocality has not yet been taken into account in the self-consistent calculations.

The main interest of the calculations based on (28.2,3) is that they do not contain any adjustable parameters; it is then possible to predict the potential from any guessed configuration of the surface atoms. This is not the case for the calculation by *Harris* and *Liebsch* [28.6] which has been very successful in predicting the diffraction pattern of (110) faces of metals [28.7] but relies upon an adjustable parameter for the corrugation.

28.2.2 Scattering Intensities Calculations

For a long time the solution of the scattering equations was restricted to a very simple model potential: the "hard corrugated wall" (HCW) defined as

$$\begin{aligned} V(R,z) &= 0 \qquad \forall\ z > \phi(\mathbf{R}) \\ &= \infty \qquad\quad\ < \phi(\mathbf{R}) \end{aligned} \tag{28.4}$$

where $\phi(\mathbf{R})$ is the corrugation function.

There are many approximate solutions [28.8,9] of this problem and the exact numerical solution is very easy to obtain [28.10,11]. However, the deficiencies of this potential are obvious:

- there is no attractive part
- the real potential is not hard but exponentially decays at the same rate as the electron density of the solid, i.e., $\exp(-\chi z)$, where $\chi \sim 2$ Å^{-1}. So this potential is not expected to give better than a qualitative description of the diffraction patterns.

Real soft potentials as given by (28.2) can be directly handled by solving numerically the integral form of the Schrödinger equation by the so-called closed-coupling method [28.12,13]. In principle this method can be applied to any kind of structure. But for a complex surface unit cell limitations can arise due either to the computing time or to the memory size required.

Another method has been developed by *Armand* and *Manson* [28.14] which is based upon an iteration of the T matrix. It has three advantages:

- it is quite cheap for computing,
- an analytical form for resonance line shapes is readily obtained which needs only the numerical calculation of one point,
- temperature effects may be introduced.

But its limitation is due to a lack of convergence for large corrugation structures.

28.3 Comparison with Experimental Data

There are now very good experimental data for helium diffraction on (110) faces of Ni [28.15], Cu [28.16], Ag [28.17], Pd [28.18] and for higher index face (stepped surfaces) of copper [28.19,20]. (The structure of these faces is shown in Fig.28.2.) The potential is determined in two steps.

1) The attractive part and the well region are adjusted in order to fit the bound state levels given by selective adsorption data. If a sufficient number of these levels is known, this can be achieved quite accurately [28.21].

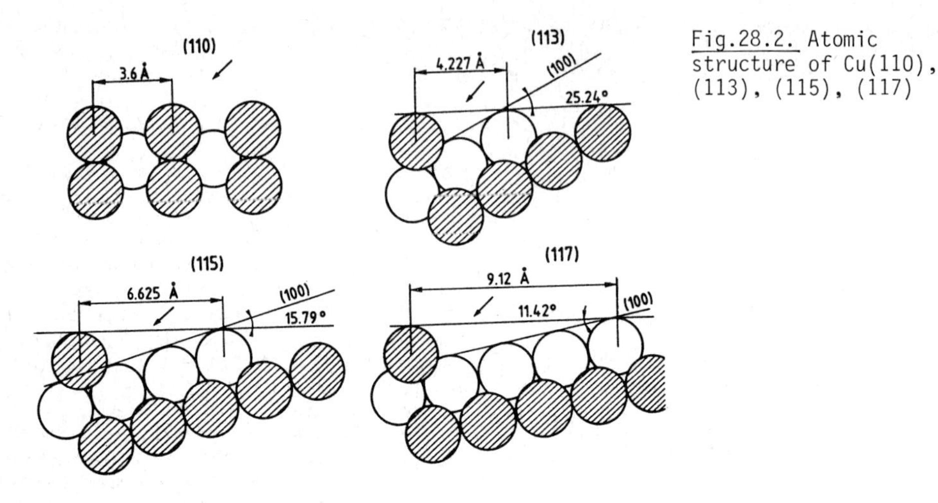

Fig.28.2. Atomic structure of Cu(110), (113), (115), (117)

2) The repulsive part (slope and corrugation) is then adjusted (generally by trial and error) in order to give the correct intensity of diffraction peaks. Note that it is essential for the uniqueness of this determination to rely upon data corresponding to a wide range of the incidence angles and incident energy. The early success of the HCW model was due only to too limited experimental data. The availability of more extensive experimental results has shown these simple models to be inadequate. For fitting the experimental data, flexible parametric forms of the potential such as the modified Morse corrugated potential of *Salanon* et al. [28.16] or the *Liebsch* and *Harris* form [28.7] are very useful.

Comparison of these results with the model of Esbjerg and Nørskov using superposition of atomic charge densities shows that this model gives very good qualitative trends but always has too large a corrugation amplitude. The discrepancy is by about a factor 1.5-2 for the various (110) faces, the agreement becoming better for the more corrugated stepped surfaces of copper.

We have assumed here that the Esbjerg and Nørskov theory includes correlation effects in the repulsive potential so that it gives the total potential. If this assumption is not accepted an attractive component has to be added, which increases the total potential corrugations and also the discrepancy with experiment. This point has been discussed by *Barker* et al. [28.22] who also pointed out that it is essential here to use atomic wave functions valid at large distances, i.e., including nonlocal exchange-correlation terms like the *Clementi* and *Roetti* data [28.23]. Using the less accurate *Herman* and *Skillman* [28.24] table would result in a larger discrepancy.

It is not yet clear whether the residual discrepancy is due to the effective medium approximation made by Esbjerg and Nørskov (a *local* assumption) or to a failure of the atomic charge density superposition to account correctly for the actual solid electron density. Nevertheless this discrepancy can be empirically removed by adjusting the Esbjerg and Nørskov constant. For instance, using $\alpha_{emp} = 600$ eV.a_0^3 gives an acceptable fit for all the copper surfaces studied [28.20]. It is important that such a fit has been achieved with copper for widely different structures. The agreement obtained for other metals for the (110) faces only is less conclusive since in this case there is only one Fourier component in the potential to be fitted. This adjustment of the Esbjerg and Nørskov constant is only an heuristic rationalization of the data which does not rely upon a physical basis. The fact that the empirical constant α_{emp} is close to the value given by *Esbjerg* and *Nørskov* in their first paper [28.1] ($\alpha = 750$ eV.a_0^3) but recognized as being in error in a subsequent publication [28.2] is certainly meaningless.

Thus some progress must still be made before the ab initio calculation of helium scattering from any atomic structure can be considered as completely accurate. However, in the present state of the theory the agreement is good enough to decide between hypothetical structures which are not too close. Note that the most difficult situation is a close-packed surface such as (100) where the smoothing of the ion core by the electrons is more pronounced. As pointed out above, the agreement is expected to be better for a more open structure. This is especially true for adsorbate structures which we have not discussed here; a good example is the O-Ni(001) system recently analyzed by *Batra* and *Barker* [28.25].

Finally we should like to emphasize here that even if a very realistic shape of the potential is needed for an accurate calculation of the diffraction pattern, it is also true that in many cases very important semiquantitative information can also be drawn from very simple models such as the HCW treated in the eikonal approximation, as shown in the following application.

28.4 Thermal Roughening of Copper Stepped Surface

In the preceding section the crystal lattice was assumed perfectly immobile which gives purely elastic scattering. In fact the thermal motion of the lattice introduces inelastic scattering through kinetic energy transfer between the incident atom and the crystal phonons. One consequence is a thermal attenuation of the elastic peaks which looks like a Debye-Waller factor effect. An example is given in Fig.28.3 for the specular scattering of helium from a close packed surface Cu(100) [28.26]. The analogy with the conventional Debye-Waller factor as used for X-ray or neutron scattering must be considered here with great care since the softness of the He surface potential introduces

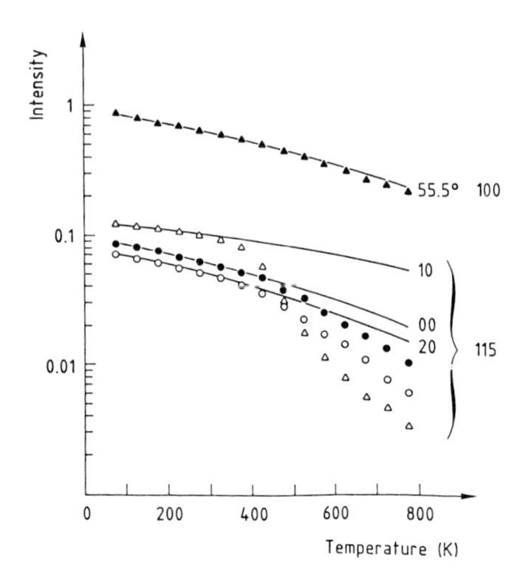

Fig.28.3. Temperature dependence of diffraction peak intensities for Cu(100) and Cu(115). Crosses are experimental points and the solid lines are theoretical calculations [28.26]

substantial complications [28.27-31]. It is not possible, within the limits of this paper, to give even an overview of this question. But even if the theory is, to date, far from complete, there are models available which are accurate enough to rationalize the thermal effect on close packed surfaces using a few adjustable parameters, as shown in Fig.28.3.

Data for a stepped surface Cu(115) [28.26] are also included in this figure. The behavior is now quite different. Above a threshold temperature the elastic intensity dramatically decreases. This change of regime cannot be explained by a sudden increase of the elastic component nor by an enhancement of the crystal vibration anharmonicity. Thus we are forced to assume some structural change.

It is well known from statistical mechanics that a linear step on a surface is not stable. The stability of an array of parallel steps is necessarily due to some repulsive interaction between the nearest steps. Then it could be expected that such a system can undergo a roughening transition above some critical temperature T_R [28.32]. This roughening occurs because kinks appear along the step in an increasing number as temperature rises. A crude model has been given recently by *Villain* et al. [28.33]. They assumed the energy for creating a pair of kinks to be 2 W_0 and the repulsive energy between the displaced step and its neighbors to be w_n per unit step length. It is then shown that the transition occurs when

$$\frac{w_n}{T_R} \exp(W_0/T_R) \sim 1 \quad .$$

As w_n decreases when the distance between steps increases, T_R is expected to behave similarly. This is confirm by the experimental data on Cu(113), (115), and (117). Treating the stepped surface as a hard wall in the eikonal approximation allows one to calculate the decrease of the elastic peak intensity due to the step roughening. One gets

$$\frac{I_{rough}}{I_{ordered}} = 1 - \exp[-2\langle u^2\rangle(1 - \cos q_x)] \quad ,$$

where

$$\langle u^2\rangle = 2\left(\frac{T}{w_n}\right)^2 \exp(-2W_0/RT)$$

is the mean-square deviation of the step position from its equilibrium and q_x is the variation of parallel momentum.

Thus, assuming that W_0 is given by a pairwise interaction potential

$$W_0 = \frac{\text{Cohesive energy}}{\text{Nb of nearest neighbors}} = \frac{3.5}{12} = 0.29 \text{ eV} \quad ,$$

w_n can be adjusted to fit the experimental data as shown in Fig.28.4. The agreement with experiment is reasonable in view of the crudeness of the model.

This is a very good example of the remarkable ability of the ABD technique to study lattice defects.

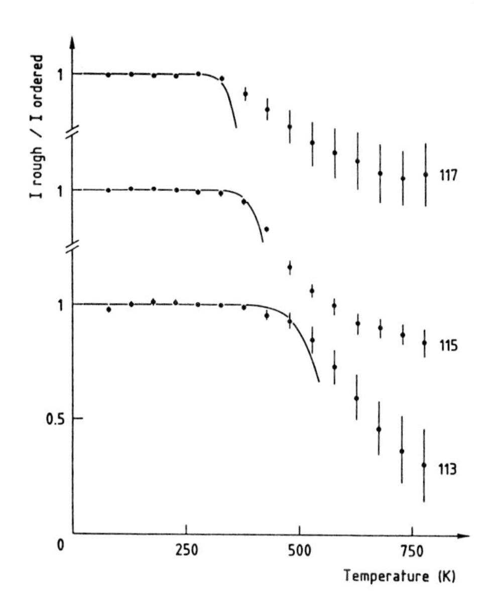

Fig.28.4. Plot of $I_{rough}/I_{ordered}$ vs T for Cu(113), (115), (117). Comparison with the model of Villain et al. [28.23]

28.5 The Reconstruction of W(100)

It has been established for at least a decade that W(100) undergoes a reversible structural change around T = 200 K [28.34]. This phase transition is now proved to be a clean surface effect [28.35,36]. LEED patterns show a reconstructed $(\sqrt{2}\times\sqrt{2})$ R45° structure at low temperature while the primitive (1×1) structure is recovered at high temperature. For further details the reader is invited to consult excellent reviews which are available [28.37,38].

In spite of much theoretical and experimental effort there is still some controversy about the structure of both the low- and high-temperature phases.

Low-T phase. The most commonly accepted model is that of *Debe* and *King* [28.39] which is schematized in Fig.28.5. It is a zigzag model with alternate tangential displacement of 0.324 Å and a common normal displacement of 0.38 Å. This model explains well the symmetries of the LEED pattern and was proposed in order to keep atoms in close contact along a hard sphere description. However LEED intensities analysis seems to indicate that the lateral displacement is not larger than 0.2 Å. Moreover, *Melmed* and *Graham* have suggested [28.40] that periodic vertical displacement can also be a convenient model.

High-T phase. Debe and King proposed on the basis of a LEED analysis that this phase is ordered with the same structure as a bulk (100) layer with simply a contraction of the interplanar spacing and eventually a slight lateral displacement. But this is questioned by *Barker* and *Estrup* [28.41], who favor a disordered structure.

Atomic beam diffraction is very well suited to cast some light onto this controversy. Indeed:

1) it is sensitive only to the topmost layer so that any disorder in this layer would easily be detected;

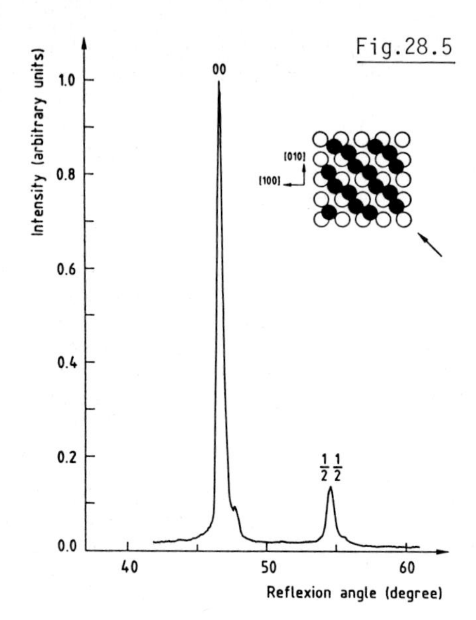

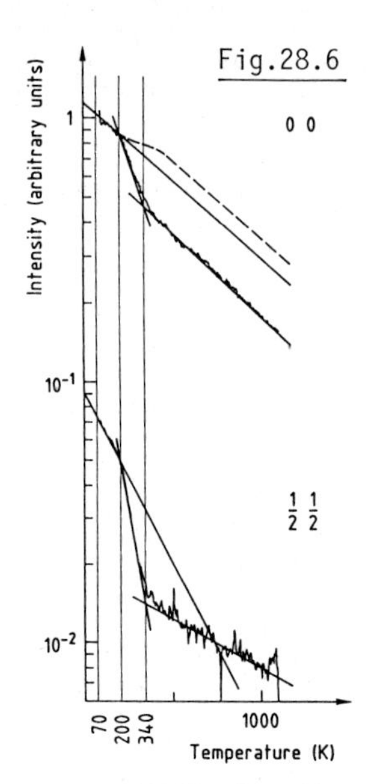

Fig.28.5. Typical diffraction pattern of He on W(100) - $E_i = 21$ meV, $\theta_i = 20.7^\circ$, T = 70 K. The insert shows the Debe and King model of W(100) reconstruction. The azimuth of the beam is indicated by the arrow

Fig.28.6. Temperature dependence of the 00 and ½ ½ diffraction peaks under the same conditions as in Fig.28.5

2) from the nature of the helium-metal interaction it is expected that vertical displacements would be more effective than horizontal ones in changing the corrugation.

Figure 28.5 shows a typical diffraction pattern taken at $E_i = 21$ meV, $\theta_i = 20.7^\circ$, T = 70 K in the azimuth shown by the diagram. This pattern displays a large specular peak and a half-order diffraction peak. Integral-order peaks and half-order peaks beyond order 1/2 are very small.

The dependence upon temperature of the specular and a half-order peak is shown in Fig.28.6. Both peaks behave quite similarly. Between 70 and 200 K they decrease according to the Debye-Waller factor. Then at 200 K a fast drop appears which ends at about 340 K. After that both peaks slowly decrease to a background value. No attempt has been made here to subtract the inelastic or incoherent background, so in this high-temperature range, the true variation of the elastic peaks is certainly faster than indicated by the curves. These results are very important since they strongly suggest that the high-temperature phase is disordered.

Indeed we have checked by an electron density calculation using the superposition model that the unreconstructed (100) face should be smoother than any reconstructed structure. Thus the specular peak should be larger for the (1×1) than for the $(\sqrt{2}\times\sqrt{2})$ R45° structure; we would then expect a behavior like the dashed line in Fig.28.5, which is actually not found.

Such behavior around 340 K is exactly the opposite of that observed by Debe and King in LEED and from which they conclude an ordered high-temperature structure.

Note also that the strong decrease of the specular peak is not compensated by the appearance of significant intensity in any other integral order peaks, which rules out the possibility of a more highly corrugated (1×1) ordered structure.

The low-temperature pattern has been analyzed in terms of the Esbjerg-Nørskov model with atomic charge density superposition for low-energy beams (20 and 63 meV). Provided the displacement is as large as 0.4 Å, a lateral displacement model explains the data but a 0.15 Å alternate vertical displacement agrees as well. Higher energy (up to 250 meV) patterns have been measured which show an enhanced corrugation as evidenced by the appearance of higher order peaks (1 and 3/2). The analysis of these patterns is in progress and we hope to be able then to decide between the possible structures.

In this case with a relatively smooth surface the choice of the empirical Esbjerg-Nørskov constant would probably affect the results. We have chosen the same constant which gives good results for copper. A change by a factor of two will not alter the above results very much but we have in progress a test of its value by a diffraction experiment on a tungsten face, W(112), which is known not to reconstruct.

28.6 Conclusion

We have shown that ABD is now a powerful tool for studying surface structures. There are still some improvements to be made in the theory of He-solid interaction but the present knowledge already allows very interesting studies.

The two examples described, the thermal roughening of copper stepped surfaces and the structure of W(100), concern results which are still preliminary. Nevertheless they give new physical insights in surface physics. Note that W(100) being a quite smooth close-packed surface is a rather severe test of the ability of ABD to elucidate surface structure.

More open surfaces such as chemisorbed layers or missing row reconstructions would be easier to study as they require less accuracy in the interaction potential knowledge [28.42].

Acknowledgment. We wish to thank very much here all those who have contributed to this paper: G. Armand, F. Fabre, A. Kara, Y. Lejay, M. Lefort, E. Maurel, from the C.E.N. Saclay, France, J. Villain from the C.E.N. Grenoble, France, and J.R. Manson, Clemson University, South Carolina, U.S.A.

References

28.1 N. Esbjerg, J.K. Nørskov: Phys. Rev. Lett. **45**, 807 (1980)
28.2 M. Manninen, J.K. Nørskov, M.J. Puska, C. Umrigar: Phys. Rev. B**29**, 2314 (1984)
28.3 D.R. Hamann: Phys. Rev. Lett. **46**, 1227 (1981)
28.4 E. Wimmer, A.J. Freeman, M. Weinert, H. Krakauer, J.R. Hiskes, A.M. Karo: Phys. Rev. Lett. **48**, 1128 (1982)
28.5 D. Haneman, R. Haydock: J. Vac. Sci. Tech. **21**, 330 (1982)

28.6 J. Harris, A. Liebsch: J. Phys. C Solid State Phys. **15**, 2275 (1982)
28.7 A. Liebsch, J. Harris, B. Salanon, J. Lajupoulade: Surf. Sci. **123**, 338 (1982)
28.8 U. Garibaldi, A.C. Levi, R. Spadacini, G.E. Tommei: Surf. Sci. **48**, 649 (1975)
28.9 N. Garcia: J. Chem. Phys. **67**, 897 (1977)
28.10 N. Garcia, N. Cabrera: Proc. 7th Int. Vac. Cong. and 3rd Int. Conf. on Sol. Surfaces, ed. by R. Dobrozemsky et al. (Vienna 1977) p.379
28.11 G. Armand, J.R. Manson: Phys. Rev. B18, 6510 (1978)
28.12 H. Chow: Surf. Sci. **62**, 487 (1977)
28.13 R.B. Laughlin: Phys. Rev. B**25**, 2222 (1982)
28.14 G. Armand, J.R. Manson: J. Physique **44**, 473 (1983)
28.15 K.H. Rieder, N. Garcia: Phys. Rev. Lett. **49**, 43 (1982)
28.16 B. Salanon, G. Armand, J. Perreau, J. Lapujoulade: Surf. Sci. **127**, 135 (1983)
28.17 A. Luntz: Surf. Sci. **126**, 695 (1983)
28.18 K.H. Rieder, W. Stocker: J. Phys. C**16**, L783 (1983)
28.19 J. Lapujoulade, Y. Lejay, N. Papanicolaou: Surf. Sci. **90**, 133 (1970)
28.20 D. Gorse, B. Salanon, F. Fabre, A. Kara, J. Perreau, G. Armand, J. Lapujoulade: Surf. Sci. (in press)
28.21 J. Perreau, J. Lapujoulade: Surf. Sci. **122**, 341 (1982)
28.22 J.A. Barker, N. Garcia, I.P. Batra, M. Baumberger: Surf. Sci. Lett. **141**, L317 (1984)
28.23 E. Clementi, C. Roetti: Atomic Data and Nuclear Data Tables **14**, 177 (1974)
28.24 F. Herman, S. Skillman: *Atomic Structure Calculations* (Prentice Hall, New York 1963)
28.25 I.P. Batra, J.A. Barker: Phys. Rev. B**29**, 5286 (1984)
28.26 J. Lapujoulade, J. Perreau, A. Kara: Surf. Sci. **129**, 59 (1983)
28.27 G. Armand, J.R. Manson: Surf. Sci. **80**, 532 (1979)
28.28 A.C. Levi, H.G. Suhl: Surf. Sci. **88**, 221 (1979)
28.29 N. Garcia, A.A. Maradudin, V. Celli: Philos. Mag. A**45**, 287 (1982)
28.30 J. Lapujoulade: Surf. Sci. Lett. **134**, L529 (1983)
28.31 G. Armand, J.R. Manson: Proceedings 17th Jerusalem Symposium: Dynamics on Surface, May 1984 (Dordrecht, The Netherlands, in press)
28.32 J.D. Weeks: In *Ordering in Strongly Fluctuating Condensed Matter*, ed. by T. Riske (Plenum, New York 1980), p.293
28.33 J. Villain, D. Grempel, J. Lapujoulade: Submitted to J. Phys. F
28.34 K. Yonehara, L.D. Schmitt: Surf. Sci. **25**, 238 (1971)
28.35 M.K. Debe, D.A. King: J. Phys. C: Solid State Phys. **10**, L303 (1977)
28.36 T.E. Felter, R.A. Barker, P.J. Estrup: Phys. Rev. Lett. **38**, 1138 (1977)
28.37 D.A. King: Phys. Scripta T**4**, 34 (1983)
28.38 J.E. Ingelsfield: In *The Chemical Physics at Solid Surfaces and Heterogeneous Catalysis*, Vol.1, Clean Solid Surfaces, ed. by D.A. King, D.P. Woodruff (Elsevier, Amsterdam 1981)
28.39 M.K. Debe, D.A. King: Surf. Sci. **81**, 193 (1979)
28.40 A.J. Melmed, W.R. Graham: Appl. Surf. Sci. **11**, 470 (1982)
28.41 R.A. Barker, P.J. Estrup: J. Chem. Phys. **74**, 1442 (1981)
28.42 T. Engel, K.H. Rieder: In *Structural Studies of Surfaces*, Vol.91, Springer Tracts in Modern Physics, ed. by G. Höhler (Springer, Berlin, Heidelberg, New York 1980) p.55

29. Structure Analysis of a Semiconductor Surface by Impact Collision Ion Scattering Spectroscopy (ICISS): Si(111) $\sqrt{3} \times \sqrt{3}$ R30°Ag

M. Aono, R. Souda, C. Oshima, and Y. Ishizawa

National Institute for Research in Inorganic Materials
Namiki 1-1, Sakura, Niihari, Ibaraki 305, Japan

A structural model for the Si(111)$\sqrt{3} \times \sqrt{3}$ R30° Ag surface is discussed on the basis of impact collision ion scattering spectroscopy (ICISS) experiments using a beam of Li^+ ions. Silver atoms at the surface are situated ~0.5 Å above the first silicon layer; their lateral positions have not yet been analyzed.

The purpose of this paper is to propose a structural model for the Si(111) $\sqrt{3} \times \sqrt{3}$ R30° Ag surface on the basis of impact collision ion scattering spectroscopy (ICISS) [29.1-7] experiments in which a beam of Li^+ ions is used.

A silicon (111) wafer, which was prepared following *Henderson* [29.8], was mounted on a two-axis sample manipulator. The wafer was cleaned in situ (~ 1×10^{-10}Torr; base pressure ~5×10^{-11}Torr) by electron-beam heating (~ 1250°C) from behind. A sharp 7×7 low-energy electron diffraction (LEED) pattern was observed. Silver was deposited onto the surface at room temperature until the intensity of an ICISS spectral peak due to the deposited silver saturated. The sample was then heated at ~300°C for ~20 s. A sharp $\sqrt{3} \times \sqrt{3}$ R30° LEED pattern was observed. Experiments of ICISS were done in the same manner as reported elsewhere [29.1-4] except that a beam of Li^+ ions was used.

Figure 29.1 shows the ICISS intensity of Li^+ ions scattered from silver atoms at the Si(111)$\sqrt{3} \times \sqrt{3}$ R30° Ag surface as a function of the polar and azimuthal angles of the ion incidence direction, α and ϕ, respectively; α is measured from the surface. The primary energy of Li^+ ions was $E_0 = 998$ eV. Hereafter, low-index azimuths will be indicated by [A], [B], [C], and [D] as defined at the top of Fig.29.1. As seen in Fig.29.1, clear shadowing effects are observed in azimuths [A], [B],[C], and [D] at α's smaller than ~ 12°.

Figure 29.2 also shows the ICISS intensity of Li^+ ions ($E_0 = 998$ eV) scattered from silver atoms at the Si(111)$\sqrt{3} \times \sqrt{3}$ R30° Ag surface, but the intensity is plotted as a function of α; measurements were made in azimuths [A], [B], [C], and [D], but the result for azimuth [D] is not shown because it was essentially the same as that for azimuth [C]. As seen in Fig.29.2, in all the azimuths [A], [B], and [C], only a single shadowing effect is observed with a shadowing critical angle of ~12° (the shadowing critical angle in azimuth [C] seems to be slightly larger than those in azimuths [A] and [B].

So far, several structural models [29.9-18] have been proposed for the Si(111)$\sqrt{3} \times \sqrt{3}$ R30° Ag surface, but a honeycomb model shown in Fig.29.3a has been regarded as the "preferred" model. This honeycomb model was first proposed by *Le Lay* et al. [29.11], on the basis of their measurement of the

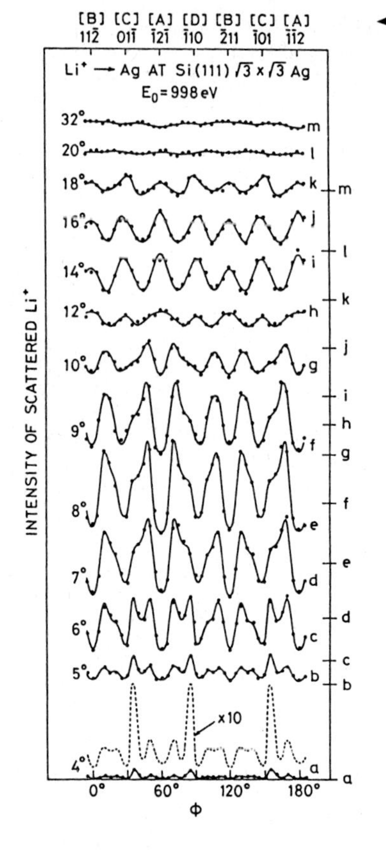

◄ Fig.29.1. The ICISS intensity of Li^+ ions scattered from silver atoms at the Si(111) $\sqrt{3}\times\sqrt{3}$ R30° Ag surface as a function of the polar and azimuthal angles of the ion incidence direction, α and ϕ, respectively; α is measured from the surface. The primary energy of Li^+ ions was 998 eV

Fig.29.2. The ICISS intensity of Li^+ ions scattered from silver atoms at the Si(111) $\sqrt{3}\times\sqrt{3}$ R30° Ag surface as a function of ion incidence direction measured from the surface, α. The measurements were made in three different azimuths [A], [B], and [C] defined in Fig.29.1

▼

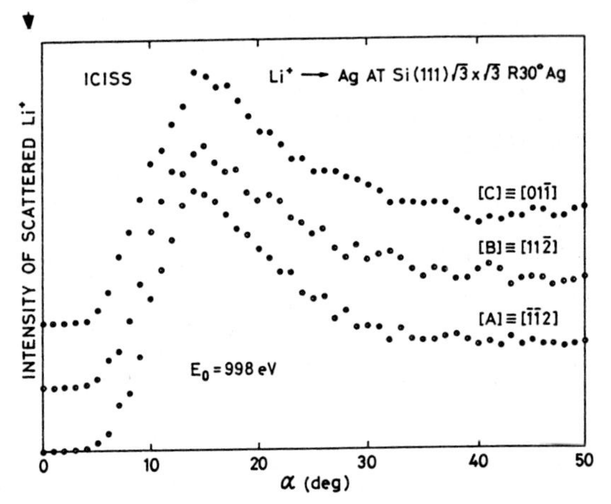

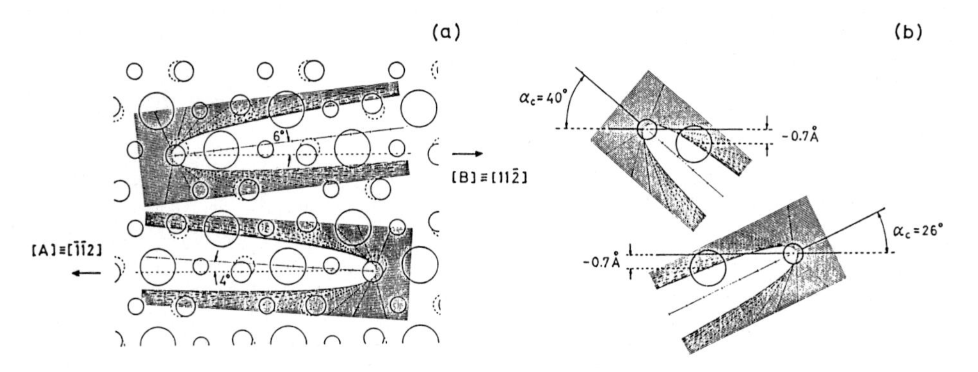

Fig.29.3a,b. Top (a) and side (b) views of the honeycomb model with buried silver atoms for the Si(111)$\sqrt{3}\times\sqrt{3}$ R30° Ag surface. The shadow cone of a silicon atom for Li^+ ions of 998 eV is also shown

coverage of silver at the surface ($\theta = 2/3$), although no detailed description was given. Later, *Saitoh* et al. [29.12,13] and *Terada* et al. [29.14] supported this honeycomb model and suggested that the silver atoms at the surface were buried under the first Si layer by ~0.7 Å, the first- and second-layer silicon atoms being displaced from their ideal positions. This honeycomb model with buried silver atoms was derived from their low-energy ion scattering (ISS) blocking experiments [29.12,13] and LEED analysis using a CMTA (constant momentum transfer averaging) approximation [29.14]. *Stöhr* et al. [29.15,16] reported that this model is consistent with the results of their surface EXAFS (extended X-ray absorption fine structure) experiments.

The honeycomb model with buried silver atoms, however, is inconsistent with the ICISS data shown in Figs.29.1,2. If the silver atoms are buried under the first silicon layer by 0.7 Å, every silver atom should be concealed by the shadow cone of its neighboring first-layer silicon atom at $\alpha < 40^{\circ}$ in directions near azimuth [B], Figs.29.3a,b. Similarly, in directions near azimuth [A], a similar shadowing effect should occur at $\alpha < 26^{\circ}$, Figs.29.3a,b. However, clear in Fig.29.2 (also Fig.29.1), no shadowing effect is observed until α decreases down to $\sim 12^{\circ}$ in both azimuths [A] and [B]. It is thus obvious that the honeycomb model with buried silver atoms is inconsistent with the ICISS data.

Recently, *Kono* et al. [29.17] modified the model on the basis of their X-ray photoelectron diffraction experiments. They claim that although the silver atoms are buried under the first silicon layer, the depth is only 0.2 ± 0.1 Å. A similar result was obtained by *Horio* and *Ichimiya* [29.18] from reflection high-energy electron diffraction experiments. However, this modified model is also inconsistent with the ICISS data; in this model, a shadowing effect should occur at $\alpha < 27^{\circ}$ (19°) in directions near azimuth [B] ([A]) in conflict with the ICISS data shown in Figs.29.1 and 2.

The ICISS data shown in Figs.29.1 and 2 and ICISS data on He^{+} ion scattering from silicon atoms at the surface (not shown) indicate that the silver atoms are situated above the first silicon layer by ~0.5 Å. However, it is not enough to displace the silver atoms upwards. In such a simple modification of the model, two shadowing effects should occur in azimuth [C] as follows. One half of the silver atoms is concealed by the nearest-neighbor silver atoms at $\alpha < 21^{\circ}$, and the remaining half of the silver atoms is concealed by the next-nearest silver atoms at $\alpha < 12^{\circ}$. Although the latter shadowing effect is observed in Fig.29.2, the former shadowing effect is not observed. This indicates that the silver atom are not arranged in the honeycomb structure.

Figure 29.4a shows the ICISS intensity of Li^{+} ions scattered from silicon atoms at the Si(111)$\sqrt{3} \times \sqrt{3}$ R30° Ag surface as a function of α. The measurements were made in azimuth [A]. In the case of Li^{+} ion scattering from silicon atoms, the intensity of scattered Li^{+} ions is contributed not only by single scattering from one or two surface layers but by multiple scattering from deeper layers. Figure 29.4b shows the atomic arrangement of the ideal silicon crystal in azimuth [A]. It is found that the several intensity minima observed in Fig.29.4a occur in directions corresponding to the crystal "aligned" axes indicated in Fig.29.4b by straight lines. Similar measurements were made also in azimuths [B] and [C], and similar results were obtained. It is thus found that in those directions which are parallel to the crystal "aligned" axes, the intensity of scattered Li^{+} ions necessarily shows a minimum. This indicates that silicon atoms at the Si(111)$\sqrt{3} \times \sqrt{3}$ R30° Ag surface are not displaced markedly from their ideal positions; this

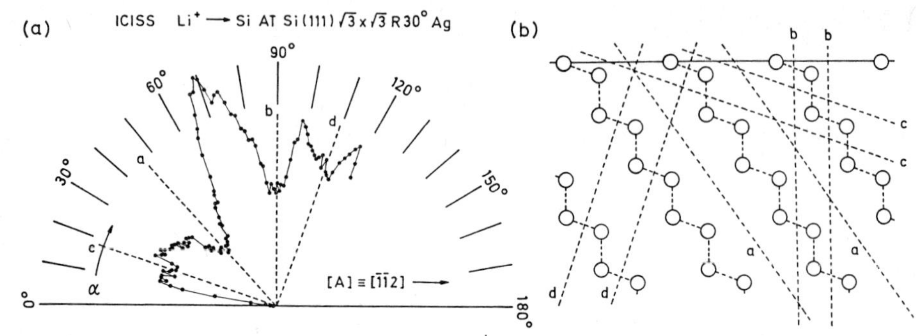

Fig.29.4. (**a**) The ICISS intensity of Li^+ ions scattered from silicon atoms at the Si(111)$\sqrt{3}\times\sqrt{3}$ R30° Ag surface as a function of the ion incidence angle measured from the surface, α. The measurements were made in azimuth [A] defined in Fig.29.1. (**b**) The atomic arrangement of the ideal silicon crystal in azimuth [A] defined in Fig.29.1

does not reject the possibility that silicon atoms are partly missing. In the case of the Si(111)7 × 7 clean surface (not shown), some of the intensity minima observed in Fig.29.4a are missing because of drastic surface reconstruction.

This work was supported in part by the Special Coordination Funds for Promoting Science and Technology, Japan.

References

29.1 M. Aono, C. Oshima, S. Zaima, S. Otani, Y. Ishizawa: Jpn. J. Appl. Phys. **20**, L829 (1981)
29.2 M. Aono, Y. Hou, C. Oshima, Y. Ishizawa: Phys. Rev. Lett. **49**, 567 (1982)
29.3 M. Aono, Y. Hou, R. Souda, C. Oshima, S. Otani, Y. Ishizawa: Phys. Rev. Lett. **50**, 1293 (1983)
29.4 M. Aono, R. Souda, C. Oshima, Y. Ishizawa: Phys. Rev. Lett. **51**, 801 (1983); Nucl. Instrum. Meth. **218**, 241 (1983)
29.5 M. Aono: Nucl. Instrum. Meth. B**2**, 374 (1984)
29.6 H. Niehus: Nucl. Instrum. Meth. **218**, 230 (1983)
29.7 H. Niehus, G. Comsa: Surf. Sci. **140**, 18 (1984)
29.8 R.C. Henderson: J. Electrochem. Soc. **119**, 772 (1972)
29.9 K. Spiegel: Surf. Sci. **7**, 125 (1967)
29.10 F. Wehking, H. Beckermann, R. Niedermayer: Surf. Sci. **71**, 364 (1978)
29.11 G. Le Lay, M. Manneville, R. Kern: Surf. Sci. **72**, 405 (1978)
29.12 M. Saitoh, F. Shoji, K. Oura, T. Hanawa: Jpn. J. Appl. Phys. **19**, L421 (1980)
29.13 M. Saitoh, F. Shoij, K. Oura, T. Hanawa: Surf. Sci. **112**, 306 (1981)
29.14 Y. Terada, T. Yoshizuka, K. Oura, T. Hanawa: Surf. Sci. **114**, 65 (1982)
29.15 J. Stöhr, R. Jeager: J. Vac. Sci. Technol. **21**, 619 (1982)
29.16 J. Stöhr, R. Jeager, G. Rossi, T. Kendelewicz, I. Lindau: Surf. Sci. **134**, 813 (1983)
29.17 S. Kono, H. Sakurai, K. Higashiyama, T. Sagawa: Surf. Sci. to be published
29.18 Y. Horio, A. Ichimiya: Surf. Sci. **133**, 393 (1983)

III.5 Photoemission

30. Surface Structure Determination with ARPEFS

J.J. Barton, S.W. Robey, C.C. Bahr, and D.A. Shirley

Materials and Molecular Research Division, Lawrence Berkeley Laboratory, and Departments of Chemistry and Physics, University of California, Berkeley CA 94720, USA

We describe a method of surface structure determination based on oscillations in core-level photoemission intensity—Angle-Resolved Photoemission Extended Fine Structure—with particular emphasis on the use of Fourier transformation. Qualitative comparisons of Fourier power spectra reveal adsorption sites and shortcomings in theoretical calculations; quantitative backtransformation analysis allows accurate bond lengths and bond angles to be determined. Examples are drawn from these similar atomic adsorption systems: c(2 × 2)S/Ni(100), p(2 × 2)S/Cu(100) and c(2 × 2)S/Ni(110).

30.1 Introduction

Recently [30.1,2], we introduced a new approach to determining surface structures: Angle-Resolved Photoemission Extended Fine Structure (ARPEFS). This technique is based on photoelectron diffraction: the interference between the direct and ion-core scattered paths for a photoelectron to enter an angular resolving detector. The key features of ARPEFS which recommend it for structure work are:

1) *Chemical Specificity*. The structural signal is contained in core-level partial cross-section oscillations. By selecting the core level observed, we select the element or even oxidation state of an element to study.

2) *Surface Sensitivity*. Using photoelectrons in the 100-500 eV energy range gives good surface sensitivity.

3) *Large Oscillation Amplitude*. The detected interference is between direct and scattered waves, giving typical oscillations of 20-50 percent.

4) *High Angular Sensitivity*. Each different emission direction yields a different view of the structure; each different combination of polarization direction and crystal orientation gives different emphasis to the scattering atoms.

5) *Simple Theoretical Model*. The above four experimental considerations combine to greatly simplify curved-wave, multiple-scattering calculations.

6) *Direct Fourier Analysis*. The Fourier transform amplitude maps out scattering power versus geometrical path-length difference. The Fourier transform provides a means of displaying the structure information directly from a measurement.

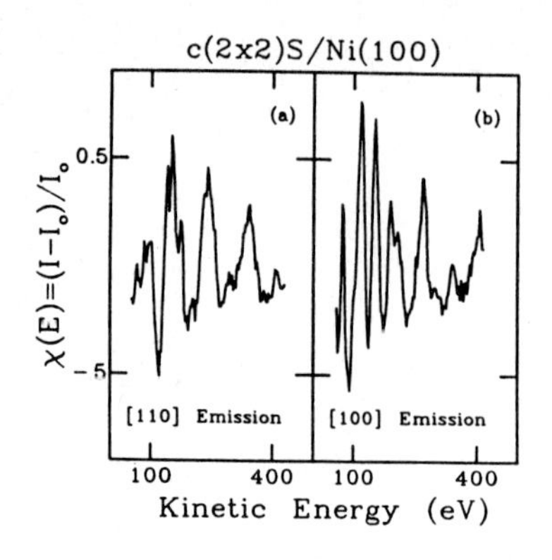

Fig.30.1. ARPEFS modulations derived from S(1s) photoemission partial cross sections. Both curves were measured from the c(2 × 2)S/Ni(100) system. (*a*) Emission along a [110] direction (45° from normal); (*b*) Emission along the crystal normal

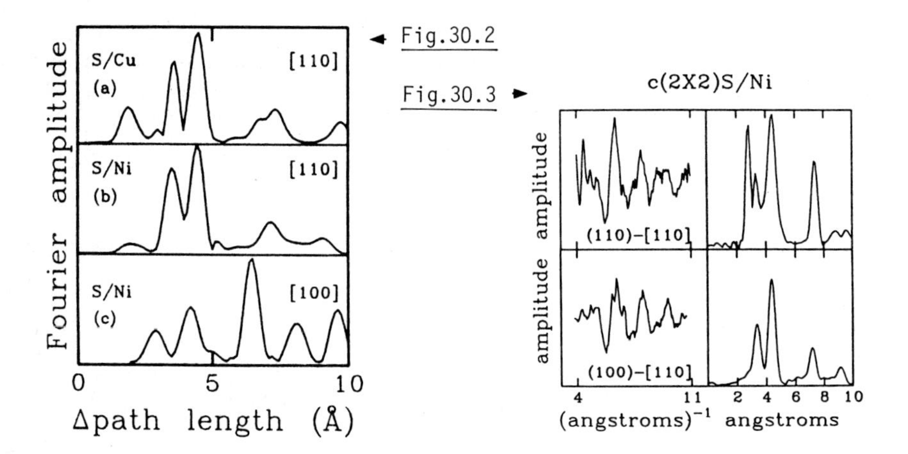

Fig.30.2. Autoregressive Fourier transforms of the three ARPEFS curves discussed in the text. Note that a path-length difference of 4.4 Å corresponds to a bond length of 2.2 Å when $\alpha_j = 173°$. (*a*) S(1s) ARPEFS from p(2 × 2)S/Cu(100); detector along [110], $\gamma = 15°$, (*b*) S(1s) ARPEFS from c(2 × 2)S/Ni(100); detector along [110], $\gamma = 0°$, (*c*) S(1s) ARPEFS from c(2 × 2)S/Ni(100); detector along [100] (normal emission), $\gamma = 20°$. In all curves the intensity below 1.5 Å varies with the background function choice and has been disregarded

Fig.30.3. ARPEFS measurements (*left panels*) and Fourier amplitudes (*right panels*) for c(2 × 2)S overlayers on Ni. Top panels were obtained in normal emission from a (110) crystal. Bottom panels were obtained by measuring along the [110] axis for a (100) crystal surface. The major peaks can be assigned by analogy. Both peaks at 4.4 Å are ~180° backscattering: a nearest neighbor Ni for (100) and a second layer Ni for (110). The 3 Å features are ~115° scattering: two nearest neighbors for (100) and four nearest neighbors for (110). The 3 Å feature for (110) is split by a generalized Ramsauer-Townsend resonance accentuated by the polarization vector position (~ 100° to the scattering vector). The 7.5 Å features indicate Ni atoms further along [110]

In this paper, we give a brief overview of ARPEFS measurement and interpretation, with an emphasis on structure determination with the Fourier transform.

30.2 Experiment

We will discuss three atomic surface structures here, c(2×2)S/Ni(100), p(2×2)S/Cu(100), and c(2×2)S/Ni(110). The first system has become the prototypical chalcogenide surface structure, and it serves to verify our methods of analysis. We reported the structure of the S/Cu system in [30.1]. The c(2×2)S/Ni(110) system provides an interesting correspondence to the S/Ni(100) system which we discuss below.

All three surfaces were prepared by exposing clean single-crystal metal samples to H_2S(g) and warming to 200°C to produce ordered overlayers.

The ARPEFS is derived by measuring a series of S(1s) core-level angle-resolved photoemission spectra for photon energies from 2575-3000 eV, typically in steps of 3 eV. The "partial cross section" is derived as the photopeak area normalized for photon flux versus the photopeak energy position. All measurements were made at the Stanford Synchrotron Radiation Laboratory's soft X-ray double crystal monochromator [30.3] with an electron spectrometer previously described [30.4].

The fractional oscillations are derived by fitting smooth curves [30.1] through the partial cross section I(E), assigning the smooth curve to I_0, and forming

$$\chi(E) = \frac{I(E) - I_0(E)}{I_0(E)} \quad . \tag{30.1}$$

To calculate a Fourier spectrum from the $\chi(E)$ curves, the abscissa is converted to momentum (k) where

$$k = \sqrt{\frac{2m}{\hbar^2}(E - E_0)} \tag{30.2}$$

with $E_0 = 11$ eV typically. The $\chi(k)$ curve is weighted by k, extrapolated with autoregressive prediction, multiplied by a Gaussian weight and Fourier transformed as described in [30.5].

The $\chi(E)$ curves for c(2×2)S/Ni(100) along the [110] (45° polar angle) and [100] (normal emission) crystal directions are shown in Fig.30.1. Notice the large oscillation size and dramatic difference in character of the two emission angles.

Fourier spectra of these curves are shown in Figs.30.2,3; they are discussed below.

30.3 Theory

Angle-resolved photoemission extended fine structure (ARPEFS) refers to oscillations in the partial cross section for photoemission due to final state interference [30.6, and ref. therein]. This interference occurs when the photoelectron can find two paths to the detector: the direct path from photoemitter to detector and a path from photoemitter to a nearby atom which can elastically scatter the electron into the detector. Consider a (1s) photoemitter at the origin of coordinates with the z axis along the electric vector $\boldsymbol{\epsilon}$. If our detector is set at a position labeled $\mathbf{R}$, then the direct probability amplitude wave will be

$$\psi_0 = M(k)\,\cos\gamma\,\frac{\exp(ikR)}{kR} \quad , \tag{30.3}$$

where M is independent of R and $\cos\gamma = (\boldsymbol{\epsilon}\cdot\mathbf{R})/|\mathbf{R}|$. If $\mathbf{a}$ denotes the "bond" vector between the photoemitter and the scatterer, then the scattered wave will be [30.6]:

$$\psi_{\mathbf{a}}(\mathbf{R}) = M(k)\,\frac{\exp(ik|\mathbf{R}-\mathbf{a}|)}{ikR}\,\frac{\exp(ika)}{|\mathbf{a}|}\,F(\mathbf{a},\mathbf{R},\boldsymbol{\epsilon},k) \quad . \tag{30.4}$$

The complex scattering factor F contains all details of the scattering process; typically, it has a large (~.3) amplitude $|F|$, and small phase ϕ, so the fractional oscillation is large and has a frequency near the path-length difference, $|\mathbf{a}|(1-\cos\alpha_j)$:

$$\psi = \sum_j \frac{\psi_0^*\psi_j + \psi_j^*\psi_0}{\psi_0^*\psi_0} = \frac{|F|}{|\mathbf{a}|}\,2\cos[ka(1-\cos\alpha)+\phi] \quad . \tag{30.5}$$

Several physical circumstances collaborate to make this simple formula useful for interpreting ARPEFS spectra:

1) The frequency of the cosine is dominated by the geometrical path-length difference. The scattering phase ϕ is usually quite linear with a $d\phi/dk \sim 0.1$ Å.
2) The scattering amplitude $|F|/|\mathbf{a}|$ (usually) contains little structure.
3) Among all of the scattering angles which can reflect electrons into the detector, backscattering ($\alpha = 180^0$) is strongly favored [30.7] in the 100-500 eV range. This selectivity is further enhanced by the polarization dependence of the final state.
4) Multiple scattering is small for all angles except forward scattering [30.8], where there is little effect on the oscillation frequency.

An important exception to this simple picture occurs for some scattering angles and energies. At these points—which we refer to as generalized Ramsauer-Townsend resonances [30.2]—the scattering amplitude $|F|$ falls to zero, and the scattering phase shift ϕ jumps by π. These resonances can complicate the ARPEFS spectrum: the amplitude drop in $|F|$ simulates a beat envelope and can split the Fourier peak for the corresponding path-length differences. On the other hand, they may also be a powerful means of determining surface structure. The resonances are sensitive indicators of surface bond angles, as we discuss below.

In spite of the simple form of (30.5), theoretical calculation of the ARPEFS curves is difficult. Single-scattering, plane-wave calculations [30.7] have contributed some valuable qualitative insights into the nature of the scattering factor F, but quantitative calculations [30.9] with reasonable model parameters reproduce neither the experimental ARPEFS, nor important features of the Fourier spectrum. We have found that such calculations agree much more closely with our measurements if only nearest neighbors and backscattering nonneighbors are included. Thus it appears that some physical mechanism not accounted for in the single-scattering plane-wave theory systematically discriminates against many of the potentially important scattering atoms. The physical origin of this enhanced selectivity is intriguing, but not crucial to the analysis we discuss below.

30.4 Structure Determination with the Fourier Transform

Given the experimental ARPEFS measurements at a number of emission angles, and an understanding of the physical origin of the oscillation, we must deduce the structure. With these simple atomic surface structures we have been developing the mechanics of the structure determination which should then be applicable to more complex systems.

The Fourier transform of the ARPEFS displays an integrated scattering amplitude versus scattering path-length difference. We must proceed with caution when we interpret the peaks in the transform for a number of reasons:

1) Any single peak may contain contributions from more than one scattering atom.
2) The extended fine structure is only approximately given by a cosine series. The oscillation frequency varies slightly with energy and, more important, the cosine envelope contributes to the Fourier peak shape.
3) The entire data range contributes to each Fourier coefficient. Inaccurate data points on the end of a spectrum are not ignored by the transform. Care must be taken when comparing two transforms to insure that the transformed range and the weighting functions are identical [30.10].
4) Transforms with different frequency resolution should be compared only with hesitation: when two Fourier peaks with different phases are merged, they need not appear as the sum of two peaks.
5) The generalized Ramsauer-Townsend effect (Sect.30.4.4) can split Fourier peaks.

With these caveats in mind Fourier analysis can be a powerful tool for studying surface structure. In the remainder of this report we discuss four ways the Fourier transform can be used to determine surface structure information from ARPEFS.

30.4.1 Adsorption Sites from Fourier Spectrum Comparisons

When confronted with an experimental Fourier spectrum our first task is assigning the major features to path-length differences. From the simple theoretical model we expect major peaks for backscattering; these path lengths must be about twice the distance from the emitter to the scatterer. Nearest neighbor scattering through angles further from 180° may also be seen since the photoemitted wave decays very little as it travels towards them. These path lengths must be less than twice a bond length.

Figure 30.2b illustrates such an assignment. The largest peak in the spectrum corresponds to backscattering ($\alpha = 173^\circ$) from a nearest neighbor (path-length difference 4.46 Å). At a slightly lower path length (3.2 Å), two other nearest neighbors with smaller scattering angles ($\alpha = 135^{\circ}$) contribute. Two higher peaks signal backscattering Ni atoms further away along [110].

When the adsorption site is unknown, the process of identifying the path-length difference will be more involved (also take note of the Ramsauer-Townsend effect discussed below). A collection of ARPEFS spectra for atomic adsorption will, however, greatly simplify the adsorption site and path-length difference assignment. For example, compare the Fourier spectra in Figs.30.2a and b, both taken along the [110] crystal axis. The striking similarity of the c(2 × 2)S/Ni(100) and p(2 × 2)S/Cu(100) spectra eliminates any doubt about the adsorption site of S/Cu. The comparison of the two S/Ni systems in Fig.30.3 is novel: the ARPEFS was measured along the same crystallographic direction for two different surfaces. If the local orientation of Ni atoms about S is the same, then the Fourier spectra from the two measurements will be the same and the site is determined. We deduce that a Ni atom must rest directly below the S on the (110) surface, at a distance close to the S-Ni bond distance for S/Ni(100). Since scattering peaks occur at shorter path lengths, we conclude that this Ni atom must be in the second layer—an atop site could not have shorter path-length distances. Hence we are drawn to the (known) fourfold geometry.

To be sure, these adsorption sites are quite simple, but the principle should apply to more complex systems. Note again two key features of ARPEFS analysis: it is elementally specific and highly angle dependent. We have the potential for measuring a great deal of information about the position of a single constituent of an adsorbate system.

30.4.2 Comparison of Theory with Experiment in Fourier Space

A second important role for Fourier analysis of ARPEFS is qualitative comparison of theoretical calculations and experimental data. The Fourier transform comparison rapidly reveals over- or under-emphasized path-length differences as an aid to correcting theoretical models. Perhaps more important, by limiting the comparison to short path-length differences, very economical calculations are possible, even for more sophisticated models for electron scattering. Since the angular momentum of the photoemission final state is restricted by dipole selection rules and since multiple scattering is important only in the forward direction, curved-wave multiple-scattering calculations can be routinely performed for comparison to experiment.

30.4.3 Empirical Backtransformation Analysis

The very close analogy between angle-resolved fine structure (ARPEFS) and the angle-integrated fine structure EXAFS leads to the third use of Fourier analysis: backtransformation analysis. The idea and its justification are drawn directly from the experiences in EXAFS [30.11]; we applied this method to c(2 × 2)S/Ni(100) and p(2 × 2)S/Cu(100) in [30.1]. The Fourier transform separates the ARPEFS into individual oscillations. By isolating a single backscattering Fourier peak and applying a complex backtransformation, the amplitude and argument of the backscattering cosine can be extracted. Then the scattering phase shift ϕ can be subtracted from the total experimental argument to give a line whose slope is the path-length difference. We have applied this analysis to the main 4.4 Å scattering peak in S/Ni(110)-[110], using the scattering phase shift derived experimentally from the main peak in the well-known S/Ni(100) system. Thus we determine that the distance between the S atom and the second layer Ni atom on the (110) surface is the same (2.23 Å) as the bond distance on the (100) surface.

30.4.4 Generalized Ramsauer-Townsend Resonance Analysis

Finally, we have been developing an entirely new method for determining surface structure information which is unique to the analysis of ARPEFS. Here we take advantage of an interesting physical feature of electron scattering: the scattering phase shift as a function of wave number can pass through the origin in the complex plane for some scattering angles. At these angles, the scattering amplitude falls to zero for some energy and rises again at higher energy. Simultaneously, the scattering phase angle jumps by pi radians. We call these points generalized Ramsauer-Townsend resonances [30.2] and they are useful for determining structure because the shape of the scattering phase argument is strongly dependent on angle.

For Ni, a resonance occurs near $k = 7$ Å^{-1} and a scattering angle of 127°. For normal emission from the (100) surface, the four Ni atoms closest to S have a scattering angle of ~127°. As the scattering amplitude dips toward zero at $k = 7$ Å^{-1}, the Fourier spectrum for this 3.2 Å path-length difference is split into two peaks as shown in Fig.30.2c. Isolating both peaks and performing the complex backtransformation give the phase jumps shown in Fig. 30.4. It appears that the scattering angle cannot exceed 127°, nor be lower than 125°.

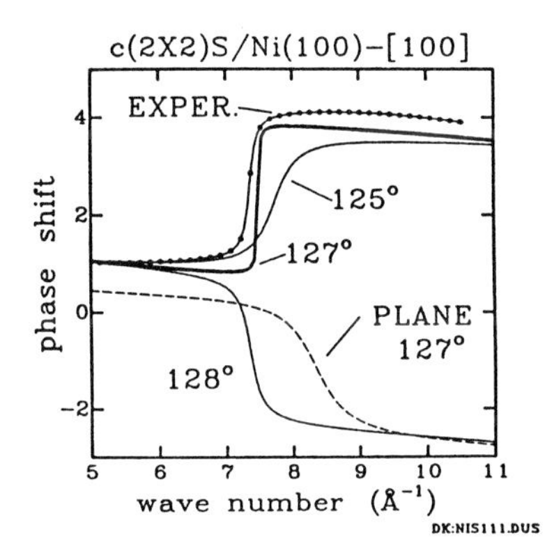

Fig.30.4. Phase shifts for scattering from Ni. (----) shows the phase shift calculated with plane wave theory $\alpha_j = 127°$. (····) is the phase shift from the experimental curve, Fig. 30.2c, where the first two Fourier peaks are backtransformed together. The zero crossing jump in phase occurs too high in wave number for the plane wave calculation. (——) are curved-wave calculations of the phase shift for the indicated scattering angles

Some important questions must be addressed before we can apply this method to unknown systems with confidence. As shown in Fig.30.4, plane wave calculations make substantial errors in the position of this resonance: theoretical calculations must have high accuracy if we rely on them for predicting the resonance. Preliminary work with changing the inner potential (E_0) shows little effect on the angle analysis, but this must be verified. Finally, we must confirm that the Fourier processing does not distort the resonance shape.

30.5 Conclusion

Angle-Resolved Photoemission Extended Fine Structure promises to be an exciting new method for examining surface structures. We see its most important early role in furthering our understanding of electron scattering and in clarifying adsorbate geometries which baffle other techniques. For both problems, the high selectivity and direct Fourier analysis features of ARPEFS should recommend its application.

Acknowledgments. This work was supported by the Director, Office of Energy Research, Office of Basic Energy Sciences, Chemical Sciences Division of the U.S. Department of Energy under Contract No. DE-AC03-76SF00098. It was performed at the Stanford Synchrotron Radiation Laboratory, which is supported by the Department of Energy, Office of Basic Energy Sciences and the National Science Foundation, Division of Materials Research.

References

30.1 J.J. Barton, C.C. Bahr, Z. Hussain, S.W. Robey, J.G. Tobin, L.E. Klebanoff, D.A. Shirley: Phys. Rev. Lett. **51**, 272 (1983)
30.2 J.J. Barton, C.C. Bahr, Z. Hussain, S.W. Robey, L.E. Klebanoff, D.A. Shirley: Proc. of the Conf. on Science with Soft X-Rays, Soc. Photo-Optical Instrum. Eng. (1984) v.447, p.82;
J.J. Barton, C.C. Bahr, Z. Hussain, S.W. Robey, L.E. Klebanoff, D.A. Shirley: J. Vac. Sci. Technol. A**2**, 847 (1984)
30.3 Z. Hussain, E. Umbach, D.A. Shirley, J. Stöhr, J. Feldhaus: Nucl. Instrum. Meth. **195**, 115 (1982)
30.4 S. Kevan: Ph.D. Thesis, University of California, Berkeley, LBL-11017 (1980)
30.5 J.J. Barton, D.A. Shirley: LBL Report, LBL-14758
30.6 P.A. Lee: Phys. Rev. B**13**, 5261 (1976)
30.7 P.J. Orders, C.S. Fadley: Phys. Rev. B**27**, 781 (1983)
30.8 S.Y. Tong, C.H. Li: In *Chemistry and Physics of Solid Surfaces*, Vol. III (CRC Press, Boca Raton, Florida 1982) p.287
30.9 E.L. Bullock, C.S. Fadley, P.J. Orders: Phys. Rev. B**28**, 4867 (1983)
30.10 The minor differences between Fig.30.2b and the bottom right-hand panel of Fig.30.3 illustrate the effect of different weighting. A Tukey weighting function was used for Fig.30.2b while a more optimal Gaussian weight gave the result in Fig.30.3. See [30.5]
30.11 P.A. Lee, P.H. Citrin, P. Eisenberger, B.M. Kincaid: Rev. Mod. Phys. **53**, 769 (1981)

31. Angle Resolved XPS of the Epitaxial Growth of Cu on Ni(100)

W.F. Egelhoff, Jr.

Surface Science Division, National Bureau of Standards
Gaithersburg, MD 20899, USA

In angle-resolved X-ray photoelectron spectroscopy (XPS) of single crystals the core level peaks exhibit enhanced intensities along major crystal axes. This phenomenon is often referred to as electron channeling (or Kikuchi beams) due to an apparent analogy with effects found in electron microscopy. The present analysis of this phenomenon for epitaxial Cu on Ni(100) demonstrates that the electron channeling (or Kikuchi beams) approach fails completely to describe the data. The actual physical basis for this phenomenon is forward scattering of photoelectrons by overlying atoms in the lattice.

31.1 Introduction

X-ray photoelectron spectra (XPS) of single crystals exhibit an angular anisotropy which takes the form of enhanced core-level-peak intensities along the major crystal axes [31.1]. This phenomenon bears a superficial resemblance to the electron channeling or Kikuchi beam phenomena observed for scattered electrons in electron microscopy [31.2]. Due to this resemblance the idea caught on that the XPS and the electron microscopy phenomena had the same physical basis, so that in current review articles the XPS phenomenon is presented in terms of an electron channeling (or Kikuchi) model [31.3]. The present work examines this model in the light of recent data on the evolution of the enhanced core level peak intensities as a function of thickness in the epitaxial growth of Cu on Ni(100).

31.2 Experimental

The work reported here was performed in an extensively modified AEI-ES200 XPS instrument[1] under ultrahigh ($\leqq 10^{-11}$Torr) vacuum conditions. The Cu was evaporated from W filaments and the deposited thicknesses were determined using two quartz-crystal thin-film-thickness monitors as well as an ion gauge calibrated for Cu flux. The quoted thicknesses are believed accurate to within ± 5%. The Ni(100) substrate was at 450 K during Cu deposition. This is high enough for complete annealing of the Cu into epitaxial overlayers but low enough to avoid interdiffusion. The angle between the X-ray source and the entrance to the analyzer is 90°. The sample is rotated so that the surface normal can vary from pointing at the X-ray source to pointing at the analyzer entrance. See [31.4] for further experimental details.

1 This commercial instrument is identified only to specify the experimental conditions and does not signify any endorsement by NBS.

31.3 Results

Figure 31.1 presents the results for the intensities of a) the Cu $2p_{3/2}$ peak and b) the Cu CVV Auger peak as a function of polar angle for various thicknesses, in monolayers (ML) $\equiv 1.6 \times 10^{15}$ atoms/cm^2, of epitaxial Cu on Ni(100). The 14 ML results are essentially converged on the thick Cu limit. Such polar intensity plots (PIPs) have also been recorded for the Cu 3d (kinetic energy, $E_k = 1247$ eV) and the Cu 3p ($E_k = 1174$ eV) for the same Cu thicknesses indicated in Fig.31.1. The 3d and 3p PIPs are almost indistinguishable from the CVV Auger PIP, indicating that the mechanism producing the peaks in the PIPs does not depend strongly on electron kinetic energy (between 917 eV and 1247 eV) or on whether the electron is an Auger electron or a photoelectron. The PIPs are recorded in the <100> surface azimuth. This azimuth is illustrated in Fig.31.1c for 3 ML Cu on Ni(100). Figure 31.1c also illustrates how electrons emitted by Cu atoms in the inner layers can scatter in the forward direction off overlying Cu atoms in the lattice at 0° and 45°.

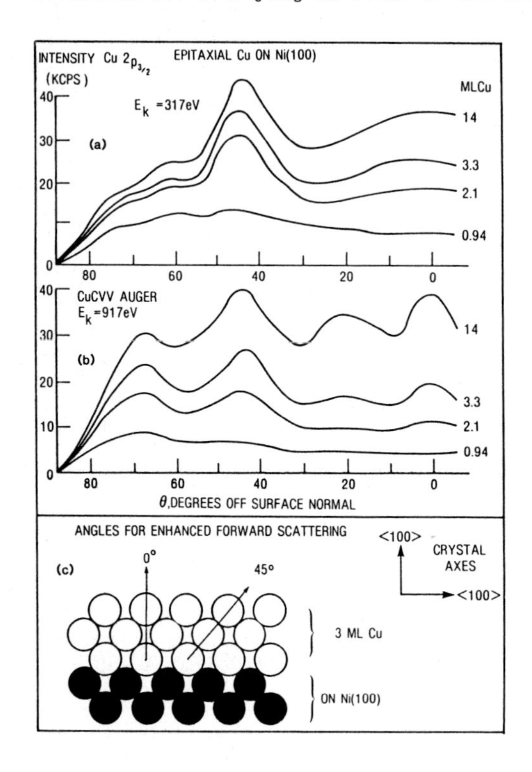

Fig.31.1a-c. The Cu $2p_{3/2}$ (**a**) and Cu CVV (**b**) intensities as a function of polar angle for epitaxial layers of Cu on Ni(100) and (**c**) an illustration of the angles at which forward scattering of the outgoing electron wave increases

31.4 Discussion

As noted in a previous publication on this topic [31.5], the peaks in PIPs of Fig.31.1 can be related to the angles at which forward scattering by overlying atoms in the lattice can occur. For example, at 0.94 ML Cu in Fig.31.1a, the Cu $2p_{3/2}$ emission is nearly isotropic (resembling the instrument response function) since there are no overlying atoms off which to scatter. At 2.1 ML a peak appears at 45°. This enhancement in the intensity at 45° is a consequence of constructive interference between the initial outgoing photoelectron from an inner layer Cu atom and the scattered wave from a top layer Cu atom. It has been demonstrated previously that such enhanced forward scattering can dramatically increase photoelectron intensity along internuclear axes [31.6].

It was consistently found in this work [which has included studies of Fe, Co, Ni, and Cu on Ni(100) and Cu(100) surfaces] that less detail is present in relatively low kinetic energy PIPs such as Fig.31.1a than in PIPs with $E_k \geqq 1000$ eV such as Fig.31.1b. This is presumably related to the escape depth increasing with kinetic energy and to scattering at high kinetic energies being more kinematic.

Were it not for the results in Fig.31.1 at lower Cu thicknesses, the 14 ML Cu results (which are essentially bulk-like) could be interpreted using an electron channeling or Kikuchi model. Figure 31.2a illustrates this basic idea, as adapted for XPS. An outgoing photoelectron wave traveling nearly parallel to a set of crystal planes can experience constructive interference for small-angle scattering. This produces two lobes of enhanced intensity (relative to isotropic emission) at angles θ on either side of the direction parallel to the crystal planes, as illustrated in Fig.31.2a. If the lobes were sufficiently close together one might think they would merge into a single peak around that particular crystal axis. This is the way in which peaks in PIPs may be viewed in an electron channeling or Kikuchi model. This picture, however, breaks down completely upon closer scrutiny.

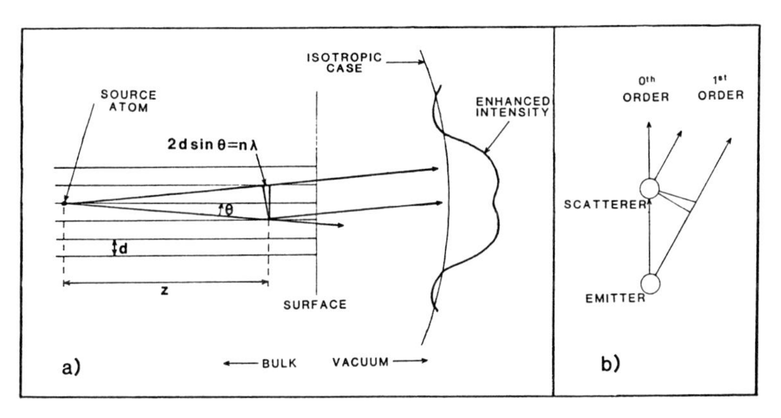

Fig.31.2. (**a**) Electron channeling or Kikuchi model for XPS intensities (the greatest enhancement is predicted for $n=1$); and (**b**), the strong zeroth and the weaker first-order contributions in the forward-scattering model

Table 31.1 presents the values for the parameters in Fig.31.2a needed to explain the data of Fig.31.1 in an electron channeling or Kikuchi model. Beginning with the 0° CVV peak, the model requires a value of 2θ (which should be a lower limit for the observed PIP peak width) of 12.8°. In the data of Fig.31.1b this peak appears narrower than this minimum but not by a wide margin. Where the model fails totally is in the value of z (which represents a lower limit on the Cu thickness at which the $n=1$ Bragg condition can be met). The 0° CVV peak begins to grow upon formation of the third Cu monolayer, just as predicted by the forward-scattering model of Fig.31.1c, but entirely inconsistent with the $z=16.2$ Å ($\approx$8 ML Cu) in the model of

Table 31.1. Parameters required to relate the model of Fig.31.2 to the data of Fig.31.1

peak	λ	hkl [7]	d_{hkl}	2θ	z
0° CVV	0.405 Å	002	1.81 Å	12.8°	16.2 Å
45° CVV	0.405 Å	022	1.28 Å	20.2°	8.0 Å
$45^\circ 2p_{3/2}$	0.689 Å	022	1.28 Å	31.2°	4.86 Å

Fig.31.2a. Thus the actual Cu thickness at which the 0° CVV peak begins to develop is considerably smaller than the minimum thickness required by the electron channeling or Kikuchi model [31.7].

In the case of the 45° CVV peak the 2θ width of 20.2° is too large and the z = 8.0 Å ($\approx$4 ML) is entirely inconsistent with the development of the peak at 2 ML. For the 45° $2p_{3/2}$ peak the 2θ of 31.2° is too large and while the z = 4.86 Å ($\approx$2.3 ML) is not too far off, the concept in Fig.31.2a of scattering from uniform continuum planes breaks down entirely at 2 ML into scattering from atomic centers.

The peaks in Fig.31.1b at 20° and 70° also cannot be explained on the basis of Fig.31.2a, the 2θ values or the z values being far too large. As suggested earlier [31.4,5] the 20° and 70° may be explained on the basis of simple 0th order forward scattering, or as suggested recently [31.8-10], may include an additional contribution from a 1st order beam within the forward scattering cone as illustrated in Fig.31.2b. At E_k = 917 eV (the data of Fig. 31.1b) the angle between the 0th and 1st order beams is $\sim 22^\circ$ [31.8-10]. This means that the 0th order beams in the 0° and 45° directions have associated with them weak 1st order beams that contribute to the 20° peak. There is also a 0th beam at 90° due to forward scattering parallel to the surface. It and the 0th order beam at 45° both contribute 1st order beams to the 70° peak. Thus the 20° and 70° peaks are composed at least partly of 1st order beams.

31.5 Conclusions

The conclusions of this work are:

1) The enhanced intensities of the Cu $2p_{3/2}$ peak at 45° and Cu CVV peaks at 45° and 0° can be fully explained on the basis of forward scattering.

2) The electron channeling or Kikuchi model fails completely to account for the data.

Acknowledgment. The author gratefully acknowledges the very helpful advice of Dr. C.S. Fadley, Dr. R.A. Armstrong, and Dr. S.Y. Tong during the course of this work.

References

31.1 K. Siegbahn, U. Gelius, H. Siegbahn, E. Olson: Phys. Scr. **1**, 272 (1970); C.S. Fadley, S.A.L. Bergström: Phys. Lett. **35**A, 375 (1971); K. Siegbahn, U. Gelius, H. Siegbahn, E. Olson: Phys. Lett. **32**A, 221 (1970);

N.E. Erickson: Phys. Scr. **16**, 462 (1977);
S. Evans, M.D. Scott: Surf. Interface Analysis **3**, 269 (1981);
S. Evans, J.M. Adams, J.M. Thomas: Philos. Trans. R. Soc. London **292**, 563 (1979)

31.2 R.D. Heidenreich: *Fundamentals of Transmission Electron Microscopy* (Interscience, New York 1964) p.191;
O.C. Wells: *Scanning Electron Microscopy* (McGraw-Hill, New York 1974) p.160

31.3 D. Briggs, M.P. Seach (eds.): *Practical Surface Analysis* (Wiley, New York 1983) p.136;
C.R. Brundle, A.D. Baker (eds.): *Electron Spectroscopy*, Vol.2 (Academic, London 1977) p.132

31.4 W.F. Egelhoff, Jr.: J. Vac. Sci. Tech. A**2**, 350 (1984)

31.5 W.F. Egelhoff, Jr.: Phys. Rev. B**30**, 1052 (1984)

31.6 S. Kono, C.S. Fadley, N.F.T. Hall, Z. Hussain: Phys. Rev. Lett. **41**, 117 (1978);
see especially Fig.3 in S. Kono, C.S. Fadley, N.F.T. Hall, Z. Hussain: Phys. Rev. B**22**, 6085 (1980)

31.7 The use of 002 and 022 planes may seem odd since they are based on atoms atoms lying above and below the plane of Fig.31.1c. This was done to give the fairest test (on the basis of the z criterion) to the electron channeling or Kikuchi model. Using 001 and 011 planes gives smaller 2θ values but larger and thus even more unrealistic z values. It may also be noted that at major crystal axes many (in principle, infinitely many) sets of crystal planes intersect and could be used to predict enhancement. However, only the lowest index ones need to be considered here since all higher index planes require much larger z values

31.8 E.L. Bullock, C.S. Fadley, B.L. Hermsmeiner, M. Sagurton, R. Saiki, B. Sinkovic, R. Trehan: To be published

31.9 R.A. Armstrong, W.F. Egelhoff, Jr.: To be published

31.10 H.C. Poon, S.Y. Tong: Phys. Rev. B**30**, 6211 (1984)

32. Evidence for Diffusion at 80 K of Gold Atoms Through Thin, Defective Oxide Layers

S. Ferrer, C. Ocal, and N. Garcia

Departamento de Física Fundamental, Universidad Autónoma de Madrid
Cantoblanco, 28049-Madrid, Spain

32.1 Introduction

During the last few years, many papers on the initial stages of the interaction of oxygen with metal surfaces have been published. In particular, the initial oxidation of aluminum surfaces has been extensively studied with most of the surface science techniques which are able to provide atomic-scale information on the interaction process [32.1]. However, microscopy studies of the three-dimensional (3D) oxidation of well-characterized metal surfaces are relatively scarce. Most of the published work concerns the study of the kinetics of the oxide growth and usually it is very difficult to infer microscopic mechanisms from measured rate laws. In general terms the most used theory to describe microscopically the growth mechanism at low temperatures of a thin oxide layer on its metal support is the Cabrera-Mott theory for oxidation of metals [32.2]. In Fig.32.1 we describe briefly the physical basis of the mechanism. When an oxygen atom approaches the oxide surface, its electron affinity level E changes by an amount W due to the interaction with the surface. If the resulting energy of the affinity level, E + W, is larger than the work function ϕ of the underlying metal, electrons from the Fermi level of the metal will tunnel through the oxide to raise the affinity level to establish thermodynamical equilibrium. Consequently, a layer of negative ions is formed on the oxide surface and another cationic layer appears at the oxide-metal interface. This is described as a plate capacitor with a potential difference V. An electric field F = V/X (X: oxide thickness) is produced that acts as a driving force for ion diffusion through the oxide and is effective only if the potential drop in one jump is comparable to the activation barrier for ion diffusion. A simple electrostatic estimation gives for V a value of a fraction of a volt. It is known that oxides containing defects have lower activation energy barriers than perfect oxides and therefore they should be more suitable to study the Cabrera-Mott mechanism.

In this work we first show the results of Ion Scattering Spectroscopy (ISS) and X-ray Photoemission Spectroscopy (XPS) studies on the 3D oxidation of Al(111) crystals. By varying the growth procedure of the oxide, ISS results show that it is possible to prepare oxides of similar thickness (about four atomic layers) but with very different surface stoichiometry. Some of

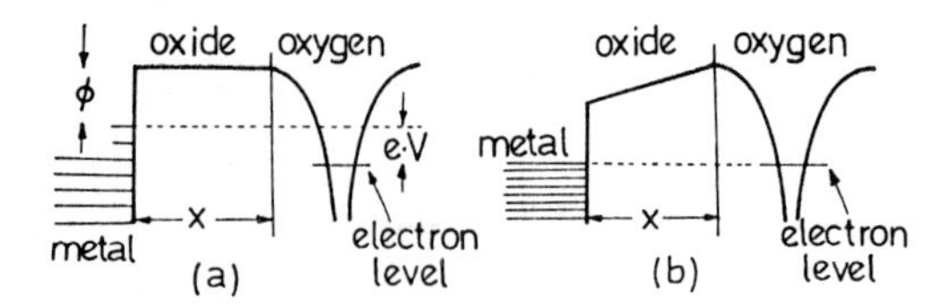

Fig.32.1a,b. Electronic levels in the metal, and adsorbed oxygen (**a**) before electrons have passed through the oxide (**b**) when equilibrium is set up (from [32.1] with permission)

them (denoted stoichiometric for brevity) exhibit an oxygen to aluminum surface concentration ratio close to 1.5 as in the Al_2O_3 formula unit, whereas others (denoted reduced) have an important deficiency in oxygen. The ulterior rate of growth of these thin oxides is noticeably different. The oxidation rate is several times faster for the reduced oxides than for the stoichiometric ones, indicating that the activation barriers for diffusion are lower in the former case.

According to the Cabrera-Mott theory, in principle, any atom with an electron affinity energy large enough to satisfy $E + W > \phi$ could be a candidate for a field-assisted diffusion process. In particular, Au atoms have larger electron affinity than oxygen atoms and therefore they could diffuse through the oxide.

We performed experiments to check this point. We deposited Au atoms from the vapor phase onto the surface of a thin (4 atomic layers), reduced aluminum oxide film epitaxially grown on the Al(111) surface. We observed diffusion at 80 K of Au atoms through the oxide layer to the underlying metal. On the other hand, gold deposited on a stoichiometric oxide of the same thickness did not diffuse, and by depositing a low electron affinity element such as K on the surface of the reduced oxide, diffusion was again not observed. These results agree with the Cabrera-Mott mechanism and they constitute, as far as we know, the first microscopic experimental confirmation of this theory.

We think that these are important results since they provide an experimental basis to understand microscopic oxidation processes. The technological importance of the area of oxidation of metals does not need to be emphasized. In addition, we also think that they may have important implications in the area of heterogeneous catalysis in the so-called strong metal-support interaction. In this case, transition metal particles supported on a reduced oxide as a substrate exhibit a catalytic activity and selectivity very different from that of similar metal particles supported on a stoichiometric oxide substrate. The so-called encapsulation effect [32.3] has been invoked as responsible for the differences between both types of catalysts. It consists in a partial encapsulation of the metal particles by the underlying oxide. We think that our results on the diffusion of metal atoms through thin reduced oxides could provide some insight in the near future to this important problem in catalysis.

We shall briefly describe the most significant experiments and results on which these statements are based. A detailed account will be published in the near future.

32.2 Experimental

The experiments were performed in a conventional ultrahigh-vacuum (UHV) system previously [32.4] described. It was equipped with a MgKα X-ray source, an ion gun, a hemispherical analyzer and leak valves for gas handling. The Au and K were evaporated in the UHV chamber to ensure cleanliness. For the ISS experiments, the scattering angle was 130° and the energy of the He ions was 500 eV. This low kinetic energy guarantees an extremely high surface sensitivity. The ion spectra are therefore fairly indicative of the composition of the topmost surface layer. We checked the possibility of sputtering due to the He beam and found it to be negligible under our operating conditions.

32.2.1 Reduced and Stoichiometric Oxides [32.5]

We now describe the procedure to grow what we call thin reduced oxides. Starting with the clean (monitored by ISS and XPS) Al(111) surface, the crystal was annealed to 700 K, kept at this temperature and exposed to 2×10^{-6}Torr of oxygen for several minutes. After this treatment the gas was pumped down and after obtaining an UHV environment the crystal was cooled to room temperature. The right side of Fig.32.2 shows the ISS (upper curve) and XPS (lower curve) spectra of the oxidized crystal after oxygen treatment for 12 minutes. In the ISS spectrum two peaks are visible which correspond to surface oxygen atoms and to aluminum atoms. Their intensity ratio is 0.5. The XPS spectrum corresponds to the photoelectrons emitted from the Al 2p level of the underlying metallic atoms (peak at 72.6 eV) and of oxidized aluminum atoms (peak at 75.9 eV). From the relative intensity of both peaks we estimate an oxide thickness of about 4 atomic layers.

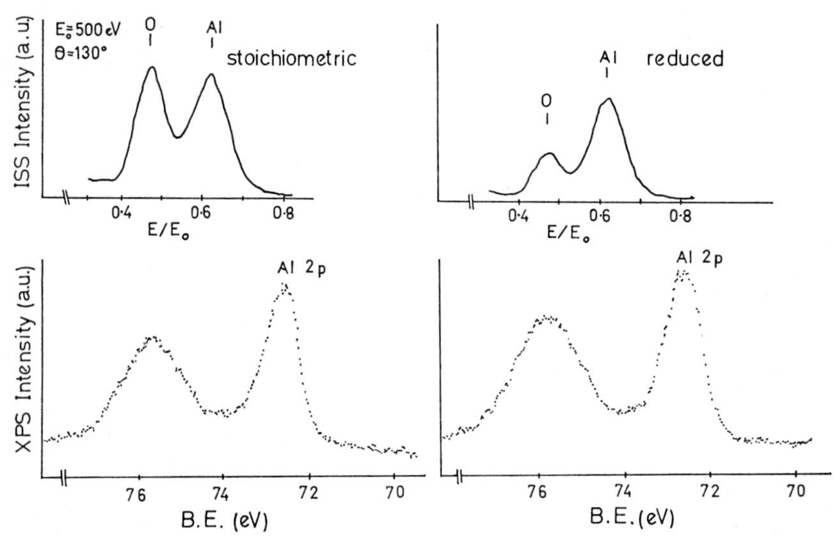

Fig.32.2. *Left*: ISS spectrum (*top*) and XPS spectrum of the Al 2p emissions (*bottom*) of the stoichiometric oxide (see text). *Right*: The same but for the reduced oxide (see text)

The procedure utilized to grow what we call stoichiometric oxides was the following. We exposed the clean Al surface at room temperature to 500 L of oxygen ($1\ \mathrm{L}=1\times10^{-6}$Torr s). After pumping the oxygen, the treatment to grow reduced oxides was repeated, i.e., anneal at 700 K and expose to 2×10^{-6}Torr of oxygen for several minutes, then pump the oxygen down and cool to room temperature. The left part of Fig.32.2 shows the spectra obtained after a 35-minute oxygen treatment. The XPS spectrum is virtually identical to the spectrum of the reduced oxide, but the ISS spectrum exhibits an O-Al peak intensity ratio of 1·1. With a heavier oxygen treatment we were able to grow thick (20 layers) oxides which displayed an O-Al ISS peak ratio of 1.5. The denominations that we have employed for these two types of oxides are clear. The stoichiometric oxide has a surface O-Al concentration ratio much closer to 1.5 (corresponding to Al_2O_3 than the reduced oxide which exhibits oxygen deficiency at the surface.

32.2.2 Gold Depositions on the Reduced and Stoichiometric Oxides

We now describe four different sets of experiments on the deposition of Au atoms on reduced and stoichiometric thin oxides. For the reduced oxides we interpreted all the experimental results as due to the diffusion of Au atoms through the oxide layer to the underlying Al metal. For stoichiometric oxides, on the contrary, our interpretation is that diffusion does not occur.

1) Starting with a thin ($\sim$4 atomic layers) reduced oxide on its metal support, characterized by ISS and XPS spectra such as the ones showed on the right side of Fig.32.2, we deposited Au from the vapor phase onto the oxide surface that was at 80 K. We monitored the intensities of the Al 2p XPS emissions of the oxide film and of the underlying metal, as a function of the deposition time. For deposition times up to $\sim$5 minutes, which corresponded to a number of deposited Au atoms equivalent to 1.5 ± 0.5 monolayers, we observed a progressive attenuation of the intensity of the XPS metal emission whereas the oxide emission was not attenuated. After 15 min of Au deposition, the intensity of the metal emission was about 80% of its initial value. If the deposited Au atoms were sitting on the oxide surface, the relative attenuation of both metal and oxide photoemission lines should be the same. As the oxide emission is not attenuated, the straightforward interpretation is that Au atoms are located at the metal-oxide interface since they cause only the attenuation of the metal emission. The same experiments performed on a stoichiometric oxide (characterized by ISS and XPS spectra such as those shown on the left side of Fig.32.2) gave quite different results. In this case, both the Al-metal and Al-oxide emissions were equally attenuated by the deposited Au, indicating that Au atoms remain on the oxide surface and do not diffuse through the oxide.

2) The binding energy (BE) of the Au 4f line is sensitive to the chemical environment. We have thus measured this BE for a dilute Au-Al alloy (obtained by depositing small amounts of Au on clean Al metal), Au clusters on oxides and Au metal foil. Relative to the bulk Au $4f_{7/2}$, the BE of the Au-Al alloy was 1.6 eV. For clusters, in our sensitivity range this BE varied from 0.7 to 0 eV with increasing cluster size. When Au was deposited at low coverages on a thin reduce oxide, we found its $4f_{7/2}$ BE to be exactly the same as that corresponding to the Au-Al alloy. At higher coverages another well-resolved line was observed starting at a BE of 0.7 eV that indicates cluster formation on the oxide surface. This is illustrated in Fig.32.3. Peak 1 corresponds to the alloyed Au, and peak 2 to Au clusters. When Au

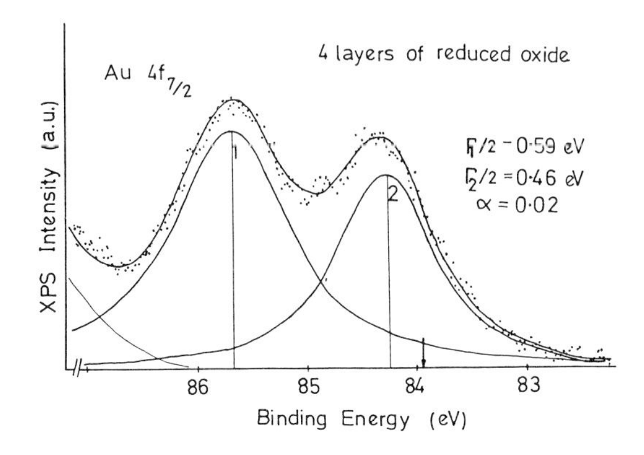

Fig.32.3. XPS spectrum of the Au $4f_{7/2}$ electron level for Au deposited on a thin reduced oxide. The experimental curve has been deconvoluted in two Doniach-Sunjic components. Component 1 corresponds to alloyed Au and component 2 to Au clusters. The arrow indicates the position of the bulk Au emission

was deposited on the thin stoichiometric oxide, the $4f_{7/2}$ line due to alloying was not observed. Only the emission due to clusters was visible even at very low coverages. Again these data indicate Au penetration through the reduced oxide at low Au coverages.

3) Starting with a surface consisting of Au deposited on a reduced oxide, we found after mild Ar sputtering that removed the oxide layer completely an XPS spectrum showing only the presence of Al metal and alloyed Au.

4) The sensitivity of ISS to small gold concentrations on the surface is noticeably larger than the sensitivity of XPS. We monitored with ISS the relative increase of the Au signal as a function of deposition time when Au was deposited on the reduced oxide surface at 80 K, Fig.32.4 (open circles). A plateau is visible, which means that after an initial adsorption of Au on the surface (not detectable by XPS), there is no buildup of the concentration of Au on the surface, indicating that Au is penetrating the crystal. The XPS showed Au emission corresponding to alloyed Au. After the plateau there is a buildup of Au on the surface which corresponds to cluster formation. The same experiment on a stoichiometric oxide resulted in a continuous increase of the Au ISS signal with no abrupt changes in slope (Fig.32.4, black circles).

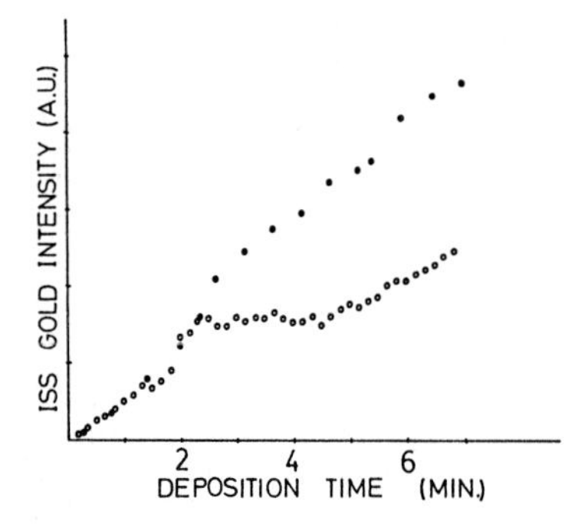

Fig.32.4. Evolution of the Au ISS intensity as a function of deposition time for 1): Au deposites on a thin reduced oxide at 80 K (open circles); 2): Au deposited on a thin stoichiometric oxide at 80 K (black circles)

From all these results we conclude that small amounts of Au deposited on thin (3 to 8 layers) reduced oxide films diffuse to the metal-oxide interface both at 300 K and at 80 K. Further Au deposition leads to the formation of Au clusters of increasing size on the surface of the oxide film.

When the experiment described in (1.) was repeated, but depositing K instead of Au on the reduced oxide at room temperature, the XPS intensities of both metal and oxide were attenuated, indicating that K atoms sit on the surface of the oxide film.

32.3 Final Remarks

The overall picture is then the following. When Au is deposited on the O-deficient oxide surface, a Cabrera-Mott capacitor forms, establishing an electric field through the layer which drives Au ions across it. Some of these ions will be trapped in the oxide matrix, building up a Coulomb repulsion that opposes the entrance of new ions. At this point the Cabrera-Mott mechanism stops and cluster formation occurs on the surface. Evidence for the presence of trapped Au will be presented elsewhere.

We thank M. Salmerón, R. Miranda, and J. Soler for encouraging discussions. This work has been supported by the CAICYT through grant number 1154-83.

References

32.1 S.A. Flodström, C.W.B. Martinsson, R.Z. Bachrach, S.B.M. Hagström, R.S. Bauer: Phys. Rev. Lett. **40**, 907 (1978);
J. Stöhr, L.I. Johansson, S. Brennan, M. Hecht, J.N. Miller: Phys. Rev. B**22**, 4052 (1980)
32.2 N. Cabrera, N.F. Mott: *Theory of the oxidation of metals*, Report on Progress in Physics, London XII, 163 (1949)
32.3 J.A. Cairns, J.E. Baglin, G.J. Clark, T.F. Ziegler: J. Catal. **83**, 301 (1983)
32.4 C. Ocal, E. Martínez, S. Ferrer: Surf. Sci. **136**, 571 (1984)
32.5 C. Ocal, S. Ferrer: Submitted to Surf. Sci.

III.6 Neutron Scattering

33. Surface Characterization by the Inelastic Scattering of Neutrons from Adsorbates

C.J. Wright

Materials Physics and Metallurgy Division, AERE Harwell, Didcot, Oxon
Great Britain

33.1 Introduction

This review illustrates how neutron inelastic scattering can be used to determine the coordination number or the local geometry of those sites at a catalyst surface at which chemisorption occurs. There are neutron diffraction methods for determining the density of different faces at the surface of a polycrystalline powder but these will not be referred to here [33.1].

The number of surface sites of different coordination which are available to a specific gas can, in principle, be deduced spectroscopically. One can compare the vibration spectrum of an adsorbed gas on the catalyst under investigation with that of the same gas adsorbed on known standards such as single-crystal surfaces. Greatest progress in this area has been made with studies of hydrogen adsorbed by nickel, palladium and platinum surfaces which can now be characterized in considerable detail. Theoretical and experimental progress now allows a vibration spectrum from hydrogen adsorbed upon a polycrystalline nickel surface, for example, to be resolved into components representative of hydrogen in four and three coordinate sites typical of those which occur on (100) and (111) surfaces. It should be emphasized, however, that the apparent understanding of nickel surfaces applies only to coverages $\leqslant 1$ monolayer. In this review we examine, in turn, our understanding of the vibration spectra of hydrogen absorbed by nickel, palladium, platinum, and MoS_2 and WS_2 surfaces.

33.2 Nickel

The scattering from hydrogen adsorbed by Raney nickel was first examined by time-of-flight [33.2] and beryllium-filter techniques [33.3]. The frequency distribution calculated from the time-of-flight data showed a local mode at $\simeq 1120\ cm^{-1}$ and surface modes whose energies were indistinguishable from those of bulk nickel. The beryllium-filter data, which covered the 200 to 2500 cm^{-1} region, contained scattering at two energies, one approximately twice the energy of the other. In the energy region where the local mode had been observed in the time-of-flight data, scattering was observed at 930 and 1130 cm^{-1}. This scattering was originally interpreted as coming from atoms which were vibrating parallel to the metal surface, in multiply coordinated sites. Scattering at 1930 and 2175 cm^{-1} was primarily ascribed to the first harmonics of the modes, together with some scattering from the metal-hydrogen stretching vibrations of singly coordinated hydrogen atoms. It was pointed out in another paper [33.4], however, that it was not possible to distinguish conclusively between singly and multiply bound hydrogen atoms in this system solely on the basis of the measured scattering data. The most interesting feature of the beryllium-filter data was the observation that the scattering at low frequencies clearly consisted of two components. Interpreters of these observations suggested three hypotheses. They suggested that the two components represented the parallel and perpendicular vibrations of hydrogen in a single site, that there was a range of different adsorption sites present at the surface, or that the hydrogen atoms were participating in collective surface excitations. Without the ability to conduct experiments in which it is possible to observe the momentum transfer dependence of the scattering intensity at these energies, it is not yet possible to rule out the last of the three possibilities. On the other hand, it is now generally accepted that it is possible to reproduce the major features of the inelastic scattering spectra solely by considering the vibrational degeneracy of hydrogen adsorbed in a single surface site. More detailed time-of-flight experiments have examined the coverage dependence of the inelastic scattering [33.5]. A high surface-area sample of nickel, supported upon alumina, was exposed to hydrogen at 300 K. Coverages between $\theta_{Ni}=0.23$ and 0.87 were examined. As the coverage increased, scattering at different energies was observed and a summary of these data, read from the figure in the original paper, is contained together with other neutron data in Table 33.1.

Table 33.1. Inelastic neutron scattering data for hydrogen adsorbed by Raney nickel

Neutron Scattering		Ni(100)		Ni(111)	
Excitations observed at low coverage cm^{-1}	Excitations observed at high coverage 33.5 cm^{-1}	Theory	EELS	Theory	EELS
		2452 [33.15] ω_{11}			
		1598 [33.12] ω_{11}		1578 [33.15] ω_{11}	1122 [33.10,11]
1130 [33.3]	1130			1210 [33.8] ω_1	1122 [33.10,11]
				1129 [33.12] ω_1	
930 [33.3]				12.33 [33.15] ω_1	
	800			1000 [33.12] ω_{11}	
640 [33.5] weak		637 [33.15] ω_1			
		602 [33.12] ω_1			
629 [33.6] weak		589 [33.8] ω_1	589 [33.9]		
	480				

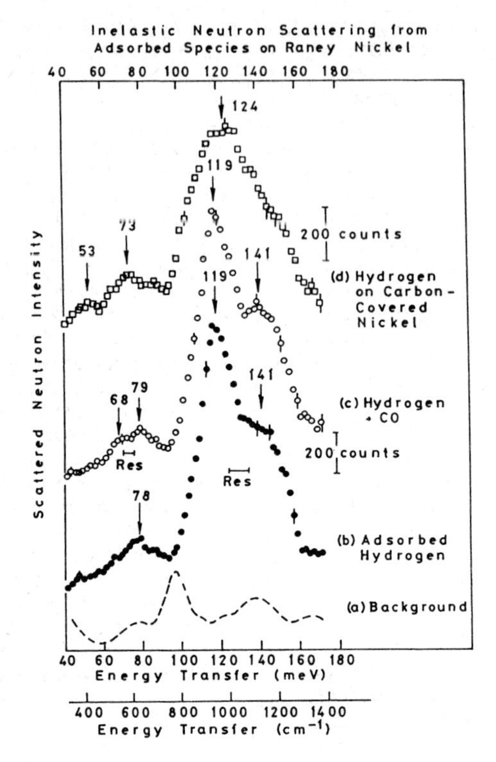

Fig.33.1. Beryllium filter spectra of (*a*) background (*b*) hydrogen adsorbed by Raney nickel (*c*) hydrogen and CO adsorbed by Raney nickel and (*d*) hydrogen adsorbed upon carbon-covered nickel

Beryllium filter results [33.6] from other authors confirmed the original findings and also provided evidence for a weaker excitation at 629 cm^{-1} (see Fig.33.1). Quantitative measurements of the intensities of the beryllium filter peaks as a function of coverage [33.7] showed that the intensities of modes at 944 and 1138 cm^{-1} were directly proportional to coverage between 10% and 100% of saturation coverage. In addition, the relative intensities of the two peaks were essentially constant and equal to two at all coverages.

The results of these neutron experiments need to be compared with single-crystal electron energy loss data and theoretical predictions before a reliable assignment can be made of the observed excitations [33.8-15]. Table 33.1 correlates this data for the 100 and 111 surfaces.

For the (110) surface it has been shown that hydrogen adsorption is very similar to that on the (111) surface since in both cases the preferred site has threefold coordination and since the coordination about a (110) site is identical to that on a facet of a (111) plane [33.14]. Comparisons of the results in Table 33.1 show that the majority of the hydrogen atoms adsorbed by Raney nickel at coverages up to saturation occupy sites of threefold coordination. The results of *Kelley* et al. [33.6] suggest that in addition a small proportion of the hydrogen atoms occupy fourfold sites.

Some of the finer details of the scattering and the adsorption process are not yet fully understood. There is disagreement between the predictions of different theoretical models [33.12,13] for the location of ω_{11} on the

100 surface and no experimental observation of this feature has yet been made. In the absence of any experimental evidence for this excitation the assignment of scattering near to 630 cm^{-1} to ω_1 must remain tentative.

Another imperfectly understood feature of the bonding is the role of singly coordinated hydrogen atoms adsorbed in the "on-top" position. It is known that hydrogen adsorbed at this position can be detected by infrared [33.16] and Raman [33.17] spectroscopies since ω_1 has been observed at 1880 cm^{-1} on alumina supported nickel. Theoretical models, however, suggest that this is the least stable of the high-symmetry sites and high-resolution neutron experiments in this energy region would be very useful if they could place an upper limit on the concentration of "on-top" hydrogen atoms. Hydrogen adsorbed in this position will also produce scattering between 600 and 850 cm^{-1} where the Ni-Ni-H angle deformation vibration will occur. In neutron scattering experiments made at higher hydrogen pressures additional scattering was observed at 484 cm^{-1}, which has yet to be satisfactorily accounted for [33.5] although it could well be due to adsorption on the sites recently proposed for platinum black [33.32].

The traditional models that have been used to explain these observations have been classical ones in which the hydrogen atoms are represented as simple harmonic oscillators adsorbed at specific sites on the surface. In contrast to this approach it has been suggested [33.18] that chemisorbed hydrogen could exhibit pronounced quantum effects. A calculation of the full potential energy surface shows that it is very anharmonic, resulting in strong coupling between hydrogen motions parallel and perpendicular to the surface. The quantitative predictions of the transition energies for this model are in reasonable agreement with the EELS data but they provide a very poor prediction of the neutron scattering spectra.

33.3 Palladium

Other metal surfaces have been investigated in much less detail than those of nickel. Measurements of the inelastic scattering from hydrogen adsorbed by palladium black [33.19] showed excitations at 823 and 921 cm^{-1} energies which are very close to those observed with infrared reflectance spectroscopy at 760 and 880 cm^{-1} for hydrogen adsorbed upon the surface of β-palladium hydride [33.20].

Since only two vibrations were observed, they were assigned to the parallel and perpendicular vibrations of hydrogen atoms in bridging sites, although it was realized that this was superficially incompatible with the observation that these two vibrations were both infrared-active. Calculations by *Muscat* [33.21] have shown that the most stable sites on the palladium surface are hollow sites in which the adsorbed hydrogen atoms are multiply bonded to atoms in and below the surface. In these sites both parallel and perpendicular excitations would be infrared-active, as suggested previously [33.19].

The spectra of hydrogen adsorbed by a relatively impure palladium black at two different coverages have been assigned to vibrations of hydrogen atoms at two coordinate binding sites by analogy with the spectra of transition metal hydridocarbonyls [33.22]. Published correlations, however, relate the ratio of ω_1 to ω_{11} to the metal-hydrogen-metal bond angle [33.23] and it has been pointed out that the assignment from [33.22] leads to a value for the palladium hydrogen bond length of 1.5 Å [33.24]. This appears to be too low when compared with the known value of the Ni-H [33.25] bond length of 1.84 Å.

It was argued [33.24] that an assignment which parallels the assignment for Raney nickel provides a much more acceptable value of the metal-hydrogen bond length.

33.4 Platinum

Time-of-flight and beryllium filter data have been published for hydrogen adsorbed by platinum black [33.26,27] but the original assignment of these spectra was difficult in the absence of any single-crystal data. Reference [33.28] compared this data and the results of more recently published EELS experiments. Table 33.2 and Fig.33.2 reproduce these results and those obtained by infrared spectroscopy for hydrogen adsorbed by MgO supported platinum [33.29].

Table 33.2. Data comparison using different methods for hydrogen adsorbed by platinum black

Neutron data cm^{-1}	EELS data cm^{-1} (111) [33.30]	EELS data cm^{-1} [6(111) × (111)] [33.31]	i.r. data [33.29] cm^{-1}	Models of *Sayers* and *Wright* [33.24] and *Sayers* [33.32]
512	550	500		$(111)_{11}$ 2-fold
616				
856				$(111)_{1}$ 3-fold
936		1130	950	$(111)_{11}$ 3-fold
1296	1230	1270		$(111)_{1}$ 2-fold
1696				
2000-2250			2100	

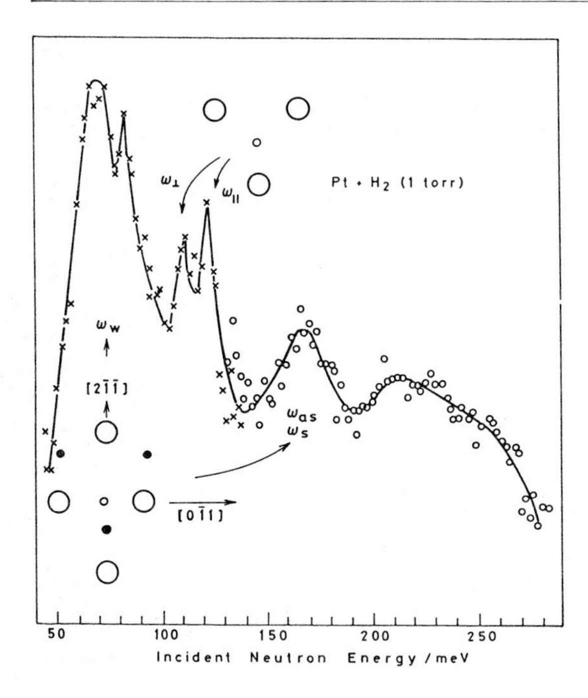

Fig.33.2. Beryllium filter spectra of hydrogen adsorbed by platinum black

The assignment of the platinum black data is more problematical than the assignment of the excitations observed for hydrogen adsorbed upon palladium or nickel. Despite similarities between some of the neutron peaks and the EELS results for the (111) surface, an assignment of those excitations to three coordinate sites may be incorrect even though it would be apparently consistent with the results for the other two metals. The uncertainty arises because one of the frequencies assigned to the hydrogen excitations is very low, and in the energy region observed in other metals for hydrogen in four sites, which should not be present on the platinum (111) surface. Using a central force model, and four coordinate sites, to interpret the excitations requires the nearest-neighbor, hydrogen-platinum distance to be 1.677 Å, a value small compared with the 1.8 Å determined for H adsorbed upon (111) nickel [33.32]. An alternative explanation [33.32] is that the two observed excitations arise from two coordinate hydrogen adsorbed on the (111) surface in a site such as that shown in Fig.33.2. Then ω_s and ω_{as} can be assigned to the vibration at 1230 cm^{-1} whilst ω_w is the vibration at 555 cm^{-1}. For this hypothesis the same central force model now leads to values for the nearest Pt-H distance of 1.93 Å. A satisfying aspect of this hypothesis is that it provides a possible explanation for the multiplicity of excitations seen in the neutron scattering experiments. The peaks at 856 and 896 cm^{-1}, for example, arise from hydrogen in the 3-fold sites on the 111 surface for which the platinum hydrogen distance can be calculated to be 1.905 Å. At the same time, however, there remains the problem of explaining why no 3-fold site excitations have been seen by EELS. INS traces have been observed for hydrogen adsorbed by highly dispersed platinum supported upon silica and alumina and they show an encouraging similarity to those for platinum black [33.33].

33.5 MoS_2 and WS_2

Sites at the surfaces of nonmetallic catalysts have also been investigated, and the adsorption of hydrogen by molybdenum and tungsten sulfides has been the subject of a number of inelastic scattering investigations which have gone a long way to clarify the sites at which adsorption occurs, and probably the sites from which hydrogen is transferred in hydrodesulfurization reactions.

Inelastic scattering from MoS_2 [33.34] data shows an intensity maximum at 153 cm^{-1}, and further subsidiary maxima near 352 cm^{-1}. The energies of these maxima correspond to those predicted for the acoustic and optic phonons [33.35,36]. The spectrum of the hydrogen-treated molybdenum sulfide differed in two respects from that of the pure material. It showed scattering beyond the cutoff in the vibrational density of states of MoS_2 and an enhancement of the scattering intensity in the region of 400 cm^{-1}. The scattering at 640 cm^{-1} was assigned to a local mode of adsorbed atomic hydrogen. Additional measurements with a beryllium-filter spectrometer detected the first harmonic of this excitation. Experiments with isostructural WS_2 [33.37] were successful in observing scattering from adsorbed hydrogen at energies up to 3000 cm^{-1} (Fig.33.3). In the energy region where the recorded spectra from hydrogen on WS_2 and MoS_2 overlapped, the similarity of the hydrogen excitations supported the hypothesis that hydrogen adsorbed on identical sites on the two surfaces.

The scattering intensities observed in the experiment with WS_2 were used to discriminate between the different sites at which the hydrogen atoms might adsorb on the tungsten sulfide surface. Predictions based on three possible models for the structure were compared with the results and the authors deduced that the model in which the hydrogen occupies an on-top site above the sulfur atoms most closely simulated the data.

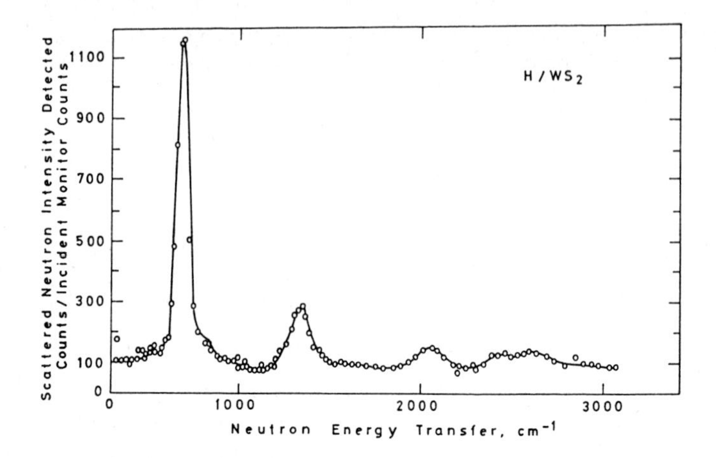

Fig.33.3. Inelastic scattering from hydrogen adsorbed by WS_2

Although this model appeared to explain the higher energy excitations of hydrogen on WS_2 in a satisfactory manner, an additional feature in the H_2/MoS_2 spectrum, the hydrogen-induced changes of the scattering intensity near to 350 cm^{-1} were more puzzling. Two explanations for this effect were suggested [33.34]. The sorption could induce changes in the lattice structure, or alternatively hydrogen sorption could selectively enhance those vibrations in which the sulfur atoms were the major participants. These would be at higher energies than those vibrations in which the molybdenum atoms were the major contributors and the observed scattering would therefore shift to higher energies on increasing sorption. It was shown 33.37 that this explanation was incompatible with calculations of the intensity of the scattering from hydrogen-covered MoS_2 and it was suggested instead that only an additional adsorption site could explain the measured intensity changes.

Experiments which have been undertaken at higher hydrogen pressures confirm this view. Measurements at up to 40 atm revealed scattering near to 400 cm^{-1}, which increased in intensity with pressure. The origin of this scattering had to be additional hydrogen sorption. Calorimetric measurements confirmed the existence of such a process, the extent of which increased with pressure until saturation occurred at $\simeq$50 atm [33.38]. The hypothesis that hydrogen sorption takes place at two sites at the MoS_2 surface, one at low, the other at high pressures, is consistent with all the available experimental data and theoretical calculations. It is not yet possible, however, to make any assignment of the site at which this high-pressure adsorption occurs. The scattering observed at 400 cm^{-1} arises from hydrogen species which adsorb at sites which become saturated at $\simeq$50 atm. Since kinetic data show that the hydrodesulfurization reaction is first order in the hydrogen partial pressure up to 50 atm, it is tempting to suggest that the species adsorbed at this second site plays an important part in the industrial process. The adsorption mechanism deduced from the scattering data is that adsorbed hydrogen molecules undergo dissociation, possibly at vacancies, to produce atoms which diffuse to sulfur sites. Hydrogen sulfide also adsorbs dissociatively since the inelastic scattering from the H_2S/MoS_2 system closely resembles that from H_2/MoS_2 [33.33]. A combination of neutron scattering with adsorption isotherm data leads to the suggestion that hydrogen sulfide poisons the molybdenum sulfide surface by adsorption and dissociation at vacancies. The hydrogen atoms produced by dissociation can diffuse over the surface to form H-S bonds, whereas the adsorbed sulfur atom remains in the original vacancy site and prevents further dissociative adsorption of hydrogen:

$$H_2S + \square + S^{--} \rightarrow 2SH^- \quad .$$

References

33.1 C.J. Wright: Specialist Periodical Reports of the Chemical Society Catalysis Vol. (1984)
33.2 R. Stockmeyer, H.M. Conrad, A.J. Renouprez, P. Fouilloux: Surf. Sci. **49**, 549 (1975)
33.3 A. Renouprez, P. Fouilloux, G. Coudurier, D. Tocchetti, R. Stockmeyer: Trans. Faraday. Soc. **73** (1977)
33.4 C.J. Wright: J. Chem. Soc. Faraday Trans. II **73**, 1497 (1977)
33.5 R. Stockmeyer, H.M. Stortnik, I. Natkaniec, J. Mayer: Ber. Bunsenges. Phys. Chem. **84**, 79 (1980)
33.6 R.D. Kelley, J.J. Rush, T.E. Madey: Chem. Phys. Lett. **66**, 159 (1979)
33.7 R.R. Cavanagh, R.D. Kelley, J.J. Rush: J. Chem. Phys. **77**, 1540 (1982)
33.8 T.H. Upton, W.A. Goddard: Crit. Rev. Solid State Mat. Sci. **10**, 261 (1981)
33.9 S. Anderson: Chem. Phys. Lett. **55**, 185 (1978)
33.10 W. Ho, N.J. Dinardo, E.W. Plummer: J. Vac. Sci. Technol. **17**, 134 (1980)
33.11 H. Ibach, D. Bruchmann: Phys. Rev. Lett. **44**, 36 (1980)
33.12 C.M. Sayers: J. Phys. C**16**, 2381 (1983)
33.13 J.E. Black: Surf. Sci. **105**, 59 (1981)
33.14 C.M. Sayers: Surf. Sci. **136**, 582 (1984)
33.15 J.E. Black, P. Bopp, K. Lutzenkirchen, M. Wolfsberg: J. Chem. Phys. **76**, 6431 (1982)
33.16 T. Nakata: J. Chem. Phys. **65**, 487 (1976)
33.17 W. Krasser, A.J. Renouprez: J. Raman Spectros. **8**, 92 (1979)
33.18 M.J. Puska, R.M. Nieminen, M. Manninen, B. Chakraborty, S. Holloway, J.K. Nørskov: Phys. Rev. Lett. **51**, 1081 (1983)
33.19 J. Howard, T.C. Waddington, C.J. Wright: Chem. Phys. Lett. **56**, 258 (1978)
33.20 I. Ratajczykowa: Surf. Sci. **48**, 549 (1975)
33.21 J.P. Muscat: Surf. Sci. **110**, 85 (1981)
33.22 I.J. Braid, J. Howard, J. Tomkinson: J. Chem. Soc. Faraday Trans. II **79**, 253 (1983)
33.23 M.W. Howard, U.A. Jayasooriya, S.F.A. Kettle, D.B. Powell, N. Sheppard: J. Chem. Soc. Chem. Comm. **18**, (1979);
U.A. Jayasooriya, M.A. Chester, M.W. Howard, S.F.A. Kettle, D.B. Powell, N. Sheppard: Surf. Sci. **93**, 526 (1980)
33.24 C.M. Sayers, C.J. Wright: J. Chem. Soc. Faraday Trans. I **80**, 1217 (1984)
33.25 K. Christmann, R.J. Behm, G. Ertl, M. Van Hove, W. Weinberg: J. Chem. Phys. **70**, 4168 (1979)
33.26 J. Howard, T.C. Waddington, C.J. Wright: J. Chem. Phys. **64**, 3897 (1976)
33.27 J. Howard, T.C. Waddington, C.J. Wright: Neutron Inelastic Scattering 1977, Vol. II (IAEA Vienna 1978) p.499
33.28 C.J. Wright, C.M. Sayers: Rep. Prog. Phys. **46**, 665 (1983)
33.29 J. Candy, P. Fouilloux, M. Primet: Surf. Sci. **72**, 167 (1978)
33.30 A.H. Baro, H. Ibach, H.D. Bruchmann: Surf. Sci. **88**, 384 (1979)
33.31 A.M. Baro, H. Ibach: Surf. Sci. **92**, 237 (1980)
33.32 C.M. Sayers: Surf. Sci. (in press)
33.33 A.J. Renouprez, J.M. Tejero, J.P. Candy: 8th Int. Congress on Catalysis Vol. III, **47** (1984)

33.34 C.J. Wright, C. Sampson, D. Fraser, R.B. Moyes, P.B. Wells, C. Riekel: J. Chem. Soc. Faraday I **76**, 1585 (1980)
33.35 R.A. Bromley: Philos. Mag. **23**, 1417 (1971)
33.36 C.M. Sayers: J. Phys. C Solid State Phys. **14**, 4969 (1981)
33.37 C.J. Wright, D. Fraser, R.B. Moyes, P.B. Wells: Appl. Catal. **1**, 49 (1981)
33.38 C. Sampson, J.M. Thomas, S. Vasudevan, C.J. Wright: Bull. Soc. Chim. Belg. **90**, 1215 (1981)

34. Infrared and Neutron-Scattering Studies of Ethene Adsorbed onto Partially Exchanged Zinc A Zeolite

J. Howard and J.M. Nicol

Department of Chemistry, University of Durham, Durham DH1 3LE, Great Britain

J. Eckert

Los Alamos National Laboratory, P.O. Box 1663/MS H805
Los Alamos, NM 87545, USA

Infrared and inelastic neutron-scattering studies of ethene adsorbed onto ZnNaA zeolite show that the adsorbed molecule occupies a single adsorption site. The C-H stretching modes are not observed in the infrared data but are seen as a broad band in the neutron spectrum. Some low-frequency adsorbate-adsorbent modes are assigned.

34.1 Introduction

In vibrational studies of species adsorbed within zeolite frameworks most attention has centered on the changes in the internal mode frequencies of the adsorbate with respect to the free molecule. On adsorption the three rotational and translational modes of the free molecule usually become hindered. These modes directly reflect the surface-adsorbate bonding, and their frequencies can therefore be used to calculate force constants and barriers to rotation. This type of analysis provides a sensitive test for any model of adsorbate-adsorbent interactions. These modes have not generally been studied because of the difficulties in observing them using optical spectroscopy, viz.:

1) these modes occur below 1000 cm^{-1} and in this region zeolite frameworks are usually extremely absorbing in the infrared;
2) many zeolites fluoresce, making Raman experiments difficult or impossible.

Neither of these difficulties applies to inelastic neutron-scattering (INS) spectroscopy for which there are also no electromagnetic selection rules [34.1]. In an INS experiment the scattering mechanism involves direct interaction between the incident neutron and the scattering nucleus. Each nucleus has a characteristic incoherent cross section (σ). Since the σ values for the atoms of a dehydrated zeolite are very small compared with the very large value ($80 \times 10^{-24} cm^2$) for the hydrogen atoms in the adsorbate [34.1], the zeolite is relatively "invisible" in an INS experiment. The INS spectrum is known to be dominated by those modes which involve large amplitude motion of the hydrogen atoms [34.1].

Both IR and INS spectra can provide information on the numbers of surface sites if multiple bands are observed for a given mode. Neutron data in particular can also readily be interpreted in terms of relative occupancies of these sites. Spectroscopic studies on ethene adsorbed on Ag_{12}-A zeolite [34.2, 3] revealed the presence of two distinct adsorption sites, while in the present case there appears to be only one site.

An adsorbate molecule may interact with a surface in a number of geometries. Group theoretical considerations in the case of IR spectra or direct intensity calculations of INS bands can then often be used to deduce the symmetry of the adsorbate-surface system as well. In addition, the position of a band, e.g., the triple band stretch of ethyne, is often a good indicator of the geometry. This has been used to show that ethyne interacts "side-on" with Ag_{12}-A zeolite [34.4].

We present INS and IR data on C_2H_4 and C_2D_4 adsorbed on a partially exchanged type A zeolite. This represents part of a program for studying the reactions of small hydrogeneous molecules over zeolite catalysts. While we observed C_2H_2 to react very slowly [34.5] with ZnNaA, C_2H_4 does not seem to do so.

34.2 Experimental

The ZnNaA was prepared by ion exchanging NaA (Linde) powder (no binder) for seven days in a 0.1 M solution of $ZnCl_2$ containing the stoichiometrically correct quantity of zinc ions. After washing, the zeolite was analyzed for Zn and Na by atomic absorption. The atomic ratio Zn/Na was determined to be 0.57.

Our IR experiments were carried out on a self-supporting ZnNaA disc in an all-metal cell. Pretreatment (heating to 450°C under a vacuum of 10^{-5}Torr) and adsorption, etc., were carried out in situ. Measurements were made using a Perkin Elmer 580B Infrared Spectrophotometer.

Neutron-scattering measurements were made using the beryllium filter spectrometer at AERE Harwell ($100 \rightarrow 800$ cm^{-1}) [34.1] and the filter difference spectrometer ($300 \rightarrow >4000$ cm^{-1}) at Los Alamos National Laboratory [34.6]. For the Harwell data the transition frequencies have been calculated from the peak maxima using standard correction factors [34.7]. The Los Alamos data were analyzed as previously described [34.6]. For the INS experiments the samples were pretreated as for the IR measurements and the samples held in thin-walled aluminum cells. The spectrum of the degassed zeolite was obtained to serve as a background. The C_2H_4 was then adsorbed and the spectrum rerun. At Harwell, coverages of 0.95 and 1.75 C_2H_4 (sequentially adsorbed onto the same sample) and of 1.75 C_2D_4 molecules per supercage were studied. The Los Alamos experiment involved an adsorption capacity of 1.03 C_2H_4 molecules per supercage. Adsorption was carried out at ambient temperature and the Harwell measurements were made at 80 K while those at Los Alamos were made at 12 K.

34.3 Results and Discussion

The IR spectra of dehydrated ZnNaA and of C_2H_4 adsorbed on it (overpressure of 5 Torr) are shown in Fig.34.1. Bands due to internal modes of the adsorbed species are clearly seen at 1326, 1451, and 1602 cm^{-1}. These transitions are readily assigned by comparison with the optical data [34.8] for gaseous C_2H_4 (Table 34.1). The change in activity of two of the modes reflects the symmetry change on adsorption. On evacuation for five minutes the intensities of the bands fall by at least a factor of 10. These observations are indicative of *relatively* weak nondissociative adsorption at a single site presumably in the α cage of the zeolite, as the ethene molecule is too large to enter the β cage. The band narrowing which takes place on adsorption

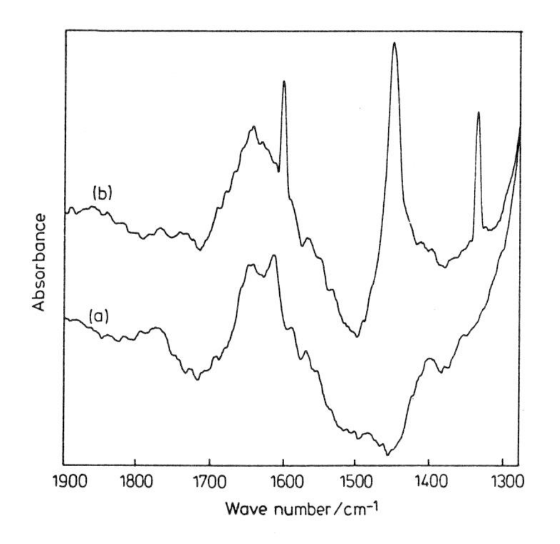

Fig.34.1. Infrared spectrum of ZnNaA (*a*) after heating under vacuum to 450°C and (*b*) after adsorption of C_2H_4 at an overpressure of 5 Torr

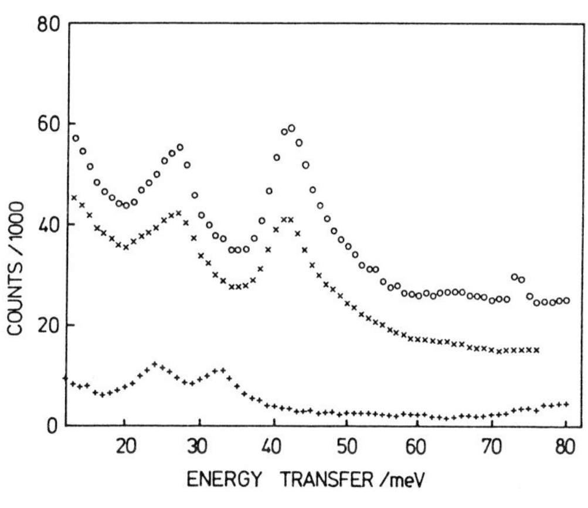

Fig.34.2. INS data (80 K) on ZnNaA obtained at AERE Harwell: (o) ZnNaA + C_2H_4 (0.95 molecules per supercage); (x) ZnNaA + C_2H_4 (1.75 molecules per supercage); (+) ZnNaA + C_2D_4 (1.75 molecules per supercage). The spectrum of the degassed zeolite is subtracted and the (o) data points have been multiplied by a factor of two

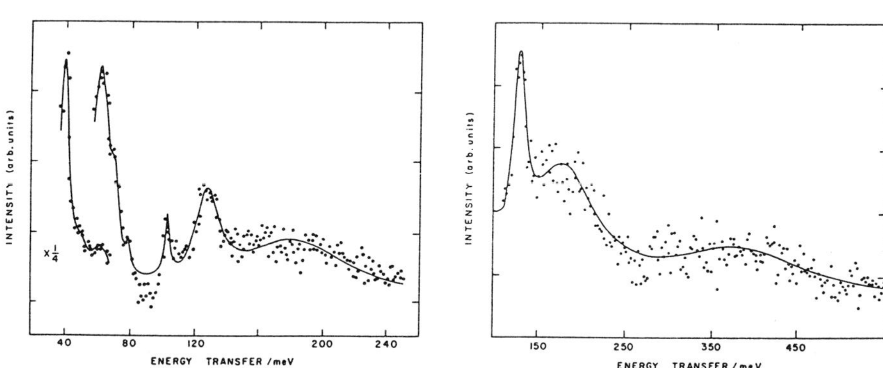

Fig.34.3. INS spectrum (12 K) of ZnNaA + C_2H_4 (1.03 molecules per supercage) obtained at Los Alamos (1 meV ~ 8.067 cm^{-1}). The spectrum of the degassed zeolite has been subtracted

Table 34.1. IR bands (cm^{-1}) and their assignments for C_2H_4 adsorbed on ZnNaA zeolite

Assignment	Activity (gas phase)	Gas Phase [34.8]	Adsorbed
ν_2 (C = C stretch)	Raman	1623	1602
ν_{12} (CH_2 deformation)	IR	1443	1451
ν_3 (CH_2 deformation)	Raman	1342	1326

Table 34.2. INS data (cm^{-1}) and assignments for C_2H_4 adsorbed on ZnNaA zeolite

C_2H_4	C_2D_4	Observed deuturation shift	Predicted deuturation shift	Assignment
183	156	0.84	0.86	τ_y
310	237	0.76	0.73	τ_x
490 566				Amplified zeolite modes
636				Overtone of τ_x
825				CH_2 Rock
1024				CH_2 Twist
1460				ν (C = C) CH_2 scissoring CH_2 rock, etc.
3050				C-H stretching modes

indicates that the adsorbed molecule is not freely rotating. In view of this it is remarkable that we do not observe any bands due to the C-H stretching modes of adsorbed C_2H_4 (expected ca. 3000 cm^{-1}). A similar observation for CdX zeolite has been reported by *Yates* and *Carter* [34.9] although these modes have been readily observed in a wide range of ion-exchanged zeolites.

The reduced intensity of these modes must result from subtle electronic changes on adsorption, and it is known that to understand these further, a good normal coordinate analysis is needed which obviously requires the frequencies of the modes to be known. The nonobservation of the C-H modes also casts some doubt on the true nature of the adsorbed species. We have studied C_2H_4 and C_2D_4 on ZnNaA using INS spectroscopy (Figs.34.2,3), summarized in Table 34.2.

34.4 INS Data Below 400 cm^{-1}

Two distinct bands are observed in the INS spectra of both adsorbed C_2H_4 and C_2D_4 in this region, and there is little change on increasing the coverage. Since there are no intramolecular modes below 586 cm^{-1}, these transitions must be hindered rotations and/or translations of the adsorbed molecule relative to the zeolite. The higher wave-number transitions occur at 310 (C_2H_4) and 237 (C_2D_4) cm^{-1} and, assuming these to be the same transition, this yields an isotopic ratio of 0.76. It is immediately evident that the only reasonable assignment of the 310 (C_2H_4) and 237 (C_2D_4) cm^{-1} bands is to the hindered rotation (τ_x) about the C = C axis. Using gas phase C_2H_4 geometry, the isotopic shift predicted is 0.72 and hence the C_2D_4 mode is expected at 223 cm^{-1}. This is within observation errors. No other mode has such a large isotopic shift since this is the only mode which involves little or no carbon motion. The corresponding mode occurs at 417 (C_2H_4) and 298 (C_2D_4) cm^{-1} for a silver exchanged A zeolite [34.2]. Since the adsorbed ethene is not easily removed on evacuation of the AgA sample, it is reasonable to expect a higher frequency in this case. The corresponding mode in Zeise's salt [$KPtCl_3(C_2H_4)H_2O$] occurs at 1180 cm^{-1} since the metal-ethene bond in this very stable complex is much stronger than in the zeolite [34.10].

The lower wave-number bands at 183 (C_2H_4) and 156 (C_2D_4) yield an isotopic ratio of 0.84. This is in good agreement with that expected (0.86) for the hindered rotation about an axis parallel to the surface but perpendicular to the C = C axis (τ_y). This corresponds to the antisymmetric metal-ethene stretching mode of organometallic complexes. In Zeise's salt τ_y occurs at 490 cm^{-1}, once again reflecting its stronger bonding [34.10].

34.5 INS Data Above 400 cm^{-1}

The INS data obtained on ZnNA + C_2H_4 using the filter difference spectrometer at Los Alamos are shown in Fig.34.3a,b and summarized in Table 34.2. Clearly the intense band at 318 cm^{-1} corresponds to τ_x observed at 310 cm^{-1} using the Harwell spectrometer. The overtone of this band is seen at 636 cm^{-1}. Above this value relatively narrow bands occur at 825 and 1023 cm^{-1} with broader features centered at ca. 1460 and 3050 cm^{-1}. The broad features clearly contain a number of unresolved features. The 3050 cm^{-1} band must be assigned to the four C-H stretching modes, which were not observed in the IR data. Similarly, the 1460 cm^{-1} band must contain a number of modes including ν(C = C), CH_2 scissoring modes and the CH_2 rock.

The assignment of the 825 (sharp) and 1024 cm^{-1} bands is more difficult. Although bands are observed in the INS spectrum of Zeise's salt [34.11] at 840 and 1030 cm^{-1}, in good agreement with the C_2H_4 + ZnNaA data, the data for the complex also contains an intense band at 720 cm^{-1}. No similar band is found for the adsorbed species. An examination of the INS data [34.11] for C_2H_4 + Ag13X also shows bands at ca. 820 and 1000 cm^{-1} with a minimum close to 700 cm^{-1}. Thus the zinc A and silver 13X data are comparable in this respect, and the adsorbed molecule is spectroscopically more similar to gaseous ethene than it is to the strongly bonded form in the complex. Although it is known that the ν(C = C) stretching mode is mixed [34.10] and so comparisons between compounds based on the value of this mode are dubious, it is worth noting that ν(C = C) occurs at 1602 cm^{-1} on ZnNaA and at 1623 cm^{-1} in gas phase C_2H_4. These similar values imply a relatively weak perturbation of the double bond on adsorption.

The INS bands at 720 and 840 cm^{-1} in Zeise's salt were assigned [34.11] to CH_2 rocking modes (ν_{17} and ν_{23}). In the gas phase these modes (ν_{10} and ν_6, respectively) occur at 826 and 1236 cm^{-1} [34.8]. It is likely, therefore, that the 825 and 1024 cm^{-1} bands we observe correspond to the CH_2 rock and CH_2 twisting modes which occur at 826 (ν_6) and 1023 (ν_4) cm^{-1} respectively in uncoordinated C_2H_4 [34.8].

We have not attempted to assign the weak bands in the region 480-650 cm^{-1} (Figs.34.2,3a). None of the hindered rotations or translations of adsorbed

C_2H_4 can occur in this region nor are there any intramolecular C_2H_4 modes. It is possible that we are observing framework vibrations enhanced in intensity by the "riding" motion of the adsorbed species, but this must be regarded as only a tentative explanation.

Acknowledgments. We should like to thank the SERC for the award of a research studentship to one of us (J.M.N.) and for the provision of neutron beam facilities in the United Kingdom. Work at Los Alamos was supported in part by the Division of Basic Energy Sciences, U.S. Department of Energy.

References

34.1 J. Howard, T.C. Waddington: In *Advances in Infrared and Raman Spectroscopy*, Vol.7, ed. by R.J.H. Clark, R.E. Hester (Heyden, 1980) Chap.3
34.2 J. Howard, K. Robson, T.C. Waddington, Z.A. Kadir: Zeolites **2**, 1 (1982)
34.3 J. Howard, Z.A. Kadir, K. Robson: Zeolites **3**, 113 (1983)
34.4 J. Howard, Z.A. Kadir: Zeolites **4**, 45 (1984)
34.5 J. Howard, J.M. Nicol: Unpublished work
34.6 A.D. Taylor, E.J. Wood, J.A. Goldstone, J. Eckert: Nucl. Instr. Method **221**, 408 (1984)
34.7 A.D. Taylor, J. Howard: J. Phys. E. Sci. Instr. **15**, 1359 (1982)
34.8 T. Shimanouchi: *Tables of Molecular Vibrational Frequencies Consolidated*, Vol.I (NSRDS 1972)
34.9 J.L. Carter, D.J.C. Yates, P. Lucchesi, J.J. Elliot, V. Kevorkian: J. Chem. Phys. **70**, 1126 (1966)
34.10 J. Hiraishi: Spectrochim. Acta **25**A, 749 (1969)
34.11 J. Howard, T.C. Waddington, C.J. Wright: J. Chem. Soc. Faraday Trans. II, **73**, 1768 (1977)

Part IV

Clean and Adsorbate-Covered Metals

IV.1 Clean Metal Surfaces

35. Theoretical Study of the Structural Stability of the Reconstructed (110) Surfaces of Ir, Pt and Au

H.-J. Brocksch and K.H. Bennemann

Institut für Theoretische Physik, WE05, Freie Universität Berlin
Arnimallee 14, D-1000 Berlin 33, FRG

Within an effective two-band model the structural stability of the $T = 0$ ground state of the (110) surfaces of Ir, Pt, and Au is compared for several reconstruction models. Band-structure contributions to the surface cohesive energy are treated in tight binding via the recursion method in the self-consistent Hartree approximation. Intersite repulsive interactions are taken into accout by a Born-Mayer potential fitted to bulk elastic data and modified at the surface. Hybridization of the 5d orbitals with a nearly free electron band of mixed sp character proves to be sufficient to discriminate between different reconstruction models for Ir and Pt. For Au, additional multilayer surface relaxation effects seem to become important in stabilizing the atomic configuration of lowest energy. For all three metals the fully relaxed missing-row structure is found to be the favored one on energetic grounds. For Au at least, this is in agreement with recent experiments.

35.1 Introduction

Although the (1×2) symmetry of the (110) surfaces of the late 5d-transition metals has long been known, the detailed local arrangement of atoms within their first few surface layers is still a highly controversial question. Only most recently, from among several reconstruction patterns conforming to the (1×2)-surface geometry observed by LEED the fully relaxed missing-row structure was confirmed by different experimental techniques for Au [35.1]. These findings can be supported by the results of our tight-binding calculations in which the ground-state cohesive energies of the missing-row surface model and of the reconstruction model proposed by *Bonzel* and *Ferrer* [35.2] (BF model) are compared to the unreconstructed (1×1) surface. Whereas from our previous results for the (110) surface of Pt [35.3] the missing-row pattern seemed to be degenerate in energy with the (1×1) phase, we are now in a position to distinguish clearly between the structures examined. This is mainly attributed to the fact that in the improved calculation the mixing of 5d electrons with sp-like states was included.

In the following a brief outline is given of the calculational procedure applied and the results obtained so far. A more complete version giving full details will be published elsewhere [35.4].

35.2 The Model

The total energy of a surface atom at site (i) is split into a band-structure part $E_{BS}(i)$ for d and sp electron cohesive energy and a Born-Mayer energy term $E_{BM}(i)$ accounting for repulsive interatomic interactions:

$$E_{coh}(i) = E_{BS}(i) + E_{BM}(i) \quad . \tag{35.1}$$

The band-structure energy is given by

$$E_{BS}(i) = \sum_{\alpha} \int_{-\infty}^{\varepsilon_F} d\varepsilon[\varepsilon - \varepsilon_0^{\alpha}(i)]\rho^{\alpha}(i;\varepsilon) \quad . \tag{35.2}$$

The sum in (35.2) extends over the five (spin-degenerate) 5d orbitals and one plane wave band of sp-like character hybridized with the 5d states. The partial local densities of states (LDOS) $\rho^{\alpha}(i;\varepsilon)$ are calculated from

$$\rho^{\alpha}(i;\varepsilon) = -\frac{1}{\pi} \lim_{\eta \to 0^+} \text{Im}\, G_{00}^{\alpha}(i;\varepsilon + i\eta) \tag{35.3}$$

via a continued fraction expansion of the local Green's functions $G_{00}^{\alpha}(i;\varepsilon)$ at (i) [35.5]. Within the effective two-band model we consider here, a set of two self-energies is assigned to each site: $\varepsilon_0^{5d}(i)$ for the 5d band and $\varepsilon_0^{sp}(i)$ for the sp band. The atomic self-energies are iteratively shifted to achieve local charge neutrality at the surface, which is the condition of self-consistency in the Hartree approximation [35.6]. The 5d tight-binding hopping parameters were taken from *Smith* and *Mattheiss* [35.7] in the two-center approximation, the ones for sp overlap and hybridization from *Harrison* and *Froyen* [35.8]. Exponential distance dependence [35.3] was assumed for the dd hopping, and canonical dependences for all other hopping matrix elements [35.8].

The repulsive energy contributions were determined from a simple ansatz [35.3] for the mutual distance dependence R_{ij} between sites (i) and (j) of the cohesive energy:

$$E_{coh}(R_{ij}) = \frac{P_s(i)}{\sqrt{Z_s(i)}} \left\{ \sum_{j=1}^{Z_s(i)} \exp[-2q(R_{ij}/R_0 - 1)] \right\}^{\frac{1}{2}}$$

$$- \frac{q}{p} \frac{R_s(i)}{Z_s(i)} \sum_{j=1}^{Z_s(i)} \exp[-p(R_{ij}/R_0 - 1)] \quad . \tag{35.4}$$

Here, $Z_S(i)$ is the coordination number at site (i). The first term in (35.4) represents the distance dependence of the band structure energy (in 2nd moment approximation for the hopping integrals), the second term gives the Born-Mayer energy. For the bulk crystal the parameters occurring in (35.4) were obtained from the standard equilibrium conditions relating the derivatives of (35.4) to experimental elastic and thermochemical data. For a surface site (i) the terms $P_S(i)$ and $R_S(i)$ in (35.4) take the form

$$\begin{aligned} P_s(i) &= E_{BS}^{5d}(i) + f_{att}[\rho_s^s(i)] \cdot E_{BS}^{sp}(i) \\ R_s(i) &= E_{BS}^{5d}(i) + f_{rep}[\rho_s^s(i)] \cdot E_{BS}^{sp}(i) \quad . \end{aligned} \tag{35.5}$$

The weight factors $f_{att} < 1$ and $f_{rep} \leq 1$ account for the reduced sp-electron charge density since the sp-wave function tails leak into the vacuum, and provide an estimate of the change in attractive as well as repulsive energy between adjacent sites due to the perturbing surface potential. They are functions [35.4] of the local sp-electron density $\rho_s^S(i)$ and d-band count and can be determined from the jellium surface model of *Lang* and *Kohn* [35.9].

Recursion method calculations were performed for 10 levels using the strictly momentum-preserving terminator procedure of *Beer* and *Pettifor* [35.10]. Preliminary calculations were performed for the central atom of a cubic bulk cluster of about 5000 atoms to obtain necessary startup information (position of Fermi level, etc.) before going to the surface [35.4]. The actual surface calculations were done for all the inequivalent surface sites (hatched atoms of the insets in Fig.35.1) of the respective structures. Multilayer surface relaxations, including row pairing, were allowed by minimizing the energy expression (35.4) with respect to the interlayer spacings by means of an iterative procedure [35.4]. With the binding energy results per site taken as input for the energy terms in (35.5), the minimization of (35.4) produced a first set of relaxation coordinates which in turn served as input for another self-consistent binding-energy calculation performed for the geometry of the relaxed surface lattice, and so on.

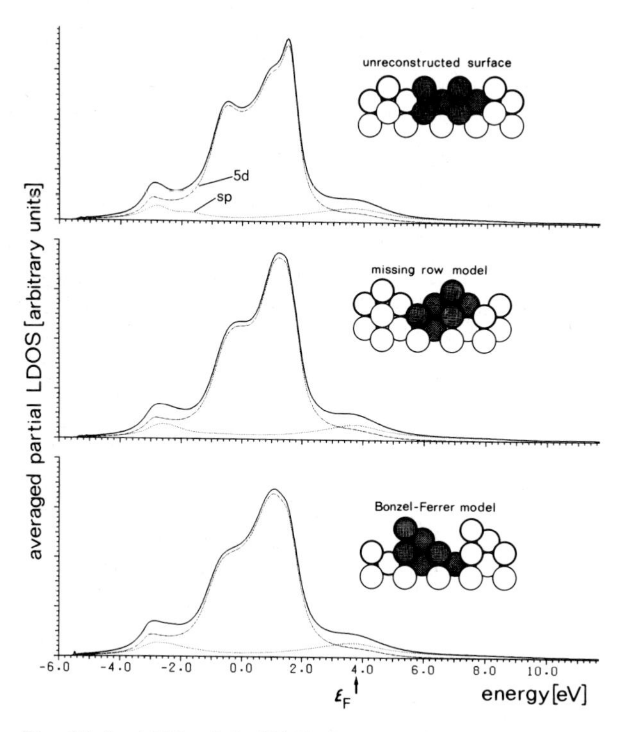

Fig.35.1. LDOS of Au(110) for the 3 surface models considered [35.4], averaged over sets of 6 atoms shown hatched in the respective insets. (——): total LDOS; the contributions from the 5d band and the sp band are marked. The arrow marks the position of the Fermi level

35.3 Results and Conclusions

As the central result of our calculations we found that among the different reconstruction models compared, the fully relaxed missing-row surface configuration proves to be the most stable one for all three metals. The ground state of the BF structure, however, always turns out to be energetically unfavored, independently of whether relaxation is included in the calculation or not. The energy differences involved [35.4] between the (1×1) phase and the missing-row one are fairly small, ranging from about 0.5 eV for Ir down to 0.1 eV for Au. (Note that the energies given refer to the subclusters of 6 inequivalent surface atoms shown hatched in the insets of Fig.35.1.) For Ir and Pt the stability of the missing-row structure is predominant even for the unrelaxed surfaces. This can be attributed mainly to an increase in binding energy effected from band hybridization. Switching on the relaxation increases this tendency slightly. The additional gain in energy due to relaxation mechanisms seems to be important, however, for the surface of gold. For all three metals and all the models studied the relaxation of the topmost layer is always found to be directed outwards. Referring to the bulk interlayer spacing of the fcc (110) geometry we got some 9% for Au, about 5% for Pt and about 4% for Ir. (The values are not much different for the three models; [35.4] for details.) The adjoining first few surface layers show an oscillatory relaxational behavior which is damped strongly on approaching the bulk. In addition, a weak tendency for row pairing in the second layer of the missing-row structure (3rd for the BF model) was obtained.

For gold these results agree qualitatively with recent experiments; *Robinson* and co-workers [35.1], however, found an even stronger outward relaxation and degree of row pairing. The physical origin of this strong outward relaxation may be found in the repulsive interaction of the (nearly) closed 5d shells in Au. The Pettifor-Liberman virial theorem shows that the 5d shells are somewhat compressed in the bulk due to the attractive sp-electron bonding. For surface sites this bonding is weaker because of the reduced sp-electron density thus allowing the d shells to relieve part of their pressure. This interplay between attractive and repulsive forces and their change in character with decreasing d-band filling is difficult to treat and may rather lead to an overestimation of the relaxation for Ir and Pt within our simple two-band model.

For the reconstruction models studied [35.4] the surface LDOS do not show any pronounced resonant surface-specific features around the Fermi level in agreement with UPS results [35.11]. As an example, Fig.35.1 gives the respective LDOS obtained for the surface of Au [35.4], averaged over the set of sites shown hatched in the insets of the figure. The behavior around the Fermi energy is governed by the small hump in the partial LDOS of the sp-like states, but it is just as smooth as in the bulk. From this it can be inferred that it is rather unlikely that a mechanism like the pseudo-Jahn-Teller effect driving a displacive reconstruction as, e.g., in the case of W(001), should be responsible for the reconstruction in this case. In summarizing, we may conclude that the origin of the reconstruction for these surfaces may be found in a delicate competition on a local scale between attractive and repulsive electronic forces, favoring slightly a missing-row-like atomic configuration with symmetric (111) facets.

Acknowledgment. This work was supported in part by the DFG and the DFG-Sonderforschungsbereich 6.

References

35.1 For a survey,see: I.K. Robinson, Y. Kuk, L.C. Feldman: Phys. Rev. B**29**, 4762 (1984) and references therein

35.2 H.P. Bonzel, S. Ferrer: Surf. Sci. **118**, L263 (1982)

35.3 D. Tomanek, H.-J. Brocksch, K.H. Bennemann: Surf. Sci. **138**, L129 (1984)

35.4 H.-J. Brocksch, K.H. Bennemann: Submitted for publication, and also in preparation for publication

35.5 See the articles by R. Haydock, V. Heine, M.J. Kelly: In *Solid State Physics*, Vol.35, ed. by Turnbull and Seitz (Academic, New York 1980)

35.6 G. Allan: Ann. Phys., Paris 5, 169 (1970); and in *Handbook of Surfaces and Interfaces*, ed. by L. Dobrzynski (Garland STPM, New York 1978)

35.7 N.V. Smith, L.F. Mattheiss: Phys. Rev. B**9**, 1341 (1974)

35.8 W.A. Harrison, S. Froyen: Phys. Rev. B**21**, 3214 (1980); B**20**, 2420 (1979)

35.9 N.D. Lang, W. Kohn: Phys. Rev. B**1**, 4555 (1970)

35.10 N. Beer, D.G. Pettifor: In *Proc. NATO Advanced Study Institute, Gent, Belgium* (Plenum, New York 1982)

35.11 For Au cf. P.O. Nilsson, L. Ilver: Solid State Commun. **17**, 667 (1975); J. Freeouf, M. Erbudak, D.E. Eastman: Solid State Commun. **13**, 771 (1973)

36. The Structure and Surface Energy of Au(110) Studied by Monte Carlo Method

T. Halicioğlu, T. Takai, and W.A. Tiller

Stanford/NASA Joint Institute for Surface and Microstructural Research,
Department of Materials Science and Engineering, Stanford University
Stanford, CA 94305, USA

The surface energy and configuration of a Au(110) surface was calculated using a Monte Carlo technique based on a potential energy function containing two-body and three-body interactions. The potential function, with the parameters calculated from experimental quantities, reproduced the average surface energy and the cohesive energy of Au correctly. Simulation calculations were performed considering discrete atoms up to the third layer from the surface. An ideal defect-free (110) surface was first considered and calculations were performed at $T = 100$ K. The relaxed surface preserved its (1×1) structure; however, the interlayer spacing d_{12} between the first and second layers displayed a contraction ($\Delta d_{12} = -9.0\%$), while the spacing d_{23} between the second and third layers expanded ($\Delta d_{23} = +4.9\%$).

For the "missing row" model of the (110) surface, calculations at $T = 100$ K produced a surface energy slightly lower than the regular (1×1) surface. This indicates that the missing row surface configuration is more stable energetically. The relaxed configuration of the (1×2) missing row model exhibited variations in the interlayer spacing somewhat less than for the (1×1) case. They were calculated as $\Delta d_{12} = -7.1\%$ and $\Delta d_{23} \approx 0.0\%$.

36.1 Introduction

In the study of solid surfaces, reconstruction plays a very important role and knowledge of the detailed surface structure is essential to a proper understanding of many reaction mechanisms taking place there. Unfortunately, a precise atomic level structural determination of reconstructed surfaces is a difficult task. In many instances, interpretation of experimental measurements is based on the choice of the "right model". Reconstruction of the Au(110) surface is one of the most thoroughly investigated surfaces among metals, and still some controversy exists [36.1-6]. At the present, two models seem to be most widely accepted for the (1×2) reconstructed Au(110) surface: i) the missing row model and ii) the saw-tooth model which has been suggested recently by *Bonzel* and *Ferrer* [36.6].

The objective of the present work is to determine the energetically most stable surface structure for the Au(110) index plane using an atomistic approach. The investigation was performed employing a Monte Carlo technique based on a potential energy function which includes both two-body and three-body interactions. For the Au(110) surface, three different models (i.e., unreconstructed, missing row and saw-tooth) were considered. The missing row model was found to be the energetically most favorable surface.

36.2 Method of Calculation

Calculations were performed using a Monte Carlo procedure based on the Metropolis approximation. This technique generates ensemble averages for desired quantities and requires a potential energy function which describes interactions among the particles in the system [36.7]. In the present work, the potential energy was calculated taking into account the two- and three-body interactions. In general, for a system of N particles, the total potential energy may be expanded as [36.8]:

$$\Phi = \frac{1}{2!}\sum_{i}^{N}\sum_{\substack{j \\ i\neq j}}^{N} u(\mathbf{r}_i,\mathbf{r}_j) + \frac{1}{3!}\sum_{i}^{N}\sum_{\substack{j \\ i\neq j\neq k}}^{N}\sum_{k}^{N} u(\mathbf{r}_i,\mathbf{r}_j,\mathbf{r}_k) + \dots \quad , \tag{36.1}$$

where $u(\mathbf{r}_i,\mathbf{r}_j)$ and $u(\mathbf{r}_i,\mathbf{r}_j,\mathbf{r}_k)$ represent the two-body and three-body interactions, respectively. The position of the ith particle was denoted by $\mathbf{r}_i$. In the numerical calculations, terms higher than three-body interactions were neglected. The two-body part of the potential is represented by a Mie-type potential

$$u(r_{ij}) = \frac{\varepsilon}{(m-n)}\left\{n\left(\frac{r_0}{r_{ij}}\right)^m - m\left(\frac{r_0}{r_{ij}}\right)^n\right\} \quad , \tag{36.2}$$

with $r_{ij} = |\mathbf{r}_i - \mathbf{r}_j|$; the equilibrium separation is denoted by r_0 and ε is the energy at $r_{ij} = r_0$. The exponents m and n account for the repulsive and attractive terms, respectively. The three-body part, on the other hand, was approximated considering the Axilrod-Teller triple dipole interactions [36.9] given by

$$u(\mathbf{r}_i,\mathbf{r}_j\cdot\mathbf{r}_k) = Z\cdot\frac{1 + 3\cos\theta_i\ \cos\theta_j\ \cos\theta_k}{(r_{ij}\cdot r_{ik}\cdot r_{jk})^3} \quad , \tag{36.3}$$

where Z is the three-body intensity parameter; $\theta_i,\theta_j,\theta_k$ and r_{ij},r_{ik},r_{jk} represent the angles and the sides of the triangle formed by the three particles i, j and k.

36.3 Determination of Parameters

To evaluate the total energy Φ for a specific material such as Au, the energy parameters (ε,r_0,m,n and Z) of (36.2,3) must be calculated. In general, this is a cumbersome procedure and requires accurate fitting to appropriate experimental data. In this work, as a first approximation, the exponents m and n of (36.2) were taken as 12 and 6, respectively. This assumption reduces (36.2) to a Lennard-Jones function. The remaining parameters (ε,r_0 and Z) were obtained from fits to reported surface energy values and the experimental bulk cohesive energy for Au considering the stability criterion expressed by

$$\frac{\partial\Phi}{\partial V} = 0 \quad , \tag{36.4}$$

where V denotes the total volume of the system. The cohesive energy was taken as -3.73 eV which is an average value over the reported data in [36.10]. The surface energy value (1550 erg/cm^2 for T = 0 K) reported by *Miedema* [36.11] was taken into consideration and fitted to a completely equilibrated Au(111) plane. For this purpose we performed a static relaxation for the top five layers as described in [36.12]. Calculations based on the experimental cohesive energy, surface energy and (36.4) produced $\varepsilon = 0.976$ eV, $r_0 = 2.6685$ Å and $Z = 2009$ eV Å^9. With these parameters, the two-body and three-body contributions to the cohesive energy are -6.55 (eV) and 2.82 (eV), respectively. Of course, now, using these parameters (36.1-3) can exactly reproduce the cohesive energy of Au and the relaxed surface energy of the Au(111) plane at the static limit (T = 0 K). This agreement (i.e., the description of the bulk and the relaxed surface energies with the same set of parameters) cannot be achieved using a semiempirical model potential based on two-body interactions alone.

36.4 Calculation of Surface Energy

Surfaces were generated in the computer as an ideal structure with an abrupt discontinuity of the bulk. This was achieved by applying periodic boundary conditions (P.B.C.) [36.7] in two directions (e.g., x and y directions) of a properly oriented system. In the third direction this provides two exposed surfaces (in the +z and -z directions) with no edge effects. The surface energy of a given surface was evaluated in two steps; first, the unrelaxed surface energy per unit area σ_{un} was calculated as

$$\sigma_{un} = \frac{(\phi_{un} - \phi_b)}{A} \quad , \qquad (36.5)$$

where ϕ_b is the total energy of the system completely equilibrated with the Monte Carlo code for the desired temperature considering P.B.C. in all three directions (this corresponds to the bulk cohesive energy of the lattice). On the other hand, ϕ_{un} represents the total energy of the same system with two exposed surfaces, on opposite sides, created by the removal of the P.B.C. in one direction. At this point no equilibration was performed after the removal of the P.B.C. The created surface area was denoted by A. The equilibrated surface energy, then, can be calculated as

$$\sigma_{eq} = \sigma_{un} + \delta_{rel} \quad , \qquad (36.6)$$

where δ_{rel} represents the additional surface excess energy per unit area due to the equilibration. Relaxation calculations were carried out only for the atoms located in the top three layers of the exposed surface. The rest of the atoms were kept fixed during the relaxation procedure, however, they contributed fully to the total energy calculations. Throughout this investigation relaxations were performed for T = 100 K, using a Monte Carlo technique based on the Metropolis approximation involving small step sizes on the order of 0.06 Å. The calculation systems contained approximately 280 particles and consisted of about 12 atomic layers (depending on the exposed surface structure). Effects associated with number of atomic layers and system size on the calculated surface energy were assessed. Relaxation of only the top two or top three layers yielded virtually the same value of surface energy as when the top five layers were relaxed. Increasing the system size to 500 particles left the surface energy unchanged.

36.5 Results and Discussions

Calculated relaxed surface energies for (1 × 2) reconstructed "missing row" and "saw-tooth" models, as well as for the (1 × 1) unreconstructed Au(110) surface are tabulated in Table 36.1, which also shows surface excess energies and the percentage variations in the first two interlayer spacings due to relaxation

Table 36.1. Calculated results for the Au(110) surface based on (1 × 1) unreconstructed, (1 × 2) reconstructed missing row and (1 × 2) reconstructed saw-tooth models. Here d_{12} and d_{23} denote interlayer spacings between the first and second, and between the second and third layers, respectively. Also included are total number of particles in the system and the step sizes used in the Monte Carlo procedure

	Au(110) surface		
	(1 × 1) unreconstructed	(1 × 2) reconstructed missing row	(1 × 2) reconstructed saw-tooth
σ_{rel} (erg/cm^2) surface energy per area (relaxed)	1545.0 ± 4.0	1515.0 ± 1.0	1965.0 ± 4.0
δ_{rel} (erg/cm^2) relaxation energy per area	-62.0 ± 4.0	-17.0 ± 1.0	-76.0 ± 4.0
% change in d_{12}	-9.0 ± 1.5	-7.5 ± 1.3	-3.0 ± 1.8
% change in d_{23}	+5.4 ± 1.6	≈0.0 ± 1.5	-6.1 ± 1.5
N_T Total number of particles	288	276	288
(N_L) (total number of layers)	(12)	(12)	(13)
Δr (Å) maximum increment in the Monte Carlo procedure	0.064	0.070	0.064

including fluctuation values. Table 36.1 also includes total number of particles in the system and corresponding incremental step sizes employed in the Monte Carlo code. These results for the relaxed surface energies indicate that the energetically most favorable surface is the (1 × 2) reconstructed missing row surface, which is followed by the (1 × 1) unreconstructed surface. Energetically, the relaxed (1 × 2) missing row surface is 30 erg/cm^2 more stable than the relaxed (1 × 1) structure. On the other hand, calculations indicate that the relaxed (1 × 2) reconstructed saw-tooth model is the least favorable one and its surface energy is about 400 erg/cm^2 higher than the missing row surface. This outcome is consistent with recent theoretical and experimental reports which agree that the missing row model is the most favorable surface structure [36.1-3,13]. Calculated surface excess energies due to relaxation, δ_{rel}, exhibit a decreasing trend on going from the missing row to the saw-tooth model. Results obtained for the relaxation of the interlayer separation also are in good agreement with the majority of the experimental reports [36.14]. The general tendency for the first interlayer spacing to contract was well reproduced in the present calculation. Comparison of these results with a recent report by *Tomanek* et al. [36.13] is quite interesting. In their calculation for the Pt(110) surface, an LCAO method incorporated with a repulsive Born-Meyer type potential was employed and

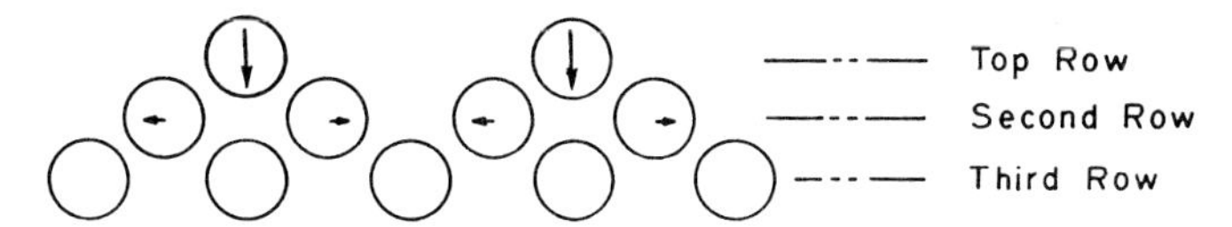

Fig.36.1. Schema of the local structure for the Au(110) surface according to the (1×2) reconstructed missing row model. The figure represents the unrelaxed structure and arrows indicate the directions of the displacement of the atoms taking place during the relaxation. While the top atoms move inward, atoms in the second row exhibit a lateral motion perpendicular to the close-packed direction

they reached the same conclusion that the missing row and the saw-tooth models provide energetically the most and the least favorable surface structures, respectively. Furthermore, their calculated values for the percentage variations in the first interlayer spacings due to relaxations for the (1×2) reconstructed missing row and for the (1×1) unreconstructed surface models agree well with our results. For the saw-tooth model the present result for the first interlayer spacing deviates somewhat, but is still consistent qualitatively with their calculations. Among the recent investigations, only *Robinson* [36.15] reports an expansion for the top surface layers of the Au(110) which disagrees with our calculations. However, his experimental observation about the lateral displacement of the atoms in the second layer was well predicted. We found about 2.3% shift in the lateral direction (perpendicular to the close packed rows) of the atoms positioned along the missing rows. Figure 36.1 displays the (1×2) reconstructed Au(110) surface according to the missing row model with the arrows indicating displacement directions of the first and second row atoms during the relaxation. At this point, the relationship between the lateral displacement of the second layer atoms and the inward displacement of the first layer atoms is easily recognized.

One of the most significant outcomes of this investigation is the understanding of the important effect exerted by three-body interactions. In many occasions, this effect (in particular, in calculating surface properties) has been emphasized [36.12,16,17]. Model calculations based only on two-body interactions predict consistently an expansion in the first interlayer spacing which is in qualitative contradiction with many experimental data for metallic surfaces [36.18]. Furthermore, as was mentioned above, pair potential-only models in general fail to provide satisfactory energy values for surfaces and for the bulk with the same set of parameters. This inherent shortcoming seems to be corrected by the introduction of three-body interactions. In these calculations, we also found that the role played by the three-body interactions is about 50% more important in the surface region than in the bulk.

The present potential energy (36.1-3) has a relatively simple functional form, therefore it can easily be used in lengthy iteration calculations. On several occasions, this same potential function has been used successfully to calculate the structure and energy-related properties for surfaces, bulk and small clusters [36.8,12,19-21]. In the present investigation, the effect of the variables m and n was not fully investigated. However, we believe that such an analysis would be quite interesting in order to understand the relationship between the curvature of the two-body part of the interactions and the final structural characteristics of the system under consideration.

36.6 Conclusions

Present calculations for the (1 × 1) unreconstructed and for two (1 × 2) reconstructed relaxed surfaces of the Au(110) plane indicate that the energetically most favorable surface is the one generated according to the missing row model. The potential energy function comprising two- and three-body interactions employed in this investigation overcomes various shortcomings of the previously used semiempirical potentials based on pair interactions only in calculating surface properties. Lateral and perpendicular displacements of atoms in the surface region were well predicted. Consideration of the three-body interactions, in that respect, seems to be the most important factor. Finally, it is believed that the present method can provide quite useful information for the interpretation of various experimental data related to the energetics and structure of surfaces.

Acknowledgments. This work was supported, in part, by the Defense Advanced Research Projects Agency under Contract Number MDA 903-83-K-0041 and, in part, by NASA Ames Research Center via Grant Number NCC2-125. The authors would like to thank Professors O. Pamuk and S. Erkoc of Middle East Technical University for their help in calculating Au parameters. Their collaboration was made possible by a Grant (074.82) provided by NATO Scientific Affairs Division.

References

36.1 L.D. Marks, D.J. Smith: Nature **303**, 316 (1983)
36.2 D.P. Jackson, T.E. Jackman, J.A. Davies, W.N. Unertl, P.R. Norton: Surf. Sci. **126**, 226 (1983)
36.3 W. Moritz, D. Wolf: Surf. Sci. **88**, L29 (1979)
36.4 J.R. Noonan, H.L. Davis: J. Vac. Sci. Technol. **16**, 587 (1979)
36.5 B. Reihl, B.I. Dunlap: Appl. Phys. Lett. **37**, 941 (1980)
36.6 H.P. Bonzel, S. Ferrer: Surf. Sci. **118**, L263 (1982)
36.7 J.J. Erpenbeck, W.W. Wood: *Modern Theoretical Chemistry*, Vol.6, Part B, ed. by B.J. Berne (Plenum, New York 1977) Chap.1
36.8 T. Halicioğlu: Phys. Stat. Solidi (b) **99**, 347 (1980)
36.9 B.M. Axilrod, E. Teller: J. Chem. Phys. **11**, 299 (1943)
36.10 R. Hultgren, R.L. Orr, P.D. Anderson, K.K. Kelley: *Selected Values of Thermodynamic Properties of Metals and Alloys* (Wiley, London 1963)
36.11 A.R. Miedema: Z. Metallkde. **69**, 287 (1978)
36.12 T. Halicioğlu, O. Pamuk, S. Erkoc: Surf. Sci. **143**, 601 (1984)
36.13 D. Tomanek, H.-J. Brocksch, K.H. Bennemann: Surf. Sci. **138**, L129 (1984); H.-J. Brocksch, K.H. Bennemann: This Volume, p. 226
36.14 H.L. Davis, J.R. Noonan: Surf. Sci. **126**, 245 (1983)
36.15 I.K. Robinson: Phys. Rev. Lett. **50**, 1145 (1983)
36.16 J.N. Schmit: Surf. Sci. **55**, 589 (1976)
36.17 U. Landman, R.N. Hill, M. Mostoller: Phys. Rev. B**21**, 448 (1980); see also, R.N. Barnett, R.G. Berrera, C.L. Cleveland, U. Landman: Phys. Rev. B**28**, 1667 (1983)
36.18 G.C. Benson, T.A. Claxton: J. Phys. Chem. Solids **25**, 367 (1964)
36.19 T. Halicioğlu, P.J. White: J. Vac. Sci. Technol. **17**, 1213 (1980); and Surf. Sci. **106**, 45 (1981)
36.20 I. Oksuz: Surf. Sci. **122**, L585 (1982)
36.21 E.E. Polymeropoulos, J. Brickmann: Chem. Phys. Lett. **96**, 273 (1983); **92**, 59 (1982)

37. Long- and Short-Range Order Fluctuations in the H/W(100) System

Roy F. Willis

Department of Physics, Cavendish Laboratory, Cambridge University
Cambridge CB3 OHE, Great Britain

The surface reconstruction phase transformations of W(100) can be understood in terms of periodic lattice distortion instabilities. A mechanism of nonlinearly coupled phonon soft modes produces a system of antiphase domain walls which can be viewed as a "soliton superlattice". Many aspects of the behavior of W(100) with variation in temperature and hydrogen coverage can be understood in terms of long- and short-range order fluctuations of this lattice dynamical model.

37.1 Introduction

Despite considerable effort over the past two decades, there is much which remains to be understood concerning the structure of the W(100) surface [37.1], in particular, the tendency to reconstruct into a lower symmetry structure, and the underlying cause and nature of the surface lattice instability. Since only the outermost atomic layers are affected, the phase transformations are truly 2-dimensional in character so that W(100) is a model system for testing our current understanding of low-dimensional critical phenomena [37.2].

Several questions need to be addressed, viz.:

1. the apparent dichotomy in behavior as regards atomic displacements within the plane of the surface or out-of-plane; LEED analysis [37.1] favors the former while field-ion microscopy [37.3] favors the latter
2. the low coverage, H-induced symmetry switching from domains with p2mg space group symmetry to domains with c2mm symmetry associated with atomic displacements along <11> and <10> directions, respectively [37.1]
3. the formation of an incommensurate phase of ordered domain walls with increasing coverage [37.4]
4. the fluctuating nature of these domain walls with increasing temperature and hydrogen coverage to produce disorder effects [37.5]
5. the role of steps and other surface defects in the "partial reconstruction" of the surface [37.6]
6. the thermodynamic driving mechanism – in particular, the inconsistent theoretical and experimental determinations of the surface electronic band structure [37.7].

In this short review, I shall present a lattice dynamical model which seeks to answer these questions. It is based on a mechanism of nonlinearly coupled phonon soft-mode distortions of the lattice drawn from ideas first introduced by *Axe* et al. [37.8] for bulk structural phase transformations, and *Fasolino* et al. [37.9] for surfaces, the latter being a special case of a more general mechanism proposed by *Heine* and *McConnel* [37.10] for incommensurate phase formation in ferroelectrics. The extension of these ideas to a model of antiphase domain walls in terms of nonlinear "soliton" behavior has been proposed by *Willis* [37.11], and recently formulated in terms of an effective lattice Hamiltonian approach by *Ying* and *Hu* [37.12].

37.2 The Model

Fasolino et al. [37.9] employed an effective lattice dynamical Hamiltonian to show that the lattice instabilities of the W(100) and Mo(100) surfaces arose from phonon soft-mode coupling effects about the $\bar{M}$ point of the surface Brillouin zone. The mechanism is illustrated in Fig.37.1.

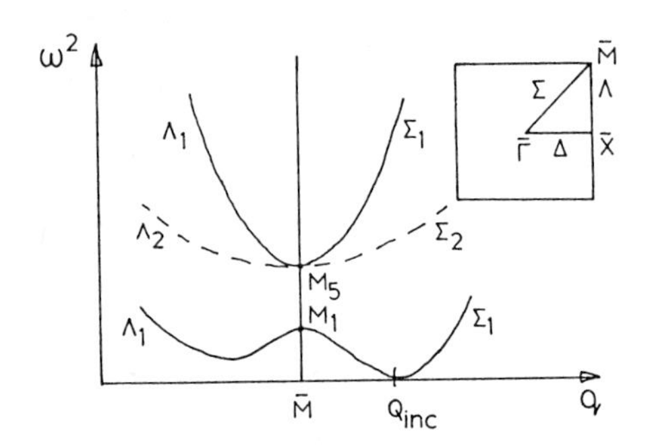

Fig.37.1. Phonon soft-mode coupling between M_5 and M_1 symmetry modes to produce incommensurate wave vector Q_{inc}. The symmetry directions in the 2-d Brillouin zone are shown inset

At the symmetry point $\bar{M}$ there is one mode M_1 with atomic displacements out of the surface plane coupling to one of the doubly degenerate modes M_5 acting along either the $\bar{\Sigma}$ or $\bar{\Lambda}$ symmetry directions with atomic displacements in directions <11> or <10> in the surface plane. Off the symmetry point $\bar{M}$, these orthogonal modes can couple and soften to low frequency at some wave vector Q_{inc} incommensurate with the undistorted (100) lattice. Thus, the high-temperature (1×1) undistorted surface has two possible modes of deformation. The M_5 modes alone can go soft at the $\bar{M}$ point to give a phase transition into the commensurate $c(2\times2)$ <11> state with displacements along <11>, as is observed for the clean W(100) surface. Alternatively, the M_5 and M_1 modes can soften and couple to give an incommensurate "$c(2\times2)$" phase as is observed for H/W(100) [37.1,4].

37.3 Nonlinear Anharmonicity

At incommensurate phase transitions, there are strongly anharmonic atomic displacements, described by "Umklapp" terms involving p^{th} order anharmonicity in a Landau expansion of the free energy [37.13]:

$$G = G_0 + \frac{1}{2}a(T - T_c)A^2 + uA^4 + VA^p \cos p\phi \quad , \qquad (37.1)$$

where A, ϕ are amplitude and phase coordinates of the periodic lattice distortion which occurs below some critical temperature T_c; a and u are Landau expansion parameters. In the simplest case of a single-component sinusoidal distortion, $u = u_0 \sin[qx + \phi(x)]$, where the atomic displacement amplitude

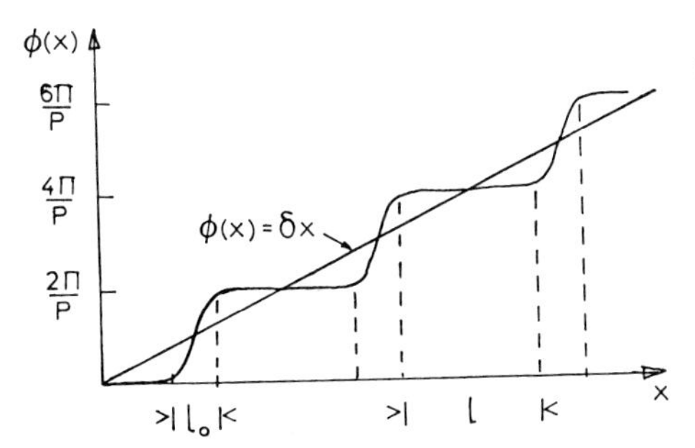

Fig.37.2. Antiphase commensurate domains of size ℓ separated by narrow regions ℓ_0 of rapidly varying phase $\phi(x)$: a "soliton superlattice"

$u = u_0 \sin qx$ is a wave commensurate with the undistorted lattice. The phase $\phi(x)$ is the shift of the atoms with respect to some potential minimum – the "lock-in" potential energy V. For $V = 0$, the unperturbed incommensurate wave distortion is simply linear in one-dimensional coordinate x; $\phi(x) = \delta x$, where δ is the relative 'misfit' or 'mismatch' between the atomic spacings in the commensurate and incommensurate sublattices. If the "lock-in" potential energy $V \neq 0$, there can be local variations in the phase which causes $\phi(x)$ to vary in a stepwise fashion, Fig.37.2.

Physically, domains of size ℓ are separated by "discommensurate" antiphase domain walls of size ℓ_0 over which the phase changes rapidly by $\pi/2$ (for $p = 4$) [37.14]. The ground-state configuration, which minimizes the free energy (37.1), is found among the solutions to the 1-d sine-Gordon equation of (nonlinearly) coupled normal modes [37.15]

$$d^2\phi/dx^2 = pV \sin p\phi \quad , \tag{37.2}$$

with solutions in the form of a solitary wave packet or "soliton" which describes the rapid phase variation over a narrow region, i.e., the domain wall discommensuration separating two commensurate domain regions

$$\phi(x) = 4/p \tan^{-1}[\exp(pV^{\frac{1}{2}}x)] \quad . \tag{37.3}$$

In general, the solutions to (37.2) are regularly spaced solitons ("soliton superlattice") the distance between solitons ℓ being equal to one-half the incommensurate wave vector Q_{inc}. The soliton superlattice is a compromise between an elastic energy term in the distorted surface layer (quadratic in amplitude A) and an interaction energy with the substrate lattice ($VA^p \cos p\phi$ term) which favors commensurate ordering, in the Landau expansion, (37.1). The order parameter of the 2nd-order (continuous) phase transition from the commensurate into this incommensurate state (Fig.37.2) is effectively the number density of domain walls, $\langle d\rangle = 2\pi/p\ell$. The theory extends to 2 dimensions where the solitons become linear boundaries in a plane [37.15]. In a diffraction experiment, the soliton superlattice produces satellite reflections about the (1/2, 1/2) order spot positions of the commensurate $c(2\times 2)$ structure, which grow in intensity and sharpness at the CI phase transition [37.4].

In the H/W(100) system, the H-induced CI phase boundary occurs at coverages in the range $H/W = 0.2$ to 0.4 [37.16,17], the actual value depending on step density on the surface; step spacings less than a critical spacing will affect soliton ordering, ℓ (Fig.37.2). As the coverage increases to $H/W \geq 0.5$, spot streaking occurs indicative of 1-dimensional disordering of the domain walls [37.11]. *Willis* [37.2] has argued that this corresponds to an effective 'roughening transition' in a striped domain configuration which occurs at a critical domain size ℓ or the order of 4 or 5 lattice spacings, i.e., 12 to 15 Å, which is in accord with recent LEED spot profile estimates of the minimum domain size [37.6]. This minimum domain size arises from wall-wall energy repulsion effects which increase exponentially with decreasing ℓ [37.5]. This value is also in close agreement with the critical distance between solitons ℓ_{cr} estimated by *Bak* and *Pokrovsky* [37.18]

$$\ell_{cr} \simeq \pi^2 \ell_0^2 \quad , \tag{37.4}$$

which for $\ell_0 \sim$ lattice constant gives a value for ℓ_{cr} of π^2. This suggests that ℓ_0, the size of the domain wall, is extremely localized in relation to its phase relation to the substrate lattice. Again this is in accord with the model of discrete lattice coupling effects between M_1 and M_5 phonon modes, proposed by *Willis* [37.11], and illustrated in Fig.37.3.

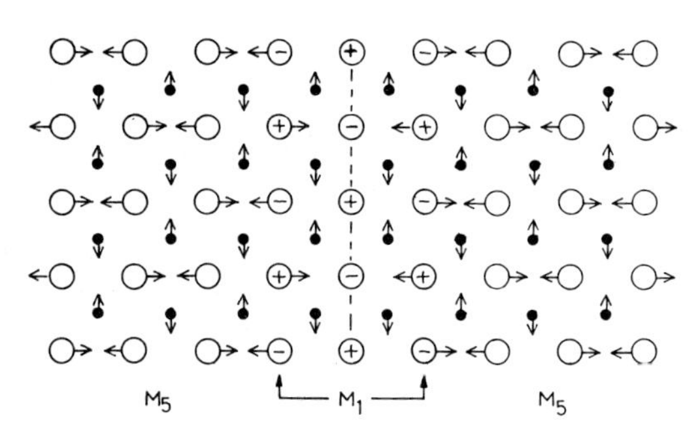

Fig.37.3. Atomic displacements along <10> in the vicinity of a domain wall produced by coupling between vertical displacements (±) of M_1 symmetry and lateral displacements (→) of M_5 symmetry. Note the 2nd layer coordinated motion (of smaller atomic displacement amplitude) which relieves stresses imposed by the top layer displacements. The H atoms sit on the dimerized bridge sites on the top layer in the M_5 domains

This concept of domain wall discommensurations and soft-phonon-mode driven lattice instabilities is a recent exciting development in the theory of structural phase transformations [37.13]. The incommensurate (and "partially disordered") phases are characterized by novel very low frequency lattice vibrations – phasons, amplitudons and wall-wall coupled modes – which characterize the normal modes of the discommensurate soliton walls and the commensurate domains. The physical picture we arrive at is one in which each superlattice unit cell can be treated as having its own normal modes with intercell interactions between modes. These phonons modes have both amplitude and phase which, at low temperature, exhibit long-range phase coherence giving the "uniform" periodic lattice distortions sensed in LEED. At a phase transition, this long-range phase coherence breaks up, i.e., the long-range order parameter goes to zero. However, there still remains a large amplitude of the *local* phonon modes in domains of much reduced size. A short range probe of these lattice vibrations therefore still senses the same local order (but with much increased amplitude fluctuations) as in the coherent phase.

There are two consequences of this picture of long- and short-range order fluctuations which are relevant to W(100): firstly, the transition from the commensurate c(2 × 2) <11> to the disordered "(1 × 1)" structure with increasing temperature is complicated by short-range order effects around the transition temperature which may be regarded as random fluctuating, small-size domains with local c(2 × 2) <11> ordering. (This is more commonly referred to as the "central peak" problem in neutron scattering cross-section measurements in bulk phase transition phenomena [37.13].) Hydrogen vibration modes observed in low coverage EELS measurements [37.19] sense this local $c(2\times2)$ "pinched" lattice site ordering so that the "(1×1)" phase observed in LEED is *not* that of the undistorted bulk lattice. This has consequences for both LEED intensity analysis [37.1] and any comparison between experimental and theoretically determined electronic surface state energy band dispersion behavior [37.7].

The second consequence is that the formation of an *ordered* array of domain walls in the incommensurate phase will affect the H-mode vibration frequencies, the latter reflecting local short-range distortion wave amplitude effects associated with the very low frequency lattice soft-mode vibrations referred to above. This being the case, it is of interest to compare the phase diagrams of the H/W(100) system deduced from LEED [37.4] (long-range coherent order effects) with that recently determined from EELS (short-range order) measurements of the H-vibration mode frequencies [37.20].

37.4 EELS Phase Diagram

The shift in EELS vibration frequency of the H/W(100) wag vibration modes with increasing hydrogen coverage is shown schematically in Fig.37.4, taken from the measurements of *Willis* [37.19]. The important point to note is that the H vibration mode frequencies are a sensitive probe of the onset of the H-induced incommensurate phase θ_{IC}, one-dimensional satellite spot streaking θ_{1D}, and two-dimensional disordering effects θ_{2D}. A low coverage switch in the atomic displacement vector from <11> to <10>, which occurs in a range below θ_{IC} $(0.05 < \theta_{switch} < 0.16)$ causes only a slight broadening in spectral line shape [37.6]. A comparison of the critical phase transition coverage points at which a sudden change in frequency occurs as a function of temperature (cf. the 300 K and 125 K data, Fig.37.4) allows the slope of phase transition boundaries to be determined, Fig.37.5).

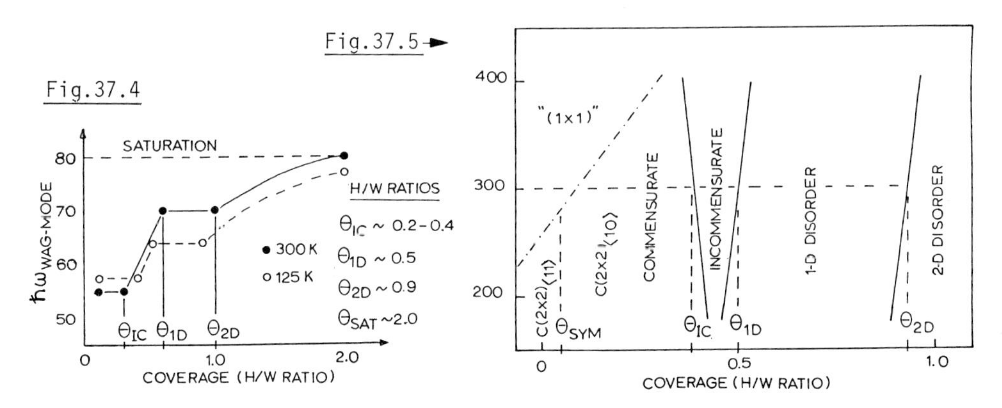

Fig.37.4. Observed EELS shift in the H wag-mode frequency with increasing coverage; the onset of incommensurate (θ_{IC}), disorder in 1-d (θ_{1D}), and disorder in 2-d (θ_{2D}) phase transitions are indicated

Fig.37.5. Phase diagram of H/W(100). The dash-dot boundary is seen only in LEED

The results, Fig.37.5, compare closely with those deduced from LEED measurements [37.17], *except that the EELS measurements do not* sense the commensurate c(2 × 2) <11> to disordered "(1 × 1)" phase transition boundary (dash-dot curve, Fig.37.5). That is, the vibration frequencies remain those characteristic of W_2 dimers, pinched together within reconstructed domains [37.19], which exist over a range of temperature above that observed in LEED: this suggests a coherence length of less than the 100 Å typical of most LEED experiments but extending over several unit cells at least so far as the H-vibrational modes are concerned.

37.5 Symmetry Switching and "Hydrogen Polaron Behavior"

To understand further the specific role of the H adlayer, we must address the problem of competing adsorbate-adsorbate and adsorbate-substrate interactions [37.4].

First, hydrogen at very low coverages stabilizes the c(2 × 2) commensurate phase by raising the transition temperature into the "(1 × 1)" phase linearly with increasing coverage (dash-dot curve, Fig.37.5). Second, at a critical coverage θ_{sym} it switches the lattice distortion displacements from <11> to <10>. Third, with further increase in coverage, it drives the system into an incommensurate phase (θ_{IC}, Fig.37.5).

In a companion paper in this volume, *Ying* and *Hu* [37.12] formulate this problem in terms of phenomenological Landau theory, treating the H/W(100) system as belonging to a 2-d XY universality class [37.21]. The symmetry switching arises because of a change in sign of the cubic anisotropy third-order term in the Landau expansion of the system's free energy [37.12]. Monte Carlo simulations and renormalization group methods, utilizing a microscopic lattice dynamical Hamiltonian, require that: a) H stabilizes the c(2 × 2) distortion into a state of lower free energy, b) it does so by occupying *specific* sites in the lattice, and c) it can do so only if its mobility across the surface is sufficiently high. The LEED measurements [37.4] suggest that this occurs only for crystal temperatures above 200 K. This is odd, since H is thought to tunnel readily across the rigid lattice surface at T = 0 K [37.22].

The low-coverage hydrogen-lattice interaction may be explained in terms of "hydrogenic polaron-like formation" in which the chemisorbed H becomes self-trapped in a deep potential well formed by *local* distortion coupling to the substrate lattice soft-phonon-mode fluctuations. Strong long-range adatom-adatom ordering effects can then occur via the soft-mode induced lattice distortions. The basic picture of a small polaron lattice distortion is shown in Fig.37.6. The polaron binding energy is the difference between the self-trapping energy and the elastic strain energy of the deformed lattice:

$$E_{binding} = E_{self\text{-}trapping} - E_{elastic\ strain} \qquad (37.5)$$

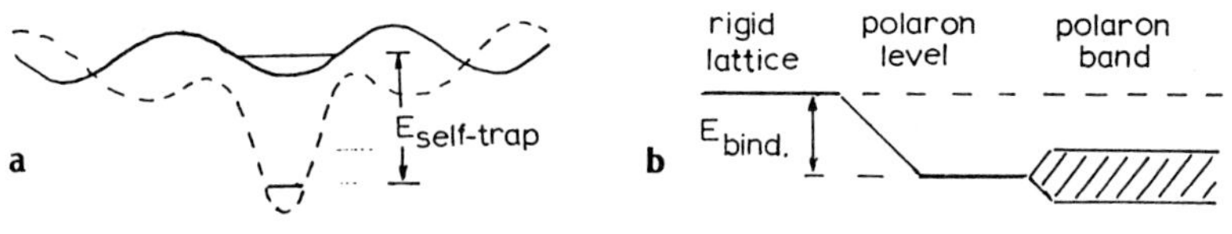

Fig.37.6. (**a**) Small polaron lattice distortion for trapped H atom. (**b**) Schematic energy level diagram showing single-particle level and multi-particle levels or narrow "polaron band"

There are thus three contributions to the system's energy: the H particle's kinetic and potential energies, and the elastic strain energy of the lattice. Following *Emin* et al. [37.23] we may treat the latter in a first approximation as a continuum, so that if we minimize the sum of these energies with respect to the dilatation energy, we get an expression for the ground-state dilatation energy of the system:

$$\Delta(\mathbf{r}) = -Z/S|\psi(\mathbf{r})|^2 \quad . \tag{37.6}$$

Here Z is an energy characterizing the strength of the H particle-continuum interaction (analogous to the local deformation potential), S is an effective "stiffness coefficient" of the elastic continuum, and $\psi(\mathbf{r})$ is the particle's wave function. That is, dilatation increases with interaction strength, decreases with increasing material stiffness and mirrors the particle's probability density distribution. At an energy minimum, the stationary wave equation

$$[-(\hbar^2/2m)\nabla^2 + Z\Delta(\mathbf{r})]\psi(\mathbf{r}) = E\psi(\mathbf{r}) \tag{37.7}$$

becomes a *nonlinear* differential equation

$$[-(\hbar^2/2m)\nabla^2 - (Z^2/S)|\psi(\mathbf{r})|^2]\psi(\mathbf{r}) = E\psi(\mathbf{r}) \quad . \tag{37.8}$$

Equation (37.8) is entirely analogous to the sine-Gordon equation (37.2) derived earlier so that the polaron model is consistent with H coupling to strongly anharmonic soft-mode lattice distortions. Furthermore, (37.8) admits both *localized* as well as *delocalized* (i.e., "polaron bands", Fig.37.6b) particle wave function solutions.

The apparent temperature dependence of the H mobility can be interpreted as lattice-activated diffusion mediated by soft-phonon mode fluctuations of the self-trapped potential wells. Lattice-mediated ordering will occur with increasing annealing temperature via a Boltzmann distribution of phonon soft-mode fluctuations. That is, small polaron transport is concerned with the motion of a light diffusing atom between adjacent self-trapped lattice positions in response to a lattice dynamical lowering of the potential barrier height. The polaron model thus explains the localized nature of the EELS H-vibration mode frequencies indicative of the H atoms remaining in $c(2\times2)$ W_2 pinched-dimer sites with increasing coverage and temperatures (Fig.37.4). The symmetry switching occurs at a critical coverage corresponding to a change in symmetry of the dilatation due to interacting polaron elastic strain fields. Increasing coverage increases the elastic strain field energy which eventually overcomes the self-trapping energy, driving the polaron binding energy to zero (37.5). At this point, it is observed experimentally that the H binding energy *suddenly* increases [37.17]. This occurs at a coverage corresponding to the spontaneous formation of domain walls (solitons) at the incommensurate phase transition. The lattice strain energy is now taken up in the coupling between the M_5 and M_1 phonon modes which minimizes the free energy by increasing the overall lattice entropy of the system.

37.6 Summary: Mechanistic Aspects

To predict the transition temperature of a phase change, one must know something about the factors which affect the difference in enthalpy ΔH and entropy ΔS between the two phases, i.e., the change in Gibbs free energy may be written as

$$\Delta G = \Delta H - T\Delta S$$
$$= \Delta E + \Delta(PV) - T\Delta S \qquad (37.9)$$
$$= \Delta E + \Delta\mu - T\Delta S \quad \text{for constant } V,P \quad .$$

That is, the system seeks to lower its internal energy ΔE or to increase its entropy ΔS balanced against $\Delta\mu$, a chemical potential term due to an increasing number density of H atoms [37.5]. Clearly, at $T = 0$ K, the driving mechanism is electronic and comes from a splitting and broadening at a prominent peak in the surface density of states in the undistorted structure which lowers the overall valence energy of the filled states. This leads to a total electronic energy stabilization of the form [37.24]:

$$\Delta E = \sum_{A,B} |\langle B|V_{AB}|A\rangle|^2/(\varepsilon_B - \varepsilon_A) \quad . \qquad (37.10)$$

Here $|A\rangle$ and $|B\rangle$ correspond to (mainly) occupied and unoccupied surface state bands, and V_{AB} is the deformation potential corresponding to deformation wave vectors (q) which couple excitations between states $|A\rangle$, $|B\rangle$. Further, ΔE will be large if the energy differences in the denominator are small (i.e., limited to a narrow energy range around the Fermi energy E_F) and if the matrix elements in the numerator are large (i.e., strong electron-phonon coupling limit [37.21]). Recent detailed electronic structure calculations [37.25,26] have demonstrated that this coupling is local in character. That is, electronic states throughout a large region of the surface Brillouin zone have their energy lowered, not just those at the Fermi energy E_F coupled by wave vectors $2k_F$. In other words, the distortion is cooperative Jahn-Teller and "bond-like" rather than charge-density-wave and "band-like" in nature [37.21,25].

In real space, this implies that: 1) the forces between the W atoms are short range in character (as assumed in the lattice dynamical calculation of *Fasolino* et al. [37.9]; 2) the forces are directional and include coupling between atoms both in the surface and to the subsurface layer (Fig.37.3); 3) the behavior at finite T is dominated by lattice (phonon) rather than electronic entropy contributions. The most important contribution in (37.10) comes from Σ_2 surface states which exist throughout a large region of the surface Brillouin zone. These states are formed mainly from x^2-y^2 orbitals, the symmetry of which favors in-plane reconstruction modes M_5 [37.27]. However, at finite temperatures (or by raising the chemical potential energy with increasing H coverage) the lattice modes become increasingly anharmonic and relieve local elastic strain fields by coupling to other anharmonic normal modes of vibrations.

It is these fluctuations in the lattice entropy terms in (37.9) which are ultimately responsible for the richness of reconstruction phase phenomena observed in W(100). These effects account for the differences observed in the electronic states' behavior at finite T compared with theoretical calculations of ground-state energies at $T = 0$ K [37.7]. They also account for the failure of mean-field theories to explain the observations. Finally, long-range order fluctuations about T_c will be sensitive to defects in the surface, i.e., will effect the actual value observed for T_c [37.28]. These fluctuations have not been taken into account in essentially *static* LEED models of the structure of the surface on either side of T_c.

Acknowledgments. The author gratefully acknowledges discussions with Profs. V. Heine and S.C. Ying, and his students E.F.J. Didham, M.S. Foster, and B.J. Hinch, which greatly helped to clarify the ideas expressed here.

References

37.1 D.A. King: Phys. Scripta T**4**, 34 (1983), and references therein
37.2 R.F. Willis: *Dynamical Phenomena in Surfaces, Interfaces and Superlattices*, Springer Ser. Solid State Sci., ed. by M. Cardona (Springer, Berlin, Heidelberg, New York, in press)
37.3 R.T. Tung, W.R. Graham, A.J. Melmed: Surf. Sci. **115**, 576 (1982)
37.4 P.J. Estrup, R.A. Barker: *Ordering in Two-Dimensions*, ed. by S.K. Sinha (Elsevier/North Holland, New York 1980)
37.5 J. Villain: *Ordering in Strongly Fluctuating Condensed Matter Systems*, ed. by T. Riste, NATO ASI Vol. 50B (Plenum, New York 1980)
37.6 J.F. Wendelken, G.C. Wang: Surf. Sci. **140**, 425 (1984), and references therein
37.7 S. Ohnishi, A.J. Freeman, E. Wimmer: Phys. Rev. B**29**, 5372 (1984); L.F. Mattheis, D.R. Hamman: Phys. Rev. B**29**, 5372 (1984)
37.8 J.D. Axe, J. Horada, G. Shirane: Phys. Rev. B**1**, 1227 (1970)
37.9 A. Fasolino, G. Santoro, E. Tosatti: Phys. Rev. Lett. **44**, 1684 (1980)
37.10 V. Heine, J.D. McConnell: J. Phys. C**17**, 1199 (1984)
37.11 R.F. Willis: *Many-Body Phenomena at Surfaces*, ed. by D. Langreth, H. Suhl (Academic, New York 1984)
37.12 S.C. Ying, G.Y. Hu: This Volume, p.341
37.13 R.A. Cowley: Adv. Phys. **29**, 1 (1980)
37.14 W.L. McMillan: Phys. Rev. B**14**, 1496 (1976)
37.15 P. Bak: Rep. Prog. Phys. **45**, 587 (1982)
37.16 D.A. King, G. Thomas: Surf. Sci. **92**, 201 (1980)
37.17 A.H. Smith, R.A. Barker, P.J. Estrup: Surf. Sci. **136**, 327 (1984)
37.18 P. Bak, V.L. Pokrovsky: Phys. Rev. Lett. **47**, 958 (1981)
37.19 R.F. Willis: Surf. Sci. **89**, 457 (1979)
37.20 E.F.J. Didham, R.F. Willis: J. Physique **45**, C5 (Suppl. no.4) (1984), and to be published
37.21 E. Tosatti: *Modern Trends in the Theory of Condensed Matter*, ed. by A. Pekalski, J. Przystawa (Springer, Berlin, Heidelberg, New York 1980)
37.22 M.J. Puska, R.M. Nieminen, M. Manninen, B. Chakraborty, S. Holloway, J.K. Nørskov: Phys. Rev. Lett. **51**, 1081 (1983)
37.23 D. Emin, M.I. Baskes, W.D. Wilson: Phys. Rev. Lett. **42**, 791 (1979)
37.24 J. Friedel: *Electron-Phonon Interactions and Phase Transitions*, ed. by T. Riste, NATO ASI Vol.29B (Plenum, New York 1977)
37.25 J.E. Inglesfield: Vacuum **31**, 663 (1981)
37.26 I. Terakura, K. Terakura, W. Hamada: Surf. Sci. **103**, 103 (1981)
37.27 M. Tomasek, S. Pick: Surf. Sci. **140**, L279 (1984)
37.28 J.F. Wendelken, G.C. Wang: J. Vac. Sci. Technol. A**2**, 886 (1984)

IV.2 Atomic Adsorption on Metal Surfaces

38. Synchrotron X-Ray Scattering Study of a Chemisorption System: Oxygen on Cu(110) Surface

K.S. Liang, P.H. Fuoss*, G.J. Hughes, and P. Eisenberger

Exxon Research and Engineering Company, Corporate Research Science Labs. Annandale, NJ 08801, USA

*AT&T Bell Laboratories, Holmdel, NJ 07733, USA

38.1 Introduction

X-ray diffraction has always been the primary means of determining the three-dimensional structure of bulk materials, just as LEED has been for the two-dimensional structure of surfaces. Although both techniques have been applied for decades, only a handful of surface-structural problems have been solved using LEED, while bulk structures are routinely solved using X-ray diffraction techniques. The main reason for this difference is that LEED analysis is based on dynamic scattering theory while X-ray scattering is usually interpretable with kinematic theory.

The main limitations in using X-ray diffraction for surface-structural study have been lack of surface sensitivity and the limited brightness of X-ray sources. However, the employment of the grazing-incidence X-ray scattering (GIXS) technique, reported first by *Marra* et al. [38.1], in conjunction with the development of high-intensity synchrotron X-ray sources, has gradually overcome this limitation. In fact, GIXS has already been applied to study surface reconstruction [38.1,2], semiconductor interfaces [38.1], and metal overlayers [38.3]. We report in this paper the first study of a chemisorption system involving a low-Z element, oxygen, on the Cu(110) surface, using the GIXS technique. The work was performed on beam line VI at the Stanford Synchrotron Radiation Laboratory (SSRL). This newly installed beam line, with a 54-pole wiggler, provides the most intense source of hard X-rays presently available [38.4].

A clean Cu(110) surface maintains the in-plane structure of the terminated bulk with an oscillatory relaxation of the surface layers [38.5]. Oxygen adsorbed on this surface is known to form a (2×1) structure with a saturation coverage at 0.5 monolayer with a doubling of the periodicity along the [011] direction. Even for such a simple system, the exact atomic arrangement of the $O(2\times1)$ surface is still a subject of dispute. Only recently was the adsorption site of oxygen directly determined to be at the long-bridge site along the [001] direction by surface EXAFS [38.6] and angle-resolved UPS [38.7].

The $O(2\times1)$ surface on Cu(110) has also been the subject of study using different scattering techniques, including LEED [38.8], low-energy ion scattering [38.9], high-energy ion scattering [38.10], and helium diffraction [38.11]. These studies agreed that the Cu(110) surface undergoes reconstruction upon the formation of the (2×1) structure. But two structural models

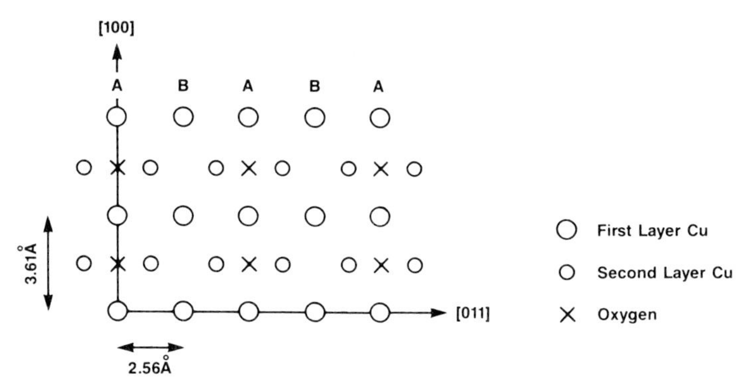

Fig.38.1. Two different structural models for the O(2 × 1) overlayer on Cu(110) surface: in the missing-row model the B-row Cu atoms are missing on the top layer [38.9,11]; in the buckled-row model the B-row Cu atoms are displaced outwards by 0.27 Å from their bulk positions [38.10]

were proposed for this reconstruction: the missing-row model [38.9,11], in which alternating [001] rows of Cu atoms are missing, and the buckled-row model [38.10], in which the non-oxygen-containing [001] Cu rows of the top layer are displaced significantly outwards from their bulk position (Fig.38.1). The results of our GIXS study favors the buckled-row model. The determination of the precise surface atom positions, as found from a complete surface crystallographic determination, is still in progress.

38.2 Experimental

Our experiments were performed on beam line VI-2 during the spring run of 1984 at SSRL. The storage ring was operated at 3 GeV with an average beam current of ~50 mA. The scattering spectrometer and its associated UHV chamber were described in detail elsewhere by *Brennan* and *Eisenberger* [38.12]. Focused, 7113 eV X-rays were produced from the wiggler source with a Pt-coated mirror and a two-crystal, parallel setting monochromator with Si(220) crystals. The diffracted beam was collimated with a 0.07° Soller slit. The measured incident X-ray flux was $\sim 5 \times 10^{10}$ photons $\cdot s^{-1} \cdot mm^{-2}$. It should be pointed out that this experiment was performed at a break-in stage of the beam line when the wiggler was operated at relatively low field (7 kG). Significant improvement of the experiment can be expected in the near future.

The Cu crystal was mechanically and electrolytically polished to a nearly optically flat surface. The surface normal was inclined about 0.5° from the [110] axis (determined from X-ray and optical reflections). The Cu surface was sputter-cleaned by 500 V Ar^+ ions followed by annealing at 500°C. During our experiment, the base pressure of the sample chamber was in the mid to high 10^{-10}Torr range. Oxygen was adsorbed at room temperature. The surface was monitored by AES and the O(2 × 1) structure was confirmed by observation of X-ray Bragg peaks.

The incident X-ray beam was kept at a mean grazing angle of 0.18° to the crystal surface (Θ_c for Cu is 0.45° at 7113 eV) to achieve maximum surface sensitivity. The sample was oriented by the (002) Bragg peaks of the bulk reflection. Only in-plane scans of the surface Bragg peaks were performed due to limited experimental time. Scans along the Bragg rods, which are crucial for a complete crystallographic determination, are in our plan for future study.

38.3 Results and Discussion

X-ray scans were carried out around the half-, first- and second-order Bragg peaks along the [011] and [100] directions. The results of the scans along the [011] direction are shown in Fig.38.2.

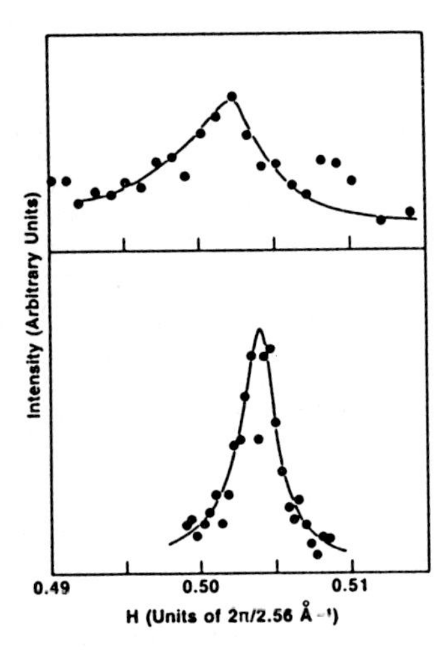

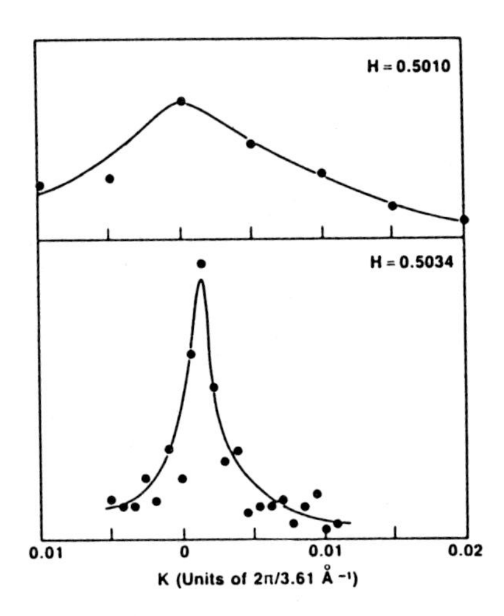

Fig.38.2. X-ray scans of the (0 1/2 1/2) Bragg reflection of O(2 × 1) structure on Cu (110) surface along the [011] direction (*left panel*) and the [100] direction (*right panel*): scans shown in the upper panel were taken after adsorption of ~1000 L oxygen at room temperature; scans shown in the lower panel were taken after annealing at 125°C for 30 min. Note the sharpening and shift of the peak after annealing

The first observation of the half-order Bragg peak (0 1/2 1/2) was made after the Cu surface was exposed to ~1000 L (1 L = 10^{-6} Torr · s) oxygen at room temperature (Fig.38.1). This exposure should produce a saturated adsorption of oxygen on the surface [38.10]. The peak was seen to be very broad for scans in both the [100] and the [011] directions with a peak intensity of about 2 counts/s. Satellite peaks were also observed but not quite reproducible in different scans. Under our experimental conditions the O(2 × 1) structure was observed to fade in a period of about 8 hours, presumably due to reaction with residual CO [38.8].

After the surface was redosed with oxygen, the sample was annealed at 125°C for 30 minutes. After the annealing, a sharp (0 1/2 1/2) peak with a shift of ~0.006 $Å^{-1}$ (or ~0.15° in 2Θ) was observed. The peak width corresponds to a correlation length of ~360 Å, which is about the maximum correlation length expected from our sample with a miscut of ~0.5° from the ideal (110) direction. The observed shift was unexpected.

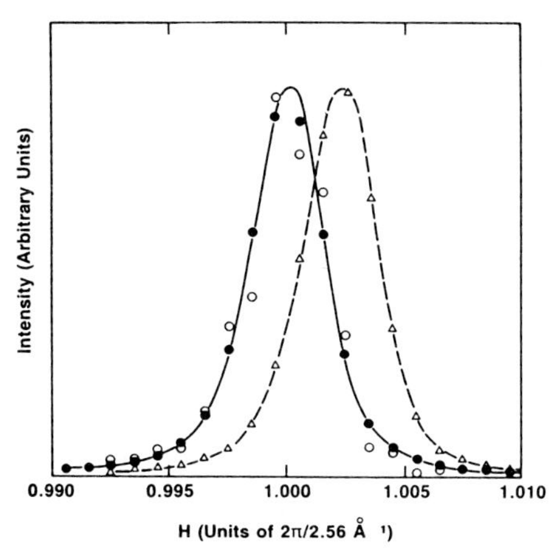

Fig.38.3. X-ray scans of the (011) Bragg reflection at different stages of the experiment: (●): before annealing of the $O(2\times1)$ surface; (△): after annealing at 125°C; (○): recleaned surface (see text)

The most surprising result in our X-ray scans is the observation of a strong first-order Bragg peak (011) (Fig.38.3). For an fcc crystal, this is a forbidden reflection from the bulk due to cancellation in the structure factor. We find that, in fact, the peak is already present on an oxygen-free clean (110) surface. Within our experimental accuracy, the peak remained unchanged after the initial oxygen adsorption at room temperature. However, the peak was observed to shift by about the same amount as the (0 1/2 1/2) reflection after annealing, and to shift back to the initial position after the surface was recleaned (Fig.38.3). The observed shift corresponds to a contraction of the d spacing of ~0.01 Å in the [011] direction.

The origin of the presence of the bulk forbidden reflection is not clear. The observation of a correlated shift of the (011) with that of the oxygen included (0 1/2 1/2) suggests a surface origin. It turns out that the other forbidden peak (100) was also observed. The measured intensity ratios between the first- and second-order peaks are about 1×10^{-3} for (011)/(022) and 4×10^{-3} for (100)/(200), respectively. The different values of the intensity ratio tend to rule out the possibility of harmonic contamination in the incoming X-rays. During the experiment, special care was taken to detune the monochromator and tune the detection system to discriminate against harmonics.

One possible explanation of the appearance of these forbidden peaks is that they are due to the surface sensitivity of the X-rays at grazing angle. Noting that the experiment was performed at 0.18° grazing incidence, the penetration of the X-rays into the surface is only about 50 Å. The observed intensity ratios seem reasonable if the first-order reflection would result from the topmost layer and the second-order reflection from all the layers in the 50 Å penetration depth. The existence of bulk forbidden reflections which are sensitive to the condition of the surface may allow studies of surfaces which do not have strong superlattice reflections.

Finally, we discuss the possible implications for the $O(2\times1)$ structural model on the basis of these observations. Noting that as the forbidden first-order peaks were followed through the formation of the $O(2\times1)$ structure, no significant change was observed in either intensity or the line shape of the (011) peak as the (0 1/2 1/2) peak developed. The measured intensity of the (011) peak is about 200 times that of (0 1/2 1/2). In the missing-row model, one would expect to observe a substantial change of the intensity of the (011) peak from a structure factor consideration. On the other hand, the change due to a buckled-row model would be observed mainly along the diffraction rods normal to the scattering plane. Since the scattering geometry used in this study had poor resolution in the normal direction, no significant change was expected to be observed on the basis of a buckled-row model. Of the two models proposed in the literature, our results from the GIXS study therefore favor the buckled-row model.

38.4 Summary

We have performed a grazing angle X-ray scattering study of the $O(2\times1)$ structure on the Cu(110) surface using the most intense X-ray source presently available, the 54-pole wiggler source at the SSRL. We observed Bragg diffraction induced by a half monolayer of oxygen with a peak intensity $\lesssim 2$ counts/s. We also observed surface diffraction at such bulk forbidden reflections as the (100) and (011). The observed shifts of these reflections at different stages of oxygen adsorptions suggest oxygen-induced surface reconstruction. Comparison of our results with the missing-row model and buckled-row model favors the buckled-row model.

We demonstrate the first X-ray diffraction study of a chemisorption system involving a low-Z element. It is clear that a complete surface crystallographic determination and phase transition study of such systems will become feasible as more intense X-ray sources from synchrotron radiation are developed.

Acknowledgment. We should like to thank R. Hewitt and S. Brennan for helpful technical discussions about the experimental setup. This work was partially done at the SSRL which is supported by the DOE, Office of Basic Energy Sciences; and the NIH, Biotechnology Resource Program, Division of Research Resources.

References

38.1 W.C. Marra, P. Eisenberger, A.Y. Cho: J. Appl. Phys. **50**, 6927 (1979); P. Eisenberg, W.C. Marra: Phys. Rev. Lett. **46**, 1081 (1981)
38.2 I.K. Robinson: Phys. Rev. Lett. **50**, 1145 (1983)
38.3 W.C. Marra, P.H. Fuoss, P. Eisenberger: Phys. Rev. Lett. **49**, 1169 (1982); S. Brennan, P.H. Fuoss, P. Eisenberger: This Volume, p.421
38.4 See SSRL Activity Report 83/01 (1983); SSRL Activity Report 84/01 (1984)
38.5 D.L. Adams, H.B. Nielsen, J.N. Andersen: Surf. Sci. **128**, 294 (1983)
38.6 U. Döbler, K. Baberschke, J. Haase, A. Puschmann: Phys. Rev. Lett. **52**, 1437 (1984)
38.7 R.A. DiDio, D.M. Zehner, E.W. Plummer: J. Vac. Sci. Technol. A**2**, 852 (1984)
38.8 G. Ertl, J. Küppers: Surf. Sci. **24**, 104 (1971); F.H.P.M. Habraken, G.A. Bootsma: Surf. Sci. **87**, 333 (1979)
38.9 R.P.N. Bronckers, A.G.J. de Wit: Surf. Sci. **112**, 133 (1981)
38.10 R. Feidenhans'l, I. Stensgaard: Surf. Sci. **133**, 453 (1983)
38.11 J. Lapujoulade, Y. Le Cruër, M. Lefort, Y. Lejay, E. Maurel: Surf. Sci. **118**, 103 (1982)
38.12 S. Brennan, P. Eisenberger: Nucl. Instrum. Meth. **222**, 164 (1984)

39. Helium Diffraction from Oxygen-Covered Nickel Surfaces

Inder P. Batra

IBM Research Laboratory K33/281, San Jose, CA 95193, USA

T. Engel

Department of Chemistry, University of Washington, Seattle, WA 98195, USA

K.H. Rieder

IBM Zurich Research Laboratory, CH-8803 Rüschlikon, Switzerland

Helium atom diffraction results are presented for oxygen-covered Ni(001) and Ni(110) surfaces. For O/Ni(001) the diffraction data has clearly favored a single vertical distance for oxygen chemisorption in both the p(2 × 2) and c(2 × 2) phases. This conclusion now seems to be quite generally accepted. For O/Ni(110) there are several competing candidates which include among others the missing row and the saw-tooth model. We have carried out simple model theoretical calculations based on the Esbjerg-Nørskov-Lang relation between the surface charge density and helium-surface repulsive potential. We have used atomic Hartree-Fock charge densities to generate surface charge densities. The corrugation coefficients have been calculated for various geometrical configurations of oxygen on Ni. These results are compared with the experimental data in order to deduce information about the chemisorption of oxygen on Ni(110).

Oxidation studies on metals have been pursued for a very long time and a few recent review articles [39.1-4] trace the developments. Essentially every surface-structural tool has been employed to investigate the oxygen-metal interaction. The most recent addition is the scanning tunneling microscope which has provided real-space images for the O-Ni(110) system [39.5]. The geometrical position of an adsorbate on a solid surface gives an important clue in understanding surface processes and reactions and hence is of interest. The oxygen chemisorption [39.4] on three close-packed planes of nickel (001), (110) and (111) takes place in well-defined super periodicities at early stages of oxidation before bulk interdiffusion takes over. The precise geometrical arrangement of oxygen atoms on Ni(001) [39.6] and Ni(110) [39.7] as deduced from helium diffraction data is the subject of this article.

On the Ni(001) surface [39.8] at low exposures (1.5 L) a p(2 × 2) structure is formed. At higher exposures (20-30 L) a c(2 × 2) structure is obtained. The Ni (110) surface shows [39.9] a more complicated sequence of structures with increasing oxygen coverage. These are in order of appearance (3 × 1), (2 × 1) and (3 × 1) at about one-third, one-half, and two-thirds monolayer coverage respectively. On the (111) plane [39.10] a p(2 × 2) and a $(\sqrt{3} \times \sqrt{3})$-R30° structure have been reported. The precise chemisorption site for the oxygen atom on these surfaces is of interest to us here. This is found by comparing the corrugation function deduced from helium diffraction data with that calculated using simple model calculations. The subject of helium diffraction has been extensively covered in a number of recent review articles [39.11,12] so shall not be discussed here at any length. Also since no helium atom diffraction data are available for O/Ni(111), no calculations are presented for this system.

To calculate the corrugation function a knowledge of the He-surface interaction potential is required. An approximate relationship between the helium-surface repulsive potential and the surface-electron density has been proposed by *Esbjerg-Nørskov* (EN) [39.13] and has the form

$$V(\mathbf{r}) = \alpha\bar{\rho}(\mathbf{r}) \quad , \qquad (39.1)$$

where $\bar{\rho}$ is the electrostatic potential (of the helium atom) averaged charge density. We note that using averaged as opposed to unaveraged charge density produces $< 15\%$ change in the charge density [39.14]. *Lang* and *Norskov* [39.15] have also provided another justification for (39.1) based on self-consistent helium-jellium model calculations. The linear relation (39.1) has also been verified [39.16] using a self-consistent field Hartree-Fock cluster model investigation. Their suggested value [39.16] of $\alpha \sim 350$ eV-bohr^3 is used in our calculations here. However, we also show the effect on the calculated corrugation when this value is reduced by one-half.

For calculating the surface charge density which enters (39.1) we superposed self-consistent Hartree-Fock atomic charge densities [39.17] arising from the two-dimensional periodic system. This approximation seems to work quite adequately [39.18-21]. The charge for each atom was generated from the self-consistent quadruple atomic zeta functions [39.17] corresponding to the configuration $3d^94s^1(^3D)$. For oxygen we assumed an ionic $O^-(1s^22s^22p^5)$ configuration. The choice of O^- is guided by cluster model calculations and has been discussed before [39.21]. We therefore used the ionic basis set consisting of three exponential functions for 2s and five functions for 2p corresponding [39.17] to the ^2P state. Oxygen was placed at various heights and in different registry patterns on Ni surfaces and the corrugation function was calculated in all cases.

A more accurate surface charge would be obtained by performing a self-consistent field (SCF) slab [39.22] calculation. However such calculations are very demanding [39.22] since the charge densities are required near the classical turning points of thermal helium particles. These classical turning points typically occur 5-7 a.u. above the atomic nuclear positions of the topmost surface layer of atoms. An SCF calculation might produce somewhat lower corrugation values than those obtained using the atomic superposition. However, another approximation made in our calculation at least partly offsets this effect. We consider only the corrugation of the repulsive part of the potential whereas the helium particles sample corrugation in the total potential. The total helium potential is probably more strongly corrugated [39.23,24] than the repulsive potential. Thus the two approximations (the lack of self-consistency and repulsive as opposed to total He potential corrugation) have the opposite effect. The net result [39.24] is that one gets sensible values of corrugations using atomic superposition. However, in view of these approximations one should treat the quantitative values with some caution.

A major issue [39.25-28] in the chemisorption of oxygen on Ni(001) has been as to whether the oxygen atoms in the $c(2\times 2)$ phase sit closer to the substrate than in the $p(2\times 2)$ phase. The experimental helium diffraction data [39.6] give a similar corrugation function for both phases and thus projects a similar distance for both structures. In particular, the corrugation function can be represented by a Gaussian of height ~ 0.55 Å and FWHM of ~ 2 Å. We calculated [39.14,21] the corrugation profile for oxygen in the fourfold sites for $p(2\times 2)$ and $c(2\times 2)$ at two different vertical distances z. The calculated values fell in the range of 0.1-0.3 Å (depending on various calculational parameters) when the oxygen atom was placed at ~ 0.2 Å

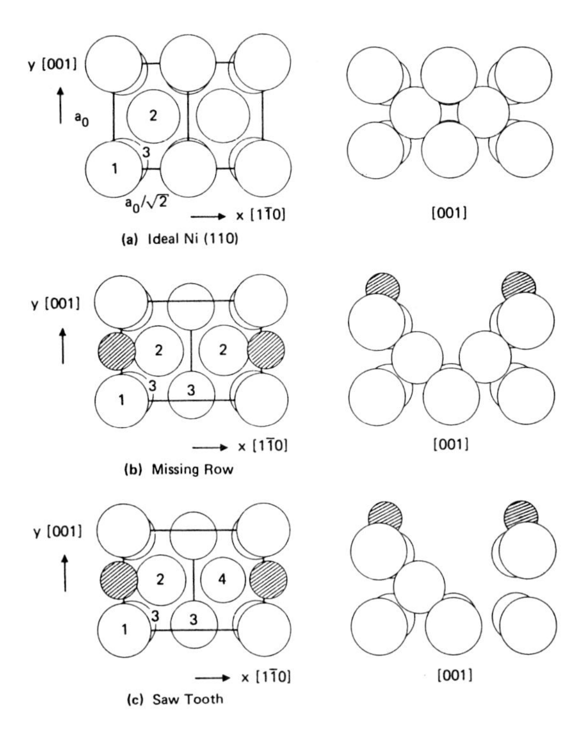

Fig.39.1a-c. Top perspective of the Ni(1$\bar{1}$0) surface: (**a**) the ideal (1×1) surface, (**b**) the missing row model for Ni(110)-O(2×1) and (**c**) the saw-tooth model for Ni(110)-O(2×1). Side perspectives parallel to the [001] direction are also shown. Cross-hatched symbols represent oxygen atoms. Numerals within the circles identify Ni atomic layers; 1 being the surface layer

above the top plane of Ni atoms. The corrugation increased when the oxygen atom was moved farther away from the substrate. At ~0.9 Å, the corrugation was calculated to be in the range of 0.3-0.6 Å depending on various parameters of the calculation. This enabled us to conclude that the helium diffraction data is consistent only with a vertical distance of ~0.9 Å and not with the shorter distance. The calculated [39.14,21] FWHM of about 2.6 Å was quite insensitive to the vertical position of oxygen and other calculational parameters.

A number of geometrical models for the adsorption of oxygen in the (2×1) phase on Ni(110) has been discussed [39.29-35]. In Fig.39.1 are shown the ideal Ni(110) and two recent models for Ni(110)-O(2×1). The x axis is in the [1$\bar{1}$0] direction and the y axis in the [001] direction. The ideal two-dimensional surface (1×1) unit cell dimensions are $a_0/\sqrt{2}$ and a_0, where $a_0 = 3.52$ Å. As early as 1962 a missing row model shown in Fig.39.1b was proposed [39.29] in which every other [001] row of the ideal (110) surface was missing. However, the oxygen atoms in that model [39.29] were supposed to occupy the positions of the missing rows. Later studies [39.30,31], however, indicated that oxygen atoms were situated in twofold long bridge sites along the [001] direction as shown in Fig.39.1b. The proposal for oxygen chemisorption on the short bridge position [39.32] has been found to be inconsistent with Rutherford backscattering data [39.33]. Another study by *Masuda* et al. [39.34] using high electron energy loss spectroscopy can be interpreted in favor of a missing row model. However, it is claimed [39.34] that in the (2×1) phase, unreconstructed surface structure as proposed by *Demuth* [39.32] is also present.

Helium-diffraction experiments [39.7] performed on Ni(110)-O(2×1) have indicated a nearly one-dimensional corrugation function with a corrugation amplitude of ~0.55 Å. The diffraction experiments were carried out on the (2×1) phase prepared by O_2 adsorption at 500 K. We calculated corrugation functions [39.7] for various models proposed for Ni(110)-O(2×1). The short bridge position model and the model where oxygen atoms replace the missing row of nickel atoms were readily eliminated. Both of these models give strong two-dimensional corrugation profiles. The experimental corrugation profile [39.7] has pronounced one-dimensional character: the missing row model shown in Fig.39.1b is a promising candidate for one-dimensional corrugation. Therefore for this model, we performed calculations for various positions of oxygen below, in and above the Ni(110) surface in the long bridge positions along the y axis.

In Fig.39.2 the calculated corrugation function ζ as a function of x for the (2×1) phase of oxygen on Ni(110) is shown. The results are for 45 meV (normal component of incident helium energy) and $\alpha = 175$ and 350 eV-bohr^3. Electrostatic potential averaging of the target charge density was performed. The results shown are for $z \simeq 0.3$ Å, where z is the vertical height of oxygen atom above the surface. We found that changing z by ±0.2 Å produced a noticeable deviation from one-dimensionality. Notice that the maximum in corrugation occurs at $x = 0$, which is above the missing row nickel atom. The corrugation profile along the y direction is flat, so there is no maximum or minimum at the oxygen site. In fact the presence of oxygen atoms along the y axis is responsible for eliminating corrugation in that direction.

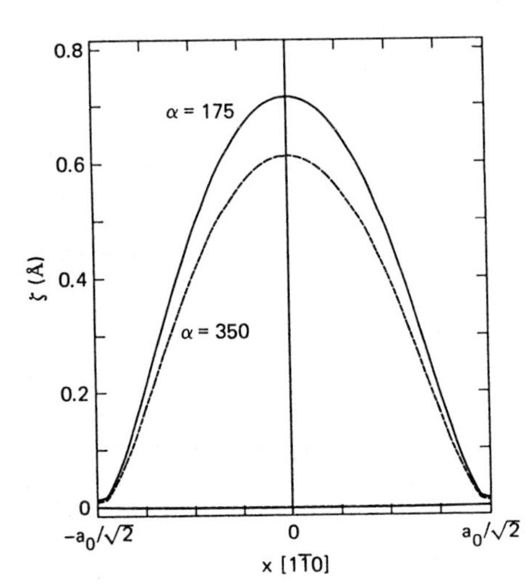

Fig.39.2. Calculated corrugation profile for Ni(110)-O(2×1) along the x[1$\bar{1}$0] direction for oxygen in long bridge sites (Fig.39.1b) at 0.3 Å above the Ni plane. The corrugation along the y axis is flat. The origin of the corrugation has been taken above a Ni atom and the zero of x axis is at the missing row atom

Our calculation favors a site for oxygen ~0.3 Å above surface Ni layer. The calculated corrugation function for $z = 0.3$ Å can be adequately represented by the following one-dimensional cosine Fourier series:

$$\zeta(x) = \frac{1}{2} \sum_m D_m \cos \frac{2\pi m}{a} x \quad , \tag{39.2}$$

where $a = \sqrt{2}a_0 = 4.98$ Å. The values of the corrugation coefficients (in Å) for $\alpha = 350$ eV-bohr^3 are: $D_1 = 0.59$, $D_2 = 0.09$, $D_3 = 0.02$. We find an average softness parameter $\kappa \approx 2.1 \pm 0.3$ Å^{-1} for the repulsive potential. The best fit of the experimental corrugation profile using the hard wall formalism [39.7] gives: $D_1 = 0.49 \pm 0.04$, $D_2 = 0.12 \pm 0.04$, $D_3 = 0.03 \pm 0.01$ Å. Similar results have recently been reported [39.12] for O/Cu(110).

A saw-tooth model which is closely related to the missing row model for O-Ni(110) has also been recently proposed [39.5,35]. In this model every other [001] row is missing in the first as well as in the second Ni atomic layer as shown in Fig.39.1c. Thus the interatomic distance along the [110] direction is doubled for the top two atomic layers. Another way of looking at this reconstruction is to imagine that every other [001] row has shifted into the position of the [001] row immediately above. The saw tooth reconstruction is quite plausible because it simply requires that some Ni atoms move by ~2.5 Å. A pure missing row model, on the other hand, requires transport of material over long distances. However, helium diffraction data alone are not able to choose conclusively between these two models because of similar corrugation profiles. A more thorough discussion is presented elsewhere [39.7].

In conclusion, we state that the observed experimental corrugation profile can conclusively pick between two different oxygen atom heights for O/Ni(001). For O/Ni(110), the corrugation function is consistent with oxygen chemisorption at a vertical distance ~0.3 Å above the top plane of Ni atoms on long bridge sites. It is interesting to note that the corrugation is primarily due to missing rows of Ni atoms. Oxygen atoms tend to smooth out the charge density contours along the [001] direction making the corrugation profile highly one-dimensional.

References

39.1 C.R. Brundle: In preparation
39.2 I.P. Batra, L. Kleinman: J. Elec. Spec. Rel. Phenom. **33**, 175 (1984)
39.3 K. Wandelt: Surf. Sci. Reports **2**, 1 (1982)
39.4 T.N. Rhodin, D.L. Adams: *Treatise on Solid State Chemistry*, Vol.6A, ed. by N.B. Hannay (Plenum, New York 1976) p.343
39.5 A.M. Baró, G. Binnig, H. Rohrer, C. Gerber, E. Stoll, A. Baratoff, F. Salvan: Phys. Rev. Lett. **52**, 1304 (1984)
39.6 K.H. Rieder: Phys. Rev. B**27**, 6978 (1983)
39.7 T. Engel, K.H. Rieder, I.P. Batra: Surf. Sci. (in press)
39.8 J.E. Demuth, T.N. Rhodin: Surf. Sci. **42**, 261 (1974)
39.9 L.H. Germer, J.W. May, R.J. Szostak: Surf. Sci. **7**, 430 (1967)
39.10 G.E. Becker, H.D. Hagstrum: Surf. Sci. **30**, 505 (1972)
39.11 T. Engel, K.H. Rieder: In *Structural Studies of Surfaces with Atomic and Molecular Beam Scattering*, Springer Tracts in Modern Physics, Vol.91 (Springer, Berlin, Heidelberg, New York 1982) p.55
J.A. Barker, D.J. Auerbach: Surf. Sci. Reports (in preparation)
39.12 I.P. Batra: Surf. Sci. (in press)
39.13 N. Esbjerg, J.K. Nørskov: Phys. Rev. **45**, 807 (1980); see also J.K. Nørskov, N.D. Lang: Phys. Rev. B**21**, 2131 (1980)
39.14 I.P. Batra, J.A. Barker: Phys. Rev. B**29**, 5286 (1984)
39.15 N.D. Lang, J.K. Nørskov: Phys. Rev. B**27**, 4612 (1983)
39.16 I.P. Batra, P.S. Bagus, J.A. Barker: Phys. Rev. B (in press)
39.17 E. Clementi, C. Roetti: Atomic Data and Nuclear Data Tables **14**, 177 (1974)
39.18 M. Manninen, J.K. Nørskov, C. Umrigar: J. Phys. F**12**, L7 (1982)
39.19 N. Garcia, J.A. Barker, I.P. Batra: Solid State Commun. **47**, 485 (1983)

39.20 R. Haydock, D. Haneman: J. Vac. Sci. Technol **21**, 330 (1982)
39.21 J.A. Barker, I.P. Batra: Phys. Rev. B**27**, 3138 (1983)
39.22 D.R. Hamann: Phys. Rev. Lett. **46**, 1227 (1981)
39.23 M.J. Cardillo, G.E. Becker, D.R. Hamann, J.A. Serri, L. Whitman, L.F. Mattheiss: Phys. Rev. B**28**, 494 (1983)
39.24 J.A. Barker, N. Garcia, I.P. Batra, M. Baumberger: Surf. Sci. **141**, L317 (1984)
39.25 J.E. Demuth, P.M. Marcus, D.W. Jepsen: Phys. Rev. Lett. **31**, 540 (1973); M. Van Hove, S.Y. Tong: J. Vac. Sci. Tech. **12**, 230 (1975); P.M. Marcus, J.E. Demuth, D.W. Jepsen: Surf. Sci. **53**, 501 (1975)
39.26 T.H. Upton, W.A. Goddard: Phys. Rev. Lett. **46**, 1635 (1981); T.S. Rahman, J.E. Black, D.L. Mills: Phys. Rev. Lett. **46**, 1469 (1981); Phys. Rev. B**25**, 883 (1982)
39.27 J. Stöhr, R. Jaeger, T. Kendelewicz: Phys. Rev. Lett. **49**, 142 (1982)
39.28 C.W. Bauschlicher, S.P. Walsh, P.S. Bagus, C.R. Brundle: Phys. Rev. Lett. **50**, 864 (1983)
39.29 L.H. Germer, A.H. MacRae: J. Appl. Phys. **33**, 2923 (1962)
39.30 L.K. Verheij, J.A. van den Berg, D.G. Armour: Surf. Sci. **84**, 408 (1979; J.A. van den Berg, L.K. Verheij, D.G. Armour: Surf. Sci. **91**, 218 (1980)
39.31 A.M. Baró, L. Ollé: Surf. Sci. **126**, 170 (1983)
39.32 J.E. Demuth: J. Colloid Interface Sci. **58**, 184 (1977)
39.33 R.G. Smeenk, R.M. Tromp, J.F. van der Veen, F.W. Saris: Surf. Sci. **95**, 156 (1980)
39.34 S. Masuda, M. Nishijima, Y. Sakisaka, M. Onchi: Phys. Rev. B**25**, 863 (1982)
39.35 M. Schuster, C. Varelas: Surf. Sci. **134**, 195 (1983)

40. Competing Reconstruction Mechanisms in H/Ni(110)

R.J. Behm, K. Christmann, G. Ertl, V. Penka, and R. Schwankner
Universität München, D-8000 München 2, FRG

The mechanisms of the hydrogen-induced reconstructions of Ni(110) have been investigated. Below ~180 K a (2 × 1) lattice gas structure with $\theta_H = 1.0$ transforms into a 2D-(1 × 2) structure during addition of hydrogen up to $\theta_H = 1.5$. The phase transition, which involves a reconstruction of the surface, exhibits first-order behavior with no apparent activation energy. In contrast, at T > 180 K and already at low coverages, an activated, local transformation into a more stable 1D structure ('streaked structure') occurs. A lattice distortion to optimize the local metal structure with respect to the metal-adsorbate bond and thus increase the binding energy is introduced as a general model for many such adsorbate-induced surface phase transformations.

40.1 Introduction

Considerable work has been performed to investigate the structure and thermodynamic stability of ordered surface phases [40.1]. Especially for reconstructed surface phases, little is yet known about the microscopic mechanism and driving force of the phase transformation. For the system H/Ni(110), we have found two competing adsorbate-induced surface reconstruction mechanisms [40.2]. A full report will be published elsewhere, including a detailed description of the apparatus and the methods applied to meet the stringent requirements for surface cleanliness [40.3]. A brief overview of the main features of this adsorption system is first presented.

Exposure of the clean surface to H_2 at 120 K leads to the formation of a continuous sequence of ordered lattice gas structures from adsorbed H atoms, which are completed by a "(2 × 1)p1g1" structure, denoted below as (2 × 1), at $\theta_H = 1.0$ [40.4]. In this coverage range He scattering [40.5], LEED structure analysis [40.6] and EELS [40.7] indicate the occupation of quasi-threefold adsorption sites for the H atoms in zigzag rows along the [1$\bar{1}$0] direction. Beyond this coverage the LEED pattern of the (2 × 1) coexists with that of a reconstructed (1 × 2) phase [40.2,3], here denoted (1 × 2), until at saturation ($\theta_H = 1.5$) the whole surface is converted into the (1 × 2). The indicated coverage assignments were recently confirmed by nuclear microanalysis [40.8].

In contrast, annealing a surface exposed to H_2 at T > 180 K or exposure at T > 180 K leads to the irreversible formation of a structure which exhibits considerable disorder in the [001] direction. In the LEED pattern streaks between the integral order beams in [01] direction are observed. The saturation coverage in this structure is roughly similar to that in the (1 × 2) [40.3,8]. The structure itself is stable and only removed by H_2 desorption.

40.2 Results

The formation of the (1×2) reconstructed phase from the unreconstructed (2×1) was followed by monitoring the intensity of the (0, 3/2) beam as a function of H_2 exposure. In Fig.40.1c also the decay of the (1/2, 1) beam, related to the (2×1), and the behavior of the integral order (1,0) beam is displayed. The maximum in the (1/2, 1) beam served to define the coverage $\theta_H = 1$ at this point [40.5]. These data allow the following two conclusions. The (1×2) grows as islands in the surrounding 'sea' of (2×1) and the onset of intensity in the (0, 3/2) beam is always delayed by ca. 0.05 L with respect to the maximum in the (1/2, 1) beam. The island growth mechanism is evident from the relationship between the (1/2, 1) and the (0, 3/2) intensity. At all coverages between $\theta = 1.1$ and $\theta = 1.45$ the sum of the two normalized intensities I/I_{max} is constant, a clear indication of coexisting, rather large islands.

The delay between the maximum in the (2×1) intensity and the first appearance of the (1×2) might suggest nucleation of the latter phase at a critical coverage, i.e., supersaturation of the (2×1) phase at $\theta_H > 1.0$. This has been ruled out, however, by measurements of the diffuse elastic background and the beam profiles in the [10] and [01] directions, as indicated in Figs.40.1b,c. The intensity of the diffuse elastic background is related to the number of uncorrelated point defects and vibrationally induced displacements in the lattice. Since the temperature in these experiments was kept constant, changes in the latter can be neglected unless new phonon modes appear. Statistically distributed H atoms in the (2×1) would present a typical example of point defects, so they should therefore lead to a possibly temporary increase in the diffuse elastic background. However, the background intensity monitored in the 'windows' 1 and 2 in the reciprocal lattice as indicated in Fig.40.1b remains at a constant level as θ_H increases from 1 and starts to grow only around $\theta = 1.25$. Obviously more random defects are created once the (1×2) starts to cover major parts of the surface, e.g., by island coalescence, at $\theta_H \geqslant 1.25$.

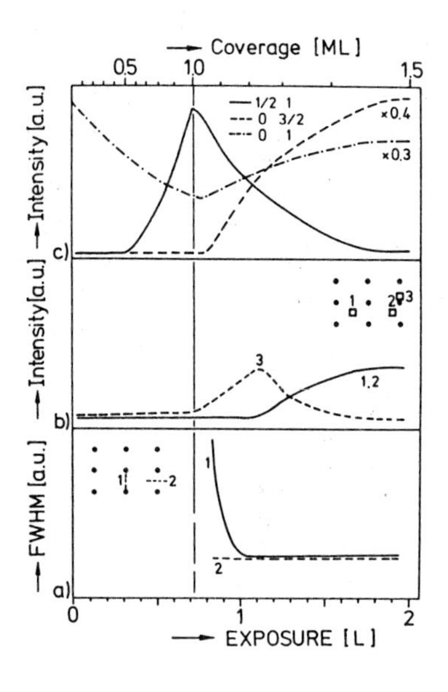

Fig.40.1. LEED during exposure to H_2 at 120 K: (*a*) profile width (FWHM) of (1, 1/2) beams as indicated; (*b*) diffuse background as indicated; (*c*) integral intensity as indicated

If, in contrast, the delay in the appearance of intensity in the (1×2) related beams were caused by the initial formation of very small nuclei of (1×2), which subsequently grow, initially broad profiles of the (1, 1/2) beam in one or both directions should be observed, depending on the shape of these nuclei. In 'window' 3 the intensity between the (1, 1/2) and (1, 0) spot was monitored, where at higher temperature the streaks are formed. Curve 3 in Fig.40.1b clearly demonstrates that there is an intensity increase in this region right at the maximum of the (2×1), thus confirming this hypothesis.

The evaluation of the beam profiles supports this conclusion. The integral order (1,0) beam does not change its shape over the entire exposure range. The FWHM of the (1/2, 1) beam reaches the instrumental resolution, given by the instrumental response function, in both directions at $\theta = 0.96$ and remains there until at around $\theta = 1.4$ when the intensity is too low for further evaluation. The corresponding behavior of the FWHM of the $(0, \bar{1}/2)$ beam is illustrated by the full line in Fig.40.1a. As soon as it can be evaluated around $\theta = 1.05$, the width of the profiles in the [10] direction is within the instrumental resolution. In the normal [01] direction, measured on the $(1, \bar{1}/2)$ beam, the FWHM is clearly larger: initially, at $\theta = 1.05$, the measured width is about double the size of that in the [10] direction. With increasing coverage the width rapidly decreases and at $\theta = 1.15$ it also reaches its final value, close to the instrumental resolution.

These results indicate that the reconstructed (1×2) grows as islands in the surrounding (2×1), starting immediately when the (2×1) is fully developed. There is no indication for a supersaturation effect in the sense that H atoms might be statistically squeezed into the (2×1) without locally initiating the reconstruction as small islands of (1×2). Initially the (1×2) islands are elongated in the $[1\bar{1}0]$ direction, with a rather small diameter in the [001] direction. The instantaneous formation of the (1×2) suggests that the reconstruction is not activated, which is supported by the results of similar LEED intensity measurements at temperatures between 120 K and 250 K [40.3].

The second reconstruction path leads to the 'streaked' structure, characterized by a LEED pattern in which the half-order spots of the (1×2) are elongated to streaks in the [01] direction. The energy and coverage dependence of the spot profiles also indicate an island growth mechanism although the islands exhibit inherent disorder. This structure is formed upon H adsorption at $T > 180$ K, independent of coverage and with a rather low critical coverage (< 5% at 300 K) [40.3], or upon annealing a hydrogen adlayer to $T > 180$ K. This transition *always* starts from the unreconstructed Ni surface, since the transformation from the reconstructed (1×2) into the streak phase proceeds through an intermediate, unreconstructed (1×1) Ni surface and partial loss of hydrogen [40.2].

This reconstruction proceeds rather slowly. To determine a possible activation barrier, a fixed amount of hydrogen was adsorbed at a low temperature which then was raised in a step and kept at the new value. The formation of the 'streaked' phase was monitored by LEED as a function of time. The rate of transformation (assumed to follow first-order kinetics) can be expressed as

$$\frac{dA_{str}}{dt} = A_{1 \times 1} \cdot \nu \cdot \exp(-E^{*}/R \cdot T) \quad ,$$

where A denotes the fraction of the surface existing in the respective phase, ν and E^* are the preexponential and the activation energy for the reconstruction. The activation barrier can be extracted from an Arrhenius plot of the integral intensities of the streaks, since for an island growth mechanism of the 'streaked' phase $A_{str} \sim I_{str}$. Analyzing the initial slopes of the intensities in this way yielded $E^* \simeq 50$ kJ/mole. In other experiments we verified that this activation energy for reconstruction also limits the adsorption rate of hydrogen at $T > 300$ K [40.3].

40.3 Discussion

The energetic situation for H/Ni, derived from thermal desorption and other data [40.3] is displayed in Fig.40.2. In the insert in the upper part the potential energy (per mole of adsorbed H_2) of the different adsorbed structures is indicated. The zero level of the energy scale corresponds to the clean unreconstructed Ni(110) surface and H_2 in the gas phase. The 'streaked' structure is more stable than H adsorbed on the (1×1) by 16 kJ/mole, but is separated by an activation barrier of 50 kJ/mole. The (1×2), however, is less stable: its net adsorption energy is only 64 kJ/mole, as outlined below.

The effect of increasing coverage becomes clearer from the main part of Fig.40.2 which shows a plot of the *integral* energy of the system as a function of coverage using the same zero level. Adsorption on the (1×1) surface [and (2×1) formation] leads to a continuous energy decrease, whose slope corresponds to the differential adsorption energy of 71 kJ/mole [40.3]. This phase obviously cannot accommodate more than one monolayer of hydrogen, which is reached in point A. Further uptake, which still leads to an energy gain of the system, is possible only by forming the reconstructed (1×2). During this process the (2×1) is locally converted into (1×2), a measured adsorption energy thus includes a contribution to raise the energy level of those H atoms already adsorbed on the (1×1)-Ni surface, e.g., in the (2×1). Since the adsorption is not activated, this adsorption energy is identical to the activation energy for desorption from the (1×2) of 51 kJ/mole [40.3], which gives the slope of the line A-B. The net adsorption energy of the (1×2) [H adsorption and displacement of the Ni atoms into the (1×2)] is given by the slope of the dashed line 0-B corresponding to 64 kJ/mole. The adsorption energy of the 'streaked' structure, symbolized by the dash-dotted line, is constant up to medium coverages at 87 kJ/mole and decreases at higher coverages, due to occupation of less favorable sites and/or repulsion between the adsorbed atoms [40.3].

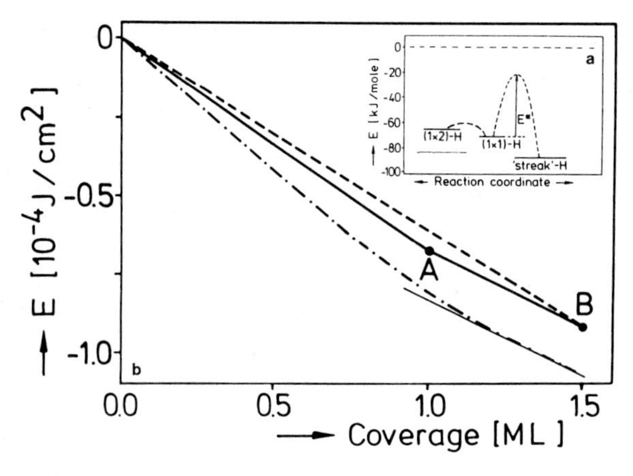

Fig.40.2. Potential energy diagram. (*a*) (Differential) adsorption energies and activation barriers of the respective reconstruction; (*b*) (integral) system energy [Ni(110) + H_2] as f(θ) for the different structures: (----) (imaginary) H/(1×2); (0——A) H/(1×1); (A——B) coexisting H/(1×1) and H/(1×2); (—·—·—) H/'streaked' structure

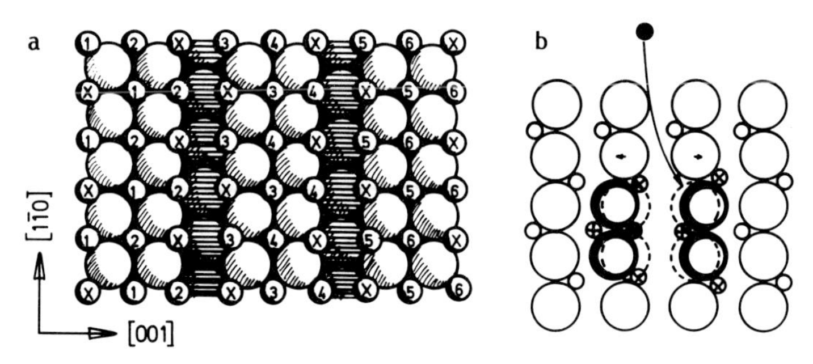

Fig.40.3. (**a**) Structure model of the (1 × 2): O: Ni atoms in the 1st layer (⊜: 2nd layer); o: H atoms, 1-6: zigzag rows in the former (2 × 1); x: additional H atoms. (**b**) Mechanism for the (1 × 1) ⇌ (1 × 2) reconstruction): O: Ni atoms (**O**: reconstructed); o ⊕ ⊗: H atoms in the (2 × 1), position I, position II; •: 'additional' H atoms to nucleate the (1 × 2)

With respect to the microscopic mechanism the data show that the (1 × 2) forms from the (2 × 1) without any noticeable supersaturation. In other words, the free enthalpy of a nucleus of (1 × 2) in a surrounding of (2 × 1) has its maximum already at a very small size and any further growth leads to a stabilization of the cluster [40.9]. Figure 40.3 illustrates a possible mechanism which would fulfill this requirement. The open dots represent H atoms in a (2 × 1) structure on the unreconstructed Ni surface. Uptake of an additional H atom (filled dot) with an adsorption energy E_{ad} causes local displacement of neighboring Ni atoms (ΔE_{rec}), as depicted. The adsorption energy of the neighboring H atoms is affected by the movement of the underlying Ni atoms ($\Delta E'_{ad}$ for H atoms in position I, $\Delta E''_{ad}$ for those in position II) and by (presumably repulsive) lateral interactions with the addition H atom (E_{aa}). The change in total energy for this nucleus can then be written as

$$E_{\Lambda} = -E_{ad} + 4 \cdot \Delta E_{rec} + 3 \cdot \Delta E'_{ad} + 4 \cdot \Delta E''_{ad} + 4 \cdot E_{aa} \quad . \tag{40.1}$$

Growth of the nucleus by adsorption of a second H atom, as indicated by the arrow, will lead to the following energy balance:

$$E_2 = -2 \cdot E_{ad} + 6 \cdot \Delta E_{rec} + 4 \cdot \Delta E''_{ad} + 8 \cdot E_{aa} \quad , \tag{40.2}$$

resulting in a stabilization of one 2-H cluster as compared to two 1-H clusters by

$$\Delta E = 2E_1 - E_2 = 2 \cdot \Delta E_{rec} + 4 \cdot \Delta E''_{ad} \quad . \tag{40.3}$$

Any further growth in the same lattice direction does not lead to an additional stabilization ΔE, barring more complicated interactions. The critical size of the cluster is thus reached already if one additional H atom is built into the (2 × 1) and locally displaces the surrounding Ni atoms. Growth in the [001] direction requires formation of a new, independent nucleus, which in turn can again easily grow in [1$\bar{1}$0] as is observed experimentally. This mechanism also predicts preferential one-dimensional growth in the [1$\bar{1}$0] direction which can formally be described as a collective phenomenon. The reconstruction will almost simultaneously affect long double rows of Ni in the [1$\bar{1}$0] direction and finally lead to the structure model of Fig.40.3a [40.2]. It is also obvious that the (2 × 1) will be a prerequisite for this mechanism.

The low observed activation barrier might be due to the fact that dissociation and subsequent chemisorption can take place at the same position and probably also simultaneously with the reconstruction. Therefore the energy gained by the former process can be used to overcome the barrier for reconstruction without any need for additional thermal activation.

In analogy the formation of the 'streaked' structure has to account for the increase in the net adsorption energy E with respect to that on the (1×1) and the observed activation barrier for its formation of ca. 50 kJ/mole. Thermal activation appears necessary, if the adsorbed particles have already completely accommodated before the reconstruction takes places. This agrees well with the experimental observation that hydrogen must first adsorb on the (1×1) surface ($\theta_{crit}\simeq5\%$) before the reconstruction can start. The increase in the net adsorption energy E must be due to an increase in the term E_{ad}. The total contribution of the other terms—especially the lattice strain ΔE_{rec}—can then be calculated to be small for the 'streaked structure', but larger for the (1×2) by using the adsorption energies on Ni(111) and Ni(100) as an upper limit for E_{ad} [40.10]. Our structure model for the (1×2), Fig.40.3, indeed postulates occupation of comparable sites [40.2].

In analogy with bulk phase transformations [40.11] surface transformations are generally characterized as being 'displacive' (gradual displacement of surface atoms) or of 'order-disorder' type (e.g., phase transitions in lattice gas systems) [40.11]. Both mechanisms seem to apply for surface reconstruction: the temperature-dependent $c(2\times2)$ like reconstructions of Mo(100) and W(100) [40.13] are unanimously attributed to periodic lattice distortions (PLD) and therefore belong to the 'displacive' type, while for the $(1\times2)\rightarrow(1\times1)$ transformation of Au(110) there is evidence of an order-disorder mechanism [40.14]. The two H-induced surface reconstructions on Ni(110) are clearly 'displacive'. Because of the island growth (i.e., constant local θ_H) a constant (not gradual) Ni displacement is expected.

The driving force for the H/Ni surface reconstruction can be identified as the gain in potential energy $N\cdot E_{ad}$ (N = number of adsorbed particles), the first term in (40.1,2). This energy gain may result from an increase in adsorbate uptake (>N) as in the (1×2) or an increase of the differential adsorption energy E_{ad} as in the 'streaked' structure. This energy increase is sufficient to compensate for all other contributions, especially the energy necessary to reconstruct the surface, ΔE_{rec}. The proposed mechanism thus consists of a local distortion of the metal surface lattice to gain the most possible substrate-adsorbate interaction energy, while simultaneously keeping the energy loss due to the metal rearrangement as low as possible.

In contrast, for W(100) and Mo(100) the geometry of the Fermi surface, which causes charge density waves at the surface, was believed to be the origin of their reconstruction. In a recent calculation, however, a more local atom-pairing model comparable to a Jahn-Teller distortion is favored as the main energy source [40.15]. This energy gain due to variations in the electronic structure of the metal may occur also for the system H/Ni(110), but it is not expected to play a dominant role. A similar increase in the adsorption energy E_{ad} was recently demonstrated to be responsible for the CO-induced removal of the hexagonal reconstruction on Pt(100). In that case the term E_{ad} could be determined separately and was found to be 155 kJ/mole on the (1×1) and 113 kJ/mole on the hex surface [40.16].

In summary, the (1 × 2) is formed by a non-activated process. The total energy gain stems from a coverage increase, while the individual Ni-H bond is weaker. The formation of the 'streaked' structure is activated by ~50 kJ/mole. It is the most stable structure. The critical coverage for nucleation amounts to ≤5% of a monolayer for the 'streaked' structure but is negligible for the (1 × 2). The proposed reconstruction mechanism accounts qualitatively for these features as well as for the preferential growth of the (1 × 2) in the [1$\bar{1}$0] direction, which simulates a collective effect. A local lattice distortion to maximize the Ni-H binding energy without creating too much strain for the Ni surface is suggested to be the energetic driving force. This model has characteristics which classify it as a displacive phase reconstruction, and is proposed as a general model for many other adsorbate-induced surface phase transformations.

Acknowledgment. We gratefully acknowledge P.R. Norton for information from [40.8] prior to publication and for valuable discussions. Financial support was obtained from the Deutsche Forschungsgemeinschaft (SFB 128).

References

40.1 L.D. Roelofs, P.J. Estrup: Surf. Sci. **125**, 51 (1983)
40.2 K. Christmann, V. Penka, R.J. Behm, F. Chehab, G. Ertl: Solid State Commun. (in press)
40.3 V. Penka, R.J. Behm, K. Christmann, G. Ertl, R. Schwankner: In preparation
40.4 V. Penka, K. Christmann, G. Ertl: Surf. Sci. **136**, 307 (1984)
40.5 K.H. Rieder, T. Engel: Phys. Rev. Lett. **43**, 373 (1979); **45**, 824 (1980); T. Engel, K.H. Rieder: Surf. Sci. **109**, 140 (1984)
40.6 W. Reimer, W. Moritz, R.J. Behm, G. Ertl, V. Penka: To be published
40.7 N.J. DiNardo, E.W. Plummer: J. Vac. Sci. Technol. **20**, 890 (1982)
40.8 T.E. Jackman, J.A. Davies, P.R. Norton, W.N. Unertl, K. Griffiths: Surf. Sci. **141**, L313 (1984)
40.9 B. Mutaftschiev (ed.): *Interfacial Aspects of Phase Transformations*, Nato Adv. Study Inst. Series, B (Reidel, Dordrecht 1982)
40.10 K. Christmann, O. Schober, G. Ertl, M. Neumann: J. Chem. Phys. **60**, 4528 (1974)
40.11 R.A. Cowley: Adv. Phys. **29**, 1 (1980)
40.12 R.F. Willis: In *Proc. Many Body Phenomena at Surfaces*, ed. by D.C. Langreth, D. Newns, H. Suhl (Academic, New York 1984)
40.13 R.A. Barker, P.J. Estrup: J. Chem. Phys. **74**, 1442 (1981); D.A. King, G. Thomas: Surf. Sci. **92**, 201 (1980) and references therein
40.14 D. Wolf, H. Jagodzinski, W. Moritz: Surf. Sci. **77**, 283 (1978)
40.15 I. Terakura, K. Terakura, N. Hamada: Surf. Sci. **111**, 479 (1981)
40.16 R.J. Behm, P.A. Thiel, P.R. Norton, G. Ertl: J. Chem. Phys. **78**, 7437, 7448 (1983)

IV.3 Molecular Adsorption on Metal Surfaces

41. The Uses and Limitations of ESDIAD for Determining the Structure of Surface Molecules

Theodore E. Madey

Surface Science Division, National Bureau of Standards
Gaithersburg, MD 20899, USA

41.1 Introduction

The principles and mechanisms of electron-stimulated desorption (ESD) and photon-stimulated desorption (PSD), as well as the utility of the electron-stimulated desorption ion angular distributions (ESDIAD) method as a tool for determining the structure of surface molecules, have been described in a recent book [41.1] and several review articles [41.2-6]. The present short paper is intended to provide a guide to the relevant literature, and to describe briefly some recent work related to the uses and limitations of ESDIAD for determining the structure of surface molecules.

41.2 Experimental Basis of ESDIAD; Mechanisms of Ion Desorption

Electronic excitation of surface molecules by a focused electron beam (20 eV to $>$1000 eV) can result in desorption of atomic and molecular ions and neutral species from the surface. The ions desorb in discrete cones of emission, in directions determined by the orientation of the bonds which are ruptured by the excitation. For example, electron-stimulated desorption (ESD) of OH bound in a standing-up configuration on a metal surface will result in desorption of H^+ in the direction of the surface normal, while ESD of H^+ from "inclined" OH, or from NH_3 adsorbed via the N atom will result in desorption of H^+ in off-normal directions.

Based on a large body of ESDIAD data [41.4], where there was a priori knowledge of the surface structure, the ion desorption angle observed in ESDIAD is related to the surface bond angle. In particular, it is found that the expected azimuthal angle of the surface bond is preserved in ESDIAD, but the polar angle is increased for the ion trajectory, due largely to image charge effects (Sect.41.3). To date, no exceptions have been found. Thus, measurements of the patterns of ion desorption provide direct information about the geometrical structure of surface molecules in the adsorbed layer. ESDIAD is sensitive to the *local molecular structure*, so that long-range order in the overlayer is not necessary for determination of adsorbate geometry.

The mechanisms of ion formation and desorption in ESD have attracted a great deal of attention from both experimentalists and theoreticians [41.1-3, 6-8]. For simplicity, we can describe ion formation in a generalized fashion. Following a fast initial electronic excitation of a surface adatom to a repulsive final state, an ion is ejected from the surface. The time required to break the surface bond is typically 10^{-14} to 10^{-15}s, corresponding to the time necessary for the ion to travel about 1 Å. There is a high probability of ion neutralization due to electron hopping from the substrate, and this can result in recapture into a bound state, or desorption of a neutral species.

It is now widely believed that the desorption of ions from both covalently bonded and ionically bonded surface species proceeds via multielectron excitations producing 2 hole or 2 hole-1 electron (2h or 2h1e) excited states [41.7,8]. These excited states can be highly repulsive, with hole localization lifetimes of the order of 10^{-14}s, so that the repulsive electronic excitation can be converted into nuclear motion. The repulsive interaction may be described as Coulombic in origin and directed mainly along the bond which is ruptured by the excitation; thus the initial ion desorption angle in ESDIAD is determined by the surface bond angle.

Various decay mechanisms of the excited states include resonant decay, valence one-hole and two-hole hopping, and core hole Auger decay [41.7,8]. Two-hole hopping and Auger decay are the slower processes, $\sim 10^{-14}$s, and states giving rise to ion desorption are generally associated with such processes. Desorption of certain neutral species, such as rare gases [41.9], are believed to proceed via 1-hole excitations.

41.3 Factors Influencing Ion Trajectories

As indicated above, it can be argued that the initial repulsive interaction for a desorbing ion is mainly along the bond direction. However, final state effects can influence the ion desorption trajectories and the resultant ESDIAD patterns. Possible final state effects include anisotropy in the ion neutralization rate, "focusing" effects due to curvature in the final state potential, and deflection of the escaping ions by the electrostatic image potential [41.4]. In general, we have no detailed knowledge of the final state potentials (although model calculations have been performed for specific systems [41.10]). There are recent calculations [41.11-13] in which the influence of the image force and of ion neutralization processes on ion desorption trajectories have been estimated. In a study by *Miskovic* et al. [41.12], analytical expressions have been obtained for the trajectories of desorbed ions, as well as ions trapped by the image field. The main conclusions from this work are the following.

1) The image potential invariably causes an increase in the polar desorption angle θ of an ion leaving a planar metal surface; the image potential does not influence the azimuthal angle ϕ. The amount of distortion of the ion trajectory by the image potential is directly related to the parameter $|V_I|/E_0$ which defines the "strength" of the image potential (V_I is the image potential at the initial ion-surface separation, and E_0 is the initial kinetic energy). Larger values of $|V_I|/E_0$ lead to large distortion of the trajectories.

2) For monoenergetic, monodirectional ions there is a critical angle for desorption θ_c, which depends on $|V_I|/E_0$. For initial desorption polar angles $\theta_0 > \theta_c$, the image potential will bend the ions back to the surface, and escape is impossible. For $\theta_0 = \theta_c$, the polar angle of the final trajectory is $\theta = 90^\circ$, parallel to the surface; as an example, $\theta_c \approx 50^\circ$ for H^+ from NH_3 on Ni [41.4].

3) For ions desorbing with a range of energies and angles (i.e., with finite energy width E_ω and angular width θ_ω), the results are more complex, and the reader is referred to the original paper [41.12].

When simple Hagstrum-type reneutralization effects are included (rate $\sim \exp(-az)$, where a is constant and z is the distance of the ion above the surface), the reneutralization rate is higher for ions having trajectories nearer the surface, i.e., for larger values of polar angle θ. Thus, the peak position of the ion angular distribution generally is shifted to smaller values of polar angle θ, toward the surface normal [41.13]. In many cases, the effects of the image potential and reneutralization on measured ion trajectories are in opposite directions, i.e., they tend to cancel one another partially.

The above results [41.11-13] based on fairly simple models (ideal planar conductors, impulsive forces) demonstrate that quantitative conclusions regarding molecular structure based on ESDIAD alone are not generally possible for large polar angle θ. Despite this, there are many useful semiquantitative, and qualitative conclusions to be drawn from ESDIAD measurements at large angle, $\theta \gtrsim \theta_c$ [41.13]. First, it is easy to observe when the polar bond angle is far away from normal rather than nearly perpendicular, and "inclined" molecules can be seen readily. Second, the presence or absence of azimuthal ordering in adsorbed molecules can be seen easily in the ESDIAD patterns independent of θ, because there is no distortion of the azimuthal angle in the image field. Finally, ESDIAD measurements at small polar angle $\theta \ll \theta_c$ are not generally as strongly perturbed by the image force, so that quantitative determinations of bond angle are possible from a measurement of ion desorption angle.

41.4 Recent Applications of ESDIAD

There are a number of cases where ESDIAD has provided new insights into structure and bonding of surface species which had not been previously recognized. Some recent examples include the following.

1) Coadsorption of polar molecules (NH_3, H_2O) with electronegative surface additives (O, Br) can lead to local azimuthal order, without dissociation of the molecule. Examples are found in the adsorption of NH_3 on Ni(111) or Ru(0001) [41.14,15]. Traces of preadsorbed oxygen (< 0.05 monolayers) will induce a high degree of azimuthal ordering in a fractional monolayer of adsorbed NH_3 molecules; the H ligands are oriented along specific azimuths. In contrast, adsorption of NH_3 on clean Ni(111) or Ru(0001) results in a random azimuthal orientation of the NH_3 molecules, which are bound to the metal via the N atom, with H pointing away from the surface. We suggest that molecular NH_3 interacts with atomic O via a hydrogen bond, leading to local azimuthal ordering in the absence of long-range order. Also, thermal desorption spectroscopy reveals an increase of NH_3 binding energy. Cases where local azimuthal ordering of polar molecules is induced include:

NH_3 + O on Ni(111), Ru(0001), Al(111)

H_2O + O on Ni(111)

H_2O + Br on Ag(110) .

2) Coadsorption of small molecules (NH_3, H_2O, CO) with an electropositive surface additive (Na) leads to molecular reorientations, i.e., "tilting" of the molecular axes, e.g., for partial monolayers of Na ($\lesssim$ 0.15 monolayers) and saturation coverages of CO on Ru(0001) at 80 K, a fraction of the CO molecules undergo a substantial change in bonding configuration; molecular CO bound perpendicular to the clean Ru(0001) surface changes to an "inclined" configuration in the presence of Na [41.16]. Other examples include [41.15,16]:

NH_3 + Na on Ru(0001), Ni(110)

H_2O + Na on Ru(0001) .

In each case, the molecular axis is "inclined" due to the electrostatic interaction of the polar molecule with adsorbed Na, which is partially ionic at low coverages. Also, the binding energies of the molecules to the surface are generally weakened by this interaction.

3) Evidence for the formation of adsorbed H_2O dimers has been found for low coverages of H_2O on Ni(110) [41.17]. It is well known that H_2O forms hydrogen-bonded clusters upon adsorption on many metal surfaces, but these are generally interpreted as large clusters having a hexagonal network. We have found recently using ESDIAD and angle-resolved UPS that H_2O adsorption on clean Ni(110) surfaces at $T \leqslant 150$ K leads at coverages below $\theta \simeq 0.5$ to the formation of chemisorbed H_2O dimers, bound to the substrate via both oxygen atoms. The linear hydrogen bond axis is orientated parallel to [001] directions. At higher H_2O coverages, a two-dimensional hydrogen bonded network forms, so that at $\theta = 1$, a slightly distorted two-dimensional ice lattice having long-range order is observed.

4) "Inclined" OH species are formed on fcc (110) surfaces. For example, the presence of oxygen on a Ni(110) surface promotes the adsorption and decomposition of H_2O at 300 K [41.18]. Angle-resolved UPS (ultraviolet photoemission spectroscopy), ESDIAD and isotope experiments all indicate that OH(ad) is formed on the surface, presumably via a hydrogen abstraction reaction, $H_2O + O(ad) \rightarrow 2\ OH(ad)$. Both ESDIAD and ARUPS indicate that the molecular axes of the OH(ad) species are inclined with respect to the surface normal, and are oriented along [001] and [00$\bar{1}$] azimuthal directions. Similar results are seen for Ag(110) [41.19].

5) We have used ESDIAD to characterize the structure and bonding of O_2 on Ag(110) [41.19]. The data indicate that both atomic oxygen ($T_{ads} > 300$ K) and molecular oxygen ($T_{ads} \cong 80$ K) are bonded in configurations which cause ion desorption in directions along [001] azimuths. Possible structures consistent with the data are at variance with existing models of the oxygen-Ag(110) system [41.20], e.g., the simplest bonding picture for molecular O_2 would involve O_2 bonded end on to single Ag atoms, slightly inclined away from the surface normal along [001] azimuths. Atomic oxygen may be bound in non-high-symmetry sites.

41.5 Summary

We summarize some of the structural information obtainable using ESDIAD as follows [41.4].

1) ESDIAD provides *direct evidence* of the structures of surface molecules and molecular complexes; in certain cases (e.g., oxides) it can provide structural information about the substrate surface.

2) ESDIAD is particularly sensitive to the orientation of hydrogen ligands in absorbed molecules. In general, electron scattering from surface H is sufficiently weak so that LEED is not very useful.

3) ESDIAD is sensitive to the *local* bonding geometry of molecules on surfaces even in the absence of long-range translational order in the overlayer.

4) The identification using ESDIAD of impurity-stabilized surface structures (H_2O and NH_3 on O-dosed surfaces) is of relevance to surface reaction mechanisms, as well as to catalyst promoters and poisons.

Acknowledgments. The author acknowledges with gratitude the contributions of a number of colleagues and visiting scientists, without whose efforts there would be little to discuss: Drs. Carsten Benndorf and C. Nöbl (Hamburg), Z. Miskovic and J. Vukanic (Belgrade), Falko Netzer (Innsbruck), Klaus Bange and Jürgen Sass (Berlin), M. Klaua (Halle), Dale Doering (U. of Florida), David Ramaker (George Washington University and NBS), Roger Stockbauer, Richard Kurtz and Steve Semancik (NBS). This work has been supported in part by the U.S. Department of Energy, Division of Basic Energy Sciences.

References

41.1 N.H. Tolk, M.M. Traum, J.C. Tully, T.E. Madey (eds.): *Desorption Induced by Electronic Transitions DIET-I*, Springer Ser. Chem. Phys., Vol.24 (Springer, Berlin, Heidelberg 1983)
41.2 M.L. Knotek: Physica Scripta T**6**, 94 (1983)
41.3 D. Menzel: J. Vac. Sci. Technol. **20**, 538 (1982)
41.4 T.E. Madey: In *Inelastic Particle-Surface Collisions*, ed. by W. Heiland, E. Taglauer, Springer Ser. Chem. Phys., Vol.17 (Springer, Berlin, Heidelberg 1981) p.80; also
T.E. Madey, F.P Netzer, J.E. Houston, D.M. Hanson, R. Stockbauer: Ref. 41.1, p.120
41.5 T.E. Madey, D.L. Doering, E. Bertel, R. Stockbauer: Ultramicroscopy **11**, 187 (1983)
41.6 T.E. Madey, D.E. Ramaker, R. Stockbauer (eds.): *Annual Reviews of Physical Chemistry* (Annual Reviews, Inc., 1984)
41.7 See papers by D.R. Jennison, J.C. Tully, R. Gomer, P.J. Feibelman, D.E. Ramaker: Ref. 41.1
41.8 D.E. Ramaker: J. Vac. Sci. Technol. A**1**, 1137 (1983)
41.9 Q.J. Zhang, R. Gomer, D.R. Bowman: Surf. Sci. **129**, 535 (1983)
41.10 E. Preuss: Surf. Sci. **94**, 249 (1980)
41.11 W.L. Clinton: Surf. Sci. **112**, L791 (1981)
41.12 Z. Miskovic, J. Vukanic, T.E. Madey: Surf. Sci. **141**, 285 (1984)
41.13 Z. Miskovic, J. Vukanic, T.E. Madey: Proc. of Symposium on the Physics of Ionized Gases, Yugoslavia, 1984 (in press)
41.14 F.P. Netzer, T.E. Madey: Surf. Sci. **119**, 422 (1982); Surf. Sci. **117**, 549 (1982)
41.15 C. Benndorf, T.E. Madey: Chem. Phys. Lett. **101**, 59 (1983); Surf. Sci. **135**, 164 (1983)
41.16 F.P. Netzer, D.L. Doering, T.E. Madey: Surf. Sci. **143** (1984);
D.L. Doering, S. Semancik, T.E. Madey: Surf. Sci. **133**, 49 (1983)
41.17 C. Nöbl, C. Benndorf, T.E. Madey: Surf. Sci. (in press)
41.18 C. Benndorf, C. Nöbl, T.E. Madey: Surf. Sci. **138**, 292 (1984)
41.19 K. Bange, T.E. Madey, J. Sass: Surf. Sci. (in press); Chem. Phys. Lett. (in press)
41.20 J.-H. Lin, B.J. Garrison: J. Chem. Phys. **80**, 2904 (1984) and references therein

42. The Study of Simple Reactions at Surfaces by High-Resolution Electron Energy Loss Spectroscopy

Neville V. Richardson, C. Damian Lackey, and Mark Surman

The Donnan Laboratories, University of Liverpool
PO Box 147, Liverpool L69 3BX, Great Britain

The use of high-resolution electron energy loss spectroscopy (HREELS) in the study of two surface reactions is described. The first reaction is the deuteration of pyridine at 350 K on a Pt(110) surface. At this temperature, pyridine may adopt one of two different orientations on the surface depending on the coverage. Close to saturation coverage, a nitrogen lone-pair bonded species is dominant, characterized by intense ring breathing modes in the 1100-1200 cm^{-1} region. The molecules bonded in this manner undergo only a limited deuteration as monitored by the growth of a band, due to C-D stretching vibrations, at 2360 cm^{-1} and the attenuation of the corresponding C-H at 3050 cm^{-1}. In contrast, at lower coverages, a π-bonded pyridine molecule predominates, characterized by an intense loss feature at 820 cm^{-1}. These molecules undergo complete H/D exchange.

As the second example, HREELS has been used to investigate the interaction of CO with a Cu(110) surface doped with potassium. Strongly interacting K/CO complexes have been identified which are stable at room temperature and exhibit vibrational losses at 1200-1250 cm^{-1} and 1400-1650 cm^{-1}. Exposure of the CO/K/Cu(110) surface to hydrogen leads to the appearance of C-H and O-H stretching vibrations at 2760 cm^{-1} and 3050 cm^{-1}, 3600 cm^{-1} respectively. This reaction is associated with a particular K/CO complex from the HREELS and thermal desorption data.

42.1 Introduction

High-resolution electron energy loss spectroscopy (HREELS) used for measuring the vibrational properties of surfaces and adsorbed species is now firmly established in the repertoire of important surface science techniques [42.1,2]. In terms of the structural aspects of solid surfaces, HREELS has made major contributions towards the determination of adsorbate molecule orientation and adsorption site [42.3-5]. In most cases, this has been possible because of the dipole selection rules which govern the scattering in the specular direction though use has also been made of the less restrictive selection rules for the impact scattering dominated regime away from specular [42.6].

In this brief report, the further developments of HREELS for the investigation of two surface reactions are described, in which the structure of the adsorbed species plays an important role. The first reaction is the deuteration of pyridine, C_5H_5N, on a Pt(110) surface at 350 K [42.7]. Previously, HREELS was used to examine the kinetics of benzene deuteration over Pt(110), where reaction proceeds to completion, i.e., C_6D_6 is formed. An isotope effect was noted and the initial rate of reaction was found to be second order in the number of CH bonds present [42.8]. The adsorption of pyridine on metal surfaces has already been demonstrated to be more involved than that of benzene, being both coverage and temperature dependent [42.9,10]. The deuteration reaction of pyridine mirrors the structural changes of the adsorbed molecule.

As the second example, the reaction between CO and H_2 on a Cu(110) surface in the presence of adsorbed potassium is briefly described [42.11]. There has been a remarkable explosion of interest in coadsorption in the presence of alkali metal atoms [42.12-14]. The stimulus, of course, is the important but poorly understood role which alkali metals play in the promotion of several important industrial catalysts: the ammonia synthesis catalyst is iron promoted with potassium whilst the Fischer-Tropsch catalysts for the production of hydrocarbons and oxygen-containing species, from CO and H_2, is also a potassium-promoted transition metal, such as Fe, Ru or Ni. Copper is not normally noted as a Fischer-Tropsch catalyst! The adsorption of CO on a potassium treated copper surface by HREELS [42.15] and angle-resolved photoelectron spectroscopy (ARUPS) [42.16] has been previously described, but in this example, the hydrogenation reaction which favors a particular potassium and CO containing adspecies is briefly discussed [42.11].

Vibrational spectroscopy provides an ideal probe of such reactions because of the ease with which specific intramolecular bonds can be identified, e.g., C-H, C-D, or C=O. HREELS, in particular, provides a means of covering a wide spectral range at good sensitivity and, at least for these reactions, the poorer resolution of HREELS compared to infrared techniques is not a disadvantage.

42.2 Deuteration of Pyridine over Pt(110)

The adsorption of benzene on transition metal surfaces invariably takes place through the π electrons of the ring such that the plane of this ring lies parallel to the metal surface [42.3,17,18]. The deuteration of benzene over Pt(110) was followed by monitoring the decrease in C-H bond stretch intensity at ca. 3025 cm^{-1} and the concomitant increase in C-D stretch intensity at 2260 cm^{-1}, Fig.42.1 [42.8]. The reaction proceeds to completion at saturation coverage within 3 hours at temperatures above 330 K and D_2 pressures of 1×10^{-7}Torr which are sufficient to ensure that D_2 arrival at the surface is not rate limiting and that the forward reaction on the surface is isolated.

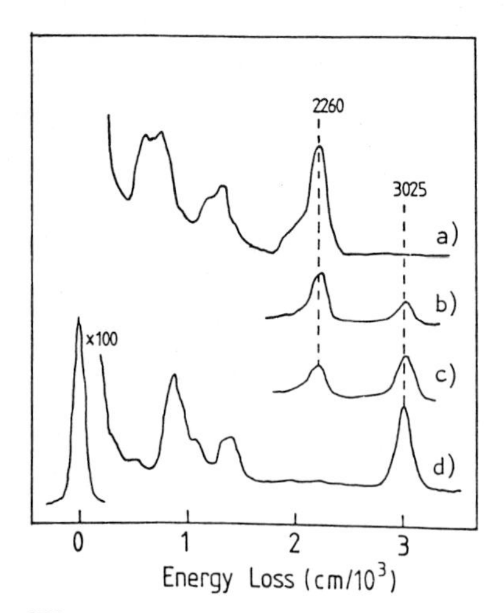

Fig.42.1. HREEL spectra, in the specular direction, for saturation coverage of (*a*) C_6H_6 on Pt(110) at 300 K; (*b*, *c*) intermediate species $C_6H_nD_{6-n}$ occurring as a result of the exchange reaction and (*d*) C_6D_6

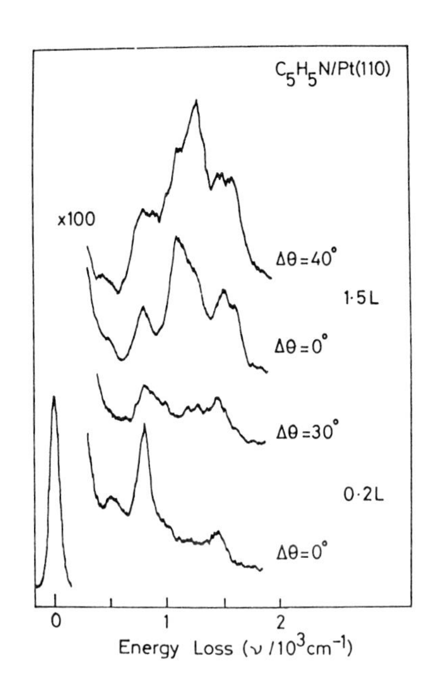

Fig.42.2. Comparison of low coverage (0.2 L exposure) and saturation coverage (1.5 L exposure) HREEL spectra, both in the specular direction ($\Delta\theta = 0^{\circ}$) and away from specular ($\Delta\theta = 30^{\circ}$), for C_5H_5N/Pt(110) at 350 K

The adsorption of pyridine of Ag(110) [42.9] and Pt(110) [42.10] surfaces at 300 K occurs with the first molecules adopting an orientation similar to that of benzene, i.e., with the aromatic ring parallel to the surface. At exposures to 0.4 L on Pt(110) this species is dominant and is readily characterized and its orientation confirmed by the HREEL spectrum in the specular direction, which is dominated by a band at 820 cm^{-1} (Fig.42.2). This band arises from the in-phase motion of the five hydrogen atoms bending out of the aromatic plane and shifted up in frequency from the value of 703 cm^{-1} found in the gas-phase species [42.17]. The energy loss occurs via a dipole mechanism as witnessed by the sharp fall in intensity away from the specular direction (Fig.42.2). The dipole selection rule demands that the dynamic dipole be perpendicular to the surface, which in turn requires, for activity in this mode, that the aromatic ring is parallel to the surface. The corresponding mode, out-of-plane hydrogen motion, also dominates the specular HREEL spectrum of C_6H_6 [42.3], C_6H_5OH [42.10,20] and $C_{10}H_8$ [42.20] species adsorbed on transition metals. It is characteristic of molecules adsorbed with their aromatic ring(s) parallel to the metal surface.

In contrast, at higher coverages modes in the range 1000-1500 cm^{-1} dominate the HREEL spectrum (Fig.42.2), those around 1100-1200 cm^{-1} correspond to ring breathing modes. *Demuth* et al. [42.9], investigating C_5H_5N/Ag(110), suggested that at this coverage molecules are much more inclined on the surface; an angle of ca. 30° between the C_2 axis of the molecule and the surface normal was determined. This is entirely plausible since, in addition to the π electrons of the ring, pyridine has the possibility of using the lone pair electrons on the nitrogen atom for bonding to the surface and ARUPS studies of pyridine adsorption on Cu(110) at full coverage were previously interpreted in terms of "upright pyridyne molecules [42.20]. At intermediate coverages, the HREELS data suggest that the two orientations can coexist on the surface. In either orientation a band is observed around 3050 cm^{-1}, corresponding to C-H stretching vibrations.

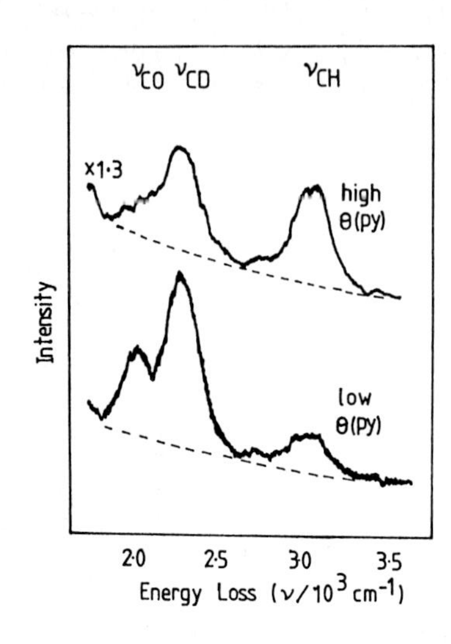

Fig.42.3. HREELS data for the deuteration of pyridine on Pt(110) at 350 K

Exposure of the *saturated* pyridine monolayer to deuterium, at a gas-phase pressure of 1×10^{-7}Torr and substrate temperature of 350 K, caused a decrease in the C-H stretching band intensity at 3070 cm^{-1} and the growth of a band at 2360 cm^{-1} arising from C-D stretching vibrations. The intensity of this latter band reached a maximum after about 1 hour and showed no further change even 12 hours later. The resultant HREEL spectrum, Fig.42.3 shows a ratio of 0.55 between the C-D band intensity I_{CD} and the total C-H and C-D band intensity $I_{CD}+I_{CH}$. *Incomplete* deuteration is clearly indicated.

If the reaction proceeds by elementary steps of the type

$$C_5H_nD_{5-n}N(ads) + D(ads) \rightarrow C_5H_{n-1}D_{6-n}(ads) + H(ads) \quad ,$$

then it seems likely that deuteration would be limited to hydrogen atoms in the α position, adjacent to the nitrogen atom, when the molecule is in the highly inclined geometry which pertains at higher coverages. The α-hydrogen atoms are then of course, those closest to the metal surface. One might perhaps have expected $I_{CD}/(I_{CH}+I_{CD})$ to be 0.4. However, this would be to assume implicitly that the C-H stretching and C-D stretching vibrations have the same scattering cross section and that all bonds contribute equally to the band intensity. Indeed, the pyridine molecule has five distinct C-H stretching vibrations with differing dynamic dipole moments, though a further complication is that the C-H and C-D bands in the HREEL spectra show both dipole and impact scattering behavior. It is suggested that the observed reaction is indeed

$$C_5H_5N(ads) + D_2(g) \rightarrow C_5H_3D_2N(ads) + H_2 \quad ,$$

in which the α-hydrogen atoms are exchanged. The other H atoms, which lie further from the metal surface, are not replaced.

The same experiment has also been performed at the much lower coverage produced by an exposure of 0.2 L. At this coverage, the HREEL spectrum, for the specular direction (Fig.42.2) indicates the presence of π-bonded molecules. The reaction proceeds more rapidly, being largely completed after 20 minutes, perhaps simply reflecting a greater D(ads) coverage of the surface.

More dramatically, the extent of reaction is much greater with $I_{CD}/(I_{CH}+I_{CD})$ reaching 0.85 (Fig.42.3). Our conclusion is that the π-bonded pyridine molecules exchange more H atoms than the lone-pair bonded molecules. By analogy with benzene, which is similarly π-bonded, it seems probable that all the H atoms are exchanged: they are all close to the metal surface.

42.3 Reaction Between CO and H_2 Over Potassium Promoted Cu(110)

The presence of potassium on copper surfaces has a profound effect on the adsorption properties of CO. The characteristic satellite features in the UPS of CO adsorbed on clean Cu surfaces (22) at temperatures <200 K, the desorption temperature, disappear when potassium is present and the 5σ orbital shifts to higher binding energy [42.16]. On warming this surface to room temperature, photoemission features remain at 5-6 eV and ca. 10.2 eV, suggesting an adsorbate species, although not molecular CO, is still present. HREELS gives more detailed information. At monolayer potassium coverages, new vibrational loss features appear centered at 1220 cm^{-1} and at 1575 cm^{-1} on CO adsorption at 170 K [42.15], Fig.42.4. Carbon monoxide on clean Cu surfaces shows a single band at 2090 cm^{-1} at temperatures below desorption, 200 K [42.23]. The thermal desorption spectra show two distinct peaks at 475 K and 545 K both at mass 28 (CO^+) (Fig.42.5a) and mass 39 (K^+) (Fig. 42.5b). These coincidences suggest that at each temperature a surface complex decomposes to produce K (or K^+) and CO species. The peak intensities suggest that the K/CO ratio in the complex giving the lower temperature peak is twice as great as that in the complex giving the higher temperature peak. It was previously suggested that the two surface species are the potassium salts of the anions

$$^{-}O-C\equiv C-O^{-}$$

I

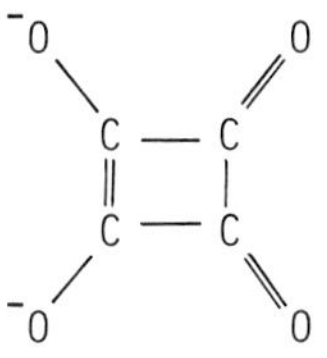

(II: four-membered ring C=C–C–C with two O^- groups on the double-bonded carbons and two =O groups on the others)

II

giving the 475 K and 545 K peaks, respectively [42.15].

Exposure of these surface species at room temperature to hydrogen at a gas phase pressure of 1×10^{-8} Torr results after warming to 400 K in the disappearance of the loss features at 1575 cm^{-1} and 1220 cm^{-1} with the simultaneous appearance of losses at 1380 cm^{-1}, 2760 cm^{-1}, 3050 cm^{-1} and 3600 cm^{-1} (Fig. 42.6) [42.11]. The latter features are attributed to C-H and O-H stretching vibrations respectively and the 2760 cm^{-1} peak to a soft C-H mode [42.1]. Also, following the reaction with hydrogen, the high-temperature peak (545 K) disappears for both CO (Fig.42.5c) and K. Thermal desorption continues to show a peak at 475 K. Of the two surface species, it seems that only the $K_2(CO)_4$ species is susceptible to hydrogenation.

In these experiments, two of the essential steps in Fischer-Tropsch reactions were demonstrated: the formation of C-C bonds and the formation of C-H and O-H bonds. The scission of C-O bonds has not been shown in this case and further experiments are required to determine the exact nature of the adsorbed species.

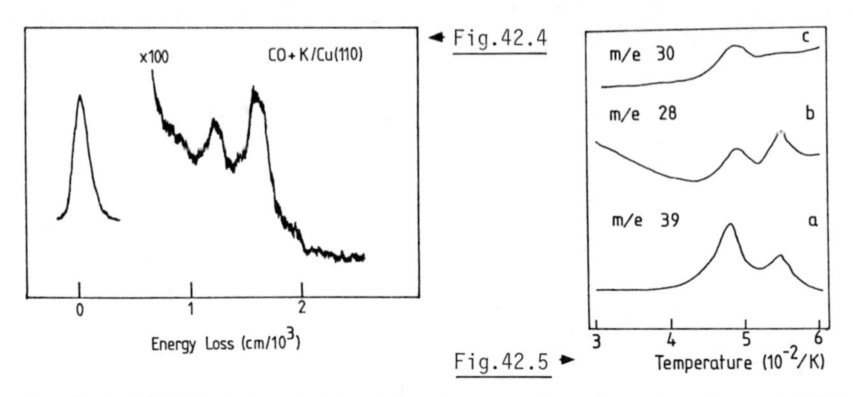

Fig.42.4. HREELS data, obtained in the specular direction for a Cu(110) surface, covered with a monolayer of potassium, then exposed to 20 L of CO at 170 K

Fig.42.5. Thermal desorption spectra (50 Ks^{-1} heating rate) for the K/CO complexes on Cu(110) (see text for details). (*a*) Potassium signal, (*b*) CO signal and (*c*) $C^{12}O^{18}$ signal following exposure of the K/CO/Cu(110) surface to 4 L of H_2

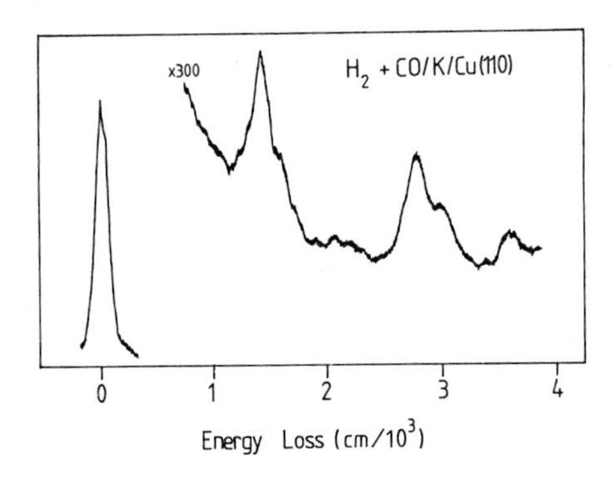

Fig.42.6. As Fig.42.4 but following exposure to about 4 L of H_2 and subsequent heating to 400 K

References

42.1 H Ibach, D.L. Mills: *Electron Energy Loss Spectroscopy and Surface Vibrations* (Academic, New York 1982)

42.2 Proceedings of the 3rd Int. Conf. on Vibrations at Surfaces: J. Elec. Spec. Rel. Phenom. **29**, **30** (1983)

42.3 H. Ibach, S. Lehwald, J.E. Demuth: Surf. Sci. **78**, 577 (1978); J.C. Bertolini, G. Dalmai-Imelik, J. Rousseau: Surf. Sci. **67**, 478 (1977)

42.4 H. Ibach, S. Lehwald: J. Vac. Sci. Technol. **15**, 407 (1978)

42.5 S. Andersson, M. Persson: Phys. Rev. B**24**, 3659 (1981)

42.6 P. Hofmann, S.R. Bare, N.V. Richardson, D.A. King: Surf. Sci. **133**, L459 (1983)

42.7 M. Surman, D.A. King, N.V. Richardson: To be published

42.8 M. Surman, S.R. Bare, P. Hofmann, D.A. King: Surf. Sci. **126**, 349 (1983)

42.9 J.E. Demuth, K. Christmann, P.N. Sanda: Chem. Phys. Lett. **76**, 201 (1980)
42.10 N.V. Richardson: Vacuum **33**, 10 (1983)
42.11 D. Lackey, M. Surman, D.A. King: To be published
42.12 G. Ertl, S.B. Lett, M. Weiss: Surf. Sci. **114**, 527 (1982)
42.13 F.M. Hoffmann, R.A. de Paola: Phys. Rev. Lett. **52**, 1697 (1984)
42.14 J.K. Norskov, S. Holloway, N.D. Lang: Surf. Sci. **137**, 65 (1984)
42.15 C.D. Lackey, M. Surman, S. Jacobs, D.E. Grider, D.A. King: Surf. Sci. (in press)
42.16 C. Somerton, C.F. McConville, D.P. Woodruff, D.E. Grider, N.V. Richardson: Surf. Sci. **138**, 31 (1984)
42.17 G.L. Nyberg, N.V. Richardson: Surf. Sci. **85**, 335 (1979)
42.18 P. Avouris, J.E. Demuth: J. Chem. Phys. **75**, 4783 (1981)
42.19 L. Corsin, B.J. Fox, R.C. Lord: J. Chem. Phys. **21**, 1170 (1953)
42.20 N.V. Richardson, P. Hofmann: Vacuum **33**, 793 (1983)
42.21 B.J. Bandy, D.R. Lloyd, N.V. Richardson: Surf. Sci. **89**, 344 (1979)
42.22 C.L. Allyn, T. Gustafsson, E.W. Plummer: Solid State Commun. **24**, 531 (1977)
42.23 J. Pritchard, T. Catterick, R.K. Gupta: Surf. Sci. **53**, 1 (1975)

Part V

Semiconductors

V.1 Elemental Semiconductors

43. Triangle-Dimer Stacking-Fault Model of the Si(111) 7×7 Surface Bonding Configuration

E.G. McRae

AT&T Bell Laboratories, Murray Hill, NJ 07974, USA

In the triangle-dimer stacking-fault (TDSF) model of the 7×7 reconstruction of Si(111) surfaces, the reconstruction extends to a depth of two double layers and consists essentially of the removal of 1/7 of the atoms belonging to the second and third layers. The resulting dislocations are dissociated and form a triangular network of partial dislocations. The unit mesh has two triangular subunits in which there are (111) stacking faults below the first and third layers, respectively. Each subunit is bordered by 18 dimers due to reconstruction of partial dislocations. The triangle-dimer geometry and the presence of stacking faults were first inferred from low-energy electron diffraction (LEED) and Rutherford backscattering (RBS) results, respectively. The TDSF model is supported strongly but not unequivocally by other observations. These include scanning tunneling microscopy (STM), H chemisorption combined with high-resolution infrared spectroscopy (HRIRS), Ar and Xe physisorption combined with electron spectroscopies, and transmission electron diffraction (TED). Specific observations that are not essential consequences of the TDSF bonding configuration include the quasi-2×2 array of "hills" in STM and relatively very intense (1 3/7) spots in TED. The possibility that these observations can be accommodated by the model is discussed.

43.1 Introduction

The nature of the 7×7 reconstruction of the Si(111) surface is not well understood and is the subject of a great deal of experimental study and speculation [43.1,2]. Much of this activity is motivated by a feeling that the discovery of the 7×7 structure might reveal an important principle underlying the long-period reconstruction of semiconductor surfaces in general. But aside from the question of its structure, the Si(111)-7×7 surface is important as a testing ground of surface experimentation. The 7×7 surface has been studied by every available method of crystal surface structure determination. The results provide an unparalleled opportunity to compare these methods and to identify methods that provide critical tests of structural models.

This paper traces the development and experimental tests of a particular model of the Si(111)-7×7 surface bonding configuration. The essence of the model is that the 7×7 unit mesh consists of two triangular, faulted subunits with dimers along their edges [43.3]. It is here called the triangle-dimer stacking-fault (TDSF) model. The presentation in this paper has three main objectives: first, to compare various experimental methods by showing how each of them reveals one or other aspect of the TDSF structure; second, to assess the TDSF model on the basis of critical experimental tests, third, to describe the principle underlying the 7×7 reconstruction as implied by the TDSF model.

43.2 Development of the Model

Experimental Basis. The TDSF model was proposed [43.3] to explain low-energy electron diffraction (LEED) and Rutherford backscattering (RBS) observations.

The 7×7 LEED pattern [43.4] shows specific formations of relatively strong fractional-order (FO) spots [43.1]. The "star" formation consists of strong FO spots on lines joining neighboring integer-order (IO) ones. It may be related through the kinematical shape transform to triangular subunits of the unit mesh (Fig.43.1). Other formations including the "3/7 hexagon" formation may be interpreted [43.1] using a kinematical forward-scattering picture [43.5] by an array of 9 dimers (or vertically separated sets of dimers) around the edge of each subunit (Fig.43.1). The observed weakening of dimer-related formations and enhancement of the star formation upon adsorption of H atoms [43.6.7] may be attributed to breaking of dimer bonds [43.7].

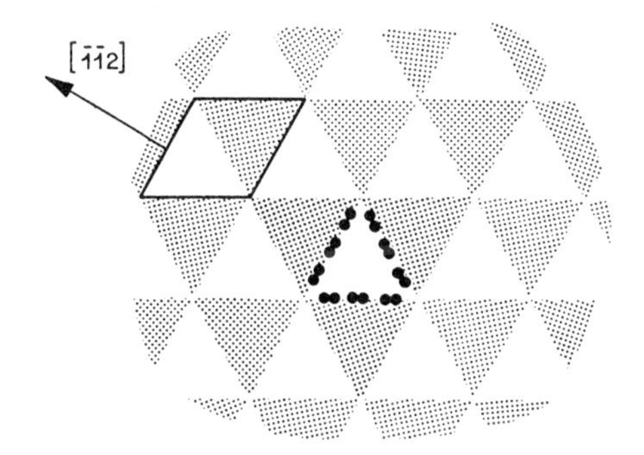

Fig.43.1. Gross properties of the TDSF model inferred (see text) from LEED and RBS. The unit mesh (outlined) contains two triangular subunits bordered by dimers as indicated for one subunit. Open and shaded subunits indicate stacking faults (a) just below the first and (b) just below the third atom layers, respectively

Other LEED observations and RBS observations [43.8,9] on the 7×7 surface may be explained by stacking faults parallel to the surface [43.10]. Stacking faults consistent with observation could be (a) just below the first or (b) just below the third atom layer, but the RBS results are best fitted by a 50-50 mixture of possibilities (a) and (b) [43.10]. All the observations are accounted for if the triangular subunits are identified as (a)- and (b)-type faulted areas, respectively [43.3]. There are two possible ways of assigning fault type to the subunits, but only the assignment shown in Fig. 43.1 is compatible with the required arrangement of dimers [43.3].

Description. In the TDSF bonding configuration (Fig.43.2) the selvedge consists of two double layers that are mirror images of each other with respect to the surface plane. As shown in Fig.43.3, the selvedge fits neatly onto the substrate which starts at the fifth atom layer and has the ideal 1×1 structure. The dimers, or alternatively the oval holes (8-membered rings, Fig.43.3a), as well as the round hole at the apex of the triangle (12-membered ring, Fig.43.3b) arise automatically from the requirements of bonding at the junctions of the faulted triangular subunits. The model selvedge is less dense than for the ideal 1×1 surface; in each unit mesh the first and

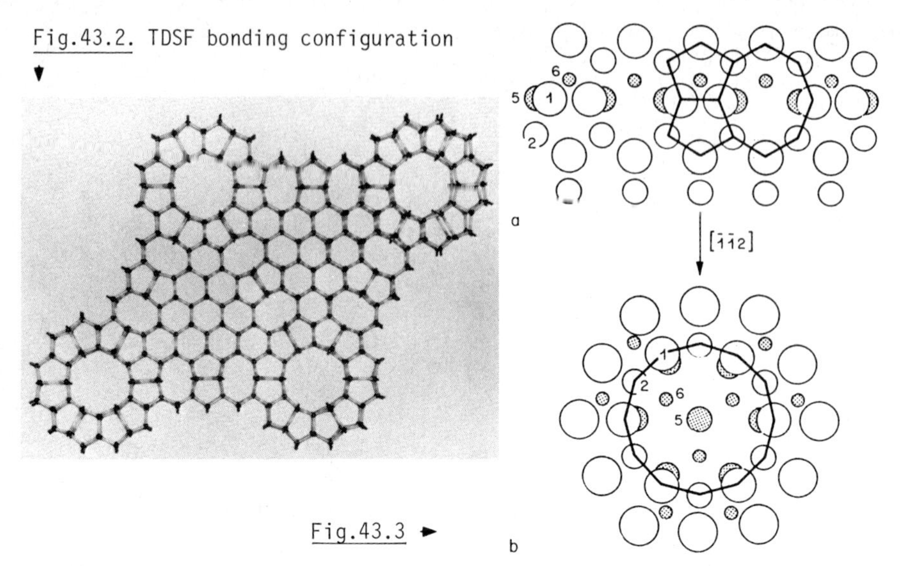

Fig.43.2. TDSF bonding configuration

Fig.43.3

Fig.43.3a,b. Detail of the TDSF bonding configuration, showing structures resulting from the joining of double-layers at (**a**) the edge and (**b**) the apex of a triangular subunit. Circles denote atoms of layers 1,2,··· as indicated. Atoms in layers 3,4 are directly below atoms in layers 1,2 and so are not shown in this view. The visible edge structure (**a**) consists of pairs of 5-membered rings alternating with 8-membered rings as outlined. The dimers are formed by pairing atoms common to each pair of 5-membered rings. The apex structure (**b**) consists of a 12-membered ring (outlined)

fourth atom layers contain 48 atoms and the third and fourth layers only 42 atoms compared to 49 per atom layer of the ideal surface. The number of dangling bonds per unit mesh is 49, the same as for the ideal surface. Of these dangling bonds, 48 come from the surface atoms and one from the fifth-layer atom at the center of the apex hole (Fig.43.3b).

43.3 Experimental Tests

Scanning Tunneling Microscopy (STM). The STM experiment by *Binnig* et al. [43.11] provides a micrograph with lateral resolution of the order of interatomic distances. A representation of one such micrograph for Si(111)-7×7 is shown in Fig.43.4a where the light and dark areas represent hills and holes, respectively. The maximum negative excursion (at the apex hole) was 2.1 Å. The model (Fig.43.4b) correctly describes both the shapes and positions of the observed holes. It accounts also for the apparent depth of the apex hole [43.12]. The observed hills are not a necessary consequence of the TDSF model bonding configuration. At present it is not clear whether they can be accommodated by the model (Sect.43.4).

Hydrogen Chemisorption—Saturation Coverage. Measurements by *Culbertson* et al. [43.13] using the ^{3}He(d,p)^{4}He nuclear reaction technique gave a saturation coverage of 1.25 ± 0.1 monolayers (ML). This figure should be increased to 1.4 ML to correct for the depletion of adsorbed H atoms by their reaction with ambient atoms [43.14]. The model indicates a saturation coverage of

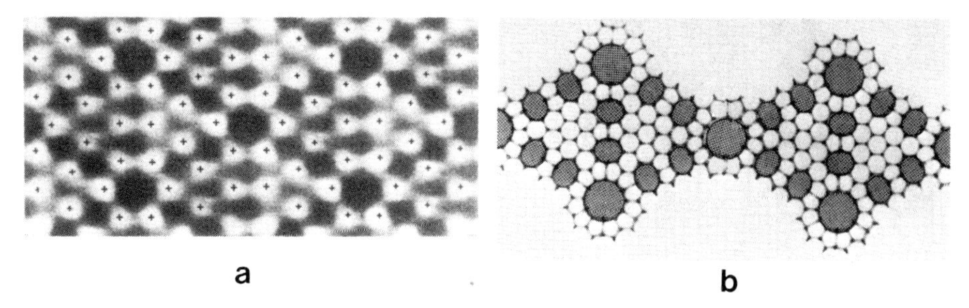

Fig.43.4. (**a**) STM image of two 7 × 7 unit meshes [43.11]. (**b**) TDSF model with "holes" shaded

(18 + 49)/49 = 1.37 ML (assuming all surface dimer bonds broken) which agrees with observation.

Hydrogen Chemisorption – High-Resolution Infrared Spectroscopy (HRIRS). HRIRS observations by *Chabal* [43.15] indicate a unique site for chemisorption of the first H atom per unit mesh. The Si-H vibration is p polarized (polarized normal to the surface) and its frequency is 4% lower than that of p-polarized peaks due to subsequent chemisorption. The unique site may be identified as the fifth-layer atom at the center of the apex hole (Fig.43.3b). The observation of a shift to lower frequency, despite the presumption that the unique site is the strongest bonding site, was explained [43.14] by the dielectric screening by the walls of the deep (6.4 Å) apex hole provided by the model.

It was also shown by *Chabal* [43.15] that the next three sites result in Si-H bonds of equal vibrational frequency. This can be explained by surface asymmetric dimers like those thought to exist on the Si(100)-2 × 1 surface. With asymmetric dimers, there are just three degenerate chemisorption sites at the edge of each apex hole.

Photomission Si(2p) Core-Level Spectroscopy. Himpsel et al. [43.16] reported that about 1/6 of the surface atoms have large 2p core level shifts in the direction expected if the atoms were negatively charged. In the model assuming asymmetric dimers, these atoms may be identified as 9 outward-displaced atoms per unit mesh from surface dimers.

Rare-Gas Physisorption – Isotherms. Conrad and *Webb* [43.17] observed that for Kr and Xe there is a unique physisorption site at which the binding energy is much greater than expected for a flat surface and is instead comparable with that expected for the binding energy to a vacancy in bulk Si. The apex hole provided by the model (Fig.43.3b) explains these observations. It is just big enough to accommodate a Xe atom and provides about the right value of the binding energy.

Rare-Gas Physisorption – X-Ray Photoemission (XPS) Level Shifts. Demuth and *Schell-Sorokin* [43.18] reported that with increasing Xe coverage θ the 3p and 5p XPS lines of Xe undergo shifts or broadening at specific values of θ. As sketched in Fig.43.5, each of these special coverage values corresponds to filling of distinct classes of Xe physisorption sites provided by the model.

Transmission Electron Diffraction (TED). TED patterns similar to each other have been reported by *Takayanagi* et al. [43.19], *Petroff* and *Wilson* [43.20] and by *Spence* [43.21]. All of these patterns were obtained near normal inci-

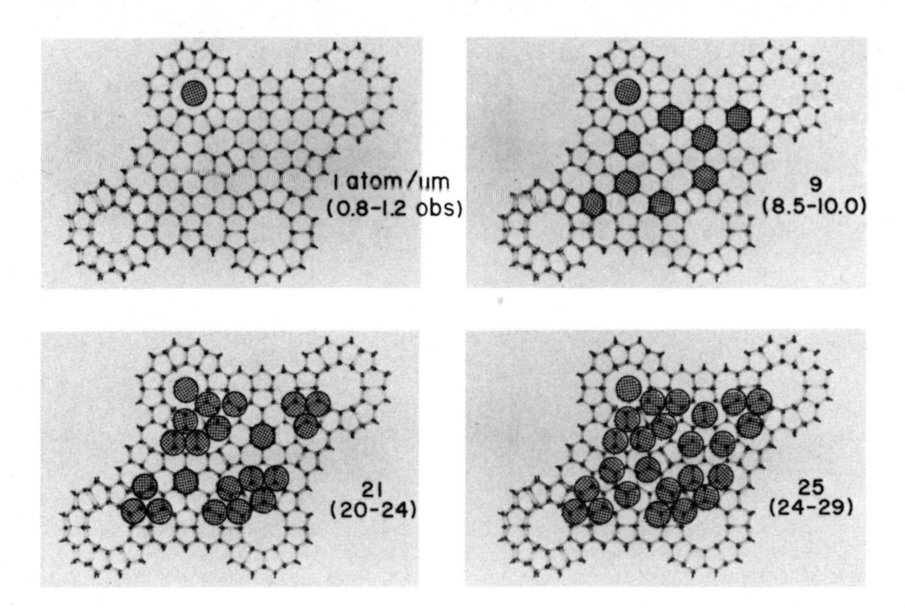

Fig.43.5. Successive stages of Xe coverage as described by the TDSF model. Shaded circles represent Xe atoms (to scale). Numbers indicate the coverage illustrated (Xe atoms per unit mesh) and the range of coverage values for which a shift or broadening of the XPS line was observed [43.18]

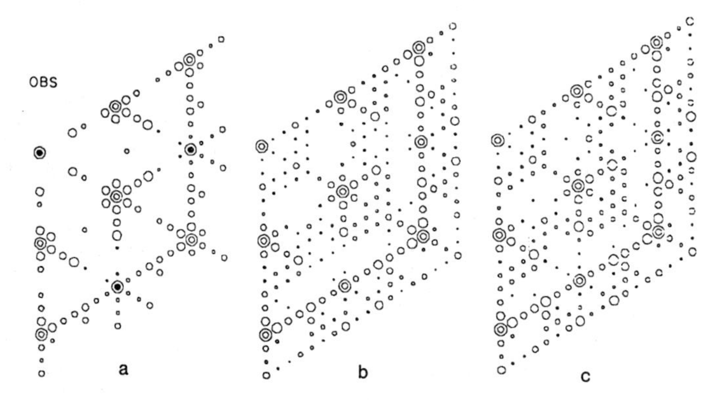

Fig.43.6a-c. Observed and calculated TED patterns. Single circles denote fractional-order beams, double circles integer-order ones. Areas of circles are proportional to estimated or calculated intensity. (**a**) Observed pattern [43.20]. Filled circles denote allowed substrate spots, (**b**) calculated pattern for TDSF model; (**c**) calculated pattern for TDSF model with two extra atoms at specific position (see text)

dence. As the forward momentum transfer is small, the FO intensities are sensitive to the lateral atom coordinates only. *Spence* [43.21] reported calculations indicating that the kinematical (single-scattering) approximation is valid for model calculations of FO spot intensities in TED. Figure 43.6 shows the observed pattern from [43.20] (Fig.43.6a) in comparison with the pattern calculated kinematically [43.22] for the TDSF model (Fig.43.6b) with an assumed dimer bond length of 2.7 Å and all other bond lengths equal to the normal Si-Si bond length 2.35 Å.

The agreement between the calculated and observed patterns is good though not perfect. As an example of the agreement achieved, the presence of triangular subunits ensures that all the intense spots lie on star formations as observed. Nearly all the spots, including "off-star" as well as star ones, have qualitatively the correct intensity relative to their neighbors.

The failures of the model show up mainly in some spots near "half-order" positions, such as {1/2 0}, {3/2 0} and {1/2 1/2}. One way to bring the calculated pattern into perfect qualitative agreement with observation is to add scattering power equivalent to between one and three extra Si atoms at the positions of the "hills" in the STM micrograph. Figure 43.6c shows the result of adding two Si atoms at these positions [43.22]. The qualitatively perfect agreement between Fig.43.6c and the observed pattern (Fig.43.6a) is apparent.

43.4 Discussion

Assessment of the TDSF Model. The model evidently succeeds in correlating a variety of experimental facts. In particular, the existence of its unique feature – the deep round apex hole – is an implication of LEED, RBS and TED and seems to be confirmed by such disparate observations as STM, HRIRS and inert-gas physisorption. However, the ability of a model to correlate observations is less important than its ability to survive critical tests, i.e., tests by experimental methods capable of eliminating specific models unequivocally.

The STM, HRIRS and TED methods have proved especially valuable by this criterion. STM [43.11] has eliminated all models with double-layer islands [43.1,23]. HRIRS [43.14] has eliminated all models that do not present a unique H-chemisorption site, such as adatom [43.11,28,29] and adatom-cluster models [43.24-27]. TED [43.19-21] has eliminated all adatom and adatom-cluster models [43.11,24-29] for which calculations have been carried out [43.22].

The TDSF model gives a good but not completely satisfactory account of STM and TED results. The need for a quasi-2×2 array of "hills" (STM) and the apparent need for extra scattering power at these positions (TED) suggest that the model should be modified by the addition of adatoms or adatom clusters at these sites. However, the model provides that these sites are top-layer atoms, so it is hard to see how adatoms or clusters could be added without generating extra dangling bonds. One possibility is a TDSF model with *one* double layer plus adatoms as suggested by *Takayanagi* [43.30]. However, *Tromp* [43.31] has shown that this model does not fit RBS data, and that extra scattering power at adatom positions is not needed to fit TED patterns if the relaxation of the TDSF model is included. New insight into this question might be provided by reflection X-ray diffraction experiments such as those by *Robinson* [43.32].

Principle Underlying the 7×7 Reconstruction. A possible driving force for reconstruction represented by the TSDF model is the backbonding of surface atoms to atoms directly beneath them [43.12]. The role of stacking faults is to shift the surface atoms into favorable positions for backbonding. Backbonding tends to cause compressional stress which is relieved by the removal of 1/7 of the atoms belonging to the second and third atom layers. A generalization of this description is that compressional stress relieved by the removal of 1/N of the second and third-layer atoms (N odd) results in an $N \times N$ TDSF structure. The $N \times N$ reconstruction is the means whereby compressional stress is relieved. The TDSF structure may be described from this viewpoint as a network of reconstructed partial dislocations of the 30° type [43.12].

The writer thanks Dr. K. Takayanagi and Dr. R.M. Tromp for making their results available prior to publication.

References

43.1 E.G. McRae: Surf. Sci. **124**, 106 (1980), and references therein
43.2 F.J. Himpsel, I.P. Batra: J. Vac. Sci. Technol. and references therein
43.3 E.G. McRae: Phys. Rev. B**28**, 2305 (1983)
43.4 J.J. Lander, G.W. Gobeli, J. Morrison: J. Appl. Phys. **34**, 2298 (1963)
43.5 T.C. Ngoc, M.B. Webb: 38th Physical Electronics Conference, Gatlinberg (1978)
43.6 T. Sakurai, H.D. Hagstrum: Phys. Rev. B**14**, 1593 (1976)
43.7 E.G. McRae, C.W. Caldwell: Phys. Rev. Lett. **46**, 1632 (1981)
43.8 R.J. Culbertson, L.C. Feldman, P.J. Silverman: Phys. Rev. Lett. **45**, 2043 (1980)
43.9 R.M. Tromp, E.J. van Loenen, M. Larami, F.W. Saris: Solid State Commun. **44**, 971 (1981)
43.10 P.A. Bennett, L.C. Feldman, Y. Kuk, E.G. McRae, J.E. Rowe: Phys. Rev. B**28**, 3656 (1983)
43.11 G. Binnig, H. Rohrer, C. Gerber, E. Weibel: Phys. Rev. Lett. **50**, 120 (1983)
43.12 E.G. McRae: Surf. Sci. (in press)
43.13 R.J. Culbertson, L.C. Feldman, P.J. Silverman, R. Haight: J. Vac. Sci. Technol. **20**, 868 (1982)
43.14 M. Henzler: Private communication
43.15 Y.J. Chabal: Phys. Rev. Lett. **50**, 1850 (1983)
43.16 F.J. Himpsel, D.E. Eastman, P. Heimann, B. Reihl, C.W. White, D.M. Zehner: Phys. Rev. B**24**, 1120 (1981);
F.J. Himpsel: Physica (Utrecht) **117**, **118**B, 767 (1983)
43.17 E. Conrad, M.B. Webb: Surf. Sci. **129**, 37 (1983)
43.18 J.E. Demuth, A.J. Schell-Sorokin: J. Vac. Sci. Technol. A**2**, 808 (1984)
43.19 K. Takayanagi, Y. Tanishiro, M. Takahashi, H. Motoyoshi, K. Yagi: Electron Microscopy **2**, 285 (1983)
43.20 P.M. Petroff, R.J. Wilson: Phys. Rev. Lett. **51**, 199 (1983)
43.21 J.C.H. Spence: Ultramicroscopy **11**, 117 (1983)
43.22 E.G. McRae, P.M. Petroff: Surf. Sci. (in press)
43.23 K.C. Pandey: Physica **117**, **118**B, 761 (1983)
43.24 L.C. Snyder, Z. Wasserman, J. Moskowitz: J. Vac. Sci. Technol. **16**, 1266 (1979)
43.25 K. Higashiyama, S. Kono, H. Sakurai: Unpublished
43.26 M. Aono, R. Souda, C. Oshima, Y. Ishizawa: Phys. Rev. Lett. **51**, 801 (1983)
43.27 L.C. Snyder: Unpublished
43.28 W.A. Harrison: Surf. Sci. 55, 1 (1976)
43.29 T. Yamaguchi: Phys. Rev. B (in press)
43.30 K. Takayanagi: Private communication
43.31 R.M. Tromp: Private communication
43.32 I.K. Robinson: This Volume, p.60

44. Structure of the Si(111) 2×1 Surface

Inder P. Batra

IBM Research Laboratory K33/281, San Jose, CA 95193, USA

F.J. Himpsel, P.M. Marcus, and R.M. Tromp

IBM T.J. Watson Research Center, Box 218, Yorktown Heights, NY 10598, USA

M.R. Cook[1]**, F. Jona, and H. Liu**[2]

Department of Materials Science and Engineering, S.U.N.Y. at Stony Brook
NY 11794, USA

A structure analysis of the Si(111)2 × 1 surface is performed using extensive new LEED data (12 beams). A modified Keating-type strain energy analysis is used to expedite search in multiparameter structural space. Although the π-bonded chain model in its original form grossly disagrees with LEED, a modification of that structure gives a more reasonable agreement. We show variation of the R factor with various important structural parameters. Our optimization so far has given us a minimum R factor of 0.42. This is not a particularly satisfying value by LEED standards, but the agreement is significantly better than for all chain models proposed to date. The major modification with respect to the earlier models is a strong intrachain buckling in the top chain. Our optimized coordinates down to six layers are given which may be useful for future electronic structure calculations.

Reconstruction of the cleaved Si(111)2 × 1 surface has been an intriguing problem for more than two decades. A number of recent review articles [44.1-4] clearly summarize the current status and provide a useful guide to the previous literature. Until recently, the buckling model proposed by *Haneman* [44.5] in 1961, where rows of the truncated bulk structure alternately move out of and into the lattice, was widely accepted. Such a reconstruction was thought to result in a net lowering of the total energy by transferring charge from the dangling bond localized on the lowered atom [44.6-8] (which becomes more p_z-like) to the dangling bond localized on the raised atom (which becomes more s-like). However, a large charge transfer is inconsistent with the small observed Si(2p) core level shift [44.9, 10]. Also, it is now generally accepted [44.4,11-14] that the corresponding increase in Coulomb energy is sufficiently large to prevent much charge transfer. Thus the expected lowering of the total energy in the buckling model does not take place [44.4].

Another challenge to the buckling model came from a disagreement between the measured [44.15,16] surface band dispersion and the calculated dispersion using one-electron models [44.6-8]. It has been suggested by several authors [44.11-14] that the eigenvalue spectrum calculated by including correlations can be made compatible with the measured surface state dispersion. However, an elegant resolution of this problem has been suggested by *Pandey*

---

1 Present address: Physics Department, University of Maine, Orono, ME 04469
2 Permanent address: Physics Department, Huanzhou University, Huanzhou, P.R.C.

[44.17,18] within the framework of a one-electron picture. He proposed a novel chain model for the reconstruction of the Si(111)2 × 1 surface. It gives the lowest total energy [44.18-20] of any structure calculated to date and also gives the best agreement with ion scattering [44.21], optical absorption [44.22] and electron energy loss spectroscopy [44.23]. The chain model involves bond breaking and rebonding to form zigzag chains in the [$\bar{1}$01] direction.

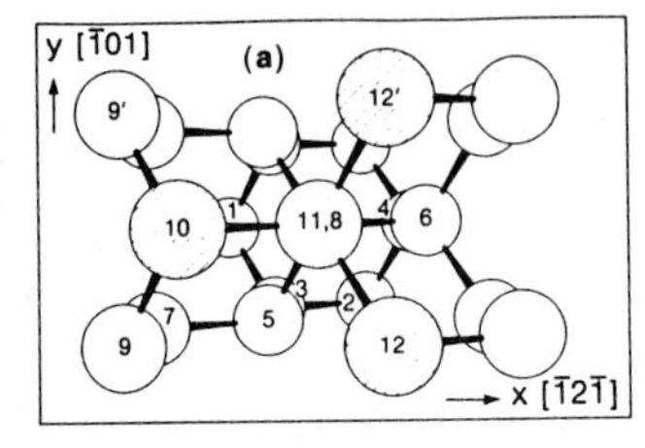

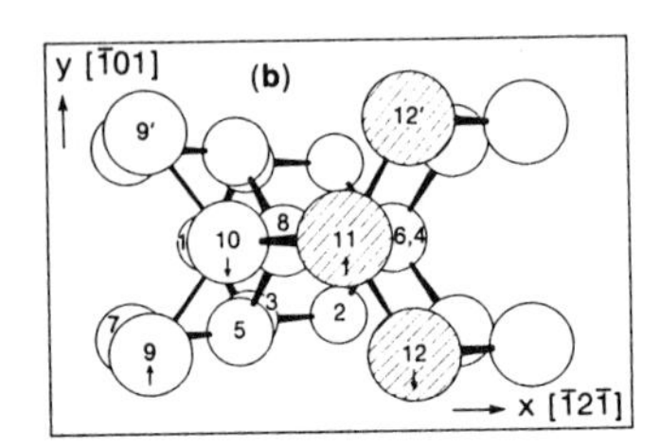

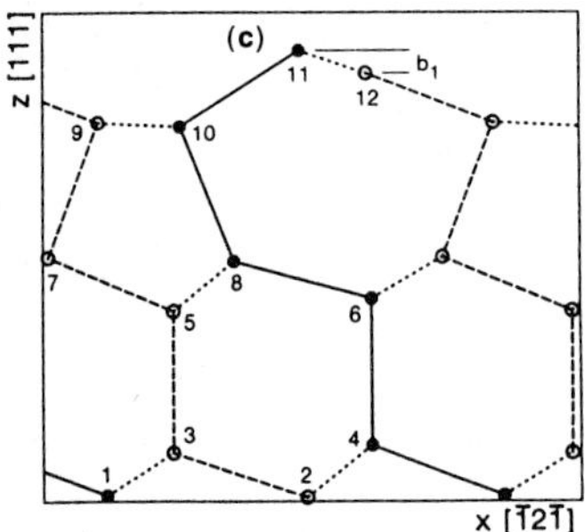

Fig.44.1a-c. Top perspective of (**a**) the ideal Si(111)2 × 1 rectangular unit cell viewed along − z [$\bar{1}\bar{1}\bar{1}$], (**b**) the LEED optimized chain model; the original Pandey [44.17,18] chain model has nearly identical top view. Atoms labeled 1-12 specify the unit cell and their coordinates are given in Table 44.1. The cross-hatched atoms have dangling bonds along the [111] direction. (**c**) Side view of our optimized structure

The formation of chains starting from the ideal truncated surface can be visualized from Fig.44.1. In Fig.44.1a we have shown a perspective of the ideal 2 × 1 rectangular unit cell viewed along the -z [$\bar{1}\bar{1}\bar{1}$] direction towards the origin. The origin is taken at the third layer atom (labeled 7 in the figure), x is in the [$\bar{1}$2$\bar{1}$] direction and y is in the [$\bar{1}$01] direction. Atoms labeled 1-12 define the surface unit cell six layers deep. The remaining atoms can be generated from these by the translation vectors: $\mathbf{a}_1 = \sqrt{3}a(1,0)$, $\mathbf{a}_2 = a(0,1)$, where $a = a_0/\sqrt{2}$ and $a_0 = 5.427$ Å is the Si lattice constant. The positions of these atoms are given in Table 44.1. It is clear that for the ideal lattice, the atoms with dangling bonds (atoms 10,12 and 12') are next nearest neighbors at a distance of 3.84 Å from each other. It is to be noted that in anticipation of showing proper numbering of the atoms for the chain model, we have purposely labeled atoms 10 and 11 in the opposite order for the ideal lattice.

In the chain model proposed by *Pandey* [44.17,18], the top layer atom 10 moves inwards towards the bulk and the second layer atom 11 moves outwards. In the process the bond between atom 11 and the third layer atom 8 (Fig. 44.1a) is broken. Atom 11 presents this as a dangling bond and a new bond is formed between atoms 10 and 8 (Fig.44.1b). Whereas for the ideal lattice the bond between atoms 10 and 11 pointed downwards, the situation is reversed in the chain model. The number of dangling bonds on the surface is identical to the ideal surface. Atoms in the top double layer are also given x displacements to maintain all bond lengths close to their bulk values. Complete displacements Δx, Δy, Δz for generating the original chain model [44.17,18] are given in Table 44.1. Two zigzag chains are formed; the top chain by atoms 12,11,12' and the bottom chain by atoms 9,10,9' both

Table 44.1. Ideal bulk truncated atomic positions (in Å) for the Si(111)2 × 1 lattice and the relative displacements required for the original π-bonded chain model[a] and the present optimized buckled chain structure. The coordinates x,y,z refer to the [$\bar{1}2\bar{1}$], [$\bar{1}01$], and [111] directions respectively (Fig.44.1). The origin is at the atom number 7 (Fig.44.1) in the third layer

Ideal Lattice				Chain Model			Optimized		
N	x	y	z	Δx	Δy	Δz	Δx	Δy	Δz
1	1.1078	1.9187	-3.9166	0.0	0.0	0.0	-0.0200	0.0	0.0200
2	4.4311	0.0	-3.9166	0.0	0.0	0.0	0.0200	0.0	-0.0100
3	2.2156	0.0	-3.1333	0.0	0.0	0.0	-0.0050	0.0	-0.0771
4	5.5389	1.9187	-3.1333	0.0	0.0	0.0	0.0045	0.0	0.0500
5	2.2156	0.0	-0.7833	0.0	0.0	0.0	0.0020	0.0	-0.1100
6	5.5389	1.9187	-0.7833	0.0	0.0	0.0	-0.0020	0.0	0.0957
7	0.0	0.0	0.0	0.0	0.0	0.0	0.0900	0.0	-0.0210
8	3.3233	1.9187	0.0	0.0	0.0	0.0	-0.0800	0.0	-0.0939
9	0.0	0.0	2.3500	0.9833	0.0	-0.2156	0.9500	0.0	-0.1700
10	1.1078	1.9187	3.1333	1.2323	0.0	-0.9990	1.2300	0.0	-1.0200
11	3.3233	1.9187	2.3500	1.0850	0.0	0.8418	1.0200	0.0	1.0200
12	4.4311	0.0	3.1333	1.1306	0.0	0.0585	1.0295	0.0	-0.1400

[a] [44.17,18]

running parallel to the y[$\bar{1}01$] direction. The interplanar distance d_{12} between these chains (which are flat) is ~1.06 Å. In the ideal structure the bilayer d_{12} ~0.78 Å. Also d_{23} ~2.13 Å for the chain model is somewhat compressed with respect to the bulk value ~2.35 Å. The energy lowering in the chain model basically arises from the fact that the dangling bonds are nearest neighbors, allowing for a substantial π-bonding among them.

The observed [44.15,16] upwards surface band dispersion follows rather naturally in the chain model. The intrachain interactions in the top chain lead to the following dispersion relation in a simple nearest-neighbor tight-binding picture:

$$E(k_y) - \lambda = -\left[\left(\frac{\varepsilon}{2}\right)^2 + 4t^2 \cos^2\left(\frac{k_y a}{2}\right)\right]^{\frac{1}{2}} . \tag{44.1}$$

Here ε is the difference in self-energies of the two inequivalent dangling bonds, t is nearest-neighbor hopping term (off-diagonal matrix element), and λ is a constant. The observed dispersion from Γ ($k_y = 0$) to J ($k_y = \pi/a$) is reproduced by setting ε ≃ 0.4 and t ≃ -0.4 eV. These values give the bandwidth for the surface band to be ~0.6 eV. The flat bands along the JK and ΓJ' directions confirm that the interchain interaction is small due to a large separation (~ 6.6 Å) between the chains with dangling bonds.

Although the chain model has successfully interpreted various experiments, the model in its original form [44.17,18] has been found to be inconsistent [44.24] with LEED. LEED is sensitive to small atomic displacements several layers deep but the original chain model [44.17,18] was designed for qualitative features. Subsequent refinement [44.19] produced a somewhat better agreement but not enough to consider the structure solved according to LEED standards. Our goal is to build on these pioneering studies by optimizing the chain model to see whether agreement with LEED data is possible. We augment our efforts towards structure determination by combining LEED and ion-scattering experiments. This utilizes the strengths of both techniques, i.e., the straightforward interpretation of ion scattering (which provides the proper region of parameter space in which to find the best-fit parameters and helps avoid false fitting minima in LEED) and the high sensitivity of LEED to small atomic displacements.

The LEED data used in our analysis consisted of 12 normal incidence beams [44.25,26] taken at room temperature. The dynamical LEED program CHANGE [44.27] was used to calculate intensity spectra of a large number of chain structures proposed in the literature [44.17-19]. The eight z coordinates of all atoms down to the fourth layer were optimized. Four of these coordinates are restricted to a relatively narrow range (±0.1 Å) to keep reasonable bond lengths between layers 2 and 3 and layers 4 and 5. The y coordinates were frozen at the mirror plane positions. To expedite the structure determination, the fifth- and sixth-layer z and all the x coordinates were determined by a modified Keating type strain energy minimization [44.28]. This consisted of minimizing the following elastic energy function:

$$E = \alpha \sum_i (d_i^2 - b_0^2)^2 + 2\beta \sum_{i<j} (\mathbf{d}_i \cdot \mathbf{d}_k - b_0^2 \cos\theta)^2 \quad , \qquad (44.2)$$

involving summation over all bonds and bond pairs. In usual applications one takes $\alpha = \alpha_0$, $\beta = \beta_0$, $b_0 = \sqrt{3}/4\ a_0$ and $\cos\theta = -1/3$ to correspond to the bulk values [44.28]. According to *Keating* [44.28] $\alpha_0 = 0.485$ and $\beta_0 = 0.138$ in units of 10^5 dyn/cm. Another useful number to remember for Si is $\alpha_0 a_0^2$, which is ~89.2 eV. Any departure of the bond length d_i from b_0 and the bond angle θ from 109.47° increases the value of E.

The essential modification we made consisted of using different values for these parameters when surface atoms were involved. These are $\alpha = 2\alpha_0$, $\beta = 10\beta_0$, $b_0 = 1/\sqrt{6}\ a_0$ and $\cos\theta = -0.5$ and we recognize the graphitic structure of the surface layer. These values of the constants also ensure that the relaxed surface has lower energy than the ideal surface and the π-bonded chain model has lower energy still. Our results, however, are not crucially dependent on these constants because our ultimate test is a comparison between the calculated and observed LEED intensities to obtain the best Zanazzi-Jona R factor [44.29]. Furthermore, on the outer 4 layers only the x coordinates were determined by the Keating calculation. The sensitivity of LEED to these coordinates is small. Other parameters of our LEED calculation were an imaginary part $\kappa = 3.5$ eV of the potential, a constant potential of -10 eV between the muffin-tin spheres, and a rms vibration amplitude of 0.1 Å (0.3 Å for the outhermost chain). The extra surface vibration amplitude did not change the R factor of our optimum structure but gave somewhat better visual agreement. Comparable vibration amplitudes were found with ion scattering [44.30] (0.28 Å for the outer chain, 0.21 Å for the inner chain and 0.13 Å in the bulk). A large enhancement of vibrations in the outer chain can be expected, since extra vibration modes exist that involve mostly bond bending and little bond stretching.

Table 44.1 gives the displacements (see last three columns) from the ideal structure for our optimized structure; the top and side views are shown in Fig.44.1b,c, respectively. The arrows in Fig.44.1b indicate the direction of motion of atoms along z to produce buckling in the chains. The calculated I(V) curves for some representative beams are shown in Fig.44.2 as full lines. Most of the major observed features are reproduced by the calculation for the optimized structure. The results for the original chain model [44.17,18] are shown by dotted lines in Fig.44.2 and are clearly unsatisfactory. This

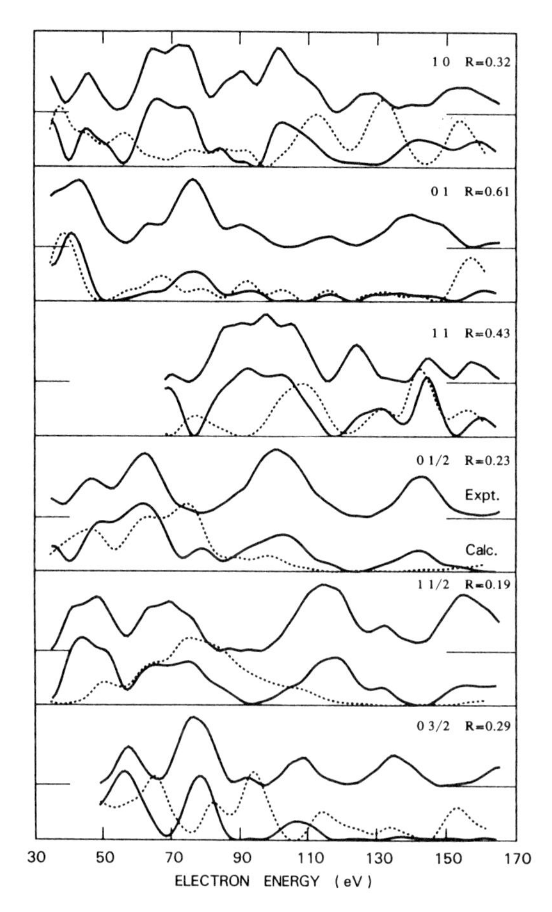

Fig.44.2. Comparison of normal incidence LEED intensity spectra for a few representative beams (*upper curves*) with calculations for the optimized Si(111)2 × 1 structure (*lower full curves*) and for the original Pandey model [44.17,18] (*dotted curves*). All curves are normalized to the same height. R factors are given for individual beams of the optimized structure

is reflected in the R factors (R = 0.42 for the optimized structure and R = 0.92 for the original chain model). The most important parameter making agreement with LEED possible is a large intrachain buckling b_1 of the top chain (Fig.44.1c). Our optimized structure has a strong buckling $b_1 = 0.38 \pm 0.08$ Å. This amount of buckling is close to the extremal case where the down atom has planar sp^2 bond configuration closely similar to the buckled (110)1 × 1 surfaces of III-V compounds. The original chain model [44.17] has $b_1 = 0.0$ and a recent optimized structure given by *Pandey* [44.31] has $b_1 = 0.09$ Å. A larger buckling ($b_1 \simeq 0.2$ Å) has been found by *Northrup* and *Cohen* [44.19,32] from total energy calculations and appears in recent LEED work [44.33].

The average vertical separation between the top and bottom chains $d_{12} \sim 1.03$ Å is quite close to the value given by *Pandey* [44.18] but due to strong intrachain buckling there is an overall compression. Finally, the average $d_{23} \simeq 2.2$ Å in our structure is intermediate to the value given by *Pandey* [44.17,18] (2.13 Å) and the bulk value (2.35 Å). The overall compression of Northrup and Cohen's model is about the same as in our optimum LEED structure. The ion-scattering analysis [44.30] also gives an optimum fit for a fair amount of buckling ($b_1 = 0.3$ Å) but the uncertainty (+0.35, -0.45 Å) is

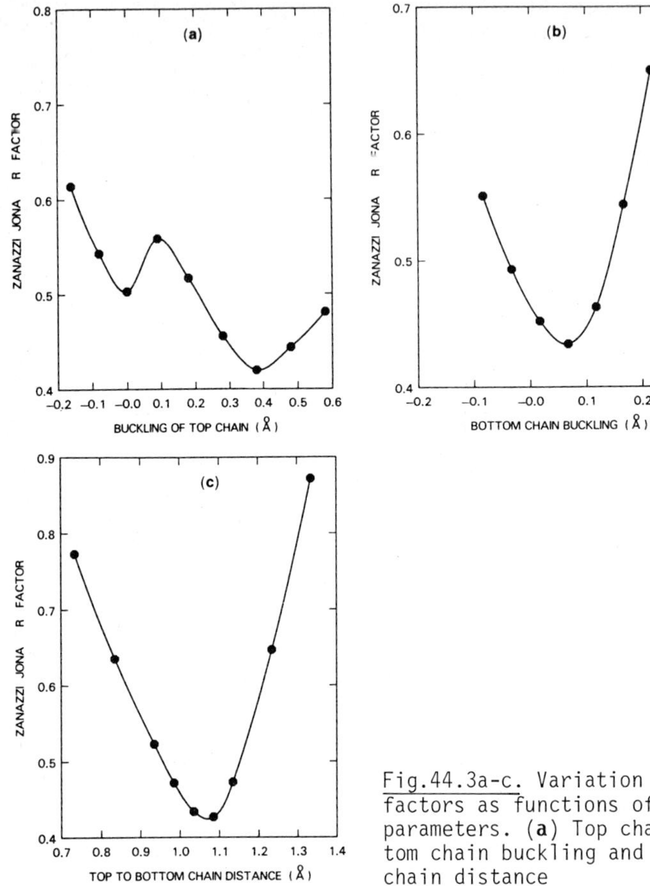

Fig.44.3a-c. Variation of Zanazzi-Jona R-factors as functions of important structural parameters. (**a**) Top chain buckling, (**b**) bottom chain buckling and (**c**) top to bottom chain distance

too large to discriminate buckled from nonbuckled structures. Subsurface relaxation plays an additional role in improving the agreement with LEED. For example, a strain energy minimization improves the R factor from 0.92 to 0.65 for the original Pandey model. The largest effect of subsurface relaxation is a buckling ~0.2 Å in the fourth layer that is driven by bond angle changes of third-layer atoms.

In Fig.44.3 we have shown the effect of key structural parameters on the Zanazzi-Jona R factor calculated using all twelve beams [44.26]. In Fig. 44.3a the effect of the top chain buckling is examined. The remaining coordinates were also optimized for $b_1 = 0$, 0.38 Å and interpolated for other values of b_1. There would be a single minimum at $b_1 = 0.38$ Å if other coordinates are fixed at their optimum values given in Table 44.1. In Fig.44.3b,c the remaining coordinates were kept fixed at their optimum values. Figures 44.3a,b suggest that the sign of buckling of the top and bottom chains is uniquely determined. A positive buckling in the top chain corresponds to atom 11 (Fig.44.1c) lying higher than atom 12. The Keating strain minimization introduces a small (less than 0.1 Å) buckling in the bottom chain where atom 10 is located higher than atom 9 (negative buckling). However, the best R factor is obtained (Fig.44.3b) when the bottom chain is buckled by +0.07 Å (atom 10 located lower than atom 9 along z). This agrees with predictions using a calculation by *Pandey* [44.31] and ion-scattering results [44.30]. Figure 44.3c shows graphically that an average d_{12} close to 1.03 Å gives a minimum in the R factor. We find that all other currently available chain models give an inferior agreement with LEED than our optimum structure. For example, R = 0.66 for the structure given by *Northrup* and *Cohen* [44.19], R = 1.01 for a recent optimized structure given by *Pandey* [44.31], and R = 0.70 for the structure that gives the best fit with ion scattering [44.30].

In summary we may state that a strong intrachain buckling of the top layer definitely improves agreement with LEED. This buckling makes the two dangling bonds inequivalent with respect to the top layer, unlike in the original chain model where inequivalence arises from the location of deeper layers. The lowest R factor we have obtained so far is about 0.42. There is definite room for improvement and we leave open the possibility that Si(111)2 × 1 structure has not been solved at the sensitivity limit level of LEED.

Acknowledgments. We are indebted to Dr. D.W. Jepsen for help with the LEED calculation and Dr. K.C. Pandey for providing unpublished results. Three of the authors (MRC, FJ and HL) were partially supported by the National Science Foundation. One of us (IPB) thanks Dr. Praveen Chaudhari for sponsoring a summer visit to Yorktown Heights where much of this work was done.

References

44.1 D.E. Eastman: J. Vac. Sci. Technol. **17**, 492 (1980)
44.2 D. Haneman: Adv. Phys. **31**, 165 (1982)
44.3 A. Kahn: Surf. Sci. Rep. **3**, 193 (1983)
44.4 D.J. Chadi: *Proc. IX Int. Vac. Cong. - V Int. Conf. Sol. Surf.*, ed. by J.L. Segovia, Madrid, Spain (1983) p.80
44.5 D. Haneman: Phys. Rev. **121**, 1093 (1961)
44.6 J.A. Appelbaum, D.R. Hamann: Phys. Rev. B**12**, 1410 (1975); M. Schlüter, J.R. Chelikowsky, S.G. Louie, M.L. Cohen: Phys. Rev. B**12**, 4200 (1975)
44.7 K.C. Pandey, J.C. Phillips: Phys. Rev. Lett. **34**, 2298 (1975)
44.8 I.P. Batra, S. Ciraci: Phys. Rev. Lett. **34**, 1337 (1975); I.P. Batra, S. Ciraci: Phys. Rev. Lett. **36**, 1707 (1976)
44.9 F.J. Himpsel, P. Heimann, T.C. Chiang, D.E. Eastman: Phys. Rev. Lett. **45**, 1112 (1980)
44.10 S. Brennan, J. Stöhr, R. Jaeger, J.E. Rowe: Phys. Rev. Lett. **45**, 1414 (1980)
44.11 C.B. Duke, W.K. Ford: Surf. Sci. **111**, L165 (1981); J. Vac. Sci. Technol. **21**, 327 (1982)
44.12 R. Del Sole, D.J. Chadi: Phys. Rev. B**24**, 7431 (1981); J. Vac. Sci. Technol. **21**, 319 (1982)
44.13 J.E. Northrup, M.L. Cohen: Phys. Rev. Lett. **47**, 1910 (1981)
44.14 A. Redondo, W.A. Goddard, T.C. McGill: J. Vac. Sci. Technol. **21**, 649 (1982)
44.15 F.J. Himpsel, P. Heimann, D.E. Eastman: Phys. Rev. B**24**, 2003 (1981)
44.16 R.I.G. Uhrberg, G.V. Hansson, J.M. Nicholls, S.A. Flodstrom, F. Houzay, G. Guichar, R. Pinchaux, G. Jezequel, F. Solal, A. Barsky, P. Steiner, Y. Petroff: Surf. Sci. **132**, 40 (1983)
44.17 K.C. Pandey: Phys. Rev. Lett. **47**, 1913 (1981)
44.18 K.C. Pandey: Phys. Rev. Lett. **49**, 223 (1982); Physica **117**B, **118**B, 761 (1983)

44.19 J.E. Northrup, M.L. Cohen: Phys. Rev. Lett. **49**, 1349 (1982); Physica **117**B, **118**B, 774 (1983)
44.20 O.H. Nielsen, R.M. Martin, D.J. Chadi, K. Kunc: J. Vac. Sci. Technol. B**1**, 714 (1983)
44.21 R.M. Tromp, L. Smit, J.F. van der Veen: Phys. Rev. Lett. **51**, 1672 (1983)
44.22 P. Chiaradia, A. Cricenti, S. Selci, G. Chiarotti: Phys. Rev. Lett. **52**, 1145 (1984);
M.A. Olmstead, N.M. Amer: Phys. Rev. Lett. **52**, 1148 (1984);
R. Del Sole, A. Selloni: To be published
44.23 R. Matz, H. Lüth, A. Ritz: Solid State Commun. **46**, 343 (1983)
44.24 R. Feder: Solid State Commun. **45**, 51 (1983);
H. Liu, M.R. Cook, F. Jona, P.M. Marcus: Phys. Rev. B**28**, 6137 (1983)
44.25 M.R. Cook et al.: To be published
44.26 F.J. Himpsel, P.M. Marcus, R. Tromp, I.P. Batra, M.R. Cook, F. Jona, H. Liu: Phys. Rev. B (in press)
44.27 D.W. Jepsen, H.D. Shih, F. Jona, P.M. Marcus: Phys. Rev. B**22**, 814 (1980)
44.28 P.N. Keating: Phys. Rev. **145**, 637 (1966);
F.J. Himpsel, I.B. Batra: J. Vac. Sci. Technol. A**2**, 952 (1984);
J.A. Appelbaum, D.R. Hamann: Surf. Sci. **74**, 21 (1978);
I.P. Batra et al.: To be published
44.29 E. Zanazzi, F. Jona: Surf. Sci. **62**, 61 (1977)
44.30 R.M. Tromp, L. Smit, J.F. van der Veen: To be published
44.31 K.C. Pandey: To be published
44.32 J.E. Northrup, M.L. Cohen: Phys. Rev. B**27**, 6553 (1983)
44.33 I.P. Batra et al.: Bull. Am. Phys. Soc. **29**, 223 (1984);
R. Feder, W. Mönch: Solid State Commun. **50**, 311 (1984)

45. Refinement of the Buckled-Dimer Model for Si(001) 2×1

Y.S. Shu and W.S. Yang

Physics Department, Peking University, Beijing, P.R.C.

F. Jona [1]

Department of Materials Science, S.U.N.Y. at Stony Brook, NY 11794, USA

P.M. Marcus

IBM Research Center, P.O. Box 218, Yorktown Heights, NY 10598, USA

A refinement (YJM2) of the four-layer-distorted buckled-dimer model for Si(001)2 × 1 presented earlier (YJM1) is described. The YJM2 model (mean R factor $R_m = 0.155$ at normal incidence) fits the experimental LEED intensity data better than YJM1 ($R_m = 0.180$), and both do so significantly better than a model recently proposed by other workers ($r_m = 0.262$).

The buckled-dimer model for the Si(001)2 × 1 surface has gained almost universal approval in its qualitative form. The idea of planar dimerization of the top-layer atoms was born in 1957 with a suggestion by *Schlier* and *Farnsworth* [45.1]; but it took more than 20 years to realize through the work of *Appelbaum* and *Hamann* [45.2] that the elastic stresses created by the formation of surface dimers cause elastic distortions which extend down to the fifth layer below the surface; the model deduced from strain energy was a symmetric unbuckled dimer. The suggestion that the dimers could be buckled (i.e., that the two dimer members have different elevations over the plane of the surface) was made by two of us and co-workers in 1979 [45.3] and thereafter quantitatively implemented in LEED calculations [45.4]. The same idea was advanced independently by several theorists: by *Verwoerd* [45.5] on the basis of chemical cluster calculations, by *Chadi* [45.6] on the basis of energy minimizations with semiempirical methods, and by *Yin* and *Cohen* [45.7] on the basis of self-consistent pseudo-potential calculations. Supporting experimental evidence was provided by photoemission [45.8], work-function [45.9] and ion-scattering [45,10,11] studies, but none of this evidence could produce a set of atomic coordinates for the surface atoms or even discriminate among the models proposed by the theorists.

Just such a set and discrimination were provided by a full-dynamical LEED intensity analysis [45.12,13] in 1982. The models proposed by *Chadi* and by *Yin* and *Cohen* were shown to fail the LEED test, and a quantitative structural model was proposed which involves asymmetric buckled dimers and distortions extending to the fourth layer below the surface, and which provides a better than adequate description of new LEED experimental data (14 spectra). In the following, we will refer to this model as the YJM1 model. Subsequently, the existence of buckled dimers was confirmed by ion scattering [45.14,15], and the failure of the Yin-Cohen model was confirmed by an independent LEED analysis [45.16]. However, the YJM1 model was criticized for not providing as "dramatic" an improvement over other models as one might expect from the right model [45.16], and for exhibiting a tilt of the dimer bond in the direction (y) in the surface plane perpendicular to the dimer axis — a tilt that

1 Sponsored in part by the National Science Foundation

seems to be inconsistent with the ion-scattering results [45.14,15]. Very recently, *Holland, Duke* and *Paton* [45.17] (HDP) produced a different structural model for Si(001)2 × 1, which has no tilt in the dimers and, in their view, improves the fit to (older) experimental data over the YJM1 model. We find this view to be incorrect: the YJM1 model provides a significantly better fit to the LEED experiment both visually and by R-factor analysis than the HDP model. We will return to this point below.

We present here a refinement (called YJM2) of the YJM1 model which was obtained with two orthogonal-experimental-design quasi-dynamical calculations [45.18] [$L_{27}(3^{13})$ and $L_9(3^9)$] followed by additional independent parameter adjustments and full-dynamical calculations. The YJM2 model has the following characteristics. It provides a better fit to experiment at normal electron incidence; it features a buckled but symmetric top-layer dimer (i.e., the shifts of the dimer atoms toward one another, along the x axis parallel to the surface, are equal); it has a dimer y tilt that is smaller than in YJM1 but still significant (0.21 Å); it features distortions involving the top four layers, with buckling of 0.26 Å in the fourth layer. The atomic coordinates in the top 4 layers are fixed by specifying the shifts from bulk positions as defined in Fig.45.1 (all numbers in Angstroms):

$$\Delta x_{11} = 0.72,\ \Delta y_{11} = 0.21,\ \Delta z_{11} = 0.06,$$

$$\Delta x_{12} = -0.72,\ \Delta y_{12} = -0.23,\ \Delta z_{12} = 0.47,$$

$$\Delta x_{21} = 0.19,\ \Delta y_{21} = 0.21,\ \Delta z_{21} = 0.03,$$

$$\Delta x_{22} = -0.21,\ \Delta y_{22} = -0.23,\ \Delta z_{22} = 0.03,$$

$$\Delta z_{31} = -0.15,\ \Delta z_{32} = 0.23,\ \Delta z_{41} = -0.13,\ \Delta z_{42} = 0.13.$$

With the data at normal electron incidence we have made a detailed beam by beam comparison between the YJM2 and the HDP models. We present all the evidence in Figs.45.2 and 3, where seven experimental LEED spectra are compared to those calculated with the YJM2 model and to those calculated with the HDP model. The minimum R factor [45.19] for the normal-incidence data

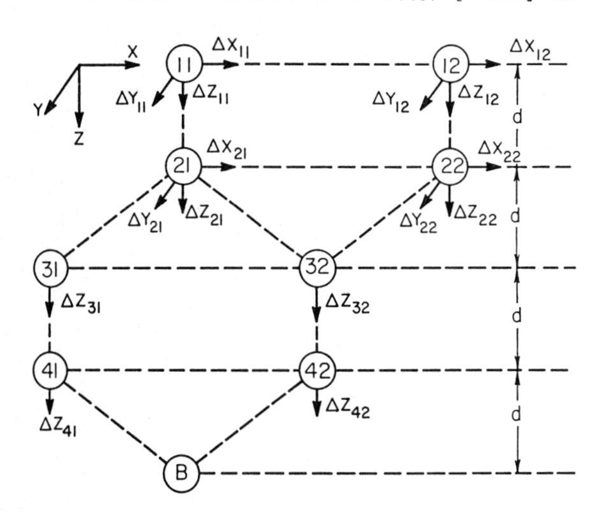

Fig.45.1. Schematic perspective drawing of the shifts of atoms from bulk positions in the four-layer-distorted buckled-dimer model of Si(001)2 × 1

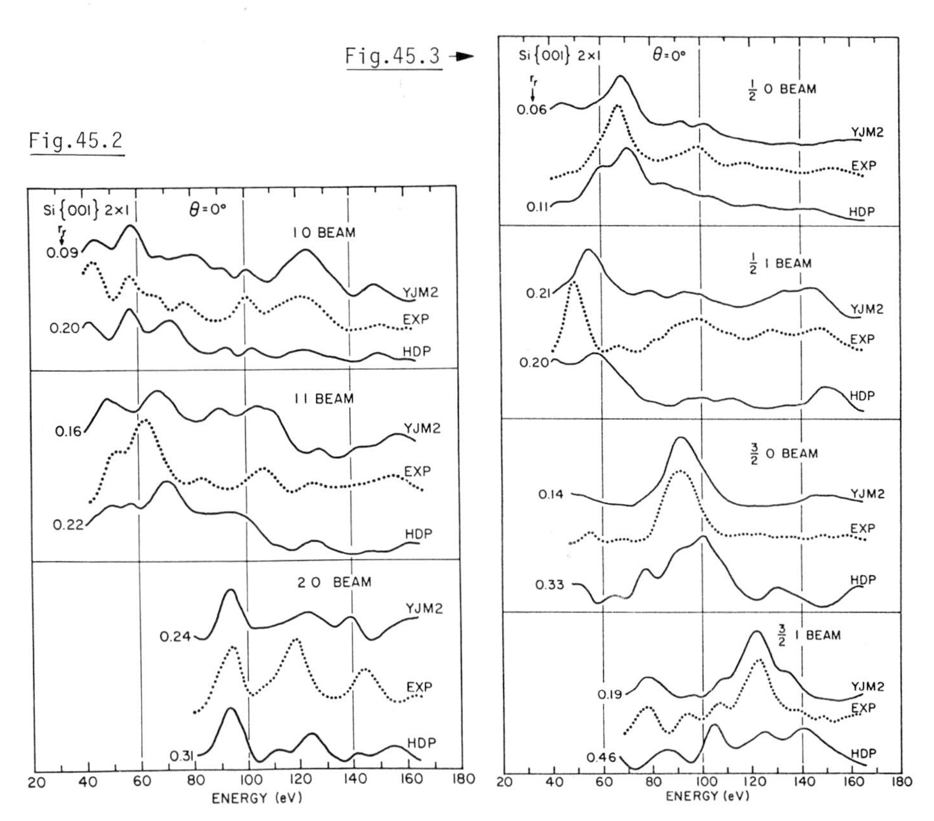

Fig.45.2. Calculated and experimental integral-order LEED spectra at normal electron incidence ($\theta = 0^{\circ}$) for Si(001)2 × 1. YJM2: model presented in this paper. HDP: model proposed by Holland, Duke and Paton [45.17]. EXP: experimental data [45.13]. Numbers on the left are R-factor values

Fig.45.3. Calculated and experimental fractiona-order LEED spectra at normal electron incidence ($\theta = 0^{\circ}$) for Si(001)2 × 1. YJM2: model presented in this paper. HDP: model proposed by Holland et al. [45.17]. EXP: experimental data [45.13]. Numbers on the left are R-factor values

is 0.155 for the YJM2 and 0.262 for the HDP model. The single-beam R factors are written near each curve in the figures. Note that the minima in R factor for both models occur at a value of the real part of the inner potential ($V_{0r} = -5 \pm 2$ eV) which is smaller than that reported for the YJM1 model ($V_{0r} = -9 \pm 2$ eV). Thus, the HDP curves in Figs.45.2,3 are shifted by about 6 eV along the energy axis with respect to those shown by HDP [45.17]. These authors show only the 11 and the 1/2 1 spectra (at normal incidence) and indeed for these two spectra a value of $V_{0r} \simeq -11$ eV provides a better fit with experiment than $V_{0r} \simeq -5$ eV implied in Figs.45.2,3. However, it is obvious from the figures that the fit with experiment of all remaining HDP spectra (see, e.g., the 10 and 20 spectra) would be considerably worsened by the choice $V_{0r} \simeq -11$ eV (the HDP curves in the figures would be shifted to the left by 6 eV). The R-factor values confirm the visual impression: for $V_{0r} \simeq -5$ eV the mean R factor for the HDP model is, as mentioned above, 0.262, while at $V_{0r} \simeq -11$ eV it would be 0.290.

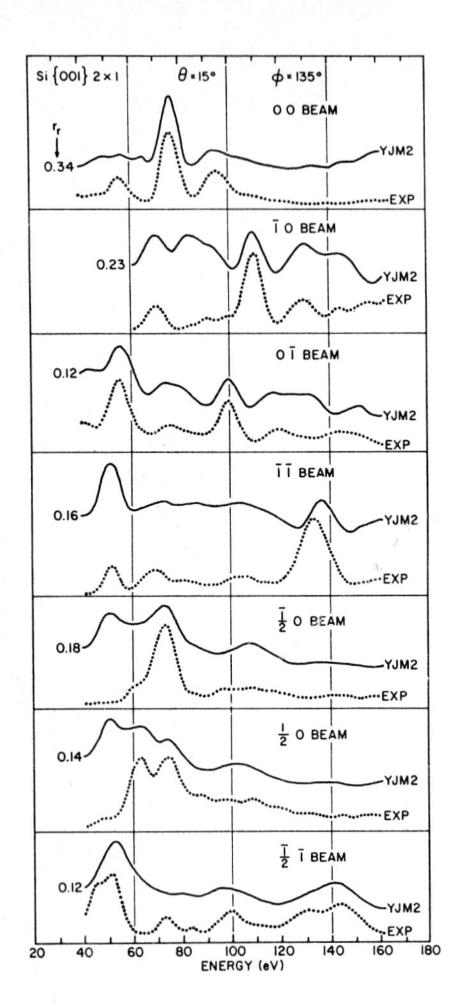

Fig.45.4. Calculated and experimental LEED spectra at $\theta = 15^\circ$, $\phi = 135^\circ$ for Si(001)2 × 1. YJM2: model presented in this paper. EXP: experimental data [45.13]. Numbers on the left are R-factor values

Scrutiny of Figs.45.2 and 3 reveals the YJM2 model yields lower R factors than the HDP model for all but one spectrum (the 1/2 1 spectrum, for which the R factors are 0.21 and 0.20, respectively). We submit that visual inspection also favors the YJM2 model consistently for all beams. Hence we conclude that the YJM2 model is substantially better than the HDP model in fitting the experimental data. The test of YJM2 at $\theta = 15^\circ$, $\phi = 135^\circ$ is depicted in Fig.45.4 (mean R factor of 0.184).

We summarize our conclusions about the Si(001)2 × 1 structure as follows:

1) The YJM2 model is a refinement and an improvement over the four-layer model YJM1 presented earlier [45.13].

2) Both the YJM1 (on the basis of comparisons between the R-factor values published in [45.13] and those quoted above for the HDP model) and the YJM2 model are substantially better than the HDP model in fitting the experimental LEED data.

3) The existence of a y tilt in our model may constitute an obstacle to its wider acceptance but, HDP notwithstanding, there exists at the present

time no other model that performs as well from the viewpoint of quantitative LEED intensity analysis. If a model without y tilt exists that fits the LEED experiment better than the YJM2 model, it was not found at the time of writing.

References

45.1 R.E. Schlier, H.E. Farnsworth: In *Semiconductor Surface Physics* (University of Pennsylvania Press, Pennsylvania 1957) pp.3-22
45.2 J.A. Appelbaum, D.R. Hamann: Surf. Sci. **74**, 21 (1978)
45.3 F. Jona, H.D. Shih, D.W. Jepsen, P.M. Marcus: J. Phys. C: Solid State Phys. on **12**, L455 (1979)
45.4 J.C. Fernandez, W.S. Yang, H.D. Shih, F. Jona, D.W. Jepsen, P.M. Marcus: J. Phys. C: Solid State Phys. **14**, L55 (1981)
45.5 W.S. Verwoerd: Surf. Sci. **99**, 581 (1980)
45.6 D.J. Chadi: Phys. Rev. Lett. **43**, 43 (1979)
45.7 M.T. Yin, M.L. Cohen: Phys. Rev. B**24**, 2303 (1981)
45.8 F.J. Himpsel, D.E. Eastman: J. Vac. Sci. Technol. **16**, 1297 (1979); F.J. Himpsel, P. Heimann, T.C. Chiang, D.E. Eastman: Phys. Rev. Lett. **45**, 1112 (1980)
45.9 P. Koke, W. Mönch: Solid St. Commun. **36**, 1007 (1980)
45.10 L.C. Feldman, P.J. Silverman, I. Stensgaard: Nucl. Instrum. Meth. **168**, 589 (1980)
45.11 R.M. Tromp, R.G. Smeenk, F.W. Saris: Phys. Rev. Lett. **46**, 939 (1981)
45.12 W.S. Yang, F. Jona, P.M. Marcus: Solid State Commun. **43**, 847 (1982)
45.13 W.S. Yang, F. Jona, P.M. Marcus: Phys. Rev. B**28**, 2049 (1983)
45.14 M. Aono, Y. Hou, C. Oshima, Y. Ishizawa: Phys. Rev. Lett. **49**, 567 (1982)
45.15 R.M. Tromp, R.G. Smeenk, F.W. Saris, D.J. Chadi: Surf. Sci. **133**, 137 (1983)
45.16 G.J.R. Jones, B.W. Holland: To be published
45.17 B.W. Holland, C.B. Duke, A. Paton: Preprint, and this conference
45.18 W.S. Yang, F. Jona, P.M. Marcus: J. Vac. Sci. Technol. B**1**, 718 (1983)
45.19 E. Zanazzi, F. Jona: Surf. Sci. **62**, 61 (1977)

46. Surface Relaxation and Vibrational Excitations on the Si(001) 2 × 1 Surface

Douglas C. Allan and E.J. Mele

Department of Physics, University of Pennsylvania
Philadelphia, PA 19104-3859, USA

We present a realistic theoretical study of surface vibrational excitations for the 2 × 1 reconstructed Si(001) surface. The extensive reconstruction of this surface is found to localize a number of interesting surface modes. These include a band whose eigenvector consists of a rocking motion of surface dimers and an unanticipated optical mode localized in the first two subsurface layers. Using a tight-binding theory for structural energies, the model incorporates the effects of bond rehybridization at the surface.

46.1 Introduction

Using a realistic but tractable theoretical model for structural energies, we have computed the surface structure and vibrational excitation spectrum of reconstructed and relaxed Si(001)2 × 1 [46.1]. Our formalism, which uses a tight-binding theory for structural energies, is described in [46.1]. As shown there and in Table 46.1, the description of the bulk phonon dispersion over the entire Brillouin zone is remarkably successful. As seen in Table 46.1, the principal deficiencies of the model are a systematic underestimate (22%-26%) of the elastic constants and a slight overestimate (6%-12%) of the zone boundary TA modes.

Table 46.1. Comparison of experimental and theoretical phonon energies (meV) at symmetry points and elastic constants (eV/Å^3) for bulk Si

	Bulk Si phonons (meV)							
	LTO(Γ)	TA(X)	LAO(X)	TO(X)	TA(L)	LA(L)	LO(L)	TO(L)
exp[a]	64.4	18.6	51.0	57.5	14.2	46.9	52.1	60.7
theory	64.4	20.8	48.1	55.5	15.1	42.4	48.7	60.3
Δ%	(FIT)	12	-6	-4	6	-10	-7	-0.7

	Bulk Si elastic constants (eV/Å^3)			
	C_{11}	C_{12}	C_{44}	B
exp[b]	1.03	0.399	0.497	0.61o
theory	0.790	0.312	0.369	0.471
Δ%	-23	-22	-26	-23

[a]G. Dolling: In *Inelastic Scattering of Neutrons in Solids and Liquids* (I.A.E.A., Vienna 1963), Vol.II, p.37; G. Nilson, G. Nelin: Phys. Rev. B**6**, 3777 (1972)
[b]H.J. McSkimin: J. Appl. Phys. **24**, 988 (1953)

46.2 Structure

We extend this model to the Si(001) surface by imposing the connectivity of the 2×1 reconstructed surface, coupling pairs of Si atoms in rows along the [110] direction on both (001) surfaces of a ten-layer slab. By following the Hellmann-Feynman forces to equilibrate the slab we produce the expected rows of asymmetric (tilted) dimers on the surface [46.2,3]. Our equilibrated structure is essentially that found by *Chadi* [46.2,3] which is very similar to one obtained in the self-consistent pseudopotential study of *Yin* and *Cohen* [46.4] and which agrees qualitatively with recent LEED experiments [46.5]. A comparison of the present and the other three equilibrium geometries is given in Table 46.2. One qualitative feature which differs among the tight-binding, pseudopotential, and LEED geometries is the dimer bond length (Table 46.2). The phonon spectrum, however, is dominated by the 2×1 asymmetric dimer connectivity of the surface and is not sensitive to the exact dimer bond length. The overall agreement of the proposed structures is more striking than their variations in detail. To assess further the reliability of energy variations computed in the present model, we also compare the asymmetric relaxation energy in the three theoretical calculations (Table 46.2). The metallic surface bands of the symmetric dimer introduce some un-

Table 46.2. Comparison of equilibrium geometries for Si(001)2×1 between three calculations and a recent LEED experiment: displacement field R for atoms near surface, energy E^{asymm} gained by asymmetric relaxation of dimers, and dimer bond length b in terms of equilibrium bond length b_0

Displacement field $\Delta\mathbf{R}$ in Å.
Ideal (k,ℓ,m): $\mathbf{R}=(k\times a/2\sqrt{2},\ \ell\times a/2\sqrt{2},\ m\times a/4)$

Layer	(k,ℓ,m)	Present ΔR_x	Present ΔR_z	Chadi[a] ΔR_x	Chadi[a] ΔR_z	YC[b] ΔR_x	YC[b] ΔR_z	LEED[c] ΔR_x	LEED[c] ΔR_z
1	(0,0,0)	0.521	0.091	0.478	0.100	0.573	-0.159	0.500	-0.250
	(2,0,0)	-1.033	-0.506	-1.071	-0.459	-1.038	-0.468	-0.900	-0.614
2	(0,1,-1)	0.103	0.079	0.094	0.075	0.093	-0.047	0.094	-0.022
	(2,1,-1)	-0.073	-0.029	-0.094	-0.011	-0.115	0.020	-0.105	0.055
3	(1,1,-2)	-0.010	-0.108	-0.025	-0.106	-0.007	-0.185	(not optimized:	
	(3,1,-2)	0.052	0.104	0.031	0.132	-0.034	0.129	average a and b)	
4	(1,0,-3)	-0.009	-0.071	-0.010	-0.088	0.061	-0.135		
	(3,0,-3)	0.002	0.070	-0.004	0.096	-0.060	0.103		
5	(2,0,-4)	0.019	0.001	0.042	0.011	*			
	(4,0,-4)	-0.028	-0.004	-0.056	0.005	*			
6	(2,1,-5)	*		0.019	0.003	*			
	(4,1,-5)	*		-0.021	0.000	*			
ΔE^{asymm} (eV/surface atom)		~0.14		~0.23		$\lesssim 0.1$			
dimer bond length $[(b/b^0-1)\times100\%]$		0.48		0.29		-4.3		+4.9	

*These atoms occur in lower half of inversion-symmetric slab unit cell
[a] [46.3]
[b] [46.4]
[c] [46.5]

certainty into part of this comparison, but it is seen that our present parameterization predicts a relaxation energy similar to the pseudopotential result. The curvature of the total energy around this equilibrium structure is obtained following the scheme outlined in [46.1]. It is worth noting that the force constants are considerably altered for surface atoms (by about 20%-30%) so that a force constant model extrapolated using bulk force constants would be unreliable for the surface.

46.3 Vibrational Excitations

A number of perturbations occurring at a surface can produce localized vibrational modes. The most obvious is naturally the reduced coordination of surface atoms. At long wavelengths, where elastic continuum theory applies, this perturbation produces the well-known Rayleigh surface wave at acoustic frequencies just below the acoustic continuum. At short wavelengths, however, the character of the acoustic modes is considerably altered and a lattice theory is necessary. The vibrational spectrum of our ten-layer slab and the projection of the bulk phonon bands onto the folded 2×1 surface Brillouin zone (hatched area) are shown in Fig.46.1. At $\underset{\sim}{q}$ near zone boundaries, reduced coordination pushes three branches (per surface) off the continuum, involving motions both normal and parallel to the surface plane. The reduced coordination also causes a general softening of modes which pushes modes into the two uppermost projected gaps.

In addition to the expected Rayleigh-type surface modes, we have found three other surface modes (labeled r, s, and sb) which are particularly informative. The first of these, a dimer rocking mode (r), extends across a projected gap at 26 meV. The displacement field for this vibration, which is almost purely the rocking motion of the surface dimer, is sketched for the K point in Fig.46.2a. Recalling that a surface of symmetric dimers has metallic electronic bands with a Fermi surface instability which drives the dimers to distort asymmetrically [46.2-4], one expects strong electronic screening to occur for the dimer rocking mode. This is precisely what we observe. Although the r mode has the character of an optical phonon, virtual transitions between dangling bond surface bands [46.6] soften the mode down to just above the acoustic continuum. The rocking motion is therefore coupled to a charge transfer oscillation of the surface dimer, making the

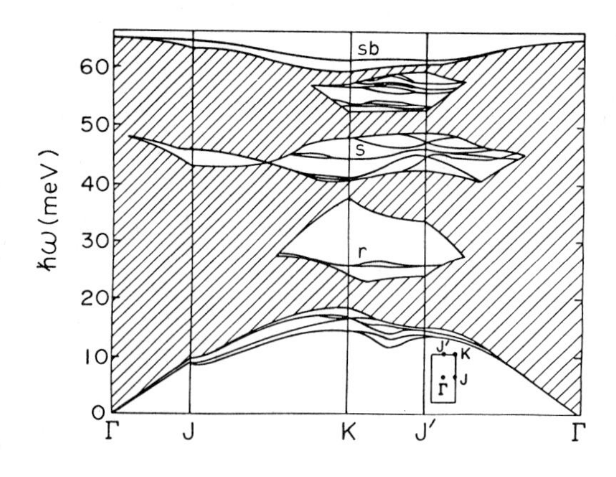

Fig.46.1. Bulk phonon band projected onto symmetry directions of the folded 2×1 surface Brillouin zone (hatched) and surface phonon bands obtained from slab calculation (solid lines)

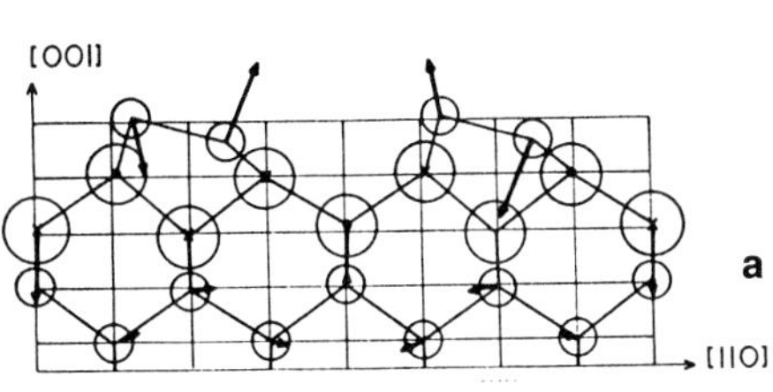

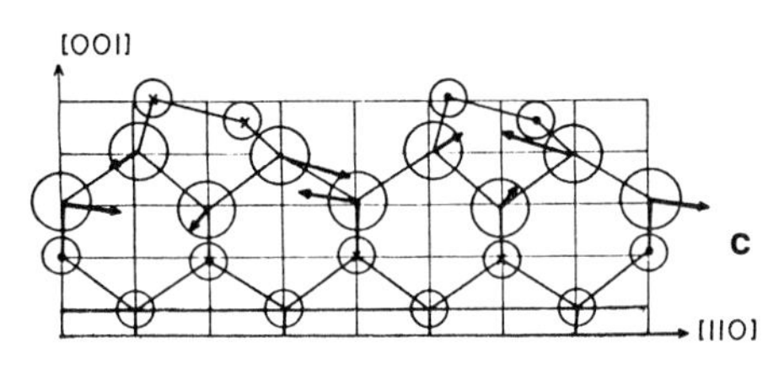

Fig.46.2a-c. Vibrational amplitudes for three localized modes (arrows are proportional to relative amplitudes) at the K point of the surface Brillouin zone. (**a**) Side (1$\bar{1}$0) view of the dimer rocking mode r ($\hbar\omega$ = 25.7 meV); (**b**) top (001) view of a zone boundary swing mode s ($\hbar\omega$ = 44.2 meV) in which adjacent dimers swing out of phase; (**c**) side (1$\bar{1}$0) view of the subsurface bond mode sb ($\hbar\omega$ = 61.2 meV) which splits off the *top* of the bulk continuum throughout the surface Brillouin zone. The atoms mainly affected are in the second layer (largest vibrational amplitude) and in the third layer (second largest amplitude)

mode strongly IR active if it persists at the zone center. As it is, the mode has a strong projection along the surface normal, and therefore should be particularly accessible to inelastic He scattering.

Another mode which involves a simple motion of surface dimers is the dimer swing mode (s) whose dispersion crosses the center of the next window in the projected spectrum at about 44 meV. Figure 46.2b shows the eigenvector of this mode at K, a swinging of alternate dimers 180° out of phase.

In addition to the expected Rayleigh modes and the suspected softened dimer rocking mode, a heretofore unexpected third kind of vibrational mode occurs on this surface. This is the subsurface bond (sb) mode, which splits off the *top* of the bulk TO continuum as a consequence of subsurface strain induced by the asymmetric dimer relaxation. This mode has the unusual property of exceeding bulk frequencies because it involves motion of fully fourfold coordinated atoms in the second and third layers (Fig.46.2c). In the relaxed geometry, with distorted bond angles, the motion of subsurface atoms projects strongly onto bond length oscillations of bonds in the (1$\bar{1}$0) plane, in a manner which is impossible in the bulk. Because the stiffening is a geometric effect, we predict that *any* Hamiltonian producing an equilibrium surface of asymmetric dimers, as shown in Fig.46.2c, will possess this high-frequency subsurface bond mode. The weak amplitude of this mode on surface atoms, as well as its relatively high energy, makes it unlikely to be seen in inelastic He scattering. It remains an intriguing possibility that EELS, however, which is not quite as surface sensitive, could be used to identify this localized subsurface vibrational mode.

46.4 Conclusion

We have shown that undercoordination, strong electronic screening, and subsurface strain produce localized vibrational modes on Si(001)2 × 1. Some of these modes (r, s, and sb) are easily related to the underlying surface geometry, potentially yielding structural information about the bonding properties of the surface. In addition, the theoretical methods we have employed here are quite efficient and flexible and should find increasing application for analyzing similar surface and interfacial vibrational problems.

Acknowledgments. This work was supported under Department of Energy contract DE-AC02-83ER45037. Early stages of this work were supported in part by the National Science Foundation under DMR 82-16718. EJM gratefully acknowledges support from a fellowship provided by the Alfred P. Sloan Foundation.

References

46.1 D.C. Allan, E.J. Mele: Phys. Rev. Lett. (1984) (in press), and references therein
46.2 D.J. Chadi: Phys. Rev. Lett. **43**, 43 (1979); J. Vac. Sci. Technol. **16**, 1290 (1979)
46.3 R.M. Tromp, R.G. Smeenk, F.W. Saris, D.J. Chadi: Surf. Sci. **133**, 137 (1983)
46.4 M.T. Yin, M.L. Cohen: Phys. Rev. B**24**, 2303 (1981)
46.5 B.W. Holland, C.B. Duke, A. Paton: Surf. Sci. **140**, L269 (1984)
46.6 Surface electronic bands before and after the asymmetric dimer relaxation are given in [46.2]

V.2 Compound Semiconductors

47. The Geometric Structure of the (2×2) GaAs(111) Surface

G. Xu

Department of Physics, Zhongshan University, Guangzhou, China

S.Y. Tong and W.N. Mei

Department of Physics and Laboratory for Surface Studies, University of Wisconsin-Milwaukee, Milwaukee, WI 53201, USA

We present results of a reconstruction model proposed for the (2 × 2) GaAs (111) surface. Our model suggests similar reconstruction mechanisms for the (111) and (110) surfaces. In both cases, surface electronic energies are lowered via orbital rehybridization between nearest-neighbor Ga and As atoms with dangling bonds.

47.1 Introduction

In the bulk of a GaAs crystal, each Ga atom is bonded to four As atoms via sp^3 bonds, and vice versa. An ideal (111) surface is terminated by a complete layer of Ga atoms. A surface Ga atom is bonded to three As neighbors in a plane 0.816 Å below. The fourth As neighbor is absent, hence each surface Ga atom has a partially filled dangling bond pointing normal to the surface. The bulk-terminated (111) configuration is unstable, because a large amount of potential energy is needed to localize electrons in these partially filled dangling bonds.

A possible reconstruction mechanism is for electrons of the surface Ga atoms to rehybridize into three sp^2 bonds, similar to the configuration in GaH_3 [47.1]. The three sp^2 bonds lie almost in a plane and are separated by approximately 120°. This rehybridization leaves an empty dangling bond on each surface Ga atom. Electrons on the As neighbors rehybridize to form three p-type orbitals separated by approximately 90° and a doubly filled s-type orbital. Thus, the rehybridized Ga and As surface orbitals resemble the respective atomic configurations closer than the sp^3 configuration of the bulk.

The electronic rehybridization described above is responsible for stabilizing the (110) surface of GaAs [47.2], resulting in a relaxed structure with a tilt angle $27° \leqslant \omega \leqslant 30°$. On a bulk-terminated (110) surface, the number of Ga and As dangling bonds are equal, one of each in a surface unit cell. Thus, the transfer of electrons between Ga and As atoms can take place without altering the (1 × 1) periodicity. However, similar rehybridization cannot take place on an ideal (111) surface, because all the unfilled dangling bonds are in the Ga atoms and each As neighbor in the plane below is fully, i.e., tetrahedrally, bonded.

Rehybridization, however, becomes likely if there are Ga vacancies on the surface [47.1-3]. Each vacancy breaks three Ga-As bonds and creates three unfilled As orbitals. These As atoms are then free to transfer electrons with their Ga neighbors. To rehybridize the surface fully, there should be one vacancy for every three remaining Ga surface atoms, and by symmetry, the vacancies form a hexagonal lattice. Thus, the model predicts a (2 × 2) periodicity. Experimentally, the GaAs (111) face is found to exhibit one stable LEED pattern: the (2 × 2) pattern, which is formed either by Ar^+ ion sputtering following by annealing at 545°C or by annealing at 545°C a MBE grown (111) surface [47.4]. Thus, the measured LEED pattern is consistent with the above reconstruction model.

47.2 Determination of Surface Geometry

We have analyzed LEED intensity-energy (IV) data by the multiple-scattering slab method to determine the atomic positions of the (2 × 2) GaAs (111) surface. We kept the same dynamical inputs on the (111) surface as on the (110) [47.2], and changed only the inputs of atomic coordinates. To analyze the GaAs(111) surface, we considered 10 beams: 5 integral-order and 5 half-order beams [47.3].

Because LEED analysis depends on comparing calculated and measured IV profiles of a large number of structural models, some of which are only very slightly different, it is extremely important that the calculation is accurate [47.5]. There have been examples where wrong structures were selected due to nonconvergent calculations. In the analysis of GaAs (111), we used the combined space method [47.6], in which multiple scattering within each atomic plane and between different planes is evaluated and summed until numerical convergence is achieved. Up to 229 beams, reduced to 47 symmetrized beams at normal incidence, are used. At typical energies, seven to ten bilayers are included in the calculation.

To help in picking the best structure, we used a reliability factor (R factor) which is a normalized average of 10 algorithms [47.7,8]. The 10 algorithms compare key features in calculated and measured IV spectra such as the position, width and area of peaks, number of peaks, energy of turning points, etc. [47.9]. A smaller R factor indicates better agreement, and a perfect match has an R factor of zero.

In Fig.47.1, we show a schematic view of the vacancy-buckling model, with the (2 × 2) unit cell bounded by broken arrows [47.1]. Atoms A, B, C are surface Ga atoms, a, c, d are As atoms in the plane below. The fourth As atom (b) in the unit cell is tetrahedrally bonded, as shown.

The structure was found by starting with a bulk-terminated (1 × 1) GaAs (111) surface and displacing the surface Ga plane vertically towards the bulk, keeping the atomic coordinates of all other atoms at those of the bulk. The R-factor value is plotted as a function of the first bilayer spacing in Fig.47.2. We found a deep minimum in the range of 0.05-0.09 Å. This very small bilayer spacing, being about only 10% of the bulk value, strongly suggests the rehybridization of surface Ga orbitals into the sp^2 type.

We then used symmetry among beams to guide us in the next step of the structural search. If the surface Ga plane, having one vacancy per (2 × 2) unit cell, is displaced vertically towards the bulk, and if the As plane below it retains the (1 × 1) bulk periodicity, then the IV spectra for the (0 1/2) and (1/2 0) beams should be very similar. The measured spectra for these two beams are, however, very different, as can be seen from Fig.47.3. From the lack of symmetry of these two beams, one can deduce that the cor-

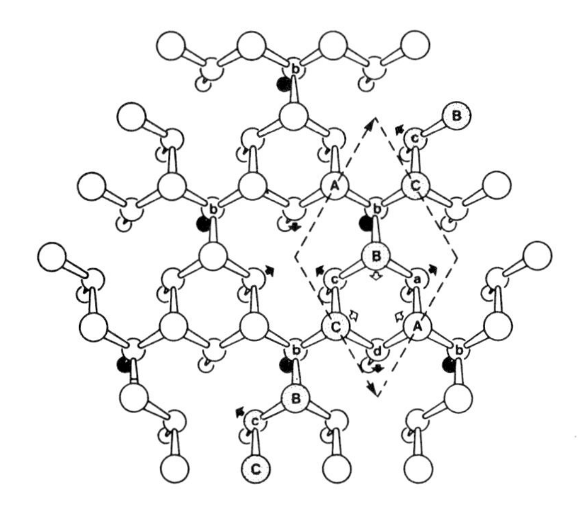

Fig.47.1. Top view of (2×2) vacancy-buckling model

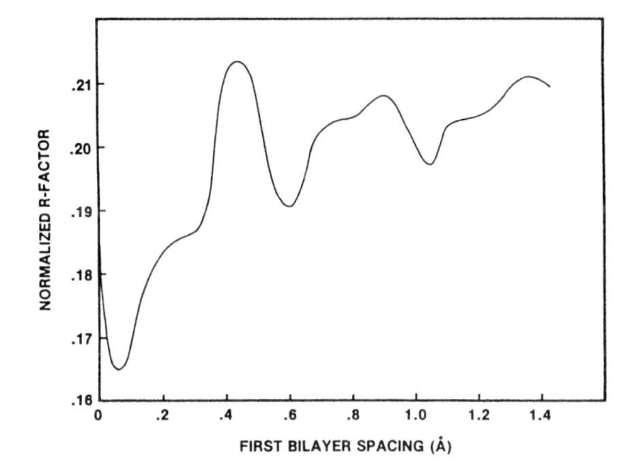

Fig.47.2. Plot of R factor vs bilayer spacing

rect structure must have lateral displacements of the surface Ga atoms, or rearrangement of the As plane into two or more subplanes having the new (2×2) periodicity, or both.

Further information can be obtained from the widths of the measured peaks. If only the surface Ga plane has (2×2) periodicity, and all deeper layers retain bulk periodicity, then the widths of peaks in the half-order beams should be substantially wider than those in the integral-order beams. The data in Figs.47.3,4 show that all peaks have about the same widths. This suggests that three or more atomic planes have (2×2) periodicity.

Following these hints, we laterally displaced the surface Ga atoms A, B and C of Fig.47.1, as well as the As atoms a, c and d. A smaller R factor was obtained. We then argued that since the fourth As atom, marked b in Fig.47.1, is tetrahedrally bonded, it may occupy a different vertical distance from the surface than the other three As atoms. Therefore, we divided the As plane into two subplanes, and varied the distance between them. We obtained a smaller R factor when the subplane containing atom b is below that containing a, c and d.

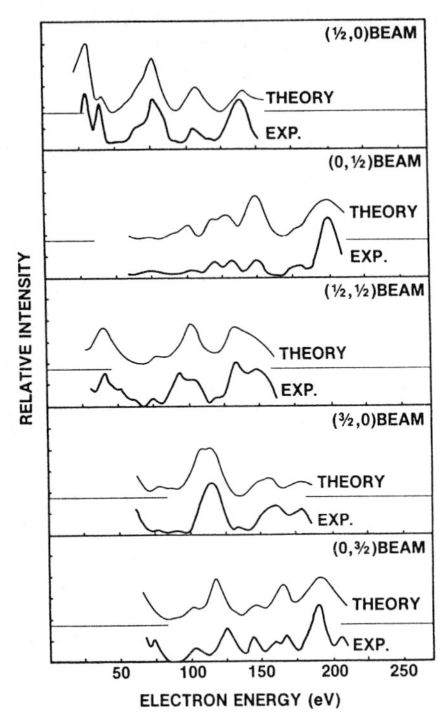

Fig.47.3. Comparison between calculaed IV curves with experiment, for half-order beams

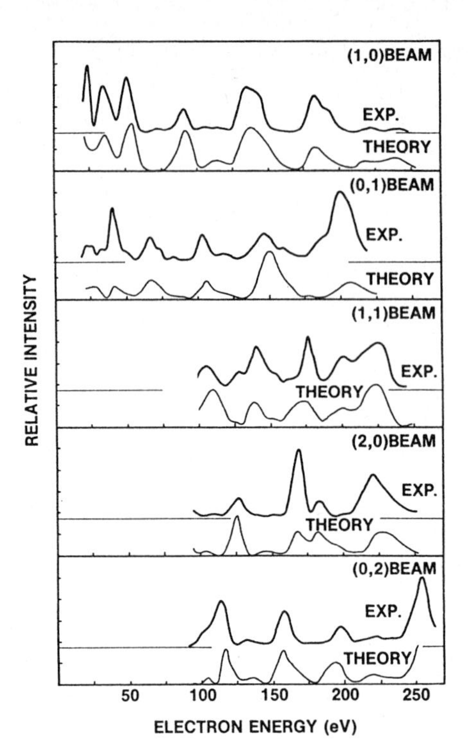

Fig.47.4. Comparison between calculated IV curves with experiment, for integral-order beams

A final improvement in the R factor was obtained when we divided the Ga plane in the second bilayer into two subplanes. The Ga atom directly below b is depressed into the bulk while the Ga atoms below a, c and d are raised towards the surface. No lateral displacements are considered for these atoms as they are probably too small to be conclusively determined (<0.03 Å). As each set of new displacements was introduced, we simultaneously varied the spacing within the first bilayer and the spacing between the first two bilayers.

47.3 Conclusion and the Atomic Positions

We have shown that the GaAs(111) surface has one vacancy per (2 × 2) unit cell. The structure determined by this study has the following properties:

i) 1/4 layer surface Ga atoms missing
ii) 3/4 layer surface Ga atoms (A,B,C) are pushed 0.706 Å towards the bulk, and 0.1 Å laterally along in the direction of the arrows in Fig.47.1
iii) 3/4 layer As atoms (a,c,d) are raised 0.04 Å towards the surface and displaced 0.28 Å laterally along the direction of the arrows of Fig. 47.1. The resulting first bilayer has a separation of 0.07 Å
iv) 1/4 layer As atoms (b), tetrahedrally bonded, are pushed 0.08 Å towards the bulk. A distance of 0.12 Å separates the two As subplanes

v) 3/4 layer Ga atoms directly below atoms (a,c,d) are raised 0.01 Å towards the surface
vi) 1/4 layer Ga atoms directly below atoms (b) are displaced 0.08 Å towards the bulk. A distance of 0.09 Å separates the two Ga subplanes
vii) All atoms in deeper layers occupy bulk sites.

The reconstruction mechanism is similar for both the (111) and (110) faces of GaAs. A recent energy-minimization calculation [47.10] determined that the vacancy-buckling model lowers the total energy by 2.3 eV per (2×2) unit cell.

This work is supported by NSF Grant No. DMR-8405049 and PRF Grant No. 1154-AC5,6.

References

47.1 S.Y. Tong, G. Xu, W.N. Mei: Phys. Rev. Lett. **52**, 1693 (1984)
47.2 S.Y. Tong, A.R. Lubinsky, B.J. Mrstik, M.A. Van Hove: Phys. Rev. B**17**, 3303 (1978)
47.3 S.Y. Tong, W.N. Mei, G. Xu: J. Vac. Sci. Tech. B**2**, No.3 (1984)
47.4 R.D. Bringans, R.Z. Bachrach: This Volume, p.308
47.5 S.Y. Tong: In *Progress in Surface Science*, Vol.7, ed. by S.G. Davisson (Pergamon, New York 1975) p.1
47.6 S.Y. Tong, M.A. Van Hove: Phys. Rev. B**16**, 1459 (1977)
47.7 M.A. Van Hove, S.Y. Tong, M.H. Elconin: Surf. Sci. **64**, 85 (1977)
47.8 S.Y. Tong, W.M. Kang, D.H. Rosenblatt, J.G. Tobin, D.A. Shirley: Phys. Rev. B**27**, 4632 (1983)
47.9 S.Y. Tong, K.H. Lau: Phys. Rev. B**25**, 7382 (1982)
47.10 D.J. Chadi: Phys. Rev. Lett. **52**, 1911 (1984)

48. A Comparison Between the Electronic Properties of GaAs(111) and GaAs($\bar{1}\bar{1}\bar{1}$)

R.D. Bringans and R.Z. Bachrach

Stanford Synchrotron Radiation Laboratory
P.O. Box 4349, Stanford University, Stanford, CA 94305, USA
and
Xerox Palo Alto Research Center
3333 Coyote Hill Rd., Palo Alto, CA 94304, USA

Results of an angle-resolved photoemission investigation of GaAs(111)(2 × 2) and GaAs($\bar{1}\bar{1}\bar{1}$)(2 × 2) are presented. Significant differences were found to exist between spectra for the two surfaces. In addition, the surface band dispersions which are derived directly from the data are distinctly different for the two surfaces.

48.1 Introduction

Along the [111] direction, GaAs consists of complete layers of As alternating with complete layers of Ga. This makes both the (111) and ($\bar{1}\bar{1}\bar{1}$) faces of GaAs polar. If only one bond per surface atom is broken then one arrives at the ideally terminated surfaces shown in Fig.48.1. It should be noted that these two surfaces are not identical: the (111) surface is stable with a final layer of Ga and the ($\bar{1}\bar{1}\bar{1}$) with a final layer of As.

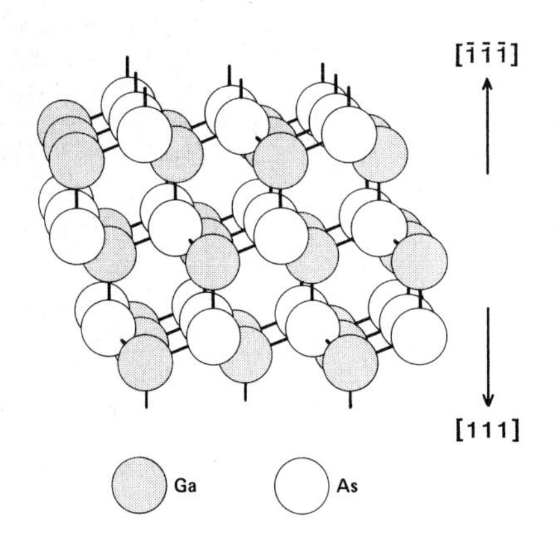

Fig.48.1. Ideal GaAs(111) and GaAs ($\bar{1}\bar{1}\bar{1}$) surfaces

The real surfaces both reconstruct, exhibiting in each case a (2 × 2) LEED pattern. Their structures, however, are expected to be different. A vacancy model (i.e., one in four missing Ga atoms) appears to be favorable for the GaAs(111) surface but not for GaAs($\bar{1}\bar{1}\bar{1}$) [48.1,2]. Having two surfaces with the same reconstruction symmetry makes an investigation of their electronic properties very useful and in this paper we present the results of an experimental comparison of GaAs(111)(2 × 2) and GaAs($\bar{1}\bar{1}\bar{1}$)(2 × 2).

We have carried out angle-resolved photoemission measurements on both surfaces and will compare photoemission spectra and also the surface state dispersions which we have obtained.

48.2 Experimental Details

The (2×2) reconstructed surfaces of GaAs($\bar{1}\bar{1}\bar{1}$) and (111) were grown in situ by molecular-beam epitaxy (MBE) from separate Ga and As effusion cells. Good (2×2) LEED patterns were also obtained for GaAs(111) surfaces produced by Ar^+ ion sputtering at 500 eV followed by annealing at 545°C. Spectra from the MBE-produced or sputter-annealed surface showed no significant differences, however, it should be noted that the MBE grown (111) surfaces were also annealed to 545°C following growth to obtain the (2×2) reconstruction.

The spectra were measured at the Stanford Synchrotron Radiation Laboratory with an experimental geometry which constrained the polarization vector of the light, the surface normal of the sample and the electron emission direction to lie in a plane. All of the data presented in this paper were collected for a fixed polarization with the photons making an angle of 45° with the sample normal. The effect of changes of the polarization direction on particular spectral features was determined and will be discussed below.

48.3 Results and Discussion

Spectra for the two surfaces are compared in Fig.48.2. With an emission angle of 35°, the spectra correspond approximately to the $\bar{K}_{1\times 1}$ point (corner) of the hexagonal surface Brillouin zone. The vertical lines in Fig.48.2 show the positions expected for bulk-derived features. These positions were determined from empirical pseudopotential calculations of bulk initial states and the assumption of k-conserving transitions to a (parabolic) final state whose zero energy was placed 7.75 eV below the top of the valence band. This method has been shown previously [48.3,4] to agree accurately between experiment and theory for the (100) surface of GaAs.

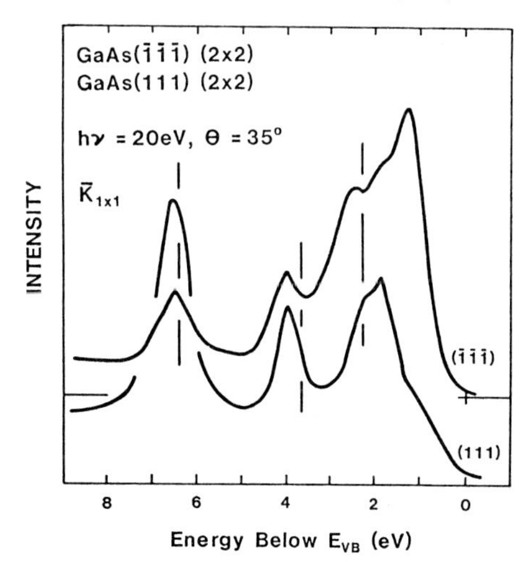

Fig.48.2. Angle-resolved photoemission spectra for GaAs(111)(2×2) (*lower curve*) and GaAs($\bar{1}\bar{1}\bar{1}$)(2×2) (*upper curve*) taken at $h\nu = 20$ eV. The spectra correspond approximately to the $\bar{K}_{1\times 1}$ point in the surface Brillouin zone

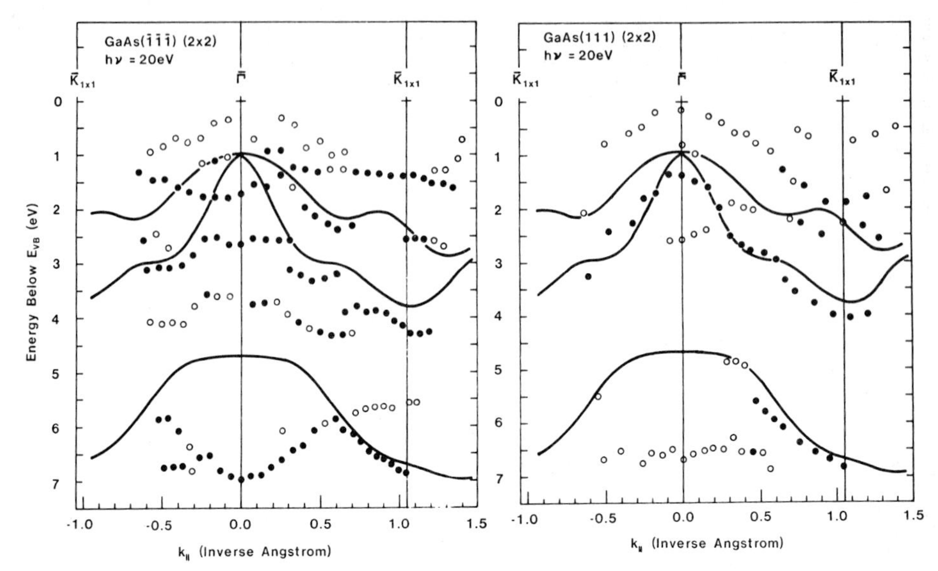

Fig.48.3. Positions in energy of strong (● ● ●) and weaker (○ ○ ○) spectral features are plotted as a function of $k_{\parallel}$ for a photon energy of 20 eV. Curves represent the positions expected for bulk initial states

Examination of the spectra in Fig.48.2 shows that there are more strong features than can be explained by bulk initial states alone and that the (111) and ($\bar{1}\bar{1}\bar{1}$) surfaces have significantly different electronic properties. Differences at other points in the surface zone occur not only in the region near the top of the valence band, but also in the region around 3 to 4 eV below the band edge. This can be seen in Fig.48.3 which collects together the data for $k_{\parallel}$ along the $\bar{\Gamma}$ to $\bar{K}$ direction. Positions of features in the spectra were located in energy and $k_{\parallel}$ and are indicated in Fig.48.3 for $h\nu = 20$ eV by filled circles (strong features) and open circles (weaker features). Also plotted are curves showing the positions expected for bulk-derived features. Plots such as this were used to determine which of the features in the spectra correspond to surface states.

Surface-related points derived from spectra taken at a variety of photon energies are plotted in Fig.48.4 for the $\bar{\Gamma}$ to $\bar{K}$ direction.

We will first examine the results for GaAs($\bar{1}\bar{1}\bar{1}$). Lines with a (2×2) symmetry have been drawn through the data points in Fig.48.4. There appear to be at least two, and possibly three, surface bands within 1.75 eV of the top of the valence band. The first of these disperses downwards from 0.3 eV at $\bar{\Gamma}$ to 0.75 eV at $\bar{K}$. A second disperses upwards from 1.75 eV at $\bar{\Gamma}$ towards the zone boundary. A third band dispersing down from 0.7 eV at $\bar{\Gamma}$ to 1.35 eV at $\bar{K}$ may be present but is not drawn in Fig.48.4. This band overlaps with the bulk features for small values of $k_{\parallel}$ and so can be seen best in the second of the (2×2) surface zones.

We will now turn to the surface features for GaAs(111)(2×2). In contrast to GaAs($\bar{1}\bar{1}\bar{1}$)(2×2), no strong surface bands were seen at binding energies greater than 3 eV below E_{VB}. The second difference is that the features appear

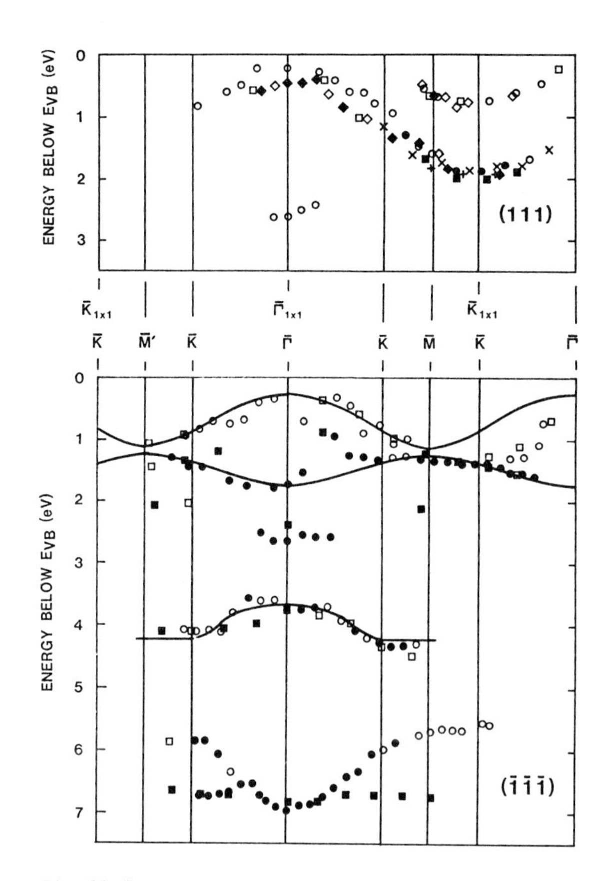

Fig.48.4. Positions in energy and $k_{\parallel}$ of surface-related features from the photoemission spectra for GaAs(111)(2 × 2) (*upper box*) and GaAs($\bar{1}\bar{1}\bar{1}$)(2 × 2) (*lower box*). Data taken at 17, 20, 22, 25 and 27 eV are shown by ♦,●, x, ■ and + symbols, respectively. Open symbols correspond to weaker features. Lines through the GaAs($\bar{1}\bar{1}\bar{1}$) data are repeated to show (2 × 2) symmetry

to have a (1 × 1) symmetry and not the (2 × 2) symmetry expected from the LEED pattern. It is possible, of course, that the surface bands do indeed have (2 × 2) symmetry, but that there is only a small photoemission probability for some portion of each band.

A buckling model for the (2 × 2) reconstructed GaAs($\bar{1}\bar{1}\bar{1}$) and GaAs(111) surfaces was introduced by *Haneman* [48.5] and was used to interpret earlier photoemission results for GaAs($\bar{1}\bar{1}\bar{1}$) [48.6]. *Jakobi* et al. [48.6] found surface states near 1.5 eV below the top of the valence band which they attributed to the occupied lone-pair orbital on the As atoms at the surface. *Harrison* [48.7] also discussed both surfaces and suggested that GaAs($\bar{1}\bar{1}\bar{1}$) (2 × 2) [GaAs(111)(2 × 2)] was stabilized by having 1/4 monolayer of Ga [As] adsorbed on top of the double layer.

Recently, a structural model for GaAs(111)(2 × 2) has been put forward [48.1] to explain the results of LEED experiments. This model proposes that one in four of the surface Ga atoms is missing and that the remaining surface atoms undergo an inward relaxation. Total energy calculations [48.2] have shown that a relaxed, vacancy surface is energetically favorable for GaAs(111)(2 × 2) but unfavorable for GaAs($\bar{1}\bar{1}\bar{1}$)(2 × 2).

Surface band-structure calculations have been carried out by *Nishida* [48.8] for (1 × 1) GaAs(111) and GaAs($\bar{1}\bar{1}\bar{1}$) surfaces with uniformly contracted [GaAs(111)] and expanded [GaAs($\bar{1}\bar{1}\bar{1}$)] surface atomic layers. *Nishida* found the p_z-like dangling bond band to be very close to the edge of the projected bulk bands for expanded GaAs($\bar{1}\bar{1}\bar{1}$) and significantly above the bulk valence band for relaxed GaAs(111). The uppermost occupied surface band in the latter case was a p_x band which was degenerate with the projected bulk bands. These findings are qualitatively consistent with the polarization dependence of our (2 × 2) data. The uppermost occupied surface states near the zone boundary showed an increase (decrease) in relative intensity as the polarization vector moved towards the surface normal for GaAs($\bar{1}\bar{1}\bar{1}$) (2 × 2) [GaAs(111)(2 × 2)].

In conclusion, the data in Figs.48.2-4 make it clear that the electronic structure of the two surfaces is remarkably different. The origin of the difference must lie in either a different surface geometry or in the effect of interchanging Ga and As atoms. It is likely that both of these are important. There is evidence that the depth of the reconstruction or the degree of bond alteration is greater in the ($\bar{1}\bar{1}\bar{1}$) case. It can be seen in Fig. 48.3 that the agreement between the bulk calculation and the data is much closer for the (111) surface, allowing a clearer separation between bulk and surface features in the spectra. The fact that the strong surface features in GaAs(111) have an apparent (1 × 1) repeat whereas the (2 × 2) symmetry is seen clearly for GaAs($\bar{1}\bar{1}\bar{1}$) may indicate that the (2 × 2) potential is weaker in the (111) case. Although the true dispersion is (2 × 2), the non-(1 × 1) dispersion may be too weak to be seen, especially when degenerate with the bulk bands.

Acknowledgments. We are grateful for useful discussions with D.J. Chadi and J. Northrup and for the skillful assistance of B.K. Krusor and L.E. Swartz. Part of this work was supported by NSF Grant DMR 81-08343, and was performed at SSRL which is supported by the Department of Energy, Office of Basic Energy Sciences; and the National Science Foundation, Division of Materials Research.

References

48.1 S.Y. Tong, G. Xu, W.N. Mei: Phys. Rev. Lett. **52**, 1693 (1984) and this Volume, p.303
48.2 D.J. Chadi: Phys. Rev. Lett. **52**, 1911 (1984)
48.3 R.D. Bringans, R.Z. Bachrach: Proc. Int. Soc. Opt. Engineering **447**, 58 (1983)
48.4 R.D. Bringans, R.Z. Bachrach: Proc. 17th Int. Conf. Phys. Semiconductors (to be published)
48.5 D. Haneman: Phys. Rev. **121**, 1063 (1961)
48.6 K. Jakobi, C. von Muschwitz, W. Ranke: Surf. Sci. **82**, 270 (1979)
48.7 W.A. Harrison: J. Vac. Sci. Technol. **16**, 1492 (1979)
48.8 M. Nishida: J. Phys. C**14**, 535 (1981)

49. X-Ray Diffraction from the (3×3) Reconstructed $(\bar{1}\bar{1}\bar{1})$B Surface of InSb

R.L. Johnson and J.H. Fock

Max-Planck-Institut für Festkörperforschung
Heisenbergstr. 1, D-7000 Stuttgart 80, FRG

I.K. Robinson

AT & T Bell Laboratories, Murray Hill, New Jersey 07974, USA

J. Bohr, R. Feidenhans'l, J. Als-Nielsen, M. Nielsen, and M. Toney

Risø National Laboratory, Risø, DK-4000 Roskilde, Denmark

X-ray diffraction measurements have been performed with synchrotron radiation under UHV conditions on the Sb rich $(\bar{1}\bar{1}\bar{1})$ surface of InSb. This InSb $(\bar{1}\bar{1}\bar{1})$ B surface has a (3×3) reconstruction [49.1]. The surface was prepared by argon-ion bombardment and annealing at 400°C and was characterized using LEED and high-resolution photoemission. Using X-rays incident at the critical angle for total reflection (0.31°) we have measured the intensities of fractional-order Bragg rods corresponding to the (3×3) reconstruction on the B surface.

49.1 Introduction

Standard X-ray crystallographic techniques can be applied to the study of surfaces. In general, scattering from the bulk of the crystal will mask the much weaker signal originating from the surface. If the X-rays are incident at the critical angle for total reflection, then the surface contribution is enhanced [49.2]. *Marra* et al. [49.3] were the first to exploit this technique to study Al-GaAs interfaces using a rotating anode source. They subsequently studied the Ge(001)-(2×1) surface [49.4] using both the rotating anode and synchrotron radiation (SR) and obtained two orders of magnitude more intensity with SR. This technique has the sensitivity and simplicity of interpretation necessary to perform crystallography on reconstructed surfaces and adsorbed monolayers.

At HASYLAB (Hamburger Synchrotronstrahlungslabor, DESY) on adjacent beamlines we have a high precision X-ray diffractometer and a high-resolution angle-resolved photoemission system (20-200 eV) with extensive sample preparation and multiple-technique analysis capabilities. By use of a transfer cell (Fig.49.1) we can take full advantage of both facilities. The sample is prepared and characterized using LEED and photoemission. It is then transferred under UHV conditions (typically 2×10^{-10}mbar) to a small cell fitted with a 0.5 mm thick cylindrical Be window which allows 360° access to X-rays. Valves on the transfer cell and on the photoemission system allow samples to be transferred back and forth easily. After removing the transfer cell from the photoemission system, it is mounted on the goniometer head of the diffractometer for the X-ray measurements.

We chose to study InSb as a prototype zinc-blende structure III-V compound semiconductor with structural properties similar to GaAs. The heavy atoms have large scattering amplitudes, which enable more accurate structure determinations. The InSb$(\bar{1}\bar{1}\bar{1})$B surface (Sb rich) has a (3×3) surface

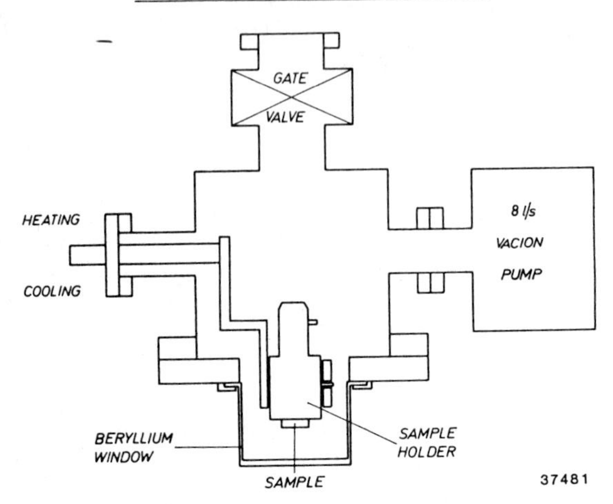

Fig.49.1. Schema of the X-ray transfer cell. The beryllium window allows 360° access for X-rays to the sample surface. The total weight of the cell is about 15 kg

reconstruction, whereas the InSb(111)A surface (In rich) has a (2 × 2) reconstruction [49.1]. High-quality crystals were available from the Max Planck Institut in Stuttgart.

49.2 Experimental

Accurately oriented, cut and polished samples were cleaned by repeated cycles of sputtering with 500 eV Ar^+ ions and annealing at ~400°C. Valence band energy distribution curves (EDCs) were recorded at $\hbar\omega = 30$ eV to determine surface cleanliness. The quality of the surface reconstruction was checked by observing LEED patterns at different points on the surface. It was found that the valence band EDCs were much more sensitive to contamination than either XPS or LEED measurements. Once a sample was clean and exhibited a reconstruction with sharp fractional order LEED spots, it was transferred to the X-ray cell and mounted on the goniometer. The pressure in the cell was monitored using the the ion-pump current and was below 1×10^{-9}mbar at all times.

The geometry used in our X-ray diffraction experiments is shown in Fig. 49.2. The nature of SR provides good (~0.005°) collimation in the vertical direction. The monochromator arrangement, set for $\lambda = 1.5$ Å, gave a similar collimation in the horizontal direction. Thus, we can choose to place the diffraction plane horizontal and achieve high resolution in both definition of incidence angle and in-plane diffraction. The position of the total reflected beam was used to check the horizontal alignment of the sample optical surface as it was rotated through 360°.

The X-ray intensities were measured with a position-sensitive detector, placed horizontally to resolve a 4° range of scattering angles simultaneously. Data were collected for 50 fractional-order in-plane surface reflections. The maximum count rate was ~180 c/min obtained at the (1/3,0) surface reflection. For several reflections the intensity along the rod was measured to check the two-dimensional character. Symmetry equivalents were used to

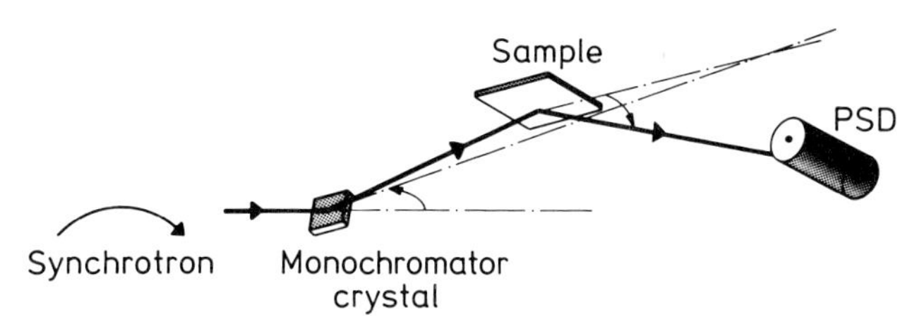

Fig.49.2. Schema of the experimental setup. The white synchrotron radiation is monochromatized by a vertical Ge(111) crystal. The sample surface is precisely horizontal and the grazing angle of incidence is controlled by tilting the monochromator. The position-sensitive detector (PSD) lies in the horizontal scattering plane. Harmonics are filtered away by a flat, gold-plated glass mirror in front of the monochromator

derive intensity error bars. The intensities were also monitored as a function of time to correct for deterioration of the surface.

49.3 Analysis

The structure factor amplitudes $|F_{\mathbf{G}}|$ derived from measured intensities were applied to a calculation of the Patterson function [49.5] (Fig.49.3a),

$$P(\mathbf{R}) = \sum_{\mathbf{G}} |F_{\mathbf{G}}|^2 \cos(\mathbf{R} \cdot \mathbf{G}) \quad ,$$

where $\mathbf{G}$ are the reciprocal lattice vectors, and $\mathbf{R}$ is the position in the surface plane. Interpreting $P(\mathbf{R})$ as the pair correlation of the atomic positions, we find strong peaks (I and II in Fig.49.3a) for interatomic vectors somewhat shorter than in the bulk, one parallel to the unit cell edge and the other at about 30° to it. A rearrangement of a layer of nine

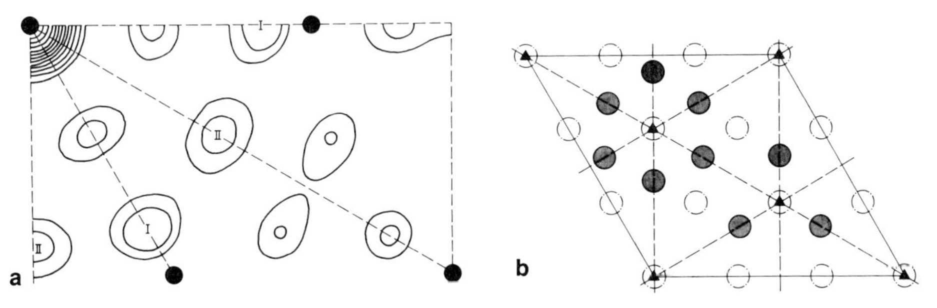

Fig.49.3. (**a**) Contour plot of the Patterson function $P(\mathbf{R})$. Positive, equally spaced contours are shown. Mirror lines in the 2D map are indicated by dashed lines. As a guide to the scale, dots mark lattice vectors of a bulk InSb unit cell. (**b**) Preliminary model of the atomic arrangement in the InSb $(\bar{1}\bar{1}\bar{1})$B (3×3) reconstructed surface. We were unable to distinguish between In and Sb atoms. Nine atoms (shaded circles), constrained by the 3m symmetry shown, have three in-plane positional parameters that have been refined. For reference, an unreconstructed second layer of atoms (broken open circles) is shown. The model is the best one obtained so far, but is only partially in agreement with the data

surface atoms in the 3×3 unit cell that is consistent with this and explains the remaining peaks in the Patterson function is shown in Fig.49.3b. Least-squares refinement of these atomic positions subject to the 3m mirror and 3-fold symmetry imposed, agreed only at the two-standard-deviation level. It is likely that a model with a different number of atoms and/or displacements in the second layer is needed to account for the discrepancy. Work is presently in progress to improve the quantity and quality of the data and to investigate other models.

References

49.1 J.T. Grant, T.W. Hass: J. Vac. Sci. Tech. **8**, 94 (1971)
49.2 G.H. Vineyard: Phys. Rev. B**26**, 4146 (1982)
49.3 W.C. Marra, P. Eisenberger, A.Y. Cho: J. Appl. Phys. **50**, 6927 (1979)
49.4 P. Eisenberger, W.C. Marra: Phys. Rev. Lett. **46**, 1081 (1981)
49.5 I.K. Robinson: Phys. Rev. Lett. **50**, 1145 (1983);
B.E. Warren: X-ray Diffraction (Addison-Wesley, Reading, MA 1969)

V.3 Adsorbate-Covered Semiconductors

50. Atomic and Electronic Structure of p(1×1) Overlayers of Sb on the (110) Surfaces of III–V Semiconductors

C.B. Duke, C. Mailhiot, and A. Paton

Xerox Webster Research Center, 800 Phillips Road-114, Webster, NY 14580, USA

K. Li, C. Bonapace, and A. Kahn

Department of Electrical Engineering and Computer Science
Princeton University, Princeton, NJ 08544, USA

The prediction of the atomic geometries of overlayers for compound semiconductors is a topic of considerable current interest, especially with regard to the mechanisms of Schottky barrier formation and the growth (e.g., by molecular-beam epitaxy) of multilayer heterojunction systems [50.1]. Moreover, such geometries are now being determined experimentally, for example by elastic low-energy electron diffraction (ELEED) intensity analyses [50.2,3]. Thus, an opportunity exists to develop and test predictive models of the geometrical and electronic structure of ordered overlayers on semiconductor surfaces. In this contribution we present a tight-binding calculation of the atomic geometries and surface-state eigenvalue spectra of p(1 × 1) overlayers of Sb on the (110) surfaces of III-V semiconductors which predicts accurately the measured structures and surface-state spectra of these surfaces. The particular systems which we examined are ordered (1 × 1) saturated monolayers of Sb on the (110) surfaces of GaP, GaAs, GaSb, InP, InAs and InSb. A schematic diagram of the experimental surface atomic geometry is given in Fig.50.1.

We determined the atomic geometries via a total-energy minimization method [50.4] and calculated the associated surface-state eigenvalue spectra using standard scattering theory [50.5]. Surface-bound states and reson-

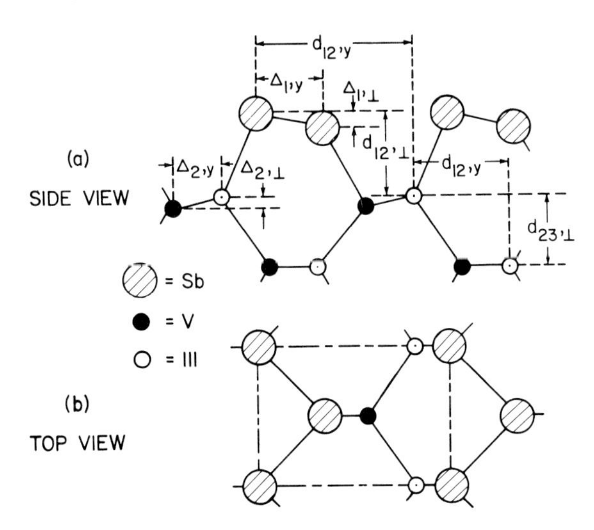

Fig.50.1a,b. Schema of the surface geometry for III-V (110)-Sb(1ML) and definition of the structural parameters. Panel (**a**): Side view. Panel (**b**): Top view [50.9]

ances are identified via an effective Hamiltonian method [50.6]. Both the total-energy minimization and the calculation of the surface-state eigenvalue spectrum were performed using the nearest-neighbor $sp^3 s^*$ empirial tight-binding model of *Vogl* et al. [50.7]. A detailed description of the model is given elsewhere [50.8].

The atomic geometries of the Sb overlayer systems are evaluated by following the procedures described by *Chadi* [50.4]. A starting structure, chosen from the ELEED structure analysis for Sb on GaAs [50.9], is utilized to initiate an automated search for the local minimum-energy surface structure. The minimum energy structures for all six overlayer systems are of the type suggested by the ELEED analysis for GaAs [50.9]. Table 50.1 gives the predicted structural parameters, defined in Fig.50.1, that characterize the atomic structure of ordered Sb monolayers on GaP, GaAs, GaSb, InP, InAs, and InSb. An important aspect of our model is its quantitative description of the ELEED structure for GaAs(110)-p(1 × 1)-Sb(1ML) reported earlier [50.9] and its successful prediction of the preliminary results [50.10] of a new ELEED structure analysis of InP(110)-p(1 × 1)-Sb(1ML). Moreover, the model predictions have proven unusually useful in the latter structure analysis because several nearly equivalent minima occur in the R-factor as a function of the structural parameters.

Table 50.1. Structural parameters predicted for the atomic geometries of III-V(110)-Sb(1ML) systems defined in Fig.50.1. The values in parentheses were determined via ELEED for GaAs(110)-Sb(1ML) [50.9] and for InP(110)-Sb(1ML) [50.10]. Units are Å and a_0 is the bulk lattice constant of the substrate

	GaP	GaAs	GaSb	InP	InAs	InSb
a_0	5.451	5.654	6.118	5.869	6.036	6.478
$\Delta_{1\perp}$	0.12	0.09 (0.10)	0.07	0.27 (0.30)	0.22	0.13
$\Delta_{1,y}$	1.91	1.87 (1.96)	1.80	1.79	1.76	1.68
$d_{12,\perp}$	2.30	2.29 (2.39)	2.25	2.22	2.23	2.21
$d_{12,y}$	4.22	4.39 (4.62)	4.73	4.16	4.34	4.74
$\Delta_{2,\perp}$	0.05	0.08 (0.10)	0.18	0.19	0.20	0.20
$\Delta_{2,y}$	1.35	1.42 (1.41)	1.57	1.48	1.54	1.68

The electronic structure of the minimum-energy p(1 × 1) Sb chains on the III-V (110) surfaces is indicated in Fig.50.2. The eigenvalue spectrum of a single isolated (p^2 bonding) Sb chain is presented in panel a. The consequences of the polarity of the III-V substrate are illustrated in panel b for GaAs(110). The lowest-energy Sb-derived surface states, labeled $S^+ = S_1$ and $S^- = S_2$ in Fig.50.2, are nonbonding s-orbital bands of states. The next two bands, designated $p_x^- = S_3$ and $p_y^- = S_4$, are the intrachain bonding p^2 orbitals within the Sb chains. Bonding of these chains to the substrate occurs via the hybrid states S_5 and S_6 which consist of bonding linear combinations

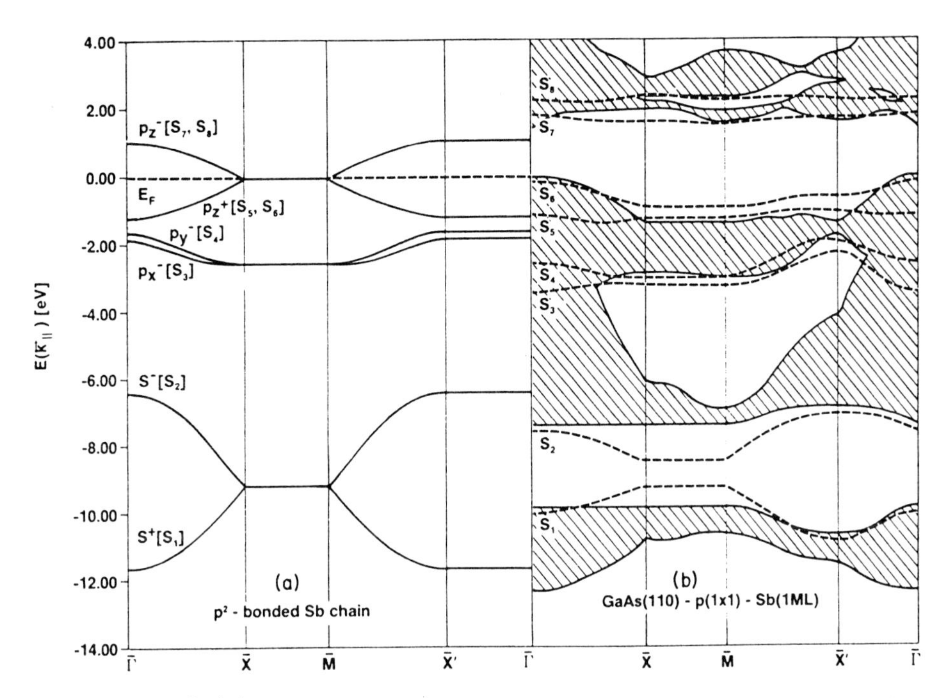

Fig.50.2.Panel (a): Energy dispersion relation for single isolated Sb chain mapped on the (110) surface Brillouin zone. Panel (b): Results for GaAs(110)-Sb(1ML). Only Sb-derived states are shown

of π (i.e., p_z) orbitals along the Sb chain with sp^3 dangling-bond orbitals of the (nearly unrelaxed) GaAs(110) substrate. These hybrid bonds are weaker than the bulk sp^3 bonds and are caused by the constraints imposed by the epitaxial growth of the large Sb species on the GaAs (110) surface. Therefore this hybrid Sb-substrate bonding is a uniquely surface phenomenon which has no analog in either bulk materials or small molecules.

In summary, we have predicted the atomic and electronic structure of ordered $p(1\times1)$ saturated monolayers of Sb on the (110) surfaces of GaP, GaAs, GaSb, InP, InAs and InSb. The atomic structures determined via total-energy minimization procedures agree with those obtained from existing ELEED data analyses for GaAs(110)-$p(1\times1)$-Sb(1ML) and InP(110)-$p(1\times1)$-Sb(1ML). The surface-state eigenvalue spectrum determined via a scattering theoretical methodology is similar to that obtained from self-consistent pseudopotential calculations [50.11] for GaAs(110)-Sb(1ML), although we have derived [50.12] a much more complete description of the Sb-substrate bonding than that available previously. In particular, our analysis reveals a novel type of hybrid Sb-substrate bonding which resolves a puzzle identified in the ELEED structure analysis of GaAs(110)-$p(1\times1)$-Sb(1ML).

References

50.1 L.J. Brillson: Surf. Sci. Repts. **2**, 123 (1982)
50.2 A. Kahn: Surf. Sci. Repts. **3**, 193 (1983)
50.3 C.B. Duke: Adv. Ceram. **6**, 1 (1983)
50.4 D.J. Chadi: Phys. Rev. Lett. **41**, 1062 (1978); Phys. Rev. B**19**, 2074 (1979)
50.5 R.E. Allen: Phys. Rev. B**19**, 917 (1979); B**20**, 1415 (1979)
50.6 R.B. Beres, R.E. Allen, J.D. Dow: Phys. Rev. B**26**, 769 (1982)
50.7 P. Vogl, H.P. Hjalmarson, J.D. Dow: J. Phys. Chem. Solids **44**, 365 (1983)
50.8 C. Mailhiot, C.B. Buke, D.J. Chadi: Surf. Sci. (submitted)
50.9 C.B. Duke, A. Paton, W.K. Ford, A. Kahn, J. Carelli: Phys. Rev. B**26**, 803 (1982)
50.10 C.B. Duke, C. Mailhiot, A. Paton, A. Kahn: Unpublished
50.11 C.M. Bertoni, C. Calandra, F. Manghi, E. Molinari: Phys. Rev. B**27**, 1251 (1983)
50.12 C. Mailhiot, C.B. Duke, D.J. Chadi: Phys. Rev. Lett. (submitted)

51. Models for Si(111) Surface upon Ge Adsorption

S.B. Zhan, John E. Northrup*, and Marvin L. Cohen

Department of Physics, University of California and Materials and Molecular Research Division, Lawrence Berkeley Laboratory Berkeley, CA 94720, USA

*Present address: Xerox PARC, 3333 Coyote Hill Road, Palo Alto CA 94304, USA

51.1 Introduction

In view of the chemical similarity between Ge and Si one would expect the surface structures observed following deposition of Ge onto the Si(111) substrate to be of complexity comparable to that of the clean Si(111)7 × 7 surface. It is therefore remarkable that a simpler reconstruction has been observed by *Chen* et al. [51.1] for room temperature deposition of Ge onto the Si(111)2 × 1 surface. They observed a $\sqrt{3} \times \sqrt{3}$ pattern corresponding to 0.2-0.4 monolayers of Ge. From the symmetry, coverage, and the fact that Ge-Si intermixing does not occur at this temperature, *Chen* et al. proposed an adatom model for the structure. Motivated by these experiments, we have carried out total energy calculations for two types of $\sqrt{3} \times \sqrt{3}$ adatom models. The results of these calculations are discussed in Sect.51.2. Ultimately, we hope to provide a definitive characterization of the Si(111)$\sqrt{3} \times \sqrt{3}$-Ge surface, since this will provide insight into the more complicated 5 × 5 and 7 × 7 reconstructions which have been observed on Si and SiGe alloys [51.2].

This paper also addresses the question of what structure occurs for one monolayer Ge deposited on the Si(111) surface. If adatom models appear to be more likely for coverages near 1/3 monolayer [51.1], and strained bulk-like double layers form at high coverage (>2 monolayers) [51.2,3], no such obvious structures exist for one monolayer coverage. We discuss possible one-monolayer structures and their energies in Sect.51.3. These results have also been discussed in an earlier publication [51.4].

The total energy is calculated using the first-principles pseudopotential method [51.5-7] within the local density functional formalism [51.8]. A repeated slab geometry [51.9] is used to model an isolated surface. The Kohn-Sham equations are solved self-consistently using a plane wave basis. Norm conserving pseudopotentials [51.10] are used to simulate the electron-ion interaction. The exchange and correlation energy is treated in the local approximation with the Ceperly and Alder functional form as parameterized by *Perdew* and *Zunger* [51.11].

51.2 Models for 1/3 Monolayer Coverage

For coverages starting around 0.2-0.4 monolayers *Chen* et al. [51.1] observed a $\sqrt{3} \times \sqrt{3}$ LEED pattern following deposition onto the Si(111)2 × 1 surface. This LEED pattern was not observed following deposition onto the Si(111)7 × 7 surface [51.1]. This suggests that the $\sqrt{3}$ structure is not in fact the most stable Ge-induced reconstruction. To explain the $\sqrt{3}$ reconstruction *Chen* et al. proposed an adatom model in which each Ge atom occupies a threefold coordinated hollow site and saturates three Si dangling bonds. Such a model

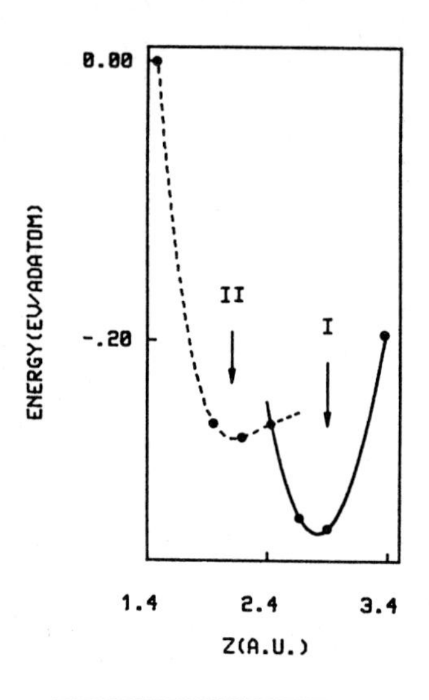

Fig.51.1. Total energy with respect to the Ge adatom position z measured from the surface-layer Si atom. Two curves are fitted separately to illustrate the two local minima

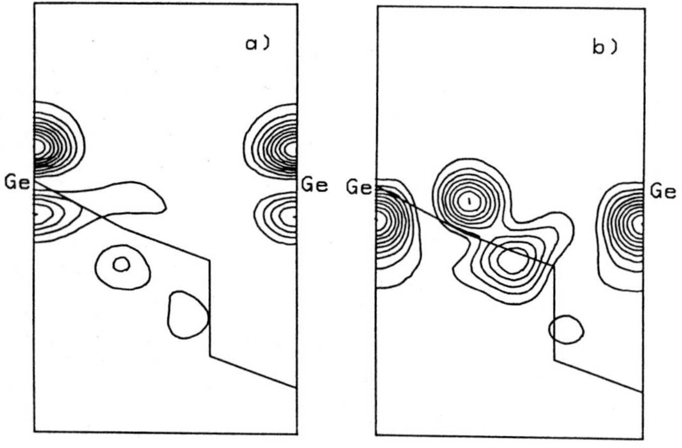

Fig.51.2a,b. Surface states in the gap of Si, (**a**) is the dangling bond, and (**b**) is the antibonding state at $\bar{M}$ point

reduces the surface dangling bond density by a factor of three, and it is this reduction in the number of broken bonds which is the physical motivation for the model. Using the supercell method to study this reconstruction is difficult technically because there are 26 atoms plus a large vacuum region in each unit cell. We used a plane wave cutoff of 6 Ry in the expansion of the wave functions, and two high-symmetry k points in the Billouin zone summations. Although substrate relaxation has not been included in the present analysis, we do not expect the results presented below to be altered significantly by its inclusion.

In Fig.51.1, we show the surface energy of the adatom model as a function of the adatom position z, above the Si(111) surface. Seven different positions were tested. There are two local minima, denoted I and II, in the energy curve. The deeper minimum occurs at the larger outward displacement. The arrows on the plot indicate the two minima. For minimum I the Ge-Si bond length is ~5.1 a.u. and for II it is ~4.7 a.u. We note that the sum of Pauling single bond covalent radii for Ge and Si is 4.53 a.u. Thus, the minimum

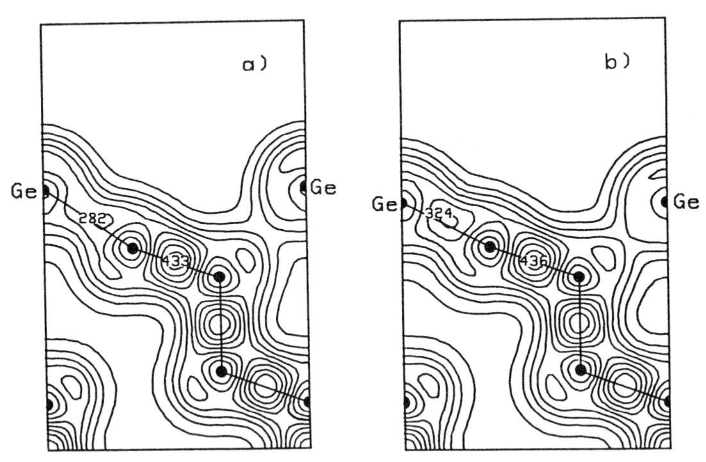

Fig.51.3a,b. Total charge density contour plots in the $(1\bar{1}0)$ plane. Atom positions are indicated by (•), (**a**) minimum I, (**b**) minimum II

energy bond length is ~0.6 a.u. larger than expected for strong covalent bonding.

The structure in the total energy curve arises from an avoided crossing of two bands of surface states which exist near the Fermi level. These states exist in the gap of Si, and the lowest energy band is half-filled. The characters of these two bands, for geometry I, is shown in Fig.51.2. For this geometry the lower energy band has dangling bond character as shown in Fig. 51.2a. The higher energy band has the character of an antibonding, conduction-band-derived surface state (Fig.51.2b). Its character is reminiscent of the surface state at the Γ point of the Ge(111) ideal topology surface [51.12]. For the inner minimum, geometry II, the characters of the upper and lower bands are reversed. Note that the energy curve (Fig.51.1) is composed of two parabola shaped curves with different origins. The outer parabola would correspond to occupation of the dangling bond state over the full range of adatom position. The inner parabola would correspond to occupation of the antibonding state. The character of the occupied band changes near the point where the total energy curve is at its local maximum. Thus the anharmonicity in the total energy is attributable to the avoided crossing.

The charge densities corresponding to geometries I and II are shown in Fig.51.3. They are noticeably different from each other. For geometry I, the charge density is reduced in the Ge-Si bonding region and increased above the Ge atom relative to the density corresponding to geometry II. The differences in the charge density reflect the different characters of the highest occupied surface states for the two geometries.

We have also considered another possible adatom geometry. Namely, we considered the possibility that the adatom is positioned in the threefold symmetric site above the second layer Si atoms. In this position the Ge is 4-fold coordinated. It is bonded to three Si surface atoms and is approximately one bond length above the second layer atom. Preliminary results indicate that this geometry is lower in energy than the hollow site adatom geometry. However, substrate relaxations for both geometries need to be included to obtain definitive results.

51.3 Models for One Monolayer Ge Adsorbed on the Si(111) Surface

In this section we discuss total energy calculations for several possible models appropriate for one monolayer of Ge bonded to the Si(111) surface. Previously considered structures include the 1×1 atop and 1×1 hollow site models [51.13-15], Fig.51.4. With our self-consistent total energy calculations, we found that the 1×1 structures are not stable. Relaxing the symmetry constraint from 1×1 to 2×1 and allowing the Ge atoms to bond to each other in Seiwatz chains [51.16] reduces the energy dramatically. Thus we think ordered 1×1 Ge covered surfaces are not likely. It is not difficult to understand why the 1×1 atop and hollow site models are energetically unfavorable. In the 1×1 atop model each Ge is bonded to a single Si atom below. Consequently the bonding potential of the Ge is unsaturated. In the 1×1 hollow site model the Si surface atoms become sixfold coordinated; and consequently sp^3 bonding, which stabilizes the bulk, is no longer effective.

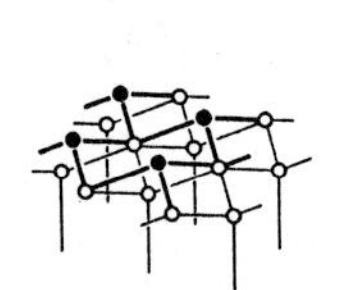

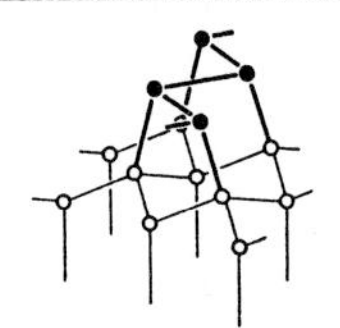

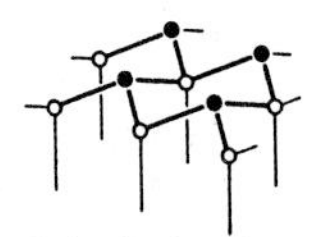

Fig.51.4. Energies (relative to the Seiwatz chain model) of four different models. (○) are Si atoms and (●) are Ge adatoms. Heavy lines are either Ge-Ge or Ge-Si bonds [51.4]

Starting from the 1×1 atop model, the 2×1 Seiwatz chain can be formed by moving the Ge atoms in adjacent rows towards each other until chains of bonds are formed between the Ge atoms. This bond formation reduces the energy by 0.7 eV/(surface atom) relative to the 1×1 atop model. Starting from the 1×1 atop model, there are many different reconstructions which can be obtained by bending the Ge-Si bond and making new bonds between Ge atoms. For example, a $\sqrt{3}\times\sqrt{3}$ structure which consists of triplets of Ge atoms bonded to each other can be formed. This model appears to be appropriate for Ge(111) $\sqrt{3}\times\sqrt{3}$-Sn [51.17]. Although we have not calculated the total energy, we expect the $\sqrt{3}\times\sqrt{3}$ triplet model to be close to the Seiwatz chain because the structures contain equal numbers of Ge-Ge and Ge-Si bonds. It is possible that in the one monolayer regime of coverage, the surface structure is similar to the triplet or Seiwatz chain models.

The discussion presented above can be valid only if the deposition temperature is kept low enough to prevent migration and intermixing of Ge and Si. If the temperature is high enough, then the Ge can occupy substitutional sites in the Si substrate and long-range migration leading to the equilibrium 7×7 surface can occur. To illustrate the possible importance of substitutions, we note that calculations show that the substitutional model, Fig.51.4, is lower in energy than the Seiwatz chain by ~0.3 eV/(surface atom). It is possible, then, that other structures which differ from the equilibrium Si(111) 7 × 7 structure by ordered substitution of Ge for Si are the most stable.

Acknowledgments. Support for this work was provided by NSF Grant No. DMR8319024 and by the Director, Office of Energy Research, Office of Basic Energy Sciences, Materials Sciences Division of the U.S. Department of Energy under Contract No. DE-AC03-76SF00098.

References

51.1 P. Chen, D. Bolmont, C.A. Sebenne: Solid State Commun. **44**, 1191 (1982); **46**, 689 (1983)
51.2 H.-J. Gossmann, J.C. Bean, L.C. Feldman, W.M. Gibson: Surf. Sci. **138**, 175 (1984)
51.3 T. Narusawa, W.M. Gibson: Phys. Rev. Lett. **47**, 1459 (1981)
51.4 S.B. Zhang, J.E. Northrup, M.L. Cohen: Surf. Sci. **144** (in press)
51.5 J. Ihm, A. Zunger, M.L. Cohen: J. Phys. C**12**, 4409 (1979)
51.6 M.T. Yin, M.L. Cohen: Phys. Rev. B**25**, 7403 (1982)
51.7 M.L. Cohen: Physica Scripta T**1**, 5 (1982) and references therein
51.8 W. Kohn, L.J. Sham: Phys. Rev. **140**A, 1133 (1965)
51.9 M. Schlüter, J.R. Chelikowsky, S.G. Louie, M.L. Cohen: Phys. Rev. B**12**, 4200 (1975)
51.10 D.R. Hamann, M. Schlüter, C. Chiang: Phys. Rev. Lett. **43**, 1494 (1979)
51.11 J.P. Perdew, A. Zunger: Phys. Rev. B**23**, 5048 (1981)
51.12 J.R. Chelikowsky: Phys. Rev. B**15**, 3236 (1977)
51.13 S. Nannarone, F. Patella, P. Perfetti, C. Quaresima, A. Savoia, C.M. Bertoni, C. Calandra, F. Manghi: Solid State Commun. **34**, 409 (1980)
51.14 P. Perfetti, S. Nannarone, P. Patella, C. Quaresima, F. Cerrina, M. Capozi, A. Savoia, I. Lindau: J. Vac. Sci. Technol. **19**, 319 (1981)
51.15 P. Perfetti, N.G. Stoffel, A.D. Katnani, G. Margaritondo, C. Quaresima, F. Patella, A. Sovoia, C.M. Bertoni, C. Calandra, F. Manghi: Phys. Rev. B**24**, 6174 (1981)
51.16 R. Seiwatz: Surf. Sci. **2**, 473 (1964)
51.17 H. Sakurai, K. Higashiyama, S. Kono, T. Sagawa: Surf. Sci. **134**, L550 (1983)

52. The Graphite (0001)-(2 × 2) K Surface Intercalated Structure

N.J. Wu

Institute of Physics, Chinese Academy of Sciences, Beijing, China

A. Ignatiev

Departments of Physics and Chemistry, University of Houston
Houston, TX 77004, USA

Further absorption of potassium (K) onto the AKAKABA...stacked surface of graphite results in the formation of a (2 × 2) superstructure. A test of seven different surface-structure models with dynamical LEED calculations indicated that an A-(2 × 2)Kα-A-(2 × 2)Kβ-AB...stacked model results in intensities in good agreement with the measured diffracted intensities. Such a model is consistent with structures observed in bulk graphite for intercalated potassium.

52.1 Introduction

The widespread interest in various adsorbed and absorbed layers onto graphite is sustained by a continual unfolding of novel predictions and observations in this system. A number of studies have been reported on physical adsorption, chemical absorption and intercalation of various atomic and molecular species with graphite [52.1-5]. Previously, our low-energy electron diffraction (LEED) investigations of the potassium/graphite system indicated that evaporation of potassium onto a graphite sample provides several surface phases [52.6,7] and, in particular, that potassium intercalates in a high-defect-density graphite surface in a layer-by-layer fashion [52.8]. The present work reports the formation and structural properties of a (2 × 2) potassium structure on graphite which results from further absorption of potassium onto a surface-intercalated graphite sample. These studies are important for studying alkali metal-graphite interactions and the eventual understanding of the rich variety of intercalate stages.

52.2 Experimental

The experiments were carried out in an ultrahigh vacuum system equipped with a four-grid LEED apparatus and quadrupole mass spectrometer. A graphite single-crystal platelet cleaved in air with scotch tape was attached onto a universal manipulator by an O-ring molybdenum foil to expose the c face (basal plane) of graphite. A pure potassium molecular-beam source shielded from the graphite by a shutter was installed in the vacuum chamber approximately 7 cm from the sample.

The surface contamination of the graphite crystal was checked by retarding field Auger electron spectroscopy (AES) before and after baking in the chamber. The diffraction beam intensities were measured via the spot photometer-fluorescent screen method at different potassium coverages. The measured LEED I-V curves obtained at normal incidence were compared with those obtained from dynamical calculations to determine potassium-graphite structures.

52.3 Results and Discussion

52.3.1 C(0001)-(2 × 2)K Formation

Evaporation of potassium onto the graphite (0001) surface results in several surface structures which depend on the initial condition of the substrate, the temperature of the substrate and of the potassium source [52.6,7]. Graphite is highly anisotropic and characterized by a layered structure with a hcp lattice (ABAB... stacking). The single layer has a honeycomb structure with two carbon atoms per unit cell. Potassium from a hot source ($T > 550$ K) deposited onto a graphite sample with large surface step densities (>1%) results in diffusion of potassium into the graphite sample. Intercalation proceeds in a layer-by-layer manner with no superstructure observed for the potassium through the first carbon layer and second carbon layer intercalation states. The absorbed potassium is therefore disordered in the near-surface region. Surface structural determinations have shown that A-A-A stacking exists for the top carbon layers with disordered potassium between the first and second carbon layers, whose spacing has been expanded to a 5.35 Å separation; thus one finds an AKAKAB... stacking [52.7,8].

Further evaporation of potassium onto this intercalated surface results in the formation of a (2 × 2) superstructure in the range of potassium exposure from 2.4 to 3.6 monolayer equivalents. Typical experimental conditions were: evaporation of potassium for about 10 seconds at a rate of approximately 2×10^{14} potassium atoms per min with background pressure in the chamber at less than 3.5×10^{-10} Torr. The temperature of the potassium source was ~500 K and the substrate temperature was varied from 80 K to 300 K.

52.3.2 LEED Analysis

Figures 52.1 and 2 show the measured LEED I-V spectra for the (10), (0 1/2) (1/2 1/2) beams of the C(0001)-(2 × 2)K structure at normal incidence. Data were obtained only for the three noted beams at normal incidence and as a result the reliability of the LEED structural determination may suffer somewhat. The good agreement obtained, however, indicates a high degree of confidence in the final structural model. In determining this surface structure, we have evaluated seven different model geometries. The simplest model consistent with the LEED pattern is that of potassium atoms in every second graphite honeycomb hollow on the top surface of graphite, i.e. (2 × 2)K-ABA... . A similar model places the potassium on top of the intercalated graphite surface, i.e. (2 × 2)K-AKAKAB... . In the dynamical calculations for these models the distances between the top two neighboring carbon layers, d_{AB} and d_{AA} respectively, were varied from 3.0 Å to 5.6 Å. The top ordered potassium-carbon layer spacing d_{12} was varied from 2.1 Å to 4.1 Å, with the two K underlayers retained as disordered potassium, as discussed previously [52.7,8]. The calculated I-V curves for $d_{12} = 2.6$ Å, $d_{AB} = 3.35$ Å and $d_{AA} = 5.35$ Å, which gave the best agreement for the two models, are shown in Fig.52.1: agreement with experiment is very poor and thus the two models are unacceptable.

Considering that the graphite substrate already has potassium atoms intercalated into the first and/or second graphite layers, the (2 × 2) structure observed may be due to potassium atoms ordering between carbon layers. We considered the possibility of such an ordered underlayer superstructure in the A-(2 × 2)K-AA... and A-(2 × 2)K-ABA... stacking models.

As in all known cases of bulk graphite intercalates, intercalation changes the stacking sequence in graphite and dilates the structure along the c axis to accommodate the intercalant. Dynamical LEED calculations were undertaken for these models with the interlayer spacing between the first two neighbor-

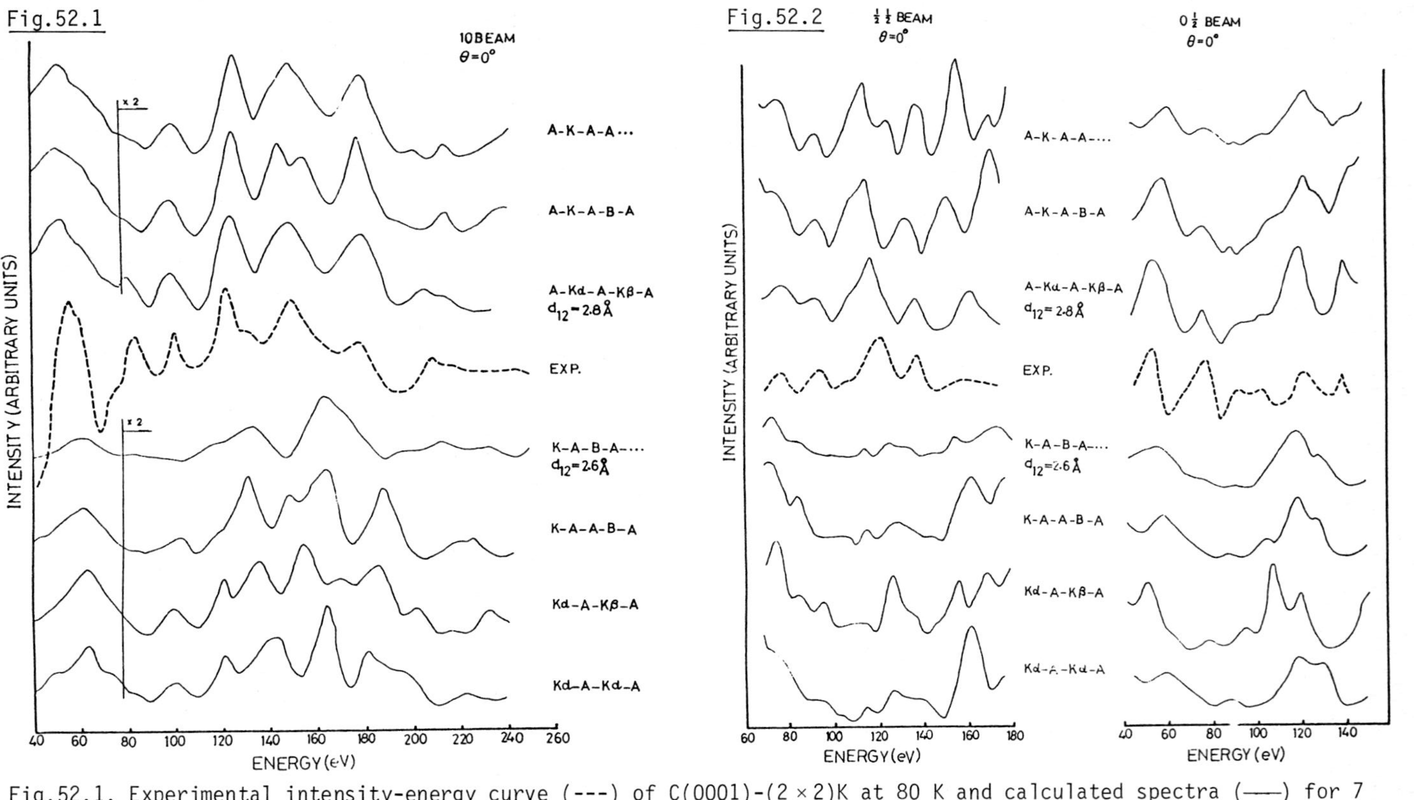

Fig.52.1. Experimental intensity-energy curve (---) of C(0001)-(2 × 2)K at 80 K and calculated spectra (——) for 7 structure models for the (10) beam at normal incidence. A and B indicate the graphite layers; K indicates potassium in (2 × 2) ordering; d_{12} denotes the distance between the first two top layers, and α and β denote the different positions occupied by the potassium atoms in the intercalate layers (see Fig.52.3)

Fig.52.2. Experimental intensity-energy spectra (---) of C(0001)-(2 × 2)K at 80 K and calculated spectra (——) for 7 structure models for the (1/2, 1/2) and (0, 1/2) beams at normal incidence. A, B, K, d_{12}, α and β are as in Fig.52.1

ing carbon layers (d_{AB} and d_{AA}, respectively) varied from 3.3 Å to 5.65 Å and the subsequent BABA... stacked layers kept at the bulk graphite separation of 3.35 Å. The d_{12} spacing between the top carbon layer and the potassium underlayer was varied from 1.9 Å to 3.3 Å. Figures 5.1,2 also show the calculated I-V curves for these two models: A-(2 × 2)K-AA... and A-(2 × 2) K-ABA... with d_{12} = 2.8 Å, d_{AA} = d_{AB} = 5.35 Å. It is obvious that the I-V spectra of the potassium underlayer models are much closer to the experimental ones than those of the potassium top layer models; however, the agreement is still poor.

A reasonable conjecture based on the crystal structure of bulk C_8K is that the potassium atom intercalant forms multilayer ordered structures [52.9,10]. The (2 × 2)Kα-A-(2 × 2)Kα-AB... , (2 × 2)Kα-A-(2 × 2)Kβ-AB... and A-(2 × 2)Kα-A-(2 × 2)Kβ-AB... stacking models based on this notion have been tested. In these structures α and β define the alternative superlattice origins as shown in Fig.52.3. The LEED I-V curves corresponding to the three structure models are shown in Figs.52.1,2. Theoretical curves for the models where top layer and interlayer potassium ordering was assumed [(2 × 2)Kα-A-(2 × 2)Kα-AB... and (2 × 2)Kα-A-(2 × 2)Kβ-AB...] again give poor agreement with experiment. Comparison between the experimental spectra and the model with interlayer potassium only [A-(2 × 2)Kα-A-(2 × 2)Kβ-AB...] for d_{AA} = 5.35 Å, d_{AB} = 3.35 Å and $d_{A\alpha}$ = 2.8 Å indicates good agreement in the (10), (0 1/2) and (1/2 1/2) beam spectra at normal incidence as shown in Figs.52.1,2. Figure 52.4 gives a detailed view of the surface-structural model for the A-(2 × 2)Kα-A-(2 × 2) Kβ-AB... structure.

Details regarding the theoretical calculation method have been described previously [52.7,11]. Briefly, reverse-scattering perturbation was used to define the multiple scattering for each layer (there are two atoms per cell in a graphite layer) and renormalized forward scattering was used to describe interlayer scattering. The calculations are similar to those recently used by *Van Hove* and *Somorjai* in studying hydrogenless benzene molecules [52.12]. Five phase shifts were used to describe the graphite and potassium. The imaginary part of the inner potential was set to -4.2 eV, and the real component of the inner potential was varied to obtain the best correspondence with the totality of the data. This variation produced a -10 eV value for the inner potential. Sixty-six reduced beams were used in the energy region of calculation. The calculated curves of Figs.52.1,2 are the average of corresponding diffracted beam required by the existence of two different atomic terminations of the substrate (A or B) as a result of monoatomic steps at the surface.

52.3.3 Discussion

The C(0001)-(2 × 2)K in A-(2 × 2)Kα-A-(2 × 2)Kβ-A... stacking is in fact the structure observed in stage-1 potassium-graphite intercalates. It is interesting to note that all of the above tested theoretical models with an ordered potassium top layer do not correlate to experiment, although the indications are that potassium does exist in a layer on top of the graphite in a disordered state. This may be due to the enhanced vibrational properties of the carbon atoms at the surface [52.13,14] which prevent ordering of the top potassium. Potassium atoms intercalating into the third and subsequent graphite layers in a layer-by-layer manner will experience an environment more like that of bulk stage-1 potassium-graphite and as a result order to the (2 × 2) structure. In our calculations ABAB... stacking is assumed for the sample after the 5th layer. This stacking assumption may not be totally correct as the additional potassium exposure required to observe the (2 × 2) structure may induce stacking changes in the 6th and possibly 7th layers. However, as LEED

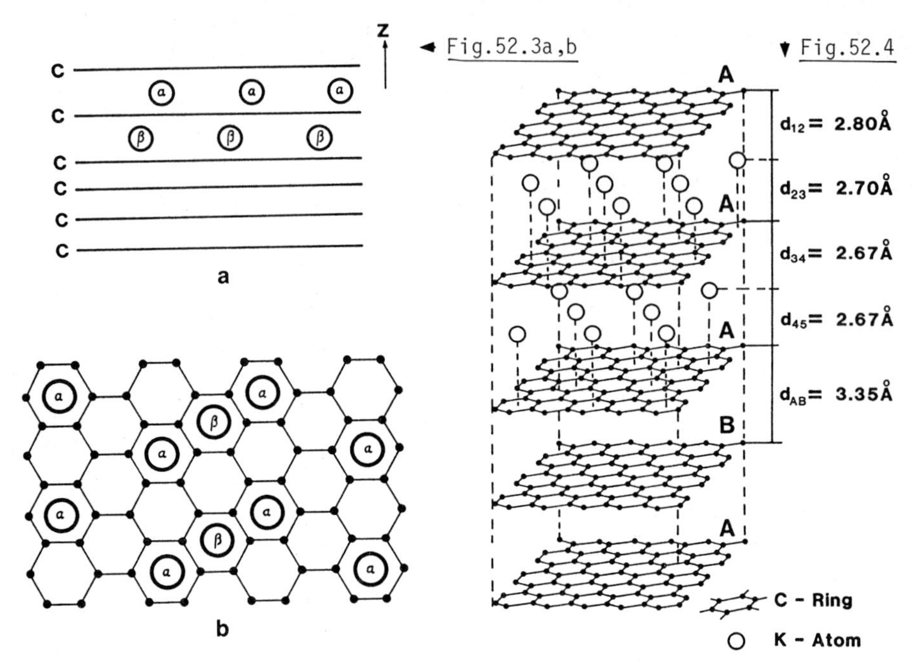

Fig.52.3a,b. Arrangement of potassium and carbon atoms in C(0001)-(2 × 2)K. (**a**) Stacking sequence of carbon and potassium layers. (——) represent carbon layers; α and β represent potassium intercalant layers. (**b**) In-plane structure of (2 × 2) structure; potassium atoms form a hexagonal (2 × 2) superlattice and occupy one quarter of the prismatic sites. α and β define the alternative superlattice origins in two subsequent potassium intercalant layers

Fig.52.4. C(0001)-(2 × 2)K surface intercalated structure

is not sensitive to such deep surface layers, we have retained ABAB... stacking beyond the 5th atomic layer to simplify the model calculations.

The C(0001)-(2 × 2)K structure exists only for 2-3 hours. With increasing time, the (2 × 2) superstructure changes into the coexistence of (2 × 2) and ($\sqrt{3} \times \sqrt{3}$)-R30° phases. Then the coexistence structure transforms into ($\sqrt{3} \times \sqrt{3}$)-R30° surface ordering without any change of potassium coverage as determined by AES. Since the ($\sqrt{3} \times \sqrt{3}$)-R30° structure has not been seen in bulk potassium-graphite intercalation compounds, it is probably a top surface structure. Studies are underway to clarify this point.

52.3.4 Conclusions

The atomic structure of a (2 × 2) superstructure observed upon further exposing a surface-potassium intercalated graphite sample to potassium has been determined by LEED measurements. The potassium does not order on top of the graphite, but instead orders only as an intercalate. The structure agreeing best for a limited number of LEED beams is that with A-(2 × 2)Kα-A-(2 × 2) Kβ-AB... stacking, with d_{AA} = 5.35 Å, d_{AB} = 3.35 Å and $d_{A\alpha}$ = 2.8 Å. This structure is in good agreement with the (2 × 2) C_8K structure of bulk potassium-intercalated graphite.

Acknowledgments. Helpful discussions with S.C. Moss and assistance of M.A. Van Hove with the computer codes are gratefully acknowledged. Partial support for work was provided by the R.A. Welch Foundation.

References

52.1 S.C. Fain: In *Chemistry and Physics of Solid Surfaces IV* ed. by R. Vanselow and R. Howe (Springer, Berlin, Heidelberg 1982) and references therein

52.2 H.J. Lanter, B. Croset, C. Marti, P. Thorel: In *Ordering in Two Dimensions*, ed. by S. Sinha (North-Holland, New York 1980)

52.3 G.K. Wertheim, S.B. Dicenzo: In *Intercalated Graphite*, ed. by M.S. Dresselhaus, G. Dresselhaus, J.E. Fischer, and M.J. Moran (North-Holland, New York 1983)

52.4 K.C. Woo, H. Mertway, J.E. Fischer, W.A. Kamitakahara, D.S. Robinson: Phys. Rev. B**27**, 7831 (1983)

52.5 N.A.W. Holzwarth, S.G. Louie, S. Rabii: Phys. Rev. B**28**, 1013 (1983)

52.6 N.J. Wu, A. Ignatiev: J. Vac. Sci. Technol. **20**, 986 (1982)

52.7 N.J. Wu, A. Ignatiev: Phys. Rev. B**28**, 7288 (1983)

52.8 N.J. Wu, A. Ignatiev: Solid State Commun. **46**, 59 (1983)

52.9 D. Nixon, G. Parry: J. Phys. C**2**, 1732 (1969)

52.10 H. Zabel: In *Ordering in Two Dimensions*, ed. by S. Sinha (North-Holland, New York 1980)

52.11 M.A. Van Hove, S.Y. Tong: *Surface Crystallography by LEED*, Springer Ser. Chem. Phys., Vol.2 (Springer, Berlin, Heidelberg 1979)

52.12 M.A. Van Hove, G.A. Somorjai: Surf. Sci. **114**, 117 (1982)

52.13 E. de Rouffignac, G. Alldredge, F. de Wette: Phys. Rev. B**23**, 4208 (1981)

52.14 B. Fyrey, F.W. de Wette, E. de Rouffignac, G. Alldredge: Phys. Rev. B (1983)

Part VI

Defects and Phase Transitions

VI.1 Theoretical Aspects

53. Energetics of the Incommensurate Phase of Krypton on Graphite: A Computer Simulation Study

M. Schöbinger and F.F. Abraham

IBM Research Laboratory, San Jose, CA 95193, USA

The energetics of the weakly incommensurate phase of a quasi two-dimensional system of 20736 krypton atoms on a graphite surface are investigated. We have chosen a krypton-graphite potential which gives the commensurate phase at the lowest-energy state. The energy increase of the incommensurate phase can be described in terms of an hexagonal domain wall network confirming the phenomenological description by *Bak* et al. [53.9]. The wall energy per unit length ξ and the energy associated with each wall intersection Λ are calculated. Since Λ is significantly negative our results provide strong evidence that the commensurate-incommensurate transition might be first order.

53.1 Introduction

A high-density surface layer of krypton physisorbed on graphite is a good model system to study the properties of the commensurate-incommensurate (C-IC) transition. The details of this transition are being extensively studied experimentally [53.1-8], theoretically [53.9-17] and by computer simulation techniques [53.18-22]. In spite of the extensive research, such basic features as the nature of the C-IC transition are still quite controversial.

The weakly IC phase has been described in terms of hexagonal or striped arrays of incommensurate domain walls separating commensurate regions [53.16]. In the domain-wall description the zero-temperature structure is essentially controlled by only two factors: the domain-wall energy per unit length ξ and the wall-crossing energy Λ. *Bak* et al. [53.9] predicted a discontinuous transition between the C phase and a hexagonal IC phase for the case when the walls attract each other ($\Lambda < 0$) and a continuous C-IC transition for $\Lambda > 0$. Since in the latter case the striped structure is energetically more favorable than the hexagonal honeycomb array, the hexagonal symmetry in the IC phase is broken near the transition. These predictions are not confirmed by the experimental findings. The C-IC transition has been interpreted experimentally to be second order. However, measurements of the atomic structure factor of the krypton film by means of synchrotron X-ray scattering cannot exclude the possibility of a weak first-order transition. Moreover, the same measurements provide no evidence for a uniaxially compressed IC phase [53.8]. *Villain* has shown [53.11] that this discrepancy between theory and experiment cannot be resolved even if one extends the theoretical description to finite temperature. The simulation results allow a direct comparison with laboratory measurements and yield additional insight into the atomic nature of the system.

It is important to choose the substrate potential appropriately and to establish that the C phase yields the lowest energy per atom [53.1,2,23]. In contrast to previous simulations [53.19-22], we employed the potential recently proposed by *Vidali* and *Cole* [53.24]. A detailed discussion of this point is given elsewhere [53.25]. It is interesting to determine whether the energetics of the weakly IC phase can be described in terms of a domain wall network. Our quasi two-dimensional (quasi 2D) simulations confirm the phenomenological description by *Bak* et al. [54.9] and we were able to compute the characteristic energies ξ and Λ of the domain wall picture. Thus, Λ was found to be significantly negative, providing evidence that the C-IC transition might be in fact first order.

In order to deal with these questions and to bridge the gap between microscopic descriptions and experimental findings further, we have performed computer simulations of a system modeling krypton on graphite, in addition to the previous ones [53.19-22]. Because of finite-size effects [53.19], we were forced to simulate a large system of 20736 atoms. To suppress thermal fluctuations which are not included in the phenomenological approach by *Bak* et al. [53.9] and to exclude the influence of thermal renormalizations [53.16, 17], we simulated the low-temperature dynamics of the system. Therefore, in the present paper we do not address second-layer effects [53.22] and dislocation formation [53.13] which might be important at temperatures where synchrotron experiments are done.

53.2 Computer Model System and Numerical Simulation Method

In accordance with the previous molecular dynamics simulations [53.19-22] we adopt the wellknown Lennard-Jones 12:6 pair potential to represent the interaction between the krypton atoms. The interatomic potentials are assumed to be pairwise additive and the empirically determined interatomic parameters are taken to be $\varepsilon_{Kr}/k_B = 150$ K, $\sigma_{Kr-C} = 3.59$ Å [53.26].

For the substrate potential we chose those parameters consistent with a lowest-energy state that is commensurate, in agreement with experiment. This turns out to be the case for the potential suggested by *Vidali* and *Cole* [53.24]. Thus we took as potential parameters for the krypton-carbon interaction $\varepsilon_{Kr-C}/k_B = 85.6$ K and $\sigma_{Kr-C} = 3.21$ Å. This potential includes anisotropy in the attractive part of the krypton-carbon pair potential. However, the potential is essentially the same as that originally suggested by *Steele* [53.27] for an isotropic interaction. The anisotropy parameter $\gamma_A = 0.4$ (c.f. [53.24]) is included for consistency reasons. (In a strictly 2D treatment this potential results in $V_g/\varepsilon_{Kr} = -0.068$ for the main Fourier component which corresponds, e.g., to $V_g/\varepsilon_{Kr} = -0.069$ in Fig.2c of [53.23].) Considering the fact that it is too early to draw definite conclusions about the actual form of the substrate potential, it seems most reasonable to choose a potential for which the resulting theoretical prediction or simulation results are consistent with experimental findings.

To exclude any spurious results due to the 2D approximation we have chosen the more time-consuming quasi 2D simulations. We used the computational cell geometry described in [53.28]. The atoms are free to move in a box with a basal plane defined by the atomic carbon positions of the graphite surface. This base is a parallelogram, compatible not only with the triangular lattice of a 2D krypton solid but also with the graphite lattice. Periodic boundary conditions were imposed at the four faces of the computational cell which passed through the sides of the basal parallelogram at normal incidence to the surface. A reflecting wall was placed at the top of the computational box at $ZL = 1.5\sigma_{Kr-C}$ since no atoms reached the second-layer mean normal distance $1.84\sigma_{Kr-C}$ at temperatures as low as $k_BT/\varepsilon_{Kr} = 0.05$. The common approximations to reduce computational time are described in [53.25] and will not be discussed here.

53.3 Simulation Results

In simulations of the weakly IC phase we fixed the temperature at $T^* \equiv k_BT/\varepsilon_{Kr} = 0.05$ and varied the coverage Θ from 1.028 to 1.089. For this series of simulations the krypton system was initialized in a perfect triangular lattice, and the atoms were assigned a velocity distribution corresponding to a temperature of $T^* = 0.05$. Equilibrium was determined by monitoring the energy, the ratio of commensurate to incommensurate atoms and the ratio of atoms in the three degenerate ground states. Since the system temperature is very low it was necessary to perform typically 20000 to 30000 time steps for a given coverage in order to obtain equilibrium properties. The time step for the simulations of this study was 0.025 ps. To visualize the result we arbitrarily define a krypton atom as being incommensurate if its averaged position is displaced from the nearest adsorption site by an amount that is greater than 20 percent of the graphite lattice constant. Representing the incommensurate atoms at various successive time steps provides a series of snapshots of the temporal evolution of the system.

Figure 53.1 shows representative snapshots for coverages $\Theta = 1.028, 1.043, 1.058, 1.073, 1.089$, respectively. For all coverages we note the honeycomb network of incommensurate domain walls separating an array of hexagonal commensurate regions which has been described in detail before [53.19-22]. Here we recapitulate only those features which are essential for the present study. Even though we find smaller and smaller commensurate domains with increasing coverage, the overall features of the system remain essentially unchanged. Especially the width of the incommensurate domain walls turns out

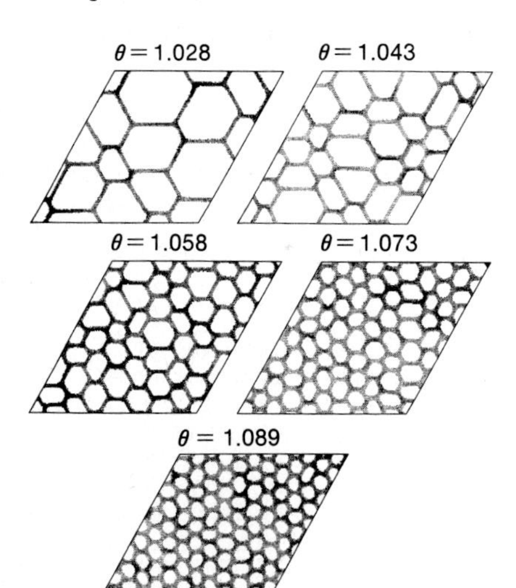

Fig.53.1. Domain-wall network for an equilibrium configuration of the weakly incommensurate phase as a function of coverage Θ at fixed temperature $T^* = 0.05$

to be invariant with respect to changes of the coverage. The walls consist of order 4 rows of atoms, which is in good agreement with the results of *Gooding* et al. [53.23] provided the above-mentioned commensurability criterion is applied to their Fig.3. As a result of a permanent expansion and contraction of commensurate regions without changing the total wall length and the number of wall intersections, the individual hexagons are not identical in size and shape. However, this reshaping confirms *Villain*'s [53.11] prediction of a 'breathing' freedom of the honeycomb array and therefore indicates that in fact the energy of the weakly IC phase is essentially determined by the total wall length and the number of wall intersections [53.9].

Figure 53.2 shows the increase $\Delta\varepsilon = \varepsilon_{IC} - \varepsilon_C$ of potential energy of the IC phase per atom compared to the commensurate phase. It is important to note that the high accuracy of the average values of the potential energy obtained from the simulations leads to an error of the energy differences which is smaller than one percent for $\Theta \geqslant 1.028$. Figure 53.2 shows that the potential energy of the weakly IC phase is a monotonous function of Θ and the lowest-energy configuration is the commensurate phase. It should be mentioned that in contrast to previous studies [53.22] we derived the energy difference between incommensurate and commensurate phase from the total potential energy of the system. Thus our results for $\Delta\varepsilon$ are independent of the arbitrary commensurability criterion explained above which we used for graphical purposes only.

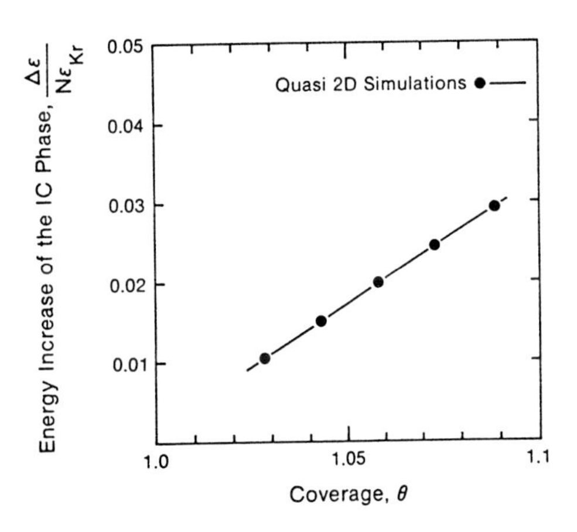

Fig.53.2. The increase $\Delta\varepsilon = \varepsilon_{IC} - \varepsilon_C$ of potential energy of the incommensurate phase per atom compared to the commensurate phase as a function of coverage

53.4 Energetic Analysis

The remarkable invariance of the wall features with respect to changes in the coverage and the indications for a 'breathing' freedom of the honeycomb array suggest a phenomenological expression for the energy increase $\Delta\varepsilon$ of the weakly IC phase in terms of these walls. Such an expression should contain a term proportional to the total wall length Lb, but it should also contain a term proportional to the number 2ν of wall intersections [53.9,16]

$$\Delta\varepsilon = \frac{2}{3}\xi L + 2\nu\Lambda + w_0 N l^{-\alpha-2} e^{-\kappa l} \quad . \tag{53.1}$$

It has been suggested in the literature that the energy should also contain an elastic interaction between the walls mediated through the domains which

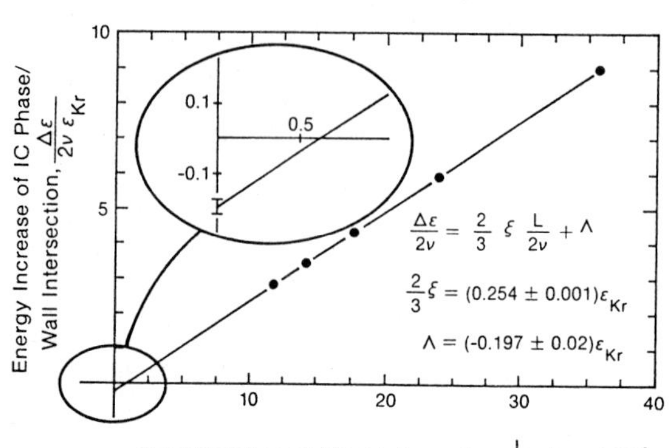

Fig.53.3. Quasi two-dimensional system: energy increase of IC phase per wall intersection plotted as a function of the total wall length per wall intersection

can be described qualitatively [53.16] by the third term in (53.1). This interaction is expected to decay exponentially when the mean distance $lb/2$ between wall intersections increases.

Figure 53.3 shows that we found no pronounced influence of the elastic wall interaction for the quasi 2D results (for details of the computation of L, see [53.25]). Instead, the results are in excellent agreement with the form given by the first two terms of the rhs of (53.1). This fact enables us to determine the wall energy per unit length $\xi = (0.254 \pm 0.001)\varepsilon_{Kr}$ and the energy associated with each wall intersection $\Lambda = (-0.197 \pm 0.02)\varepsilon_{Kr}$ although the finite size of the computational cell determines an upper limit for $L/2\nu$ and the mean distance between wall intersections (a lower limit for $\Theta - 1$), respectively. Note that the absolute values of ξ and Λ are of the same order of magnitude and Λ is significantly negative.

To analyze the implications of our result for the nature of the C-IC transition we have to realize that in a typical adsorption experiment the vapor of the adsorbate is always present and acts like a reservoir of particles. *Villain* and *Gordon* [53.16] have modified the results of *Bak* et al. [53.9] to take this situation into account. The fixed quantity is the number M of substrate sites, not the number N of adatoms. Therefore a term proportional to the chemical potential μ has to be taken into account:

$$W = \Delta\varepsilon - \mu N \quad . \tag{53.2}$$

It is sufficient to discuss the limit of large wall separations ($l \gg 1$) where one finds [53.16] $L = 2N/l$, $\nu = 4N/3l^2$ and the quantity W can be simply expressed as a function of the average distance $lb/2$ between wall intersections:

$$W = \frac{4}{3}\xi\frac{N}{l} + \frac{4}{9}6\Lambda\frac{N}{l^2} + w_0 N l^{-\alpha-2} e^{-\kappa l} \quad . \tag{53.3}$$

A necessary condition of the stability of the commensurate phase is that its energy is lower than W, $W > -\mu M$, i.e.,

$$\frac{4}{3}(\xi - \mu)\frac{N}{l} + \frac{4}{9}[6\Lambda + \xi - (\xi - \mu)]\frac{N}{l^2} + w_0 N l^{-\alpha-2} e^{-\kappa l} > 0 \quad . \tag{53.4}$$

The critical quantity in this expression is $\tilde{\Lambda} = \Lambda + \varepsilon/6$, which substitutes the bare value of the wall intersection energy Λ discussed by *Bak* et al. [53.9]. In the quasi 2D simulations Λ is sufficiently negative, resulting in a negative value for the effective wall intersection energy $\tilde{\Lambda}$. For $\tilde{\Lambda} < 0$ the transition is first order and the hexagonal symmetry is preserved [53.16] in the IC phase.

53.5 Conclusions

We found that the difference in potential energy between IC and C phase consists of two contributions which are proportional to the total length of the incommensurate domain walls L and the number of wall intersections 2ν, respectively. This result confirms the phenomenological description by *Bak* et al. [53.9]. Since the energy Λ associated with each wall intersection turns out to be significantly [53.16] negative, the simulations provide strong evidence that the C-IC transition might, in fact, be first order. This point is important since, due to the limited size of the computational cell, we were not able to investigate the phase transition directly. In contrast to earlier calculations by *Talapov* [53.29], additional 2D simulations indicate that the wall intersection energy is significantly negative, even in a 2D system [53.25]. For finite temperature, *Villain* has shown [53.11] that the "breathing entropy" results in a systematically negative contribution to (53.1). However, employing the estimate for the pertinent change in the effective wall intersection energy [Eq. (4.30) of Ref.53.16] for the present parameter choice we find that finite temperature effects might account for at most 20 percent of Λ. Thus this contradiction seems to be a result of the fact that Talapov employs the "harmonic" approximation for the krypton-krypton pair interaction.

References

53.1 O.E. Vilches: Ann. Rev. Phys. Chem. **31**, 463 (1980); A. Thomy, X. Duval, J. Regnier: Surf. Sci. Rep. **1**, 1 (1981)
53.2 M. Nielsen, J. Als-Nielsen, J. Bohr, J.P. McTague: Phys. Rev. Lett. **47**, 582 (1981)
53.3 M.D. Chinn, S.C. Fain: Phys. Rev. Lett. **39**, 146 (1977)
53.4 S.C. Fain, M.D. Chinn, R.D. Diehl: Phys. Rev. B**21**, 4170 (1980)
53.5 P.W. Stephens, P. Heiney, R.J. Birgeneau, P.M. Horn: Phys. Rev. Lett. **43**, 47 (1979)
53.6 R.J. Birgeneau, E.M. Hammonds, P. Heiney, P.W. Stephens, P.M. Horn: In *Ordering in Two Dimensions*, ed. by S.K. Sinha (Plenum, New York 1980)
53.7 D.E. Moncton, P.W. Stephens, R.J. Birgeneau, P.M. Horn, G.S. Brown: Phys. Rev. Lett. **46**, 1533 (1981)
53.8 P.W. Stephens, P.A. Heiney, R.J. Birgeneau, P.M. Horn, D.E. Moncton, G.S. Brown: Phys. Rev. B**29**, 3512 (1984)
53.9 P. Bak, D. Mukamel, J. Villain, K. Wentowska: Phys. Rev. B**19**, 1610 (1979)
53.10 H. Shiba: J. Phys. Soc. Jpn. **46**, 1852 (1979); **48**, 211 (1980)
53.11 J. Villain: In *Ordering in Strongly Fluctuating Condensed Matter Systems*, ed. by T. Riste (Plenum, New York 1980) p.221
53.12 P. Bak: Rep. Prog. Theor. Phys. **45**, 587 (1982)
53.13 S.N. Coppersmith, D.S. Fisher, B.I. Halperin, P.A. Lee, W.F. Brinkman: Phys. Rev. Lett. **46**, 549 (1981); Phys. Rev. B**25**, 349 (1982)
53.14 D.A. Huse, M.E. Fisher: Phys. Rev. Lett. **49**, 793 (1982)
53.15 M. Kardar, A.N. Berker: Phys. Rev. Lett. **48**, 1552 (1982)
53.16 For the most recent review of the theory including comparison with experiment, see J. Villain, M.B. Gordon: Surf. Sci. **125**, 1 (1983)
53.17 M. Schöbinger, S.W. Koch: Z. Phys. B**53**, 233 (1983)

53.18 F. Hanson, J.P. McTague: J. Chem. Phys. **72**, 6363 (1980)
53.19 F.F. Abraham, S.W. Koch, W.E. Rudge: Phys. Rev. Lett. **49**, 1830 (1982)
53.20 S.W. Koch, F.F. Abraham: Helv. Phys. Acta **56**, 755 (1983)
53.21 F.F. Abraham, W.E. Rudge, D.J. Auerbach, S.W. Koch: Phys. Rev. Lett. **52**, 445 (1984)
53.22 S.W. Koch, W.E. Rudge, F.F. Abraham: Preprint 1983
53.23 R.J. Gooding, B. Joos, B. Bergersen: Phys. Rev. B**27**, 7669 (1983)
53.24 G. Vidali, M.W. Cole: Phys. Rev. B**29**, 6736 (1984)
53.25 M. Schöbinger, F.F. Abraham: To be published
53.26 S. Rauber, J.R. Klein, M.W. Cole: Phys. Rev. B**27**, 1314 (1983)
53.27 W.A. Steele: Surf. Sci. **36**, 317 (1973)
53.28 S.W. Koch, F.F. Abraham: Phys. Rev. B**27**, 2964 (1983)
53.29 A.L. Talapov: Phys. Rev. B**24**, 6703 (1981)

54. Theory of Commensurate-Incommensurate Phase Transitions in W(001)

S.C. Ying and G.Y. Hu[1]

Department of Physics, Brown University, Providence, RI 02912, USA

A theoretical model is proposed to explain the commensurate-incommensurate phase transitions on the W(001) surface. The incommensurate phase on W(001) arises from the coupling of the in-plane distortions with the component of displacement normal to the surface. Near the commensurate-incommensurate transition, the surface is composed of domains with in-plane displacements forming a c(2 × 2) structure and walls separating the domains in which the surface atom displacements have mixed in-plane and out-of-plane components. The H adsorption can increase the coupling of the two distortion modes, so enhancing the incommensurate phase. This can account for the observed LEED data on H/W(001). Finally, finite temperature effects are also studied, enabling a theoretical prediction for the entire phase diagram. The general model used in the present approach can also be applied to the study of commensurate-incommensurate phase transitions on other surfaces and chemisorption systems.

54.1 Introduction

The reconstruction of a W(001) surface from a (1 × 1) to a c(2 × 2) structure has been intensively studied both theoretically and experimentally [54.1]. Despite some controversies, by now we have arrived at a good understanding of this reconstruction. On the hydrogen chemisorbed surface W(001), a further commensurate (C)- incommensurate (IC) phase transition occurs when the coverage and/or temperature is increased. This transition is far less understood than the (1 × 1) to c(2 × 2) reconstruction. Various mechanisms have been proposed for the C-IC transition [54.2-4]. To date, experimental data are not conclusive enough to favor any particular model. Among these models, perhaps the most plausible and interesting is that proposed by *Fasolino* et al. [54.4]. In this model, the incommensurate phase arises not because of long-range interactions but rather from the coupling of two vibrational modes, one perpendicular to the surface and one in the plane of the surface. This coupling only exists when one moves away from symmetric (commensurate) wave vectors for the reconstruction. This then is a special case of the general mechanism proposed by *Heine* and *McDonnell* for the incommensurate phase in ferroelectrics [54.5]. Recently, *Willis* also discussed how the incommensurate phase can be viewed as a periodic structure of antiphase domain walls or "solitons" [54.6]. While all these ideas are very intriguing they have been pursued only at a phenomenological level. In this note we show that starting from a microscopic lattice dynamical Hamiltonian for W(001) and using the "phase modulation only" concept, we can arrive at a soliton Hamiltonian for the C-IC transition. Through extensive studies in the last few years, the details of the C-IC transition for this

1 Research supported in part by the MRL program at Brown University funded through the National Science Foundation.

soliton Hamiltonian including finite-temperature properties are well understood [54.7,8]. Previously we showed that H adsorption can be interpreted with the same lattice dynamical Hamiltonian with renormalized coverage and temperature-dependent force constants [54.9]. In the present case, H adsorption therefore leads to coverage (θ) and temperature (T) dependent parameters in the soliton Hamiltonian, from which we can obtain the phase diagram in the θ-T plane.

54.2 Soliton Hamiltonian

We have shown in a separate paper that for the surface phase transition on W(001) the subsurface degrees of freedom can be eliminated to yield an effective two-dimensional surface lattice Hamiltonian [54.10]. As stated above, we interpret the incommensurate phase on W(001) as arising from the coupling of an in-plane mode which will be taken to be in the y direction and a mode normal to surface along the z direction. Thus the simplest lattice Hamiltonian describing these two modes and their mutual coupling is

$$H = \sum_{\boldsymbol{\ell}} \left\{(1/2)A_y u^2_{\boldsymbol{\ell}y} + (1/2)A_z u^2_{\boldsymbol{\ell}z} + (1/4)B_y u^4_{\boldsymbol{\ell}y} + (1/4)B_z u^4_{\boldsymbol{\ell}z} + (1/4)B_{yz} u^2_{\boldsymbol{\ell}y} u^2_{\boldsymbol{\ell}z}\right\}$$

$$+ (1/2) \sum_{\langle\boldsymbol{\ell},\boldsymbol{\ell}'\rangle} [J_\ell(u_{\boldsymbol{\ell}y} - u_{\boldsymbol{\ell}'y})^2 + J_z(u_{\boldsymbol{\ell}z} - u_{\boldsymbol{\ell}'z})^2]$$

$$+ (1/2) \sum_{\langle\boldsymbol{\ell},\boldsymbol{\ell}'\rangle} [J_t(u_{\boldsymbol{\ell}y} - u_{\boldsymbol{\ell}'y})^2 + J_z(u_{\boldsymbol{\ell}z} - u_{\boldsymbol{\ell}'z})^2]$$

$$- (J/2) \sum_{\boldsymbol{\ell}} (u_{\boldsymbol{\ell}y} u_{\boldsymbol{\ell}+a\hat{y},z} - u_{\boldsymbol{\ell}+a\hat{y},y} u_{\boldsymbol{\ell}z}) \quad . \tag{54.1}$$

Here, the sum over $\langle\boldsymbol{\ell},\boldsymbol{\ell}'\rangle$ is over nearest neighbors only:

$$\boldsymbol{\ell} = \ell_x a\hat{y} + \ell_y a\hat{x}$$

and $u_\ell = (-\ell)^{\ell_x+\ell_y} \mathbf{v}_\ell$ is related to the real displacement vector $\mathbf{v}_\ell$ by an alternating phase factor. Next we change from $\mathbf{u}_\ell$ to the phase and amplitude variables

$$u_{\ell y} = \rho_1(\ell)\cos\Phi(\ell) \quad ,$$

$$u_{\ell z} = \rho_2(\ell)\cos[\Phi(\ell) + \delta] \quad . \tag{54.2}$$

In the commensurate phase $\Phi(\ell) = 0$ or multiples of π and $\rho_1(\ell) = \rho_0 = -A_y/B_y$. Near the C-IC transition, amplitude fluctuations are costly in energy and we can used the fixed amplitude approximation

$$\rho_1(\ell) = \rho_0 = -A_y/B_y \tag{54.3}$$

and $\rho_2(\ell) = \lambda\rho_0$ with λ to be determined below. Also, to maximize the coupling between y-z modes, δ should take the value $\pi/2$. If we now go to the continuum limit and keep only the lowest-order gradient terms, the following Hamiltonian results

$$H = \int d\mathbf{r}\left\{V(\Phi) + (1/2)f_x(\Phi)(\partial\Phi/\partial x)^2 + (1/2)f_y(\Phi)(\partial\Phi/\partial y)^2 - J\lambda\rho_0^2(\partial\Phi/\partial y)\right\} \quad , \tag{54.4}$$

where

$$V(\Phi) = V_0(1 - \cos 2\Phi) - B_0(1 - \cos 4\Phi)$$

$$V_0 = (\rho_0^2/4)(-A_y/2 + A_z\lambda^2 + B_z\lambda^4\rho_0^2/2)$$

$$B_0 = (\rho_0^4/32)(B_y + B_z\lambda^4 - B_{yz}\lambda^2)$$

$$f_y(\Phi) = 2J_\ell\rho_0^2(1 + \varsigma \cos^2\Phi)$$

$$f_x(\Phi) = 2J_t\rho_0^2(1 + \varsigma \cos^2\Phi)$$

$$\varsigma = \lambda^2 J_z/J_\ell - 1 \quad . \tag{54.5}$$

The Hamiltonian in (54.4) is a typical 2D anisotropic soliton model. In the x direction only commensurate terms appear. The linear term in $\partial\phi/\partial y$ provides the driving force for the incommensurate phase. It differs slightly from the standard Hamiltonian for C-IC transitions through the Φ dependence of the elastic coefficients $f_x(\Phi)$ and $f_y(\Phi)$.

54.3 Commensurate-Incommensurate Transition at Zero Temperature

At $T=0$, we only need to find the ground state configuration $\Phi(x,y)$ that minimizes the total energy. Obviously $\partial_x\Phi = 0$. Minimization of the energy leads to the solution

$$V(\Phi) = f_y(\Phi)(\partial\Phi/\partial y)^2 + c \quad . \tag{54.6}$$

Now we consider a single soliton or antiphase domain wall where the phase angle Φ varies from 0 to π. For this case, the integration constant $c=0$. The energy of the soliton per unit length is

$$\varepsilon_s = \varepsilon_s^0 - \pi J\lambda\rho_0^2 \quad , \quad \text{where}$$

$$\varepsilon_s^0 = \int_0^\pi d\Phi\sqrt{[2f_y(\Phi)V(\Phi)]} \quad . \tag{54.7}$$

The C-I phase transition occurs at $\varepsilon_s = 0$, which is

$$\sqrt{(2V_0J_\ell/\rho_0^2)}\,I(\varsigma,b) = \pi\lambda J/4 \quad . \tag{54.8}$$

Here

$$I(\varsigma,b) = \int_0^1 dx[(1 + \varsigma x^2)(1 - bx^2)]^{\frac{1}{2}}$$

and

$$b = 4B_0/V_0 \quad .$$

The value of λ is determined by minimizing the soliton energy. To illustrate the nature of the C-IC transition from this model, we have generated the

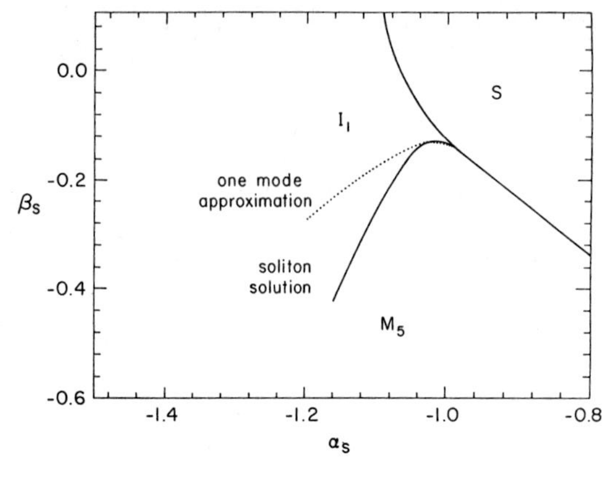

Fig.54.1. Soliton solution for C-IC phase diagram at T = 0 as a function of surface force constants α_S and β_S (in units eV/Å^2) for W(001). M_5 denotes commensurate region, I_1 incommensurate region and S the stable (1 × 1) region. Boundaries from one-mode solution are for comparison

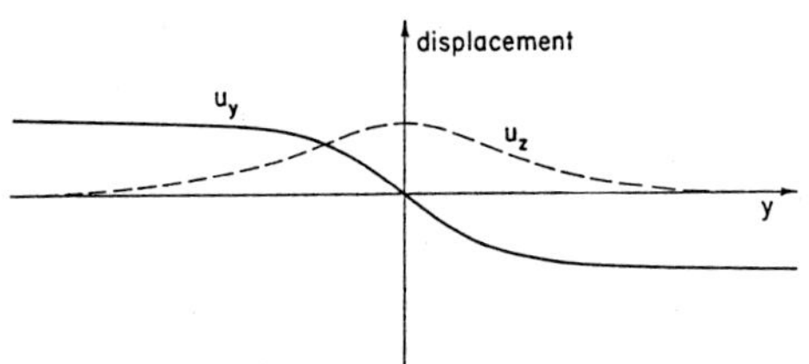

Fig.54.2. Phase relation between lateral displacement u_y and vertical displacement u_z at an antiphase domain wall

soliton Hamiltonian from the same lattice dynamic Hamiltonian for W(001) as adopted in [54.10]. The coefficients of $A_y, A_z, J_\ell, J_t, J_z$ are then functions of the surface force constants α_S, β_S. The magnitude of the anharmonic terms in (54.5) are determined mainly by bulk force constants, so they can be taken approximately as constants. In our calculation they are chosen to be $B_y = 44$, $B_z = 15$ and $B_{yz} = 30$ (in units of eV/Å^4). We noted that the phase boundary is not very sensitive to the magnitude of the anharmonic term. In Fig.54.1 we present the phase boundary separating the M_5 (commensurate in plane distortion) and I_1 (incommensurate region) from the present soliton solution. By comparison, we also present the phase boundary for the one-mode solution where $\Phi(y)$ is taken to be the linear form qy. The incommensurate region corresponding to the soliton picture is wider in the phase space, indicating that the one-mode picture overestimates the energy of the incommensurate phase. This is true except in a small region where $\rho \to 0$ and one expects the "phase modulation only" assumption to be a poor description. The physical picture of the incommensurate region I_1 here consists of domains of commensurate regions with displacement in y (or equivalently x) direction separated by solitons (antiphase domain walls) in which y and z displacements are mixed. This is similar to the picture hypothesized by *Willis* [54.6] as illustrated in Fig.54.2. The exact thickness of the antiphase domain walls and their separations depends on the location in the parameter space (α_S, β_S).

54.5 Finite Temperature Solution

At finite temperature, the soliton lines in two dimensions start to bend and collide and these thermal fluctuations must be included to determine the boundary for C-IC transitions. Here we follow the treatment of *Pokrovsky* et al. [54.8] in discussing finite-temperature properties. The basic idea is to integrate out the short-wavelength fluctuations around the equilibrium soliton configuration. The elastic constant controlling the short-wavelength fluctuations is

$$\kappa = \rho_0^2\sqrt{(J_\ell + J_z\lambda^2)(J_t + J_z\lambda^2)} \quad . \tag{54.9}$$

After this coarse-graining procedure the coefficient of the $(1-\cos 4\phi)$ term in the potential $V(\phi)$ becomes negligible, and the coefficient of the $(1-\cos 2\phi)$ term is renormalized from V_0 to V_{eff}, given by

$$V_{eff}/\kappa = [V_0/\kappa]^{1/(1-T/T_t)} \tag{54.10}$$

with

$$T_1 = 2\pi\kappa \quad .$$

Now we can apply the same considerations as in the $T=0$ case to determine the C-IC boundary but with V_{eff} replacing V_0. This leads to the equation for the boundary as

$$[V_0/\kappa]^{1/2(1-T/T_t)} I(\varsigma,b)\sqrt{(2J_\ell k/\rho_0^2)} = \lambda\pi J/4 \quad , \tag{54.11}$$

shown in Fig.54.3. The incommensurate region in the parameter space (α_S, β_S) broadens as the temperature increases because of its relatively larger entropy.

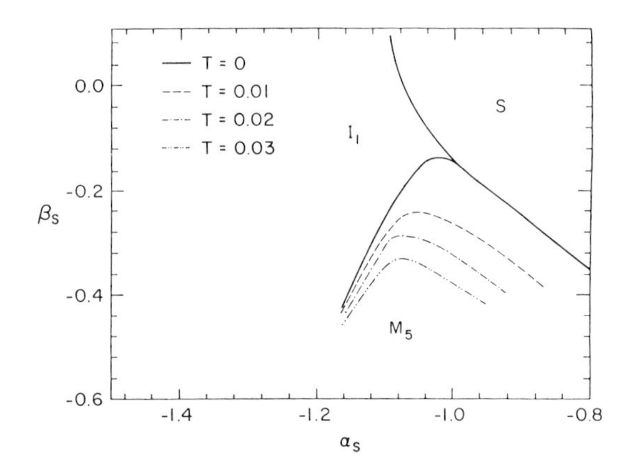

Fig.54.3. C-IC phase boundaries at different temperatures (in units of eV) in the surface force constants (α_S, β_S) plane for W(001) from the solution of the soliton Hamiltonian

54.5 Discussions

In this paper, we have discussed how a soliton Hamiltonian can be derived from a microscopic lattice dynamic Hamiltonian for W(001). The key point is that near the C-IC boundary, the incommensurate phase is not described by a sinusoidal displacement of a single incommensurate wave vector but rather by a periodic structure of antiphase domain walls (solitons) separating commensurate regions. With the help of our soliton Hamiltonian, we can calculate quantitatively properties of the antiphase domain wall in terms of original lattice dynamic parameters. The C-IC boundary at zero temperature is determined by the condition that the energy required to create a single antiphase domain wall vanishes. As our results in Fig.54.1 indicate, this provides a much better description of the incommensurate phase than a single wave-vector sinusoidal distortion. At finite temperatures, it is the free energy rather than the energy of the soliton that needs to be considered. Since there are many configurations for which a soliton line can bend or deform, it leads to a large entropy and lower free energy at finite temperatures. Thus an increase in temperature always favors the incommensurate phase. In fact, our result in (54.11) indicates that beyond a certain temperature T_1, the commensurate phase ceases to exist. In the present treatment, dislocations have not been taken into consideration. Actually, for the two-dimensional soliton lattice they are crucial [54.11]. When dislocations are included properly, a liquid phase separates the commensurate and incommensurate phases. However, it is expected that for short- to medium-range order, such as that studied in typical LEED experiments, the liquid phase differs little from the solid phase [54.12] and the C-IC boundary determined here still provides a useful description. As mentioned earlier, H-adsorption renormalizes the effective soliton Hamiltonian. Whether it actually stabilizes or destabilizes the incommensurate phase depends sensitively on the H-W interaction potential. The details of the effect of H on C-IC transition will be described in a future publication.

References

54.1 For comprehensive review articles, see the Proceedings of the Symposium on Statistical Mechanisms of Adsorption, Surf. Sci. **125** (1983)
54.2 K.H. Lau, S.C. Ying: In *Ordering in Two Dimensions*, ed. by S.K. Sinha (North-Holland, New York 1980)
54.3 T. Inaoha, A. Yoshimori: Surf. Sci. **115**, 301 (1982)
54.4 A. Fasolino, G. Santoro, E. Tosatti: Surf. Sci. **125**, 317 (1983)
54.5 V. Heine, J.D.C. McConell: Phys. Rev. Lett. **46**, 1092 (1981)
54.6 R.F. Willis: This Volume, p.237
54.7 W.L. McMillan: Phys. Rev. B**14**, 1496 (1976); P. Bak, V.J. Emery: Phys. Rev. Lett. **36**, 978 (1976)
54.8 V.L. Pokrovsky, A.L. Talapov: Sov. Phys. JETP **51** (1), 134 (1980); V.L. Pokrovsky, A.L. Talapov, P. Bak: Preprint (1983)
54.9 L.D. Roelofs, S.C. Ying: Submitted to Surf. Sci.
54.10 G.Y. Hu, S.C. Ying: Submitted to Surf. Sci.
54.11 J.M. Kosterlitz, D.J. Thouless: J. Phys. C**6**, 1181 (1973); S.N. Coppersmith, D.S. Fisher, B.I. Halperin, P.A. Lee, W.F. Brinkman: Phys. Rev. B**25**, 349 (1982)
54.12 R.A. Barker, P.J. Estrup: J. Chem. Phys. **74**, 1442 (1981)

55. Molecular Dynamics Investigation of Dislocation-Depinning Transitions in Mismatched Overlayers

K.M. Martini, S. Burdick, and M. El-Batanouny
Department of Physics, Boston University, Boston, MA 02215, USA

G. Kirczenow
Department of Physics, Simon Fraser University, Burnaby, B.C. V5A 1S6, Canada

55.1 Introduction

In recent years great strides have been made towards understanding the structural properties of incommensurate systems [55.1]. This is largely due to the advent of novel experimental methods and the discovery of systems amenable to these methods [55.1,2]. In an effort to explain the experimental results, theoretical investigations have concentrated mainly on determining the phase diagrams and the order of the structural phase transitions involved [55.1,3-5]. Although understanding the dynamical behavior of these systems is an essential ingredient for interpreting the experimental results, we find very little work done towards that end [55.6,7]. This stems from the inherent complexities involved: nonlinearity, discreteness, and incommensurability.

We report here on a new computational approach that deals with these complexities simultaneously, while maintaining simplicity of analysis. The results are then directly amenable to comparison with experiment. In this approach a combination of molecular dynamics and Fourier analysis (space and time) is implemented. The results include detailed phonon and collective dislocation (kink)-mode dispersion relations, including zone-folding effects due to phonon-dislocation scattering, local mode structure, relaxation and finite-size effects. In the light of the these results we propose a mechanism, alternative to a previously proposed entropic process [55.8], that can explain the fluid-like phase observed recently in krypton on graphite [55.2], which is believed to be a manifestation of dislocation motion [55.9].

55.2 Method

We used the Frenkel-Kontorova model of a rigid sinusoidal substrate potential with a harmonic coupled chain overlayer, since it has been widely used to study and interpret surface phenomena. The potential energy in this model is

$$V = C/2 \sum_i (x_i - x_{i+1})^2 + W/2 \sum_i \{1 - \cos 2\pi[(i-1)b + x_i]/a\} \quad , \qquad (55.1)$$

where a is the period of the substrate potential, x_i is the deviations of the ith overlayer particle measured from its ideal position in absence of the substrate, W the substrate potential depth, b the length of an unstretched overlayer bond, and C the spring constant. We introduced a quantity $SR = (2\pi/a)^2 W/8C$, the *strength ratio*, which serves as a parameter of the relative importance of the substrate potential with respect to the overlayer spring stiffness. In this study we used the nearly incommensurate system with $a/b = 9/8$. The system size used was 128 overlayer particles and a substrate of infinite extension. We treated a chain with free ends in order to permit the system to assume its natural length in the presence of the substrate, and to study end effects.

We started the overlayer particles from some arbitrary positions and then allowed them to relax to their $T = 0$ equilibrium positions. Due to the free ends, the equilibrium length of the overlayer chain can differ markedly from its value in absence of the substrate. By employing a large number of initial configurations, we could sample all the static local minima of the system (metastable and ground states). This procedure was therefore, used to construct the phase diagram. After equlibrium was reached a small random displacement ($<10^{-3}a$) was added to the position of each particle to generate the initial conditions for our molecular dynamics simulation. We set the velocity of each atom equal to zero initially, and integrated Newton's equations of motion numerically to find the subsequent motion. The vibrational spectrum is obtained directly from the Fourier transformation

$$G(k,w) = \int dt \sum_j u_j(t) \exp[-i(jbk + wt)] \quad , \tag{55.2}$$

where $u_j(t)$ is the displacement of atom j from its equilibrium position. We identified modes with the local maxima of $[G(k,w)]^2$ [55.10].

55.3 Results and Discussion

In the chain described above, the metastable states sampled involved configurations with a dislocation number between 0 and 14. These metastable states persist throughout most of the phase diagram while the ground state samples all configurations as SR is varied [$SR = 0.076$ marks the onset of the commensurate-incommensurate (IC) transition]. These results agree with previous work [55.3,4].

At high strength ratios the dislocations were found to be pinned. The pinning is due to the discreteness of the chain [55.11]. However, the repulsive interaction between the dislocations, which was confirmed in our studies, can lead to their depinning. The depinning threshold depends on the separation of the dislocations and, therefore, on their local density.

The pinning of the dislocations [55.12] gives rise to a large number of additional metastable states: disordered configurations. These states disappear at low SRs where the repulsion between the dislocations leads to their ordering. We can, therefore, refer to this transition as either a *pinning-depinning* transition, or a *disorder-order* transition. This transition takes place, in the present case, between $SR = 0.2$ and $SR = 0.1$, above the IC transition [55.13].

We have studied the vibrational features of this system at SR values representative of the different regions in the phase diagram. Here we present only results for the well-ordered state with 14 dislocations at $SR = 0.1$. The spectrum is depicted in Fig.55.1. Three types of modes can be identified. The phonon-like modes are confined to the well-defined upper branches: a main branch and several zonefolded ones. The zonefolding wave vector is the inverse of the dislocation spacing and not related to the substrate-overlayer mismatch (originating from phonon-dislocation scattering). The low-lying branch is comprised of collective dislocation modes and its dispersion originates from their interactions. In these modes the dislocations behave like a chain of coupled quasiparticles so that their dispersion relation can be described

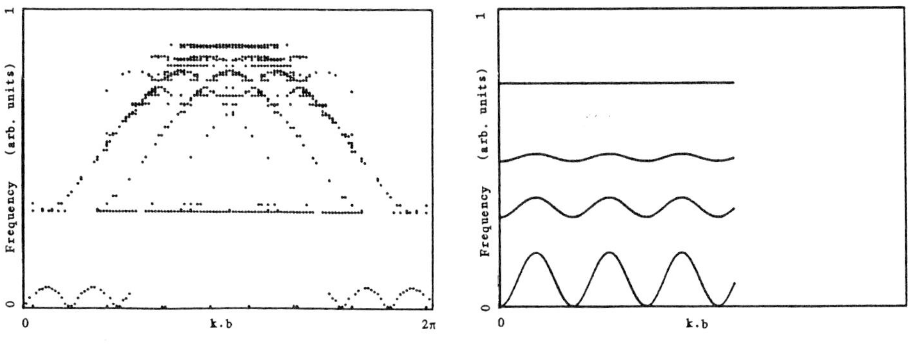

Fig.55.1 Fig.55.2

Fig.55.1. Dispersion relation of the incommensurate overlayer as a function of the renormalized wave vector k (mapped onto the extended Brillouin zone of the unstretched chain) with $a/b = 1.125$ and $SR = 0.1$

Fig.55.2. Schema of the collective dislocation-mode dispersion at different SR values

in terms of an extended Brillouin zone (BZ) scheme corresponding to their spacing. The reciprocal lattice vector in this case is $2\pi/9.2$ as it should be for a dislocation lattice spacing of 9.2 atomic lattice units. We have verified this relation between dislocation spacing and modulation wave vector, as a function of dislocation number (from 5 to 14). The maximum frequency of this branch occurs at the BZ boundary, while its lowest frequency occurs at the BZ center. The lowest frequency is, therefore, a measure of the pinning forces, while the bandwidth is a measure of the interdislocation (ID) interaction and depends on their density and SR. Figure 55.2 schematizes the changes in the dislocation branch as a function of SR. The flat dispersion at high SR indicates that the dislocations act like Einstein oscillators. At intermediate SR the ID interaction is turned on and the associated branch exhibits a bandwidth. The zone center frequency ($k = 0$) vanishes at SR values corresponding to the depinning transition (SR_d). It stays zero for $SR < SR_d$ if the system is infinite. For finite systems, however, it will rise again and assume very small values because of end effects [55.14, 15]. A careful study of the static configuration of the end atoms revealed that they assume a partial-kink-like profile [55.14].

We can further identify in Fig.55.1 several local modes (extended modes in k space). The one just below the phonon branch is due to an end mode. Furthermore, we interpret the spread in k in the main phonon branch as being caused by a mild localization of the phonon modes.

55.4 Conclusion

This molecular dynamics study has yielded for the first time the full vibrational spectra of the finite, free end Frenkel-Kontorova chain. Extensions of this technique to more realistic models are obvious and in progress. Experimental results for the vibrational features are not available, but we hope that high-resolution inelastic electron scattering or neutral atom beam scattering will soon provide some answers. Moreover, we have demonstrated that dislocation pinning due to discreteness, which is manifested in an Einstein-like mode, can be removed by ID interactions leading to a gapless dispersion branch. The process of depinning through ID interactions can explain the nature of the fluid-like phase, mentioned above, since local dislocation depinning can give rise to this effect.

This work was supported in part by the Department of Energy, contract No. DE-AC02-83ER45019.M001.

References

55.1 S.K. Sinha (ed.): *Ordering in Two Dimensions* (North-Holland, Amsterdam 1980);
P. Bak: Rep. Prog. Phys. **45**, 587 (1982);
J. Villain: In *Ordering in Strongly Fluctuating Condensed Matter Systems*, ed. by T. Riste, NATO B**50**, 222 (Plenum, New York 1980)
all give an overview
55.2 D.E. Moncton, P.W. Stephens, R.J. Birgenau, P.M. Horn, G.S. Brown: Phys. Rev. Lett. **46**, 1533 (1981)
55.3 F.C. Frank, J.H. van der Merwe: Proc. R. Soc. (London) A**198**, 205 (1949)
55.4 J.A. Snyman, J.H. van der Merwe: Surf. Sci. **42**, 190 (1974)
55.5 J.E. Sacco, J.B. Sokoloff: Phys. Rev. B**18**, 6549 (1978)
55.6 T. Schneider, E. Stoll: Phys. Rev. B**13**, 1216 (1976)
55.7 A.D. Novaco: Phys. Rev. B**22**, 1645 (1980)
55.8 M.E. Fisher, D.S. Fisher: Phys. Rev. B**25**, 3192 (1982);
V.L. Pokrovsky, A.L. Talapov: Zh. Eksp. Teor. Fiz. **78**, 269 (1980)
55.9 F.F. Abraham, S.W. Koch, W.E. Rudge: Phys. Rev. Lett. **49**, 1830 (1982)
55.10 As a test of our technique we have been able to reproduce the results obtained theoretically by Novaco [55.7] for the simpler case of an infinite periodic chain
55.11 M. Peyrard, S. Aubry: J. Phys. C**16**, 1593 (1983) and references therein
55.12 P. Bak: Phys. Rev. Lett. **46**, 791 (1980);
V.L. Pokrovsky: J. Phys. **42**, 761 (1981)
55.13 Note that the ends of the chain always remain pinned!
55.14 K.M. Martini, S. Burdick, M. El-Batanouny, G. Kirczenow: To be published
55.15 This result was also confirmed recently by S.R. Sharma, B. Bergensen, B. Joos (private communication)

56. Quantitative Analysis of LEED Spot Profiles

M. Henzler

Institut für Festkörperphysik der Universität Hannover
Appelstraße 2, 3000 Hannover, Fed. Rep. of Germany

56.1 Introduction

Since the discovery of electron diffraction [56.1] it has been known that surfaces even of single-crystal substrates frequently are not as perfect as expected from a simple periodic structure. In most cases, therefore, the surface has been treated again and again, until a "good" pattern with spots as sharp as possible was obtained, and defects neglected. There was, however, a growing awareness, that less than ideal diffraction patterns contain valuable information on nonperiodic features [56.2,3]. The evaluation of these diffraction features opens a new field of study of associated phenomena, which include nonperiodic phenomena such as epitaxial growth, surface reactions, random adsorption, phase transitions or thermal disorder. Since several recent reviews are available [56.4-8], the procedures, results and problems are discussed here only briefly. The general features apply to all diffraction geometries (such as LEED, RHEED, atom diffraction or TEM) if the different ranges of reciprocal space are taken into account with the help of the appropriate Ewald sphere.

56.2 Qualitative Evaluation

As demonstrated in the literature [56.6,7], a qualitative evaluation is possible with respect to periodicity by observing the existence of spot position and its variation with electron energy. A reconstruction of spot position in reciprocal space enables structures such as superstructures, step arrays or facets to be distinguished. The observation of spot shape (such as splitting or broadening) [56.7,8] indicates the kind of nonperiodic structure-like variations in size or distances. The variation with $K_{\parallel}$ (scattering vector parallel to the surface) yields the shape or the symmetry with respect to substrate, the dependence on energy, (or) provides an easy distinction between structures within one layer (no dependence), two layers (such as epitaxial layers) or many layers.

56.3 Scheme of Quantitative Evaluation

For periodic arrangements, all spots are as sharp as given by the instrumental limit. Here the spot position is already the full quantitative information with respect to the unit arrangement. It provides terrace width or facet orientation or mosaic orientation as described in the qualitative evaluation (Sect.56.2). The arrangement within the unit may be derived from the intensities, which is not considered in this paper [56.17-20]. For nonperiodic structures, the spots are broadened due to finite size or finite range of periodicity [56.4-8]. Those sizes or distribution of sizes may be derived

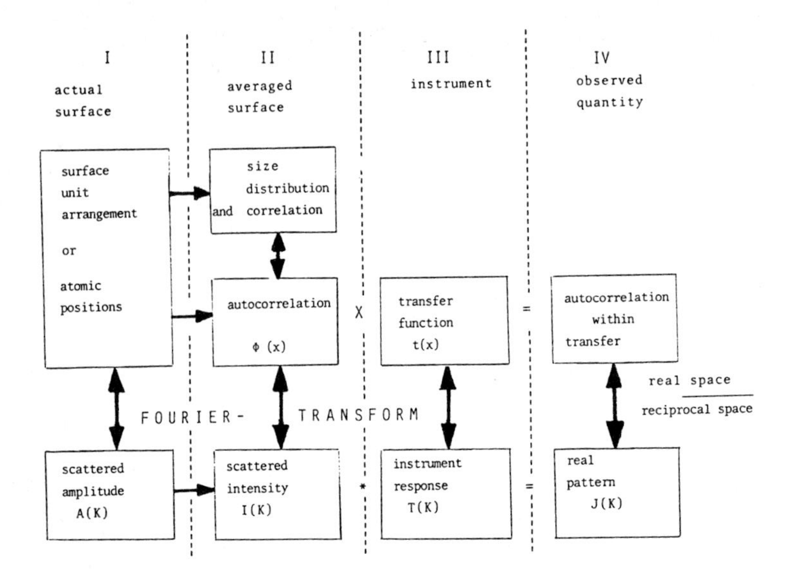

Fig.56.1. Scheme for quantitative evaluation of diffraction spot profiles

from the details of the spot profile. A schematic procedure of evaluation is shown in Fig.56.1. The actual surface is represented in the upper left corner, the observed pattern in the lower right corner. There are several ways to calculate the expected diffraction pattern or to extract parameters of the actual surface from the diffraction pattern. Starting from the description of the surface by atomic positions or with a probability distribution of domain sizes or terrace sizes, first the autocorrelation may be calculated either within an approximation [56.9] or numerically [56.10,11] or in closed form [56.12,13]. After Fourier transforming, the intensity is obtained as expected from an ideal instrument (the kinematic approximation is discussed in the next section). The instrumental parameters may be introduced via the transfer function in real space or the instrument response function in reciprocal space [56.9], Sect.56.6. Starting from the observed LEED pattern (Fig.56.1), all steps are reversible (convolution and Fourier transform) except those from the second to the first column (such as autocorrelation and absolute square). Notably it has been shown that the step from probability distribution to autocorrelation is reversible in closed form [56.13].

56.4 Kinematic Approximation

For the spot profile calculations done so far the kinematical approximation has usually been used. That means that the surface is described by identical units consisting of one or several surface atoms and all underlying atoms [56.6]. All dynamic effects are included except those due to different composition and neighborhood of the units. Therefore the approximation in its basic form is restricted to cases where all units have the same composition such as stepped surfaces, homoepitaxial layers, adlayer structures with superstructure (only with respect to extra spots). The effect of different neighborhoods is less important, the larger the number of units forming an entity (e.g., broad terraces or large superstructure domains). To date all published reports indicate that deviations due to small numbers of units are not detectable [56.14-16].

56.5 Dynamic Extensions

The kinematic approximation assumes that all units provide the same scattered amplitude within the (broadened) angular range of a diffraction spot. For an improvement the amplitudes have to be calculated separately for all existing units according to the now established standards of dynamic calculation [56.17-20] for the wanted energy and angle range. The spot profile is calculated from the autocorrelation using those dynamic amplitudes [56.14,15]. So far there are only a few theoretical approaches to the problem [56.14,15,20,21] or comparisons of experiment and calculation [56.14,21].

56.6 Instruments and Instrumental Limits

The ideal instrument should provide a diffraction spot of zero diameter, that is, the instrument response function should be a δ function and the transfer function should be a constant (zero half-width, infinite transfer width). There are, however, a lot of reasons to give a finite spot size even for an ideal crystal [56.6,7,9,22,23]: finite source diameter and lens aberration produce a finite spot size. The larger the electron energy, the easier it is to focus the beam (reduced space charge and magnetic and electrostatic disturbances). On the other hand, it is also important to have a better focus due to the smaller separation between the diffraction beams [56.6]. The energy spread is important only for nonspecular beams [56.6,23]. Other important factors are the divergence of the beam [56.23] and the inherent uncertainty in $K_\perp$ due to the deficiencies in $\Delta K_\parallel$ listed above [56.6]. The appropriate figure of merit for a given system is the transfer width (= half-width of the transfer function), although details of the transfer function like base width may also be important [56.9,22]. The performance of the usual and special instruments is described in [56.6,23]; transfer widths of 300 nm have been obtained, and improvements with transfer widths up to several microns may work soon. The usual 4-grid LEED optics provide a transfer width of 10-20 nm, so that depending on accuracy of data collection, distances up to 50 nm may be analyzed. A Faraday cup instead of the phosphoric screens (with grids) usually gives an appreciable improvement.

The advantage of a high-resolution system is not only a larger range of detectable defect sizes (so far up to about 1 micron), but it also considerably improves the range of qualitative detection. For example, in epitaxial and overlayer studies a central spike together with a broad shoulder is observed [56.24,25]. Detection of the central spike, which is important for the description of the surface, may be possible in many cases only with high-resolution systems [56.6]. With low-resolution systems at most a variation of half-width may be observed. Another important parameter is the dynamic range between maximum and minimum intensity, so that the spot profile may be measured and evaluated over most of the Brillouin zone.

56.7 Experiments

Most relevant experiments deal with steps on clean surfaces or overlayers on step-free surfaces. The steps on ion-bombarded and partially annealed surfaces have been measured especially for semiconductors [56.26-29]. The first stages of epitaxy have been studied with diffraction for both metals and semiconductors [56.24,25,30-36]. Several calculations concerning the spot profiles during epitaxy have been published [56.9,10,30,31,37-42]. Atomic steps have also been found after surface reactions: atomic hydrogen etches the silicon surface [56.43]. Steps at the interface formed during oxidation of silicon have been detected both with high-resolution electron microscopy [56.44-46] and with LEED [56.47-51]. The immediate influence of these steps on electrical properties has been shown [56.44-50]. There are many studies on overlayers on nearly perfect substrates, some of them showing distinct defect features like steps induced by oxygen adsorption on tungsten [56.52] and domain formation of oxygen or CO on different substrates [56.53-57]; more examples are discussed in the papers on phase transitions [56.58-59].

Metal overlayers on metals [56.60] or on semiconductors [56.61] show a wide variety of structural features, forming in part fairly perfect overlayers [56.62,63] or distinct defect structures. The time dependence of island growth has been studied [56.54,55,64]. The substrate may be affected by overlayer arrangements [56.52,65-67] forming steps or facets. The detection of mosaic defects of the substrate has been described in [56.51]. The contribution of thermal diffuse scattering to spot profile and background is discussed in [56.68,69]. Improvements are made by measuring the spot profile throughout the Brillouin zone with either high lateral [56.13] or energy resolution [56.70].

56.8 Conclusion

The wide range of applications shows that spot profile analysis provides a lot of additional information beyond the currently more usual intensity analysis for the study of periodic surfaces. Recent developments in theory provide the possibility of including dynamical features into profile analysis, the instruments available now provide more qualitative and quantitative details on spot profiles. The combination of these two features promises further progress in the near future.

References

56.1 C.J. Davison, L.H. Germer: Phys. Rev. **30**, 705 (1927)
56.2 R.L. Park: J. Appl. Phys. **37**, 295 (1966)
56.3 P. Estrup, E.G. McRae: Surf. Sci. **25**, 1 (1971)
56.4 M. Henzler: In *Electron Spectroscopy for Surface Analysis*, ed. by H. Ibach, Topics Curr. Phys., Vol.4 (Springer, Berlin, Heidelberg 1977)
56.5 M.G. Lagally: Appl. Surf. Sci. **13**, 260 (1982)
56.6 M. Henzler: Appl. Phys. A**34** (in press), and Surf. Sci. (in press)
56.7 M.G. Lagally: "Diffraction Techniques", in *Methods of Experimental Physics: Surfaces* (Academy, New York 1984)
56.8 M. Henzler: Appl. Surf. Sci. **11/12**, 450 (1982)
56.9 R.L. Park, J.E. Houston: Surf. Sci. **21**, 205 (1970); **26**, 269 (1971)
56.10 M. Henzler: Surf. Sci. **73**, 240 (1978)
56.11 J.M. Pimbley, T.-M. Lu: J. Appl. Phys. **55**, 182 (1984)
56.12 P.R. Pukite, C.S. Lent, P.I. Cohen: To be published
56.13 H. Busch: Master Thesis, University of Hannover, Fed. Rep. of Germany (1984), and to be published
56.14 H. Jagodzinski, W. Moritz, D. Wolf: Surf. Sci. **77**, 233, 249, 265, 283 (1978)
56.15 W. Moritz: Inst. Phys. Ser. **41**, Chap.4 (London 1978)

56.16 G.-C. Wang, T.-M. Lu: Phys. Rev. Lett. **50**, 2014 (1983)
56.17 F. Jona, J.A. Strozier, Jr., P.M. Marcus: This Volume, p.92
56.18 H.L. Davis, J.R. Noonan: To be published
56.19 M.A. Van Hove: This Volume, p.100
56.20 J.B. Pendry: This Volume, p.124, and Surf. Sci. (in press)
56.21 D. Saloner, M.G. Lagally: J.V.S.T. A**2**, 935 (1984), and this Volume, p.366
56.22 R.L. Park, J.E. Houston, D.G. Schreiner: Rev. Sci. Instrum.**42**, 60 (1971)
56.23 M.G. Lagally, J.A. Martin: Rev. Sci. Instrum. **54**, 1273 (1983)
56.24 P.O. Hahn, J. Clabes, M. Henzler: J. Appl. Phys. **51**, 1273 (1983)
56.25 K.D. Gronwald, M. Henzler: Surf. Sci. **117**, 180 (1982)
56.26 M. Henzler: Surf. Sci. **22**, 12 (1970)
56.27 G. Schulze, M. Henzler: Surf. Sci. **73**, 553 (1978)
56.28 D.G. Welkie, M.G. Lagally: J.V.S.T. **16**, 784 (1979); H.M. Clearfield, M.G. Lagally: J.V.S.T. A**2**, 844 (1984)
56.29 F.W. Wulfert: Ph.D. Thesis, Hannover (1982); M. Henzler: Surf. Sci. **132**, 82 (1983)
56.30 J.M. Van Hove, P. Pukite, P.I. Cohen, C.S. Lent: J.V.S.T. A**1**, 609 (1983); B**1**, 741 (1983)
56.31 P.R. Pukite, J.M. Van Hove, P.I. Cohen: J.V.S.T. B**2**, 243 (1984)
56.32 P.J. Dobson, J.H. Neave, B.A. Joyce: Surf. Sci. **119**, L339 (1982); Appl. Phys. A**34**, 179 (1984)
56.33 J.J. Harris, B.A. Joyce, P.J. Dobson: Surf. Sci. **108**, L444 (1981)
56.34 P.J. Dobson, N.G. Norton, J.H. Neave, B.A. Joyce: Vacuum **33**, 593 (1983)
56.35 K. Rudnick, M. Henzler: Phys. Electronic Conf. 1984 and to be published
56.36 C.S. Lent, P.I. Cohen: Surf. Sci. (in press)
56.37 C.S. Lent, P.I. Cohen: J.V.S.T. (in press)
56.38 J.M. Pimbley, T.-M. Lu: J. Appl. Phys. **55**, 182 (1984); J.V.S.T. (in press)
56.39 J.M. Cowley, H. Schuman: Surf. Sci. **38**, 53 (1973)
56.40 J. Lapujoulade: Surf. Sci. **108**, 526 (1981), and these Proceedings
56.41 T.-M. Lu, M.G. Lagally: Surf. Sci. **120**, 47 (1982); J.V.S.T. **17**, 207 (1980)
56.42 M.R. Presicci, T.-M. Lu: Surf. Sci. (in press)
56.43 G. Schulze, M. Henzler: Surf. Sci. **124**, 336 (1983)
56.44 S.M. Goodnick, R.G. Gann, J.R. Sites, D.K. Ferry, C.W. Wilmsen, D. Fathy, O.L. Krivanek: J.V.S.T. B**11**, 803 (1983)
56.45 D.K. Ferry: J.V.S.T. (in press)
56.46 T. Sugano, J.J. Chen, T. Hamano: Surf. Sci. **98**, 154 (1980)
56.47 P.O. Hahn, M. Henzler: J. Appl. Phys. **52**, 4122 (1981)
56.48 P.O. Hahn, M. Henzler: J. Appl. Phys. **54**, 6492 (1983)
56.49 P.O. Hahn, S. Yokohama, M. Henzler: Surf. Sci. (in press)
56.50 P.O. Hahn, M. Henzler: J.V.S.T. A**2**, 574 (1984)
56.51 M. Henzler, P. Marienhoff: J.V.S.T. (in press)
56.52 H.M. Kramer, E.G. Bauer: Surf. Sci. **92**, 53 (1980); **93**, 407 (1980)
56.53 M.G. Lagally, G.-C. Wang, T.-M. Lu: In *Ordering in Two Dimensions*, ed. by S.K. Sinha (North-Holland, New York 1980) p.113
56.54 G.-C. Wang, T.-M. Lu: Phys. Rev. Lett. **50**, 2014 (1983)
56.55 J.H. Perepezko, J.T. McKinney, P.K. Wu, M.G. Lagally: Phys. Rev. Lett. **51**, 1577 (1983)
56.56 E.D. Williams, W.H. Weinberg: Surf. Sci. **109**, 574 (1981)
56.57 F.M. Hoffmann, A. Ortega, H. Pfnür, D. Menzel, A.M. Bradshaw: J.V.S.T. **17**, 239 (1980)
56.58 R.F. Willis: This Volume, p.237
56.59 S.C. Fain: This Volume, p.413
56.60 E.G. Bauer: Appl. Surf. Sci. **11/12**, 479 (1982)
56.61 P.S. Ho, K.N. Tu: *Thin Films and Interfaces* (North Holland, Amsterdam 1982)

56.62 R.T. Tung, J.M. Gibson, J.M. Poate: Phys. Rev. Lett. **50**, 429 (1983)
56.63 R.T. Tung, J.M. Poate: Surf. Sci. (in press)
56.64 E. Suliga, M. Henzler: J. Phys. C**16**, 1543 (1983)
56.65 E. Suliga, M. Henzler: J.V.S.T. A**1**, 1507 (1983)
56.66 J. Clabes, W. König: To be published
56.67 J. Clabes: Surf. Sci. (in press)
56.68 R.L. Dennis, M.B. Webb: Surf. Sci. **58**, 429 (1976)
56.69 M.G. Lagally: In *Surface Physics of Materials*, ed. by J.M. Blakely (Academic, New York 1975)
56.70 M. Rocca, H. Ibach, S. Lehwald, B.M. Hall, M.L. Xu, S.Y. Tong: This Volume, p.156

57. Measurement of the Specific Heat Critical Exponent Using LEED*

N.C. Bartelt and T.L. Einstein

Department of Physics and Astronomy, University of Maryland
College Park, MD 20742, USA

L.D. Roelofs

Department of Physics and Astronomy, University of Maryland, College Park, MD 20742 and Physics Department, Haverford College, Haverford, PA 19041, USA

Near a second-order phase boundary, integrated intensities of "extra" LEED beams exhibit a $|T - T_c|^{1-\alpha}$ singularity, where α is the specific heat critical exponent. We discuss the origin of this effect, apply it to real and Monte-Carlo-generated data, and comment on generalizations.

Interest in using low-energy electron diffraction (LEED) to study the critical behavior of surface phase transitions has, in part, been inspired by analogies with X-ray and neutron-diffraction experiments. These analogies, however, have limits. The resolution of typical LEED instruments is significantly smaller than that of X-ray (but not neutron) experiments. Multiple scattering is generally important in LEED (though the new length scales it introduces are small). These problems make it more difficult to study the diverging correlation length near a second-order phase transition using LEED and have probably hindered the widespread use of LEED as a probe of critical phenomena. Here we point out that an alternative application of LEED can be used to study the energy singularities associated with a phase transition.

Scattered electron intensities in LEED are determined by correlation functions of the atoms at the surface. The finite resolution of LEED instruments means that these intensities are sensitive only to correlations of finite range. As *Fisher*, *Langer* [57.1] and others [57.2] pointed out long ago, and as we have recently emphasized [57.3,4], finite-length correlations can be expected to have the same singularity (viz. $|t|^{1-\alpha}$) as the energy at a critical point. Thus, if I(T) represents the temperature-dependent *integrated* LEED intensity, then near a critical point

$$I(T) = \begin{cases} A + B_-|t|^{1-\alpha'} + C_-|t| + \ldots & T < T_c \\ A - B_+|t|^{1-\alpha} - C_+|t| + \ldots & T > T_c \end{cases} , \qquad (57.1)$$

where α, α' are the exponents governing the divergence of the specific heat, and t is the reduced temperature $(T - T_c)/T_c$. This statement assumes I(T) is insensitive to the phase of the order parameter.[1] Scaling predicts $\alpha' = \alpha$ [57.5]. By universality, the ratio B_+/B_- will be independent of experimental details and the degree of multiple scattering [57.6].

1 The phase of the order parameter refers to which *component* is present in a particular ordered state. For example, for $c(2\times2)$ order the phase of the order-parameter determines which $c(2\times2)$ sublattice is occupied

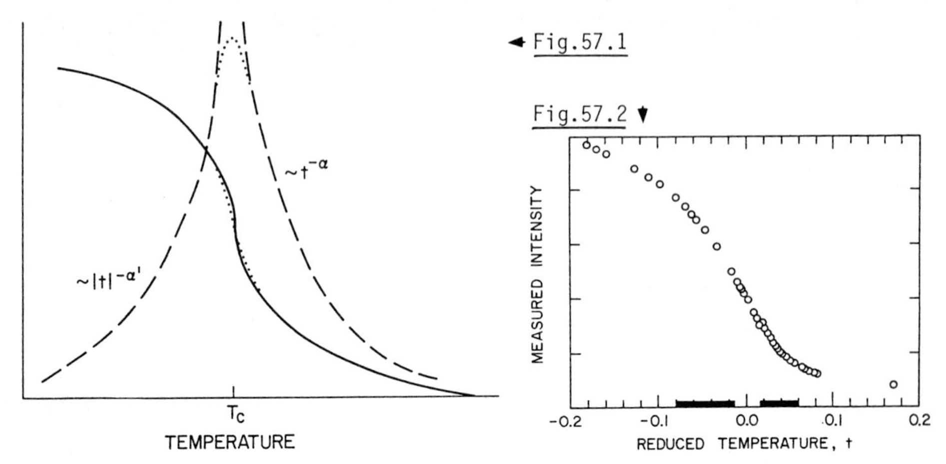

Fig.57.1. Schema of the energy-like singularity associated with positive α, seen by monitoring the intensity of an "extra" spot (——) in a diffraction experiment. The derivative with respect to temperature (---) diverges like the specific heat. (····) indicates the finite-size rounding which occurs in real systems when the correlation length exceeds the size of defect-free regions

Fig.57.2. Intensity of an adlayer-induced LEED spot vs T for p(2 × 2) O/Ni (111). The Debye-Waller factor was removed (with negligible effect here). The data is from the undeconvoluted center of the spot, corresponding to an integration ~3% of the way to the nearest integer-order beam. Similar curves are obtained by deconvolution followed by explicit integration. The heavy bars on the abscissa indicate the thermal range used to obtain β, γ, and ν in [57.7]

To be explicit, consider a system of gas atoms adsorbed on a single-crystal substrate. At low temperatures this system might have a long-range order which vanishes continuously as T is raised. If in a LEED experiment the intensity of a diffraction feature associated with this order (an integrated spot intensity, for example) is monitored, then one expects to see a singularity of the type shown schematically in Fig.57.1. Equation (57.1) applies to the regime in which $k_I\xi(t)$ is large, where k_I is the instrumental resolution (essentially the radius from the center of the spot over which one integrates) and ξ is the correlation length, which diverges like $|t|^{-\nu}$. Farther from T_C, when $k_I\xi$ is small, one recovers the more familiar diffraction picture adopted in [57.7,8]. In this language, deconvolution is an attempt to decrease k_I numerically. Notice that the *point of inflection* of the integrated intensity curve corresponds to the maximum of the specific heat, and offers a *natural estimate* of the position of the critical temperature T_C.

As a sample application of these ideas, Fig.57.2 contains the intensity of a half-order LEED spot as a function of temperature for the p(2 × 2)O/Ni (111) order-disorder transition [57.7]. There is no temperature where the slope of this function appears to diverge. Thus either α is negative or small [how small depends upon how small a number one can accept for the coefficient $B_\pm$ of (57.1)], or finite-size rounding (Fig.57.1) obscures most of the critical behavior. The value $\alpha \to 0$ is consistent by exponent scaling relations

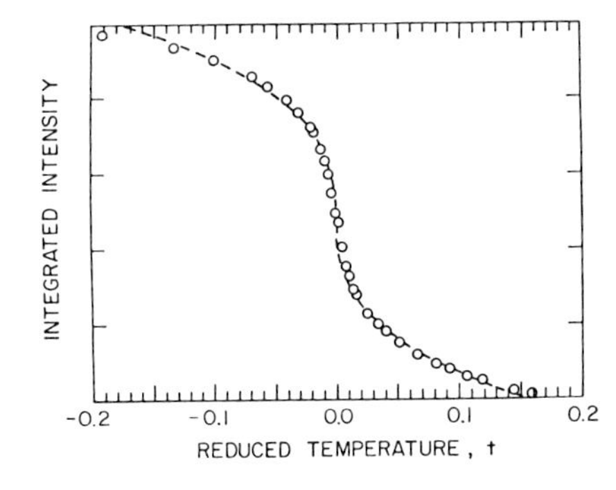

Fig.57.3. Simulated LEED intensity (o) for a p(2 × 2) overlayer on a triangular substrate vs reduced temperature. The energy-like singularity is particularly evident for this universality class (which has an unusually large α of 2/3). (---) is the result of a 6-parameter fit based on (57.1). See text for details

[57.8] with the Ising-like exponents β, γ, and ν found previously using the diffraction picture [57.7]. For comparison, the results of a Monte Carlo calculation of the structure factor, i.e., the kinematic LEED pattern, of a p(2 × 2) structure on a triangular lattice is shown in Fig.57.3. The size lattice used (3888 sites) is comparable to the expected average plateau size on the Ni(111) surface used in the experiment. The structure factor was integrated 12% of the way to the nearest integer order position to mimic poor instrumental resolution. Only data within 15% of T_c was used, while points within 1% of T_c were discarded because of finite-size rounding. Fitting this data to the form of (57.1) (with $C_\pm = 0$, however) yields $\alpha = 0.64 \pm 0.06$, $\alpha' = 0.61 \pm 0.06$, and $B_+/B_- = 1.0 \pm 0.1$. These values are consistent with those ($\alpha = 2/3$, $B_+/B_- = 1$) of the 4-state Potts model universality class, as expected [57.8.9], corroborating the previous conclusion that there is some complication in the case of O/Ni(111). It is noteworthy that α distinguishes more clearly between different universality classes than the other three exponents, especially the often-sought β.

An application of these ideas is that one can study critical phenomena by collecting the current from the *entire* LEED screen. At Maryland, *Iwasaki* and *Zhu* are studying the (7 × 7) reconstruction of Si(111) with this method [57.10].

A natural way to study the wave-number dependence of the critical scattering in a LEED experiment in the above framework is to observe that if the instrumental resolution k_I is changed, measured LEED intensities should satisfy a scaling relation

$$I(k_I, T) = |t|^{2\beta} Y\, k_I \xi(t) \quad , \tag{57.2}$$

where the generally unknown scaling function Y depends on the universality class. The exponents β and ν can thus be obtained by requiring that intensities for different instrumental resolutions satisfy this relation. This technique is analogous to finite-size scaling [57.11], which is often used in statistical mechanics [57.12]. The method takes advantage of the experimenter's control over instrumental resolution to study critical phenomena.

As final comments, we note that any probe of short-range order which is sensitive to a surface transition can be similarly used to study the exponent α or the order of the phase transition. (Examples might include frequency shifts in electron energy loss spectroscopy or infrared spectroscopy or level shifts in photoemission [57.3].) If the transition is first order, this method will detect the associated discontinuity [57.13], perhaps rounded by finite-size effects, in the energy. To avoid the added complication of Fisher renormalization [57.14], one should perform the experiment at an extremum of the temperature vs coverage boundary ($dT_c/d\theta = 0$), as was done for O/Ni(111).

References

*Work supported by the Department of Energy under grant DE AS05-79ER-10427. Computer facilities supplied by U. of Maryland Computer Science Center.

57.1 M.E. Fisher, J.S. Langer: Phys. Rev. Lett. **20**, 665 (1968)
57.2 M.E. Fisher: In *Critical Phenomena*, ed. by M.S. Green, J.V. Sengers (National Bureau of Standards, Washington 1966)
E. Riedel, F. Wegner: Phys. Lett. **29**A, 77 (1969)
57.3 N.C. Bartelt, T.L. Einstein, L.D. Roelofs: Surf. Sci. Lett. (to be published)
57.4 N.C. Bartelt, L.D. Roelofs, T.L. Einstein: Preprint
57.5 H.E. Stanley: *Introduction to Phase Transitions and Critical Phenomena* (Oxford University Press, New York 1971)
57.6 A. Aharony, P.C. Hohenberg: Phys. Rev. B**13**, 3081 (1976)
57.7 L.D. Roelofs, A.R. Kortan, T.L. Einstein, R.L. Park: Phys. Rev. Lett. **46**, 1465 (1981)
57.8 T.L. Einstein: In *Chemistry and Physics of Solid Surfaces IV*, ed. by R. Vanselow, Springer Ser. Chem. Phys., Vol.20 (Springer, Berlin, Heidelberg 1982) Chap.11 and references therein
57.9 M. Kaufman, D. Andelman: Phys. Rev. B**20**, 413 (1984)
57.10 H. Iwasaki, Q.-G. Zhu, N.C. Bartelt, R.L. Park: Preprint
57.11 M.N. Barber: In *Phase Transitions and Critical Phenomena*, Vol.8, ed. by C. Domb, J.L. Lebowitz (Academic, London 1983) Chap.2;
M.E. Fisher, M.N. Barber: Phys. Rev. Lett. **28**, 1516 (1972)
57.12 D.P. Landau: Phys. Rev. B**13**, 2997 (1976); **14**, 255 (1976);
K. Binder: In *Monte Carlo Methods in Statistical Physics*, ed. by K. Binder (Springer, Berlin, Heidelberg 1979) Chap.1
57.13 Associated with the discontinuity fixed point is the exponent $\alpha = 1$, M.E. Fisher, A.N. Berker: Phys. Rev. B**26**, 2507 (1982)
57.14 M.E. Fisher: Phys. Rev. **176**, 257 (1968)

58. Short-Range Correlations in Imperfect Surfaces and Overlayers

J.M. Pimbley and T.-M. Lu

Center for Integrated Electronics and Physics Department, Rensselaer Polytechnic Institute, Troy, NY 12181, USA
and General Electric Corporate Research and Development Center, Schenectady, NY 12301, USA

Extended defects (such as steps) on imperfect, periodic surfaces and relative positions of surface adsorbates may both exhibit short-range order. One powerful technique for studying this surface ordering is spot profile analysis of diffraction (low-energy and high-energy electron, atom and X-ray) measurements. This analysis requires only kinematic (single-scattering) diffraction calculations and thus is more tractable than conventional multiple-scattering problems. Here we discuss kinematic surface scattering results and their applicability to surface ordering studies.

Many techniques have been used to study surface and overlayer imperfections. Direct imaging techniques include scanning electron microscopy (SEM), transmission electron microscopy (TEM), and scanning tunneling microscopy (STM). Examples of indirect methods include ion-scattering techniques such as Rutherford backscattering spectroscopy (RBS) and low-energy ion scattering (LEIS) as well as atom, X-ray and electron diffraction. Although many of the direct imaging techniques have now achieved atomic resolution, the indirect methods still play an important role which cannot be replaced by the imaging techniques. Among them, the diffraction methods, notably electron diffraction, have been shown to be particularly useful for in situ, quantitative study of surface extended defects and the imperfections associated with the ordering and growth of overlayers on the surface [58.1,2]. In this paper, we report the progress we have made on the theory of diffraction from imperfect surfaces and overlayers.

If the surface or overlayer is perfect, the diffraction will give very sharp beams in which the width is controlled by the instrumental broadening [58.3]. For an imperfect surface or overlayer, due to the lack of long-range order, the beam will broaden, split, or develop shouldered structures depending on the nature and density of imperfections. In order to interpret the beam shape, one must assume a distribution of defects and evaluate the atomic pair correlation function $p(\mathbf{r})$ for the surface, overlayer, or the combined overlayer/substrate system [58.1,2]. The angular distribution of the diffracted intensity, or simply the "angular profile", is then calculated within the kinematic approximation by taking the Fourier transform of $p(\mathbf{r})$ [58.4]. The calculated angular profile is then compared to the measured angular profile. In the kinematic approximation, all diffracted electrons (or X-rays) suffer only one scattering event. Even for electrons, this approximation describes very well the diffraction beam shape at any energy. The energy dependence of the peak intensity, however, requires a complete multiple-scattering calculation. We shall begin by discussing the one-dimensional (1D) diffraction problem and treat the more difficult two-dimensional (2D) problem later.

The most common extended defects in surfaces or overlayers are perhaps steps and antiphase boundaries. Here we shall consider stepped surfaces as our example and the treatment can be generalized to surfaces or overlayers with antiphase boundaries. Steps occur on clean surfaces [58.5,6] as well as in adsorbed overlayers [58.7,8]. Quantitative study of the distribution of steps can give information on the force laws governing the clean surface and the films grown on the surface.

Several classes of model stepped surfaces have attracted extensive discussion in the literature. The vicinal surface [58.5,6], in which steps occur at well-defined, regular intervals, reflects very strong long-range step-step interaction. The diffraction pattern shows well-defined split Bragg peaks. The width of each split beam is determined by the instrument response function. The separation of the split beams is inversely proportional to the terrace width and is independent of the instrumental broadening effect.

Another class of stepped surfaces with important applications are those with a random distribution of steps [58.9-13]. In this case, step-step interactions are absent. It has been shown that a random distribution of steps leads to a geometric distribution of terrace widths [58.10]. The diffraction beams generally show broadening. Special techniques have been developed [58.9-13] to solve the diffraction problem for this particular terrace width distribution and the angular profile can be solved exactly in closed form. Recent applications of this model include the epitaxial growth of adsorbed islands [58.12,13] and the incommensurate domain walls in a reconstructed surface [58.14].

The diffraction problem for an *arbitrary* distribution (1D) of terrace widths is more difficult and very often does not lead to a simple closed-form solution. In the past, a number of general theoretical approaches have been introduced to evaluate the atomic pair correlation for an arbitrary distribution of terrace widths. These theories include an approximate analytical method [58.15], schemes for computer calculation [58.16,17], and an exact and analytical method [58.13,18] developed by us recently. For illustration, we shall use the latter method to solve a diffraction problem involving a cutoff geometric distribution of terrace widths. This distribution represents a result of very strong short-range step-step interactions. The distribution can be written as

$$P(N) = \gamma(1-\gamma)^{N-N_0} \quad , \quad N \geqslant N_0$$

$$= \quad 0 \quad , \quad N < N_0 \quad ,$$

where N_0 is the minimum existing terrace width and γ is the probability of finding a surface atom located at a step edge with the restriction that the distance between two adjacent steps cannot be smaller than $N_0 a$, where a is the lattice constant perpendicular to the step edge. The average terrace size $\bar{N}a$ is given by $(1/\gamma + N_0 - 1)a$. We specify the probability that a surface (vertical) step is either "up" or "down". The atomic pair correlation function for this step distribution can be evaluated exactly using the method outlined in [58.18]. The angular profile is then obtained by taking the Fourier transform of the pair correlation function.

To study the effect of N_0 and γ on the shape of the angular profile, we calculated the angular profile as a function of N_0 and γ for a fixed value of $\bar{N}$, assuming $\bar{N} = 7$. For $N_0 = 1$ ($\gamma = 1/7$), the model reduces to the case of a random distribution of steps which has been solved previously using other

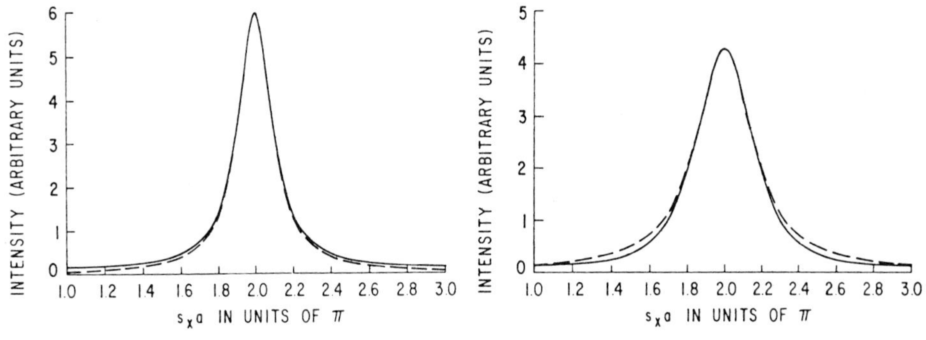

Fig.58.1. The angular distribution of diffraction intensity for a cutoff geometric distribution of terrace widths evaluated at the out-of-phase (between adjacent terraces) scattering condition, $\bar{N} = 7$, $N_0 = 1$, and $\gamma = 1/7$. (---) is a Lorentzian fit

Fig.58.2. The angular distribution of diffraction intensity for a cutoff geometric distribution of terrace widths evaluated at the out-of-phase (between adjacent terraces) scattering condition. $\bar{N} = 7$, $N_0 = 2$, $\gamma = 1/6$. (---) is a Lorentzian fit

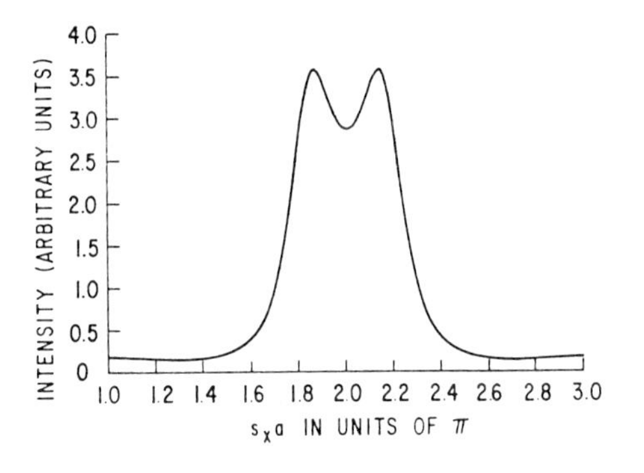

Fig.58.3. The angular distribution of diffraction intensity for a cutoff geometric distribution of terrace widths evaluated at the out-of-phase (between adjacent terraces) scattering condition. $\bar{N} = 7$, $N_0 = 3$, and $\gamma = 1/5$

techniques [58.11,18]. The angular profile is broadened and the shape is close to a Lorentzian function. Figure 58.1 is a plot of the (10) beam evaluated at the out-of-phase (between adjacent terraces) scattering condition. No instrumental broadening has been included. The dashed curve is a Lorentzian profile with the same width as the calculated profile. The momentum transfer parallel to the surface is given by $s_x a$. For $N_0 = 2$ ($\gamma = 1/6$), which corresponds to the case of no single-atom terraces, the profile deviates slightly from the Lorentzian shape as shown in Fig.58.2. Since the deviation is small, it is probably difficult to differentiate experimentally.

A drastic change in the angular profile shape occurs when $N_0 = 3$ ($\gamma = 1/5$), which corresponds to the case where no terrace has a width equal to or less than two atoms. As is shown in Fig.58.3, the profile splits and this splitting should be measurable using a conventional electron diffractometer [58.3]. For $N_0 > 3$, the splitting becomes sharper. The broadening of split beams becomes narrower as $\bar{N} \to N_0$ ($\gamma \to 1$). When $\bar{N} = N_0$, the model reduces to the vicinal surface and the width of the split beams reduces to zero. (In an experiment, the width of a beam can never be zero, but is limited by the instrumental broadening). All profiles are evaluated at the out-of-phase scattering condition. Qualitatively, the separation of the splitting is a measure of the average terrace size, and the width of the individual beams is a measure of the spread in the terrace width distribution.

Other model surfaces representing short-range step-step interactions have also been considered by *Henzler* [58.16] and by *Saloner* and *Lagally* [58.17]. The pair correlation functions are generated with the aid of a computer. In these models, the sharp cutoff is replaced by smoother and rounded distributions which are, perhaps, more realistic. In a sense, our cutoff geometric distribution represents an extreme example of strong, short-range step-step interactions.

Step-step interactions are not the only source that determines the terrace width distribution. For example, during the epitaxial growth of overlayer films [58.7,8], the terrace width distribution can be controlled by the kinetics of ordering (growth modes). Recent in situ electron diffraction studies of molecular beam epitaxy showed that both geometric and nongeometric terrace width distributions can occur at the initial stages of epitaxy [58.12,13,19, 20].

The approaches mentioned above for obtaining the angular profile are restricted to one dimension. Extension to two dimensions can be made trivially [58.1,21] under the assumption that the step probability for the two lateral directions, say **a** and **b** on the surface, are independent and are for an unrestricted (unspecified) number of layers. Of course, these assumptions may not be realistic in practice.

Recently, we developed a technique which allows us to calculate the nontrivial two-dimensional pair correlation function for a random distribution of steps in which the step probabilities in the **a** and **b** directions are mutually dependent. It is applicable to any specified number of layers. This model is a two-dimensional extension of the one-dimensional random step model with a geometric distribution of terrace widths. However, we are not able to write down the corresponding 2D terrace width distribution for this particular model. This is due to the infinite possibilities of the terrace shape. The pair correlation function is a function of the elements of a third rank tensor T_{prq}. For a two layer system, $p,q,r = 1$ or 2. Further, T_{111} is defined as the probability that lattice site (i,j) is occupied when sites $(i-1,j)$ and $(i,j-1)$ are occupied: T_{211}, T_{112} and T_{212} have analogous interpretations. It can be shown that only four of the eight tensor elements T_{prq} are independent. The pair correlations can be calculated for any coverage. The angular profile consists of a Dirac delta function at exactly the Bragg peak position and a broader background profile (an approximate 2D Lorentzian function). These features are very similar to the result obtained from the one-dimensional random step model for a two-layer system with a geometric distribution of terrace widths [58.12,13]. The 2D angular profile also resembles the measured angular profile obtained by *Gronwald* and *Henzler* [58.19] in their Si/Si(111) molecular beam epitaxy experiment.

As a special case, consider $T_{111} = T_{112} = T_{211} = T_{212} = \theta$. This means that the occupation probability at lattice sites (i,j) is always equal to the coverage θ and is independent of the occupation at neighboring lattice sites. The model then reduces to the 2D random adsorption model which we have solved previously [58.22].

So far, almost all the surface diffraction results were interpreted either using 1D analytical techniques or using 2D optical simulation [58.23]. With the 1D analytical techniques, quantitative results can be obtained if the defects can be represented by 1D models such as vicinal surfaces. If the defects are of 2D nature, such as 2D random steps, the 1D analysis works only semiqualitatively. Unlike the 1D case in which the defect distribution is controlled by a well-defined 1D terrace width distribution for each layer, the 2D defect distribution is controlled by both the shape and size of the "2D terraces" in each layer. Because of this complication, the 2D pair correlation function for an *arbitrary* distribution of defects (steps, antiphase boundaries) has not been solved so far. Our 2D random step model outlined above represents the first nontrivial 2D step model that has been solved analytically. With the recent advances of high-resolution electron diffractometers [58.24,25], a more rigorous and quantitative treatment of the 2D diffraction problems is certainly very desirable.

References

58.1 M. Henzler: In *Electron Spectroscopy for Surface Analysis*, ed. by H. Ibach, Topics Curr. Phys., Vol.4 (Springer, Berlin, Heidelberg 1977)
58.2 M.G. Lagally: In *Chemistry and Physics of Solid Surfaces IV*, ed. by R. Vanselow, R. Howe, Springer Ser. Chem. Phys., Vol.20 (Springer, Berlin, Heidelberg 1982)
58.3 For a review, see T.-M. Lu, M.G. Lagally: Surf. Sci. **99**, 695 (1980)
58.4 A. Guinier: *X-ray Diffraction* (Freeman, San Francisco 1963)
58.5 H. Wagner: In *Solid Surface Physics*, ed. by G. Höhler, Springer Tracts Mod. Phys., Vol.85 (Springer, Berlin, Heidelberg 1979) p.151
58.6 G.A. Somorjai: In *Treatise on Solid State Chemistry*, Vol.6A, ed. by N.B. Hannay (Plenum, New York 1976)
58.7 J.D. Weeks, G.H. Gilmer: In *Advances in Chemical Physics*, Vol.40, ed. by I. Prigogine, S.A. Rice (Wiley, New York 1979)
58.8 A.A. Chernov: In *Crystal Growth and Characterization*, ed. by R. Ueda, J.B. Mullin (North-Holland, Amsterdam 1975)
58.9 D. Wolf, H. Jagodzinski, W. Moritz: Surf. Sci. **77**, 283 (1978)
58.10 T.-M. Lu, M.G. Lagally: Surf. Sci. **120**, 47 (1982)
58.11 M.R. Presicci, T.-M. Lu: Surf. Sci. **141**, 233 (1984)
58.12 C.S. Lent, P.I. Cohen: Surf. Sci. **139**, 122 (1984); J. Vac. Sci. Technol. A**2** (2), 861 (1984)
58.13 J.M. Pimbley, T.-M. Lu: J. Vac. Sci. Technol. A**2** (2), 457 (1984); J. Appl. Phys. (in press)
58.14 P. Fenter, T.-M. Lu: To be published
58.15 J.E. Houston, R.L. Park: Surf. Sci. **21**, 209 (197o); **26**, 269 (1971)
58.16 M. Henzler: Surf. Sci. **73**, 240 (1978)
58.17 D. Saloner, M.G. Lagally: J. Vac. Sci. Technol. A**2** (2), 935 (1984)
58.18 J.M. Pimbley, T.-M. Lu: J. Appl. Phys. **55** (1), 182 (1984)
58.19 K.D. Gronwald, M. Henzler: Surf. Sci. **117**, 180 (1982)
58.20 P. Hahn, J. Clabes, M. Henzler: J. Appl. Phys. **51** (4), 2079 (1980)
58.21 T.-M. Lu, M.G. Lagally: J. Vac. Sci. Technol. **17** (1), 207 (1980)
58.22 J.M. Pimbley, T.-M. Lu: Surf. Sci. **139**, 360 (1984)
58.23 H. Lipson: *Optical Transforms* (Academic, New York 1972)
58.24 M. Henzler: Appl. Surf. Sci. **11/12**, 450 (1982)
58.25 J. Martin, M. Lagally: J. Vac. Sci. Technol. A**1**, 1210 (1983)

59. Domain Size Determination in Heteroepitaxial Systems from LEED Angular Profiles

D. Saloner and M.G. Lagally

Department of Metallurgical and Mineral Engineering and Materials Science Center, University of Wisconsin, Madison, WI 53706, USA

The angular distribution of intensity in the fundamental beams diffracted from a surface with a lattice gas overlayer consists of two components, a broad "diffuse intensity" distribution reflecting the part of the correlation function resulting from the size distribution of ordered domains at the surface, and a delta-function component reflecting the part of the correlation function due to the order of the clean substrate. The magnitude of the diffuse component depends on electron-beam energy, degree of order, and coverage. We present an investigation of the relative weighting of the two components as a function of these parameters. Results of kinematic and multiple-scattering models are compared with each other and with measurements for GaAs(110) p(1 × 1)-Sb.

Under suitable conditions, the deposition of a submonolayer of atoms on a substrate surface can lead to the growth of ordered two-dimensional islands. These conditions include a net cohesive adatom-adatom interaction, relative strengths of adatom and substrate interactions that allow the formation of a lattice-gas overlayer, and a high enough temperature to permit ordering to occur. The shape, size distribution, and location of two-dimensional islands are determined by the nature of the defects on the substrate, by interatomic forces, and by externally adjustable parameters such as temperature and coverage. Investigation of the growth of islands in submonolayer films will provide insight into ordering processes in general. For example, different growth modes, resulting from different physical mechanisms by which the ordering occurs, lead to different distributions of ordered-domain sizes. A determination of these size distributions can identify the growth mode and thus provide information on atomic interactions, influence of defects, and the ordering mechanisms. This type of information is essential in understanding epitaxy or the properties of epitaxial films, as well as in studies of phase transitions in submonolayer films and of ordering kinetics.

An analysis of the diffracted-intensity distribution in angular profiles measured in surface-sensitive diffraction techniques such as LEED, RHEED, atomic-beam diffraction, or grazing X-ray diffraction, provides information on the domain size distribution [59.1,2]. We report here on calculations directed at determining island size distributions from the distribution of intensity in profiles of fundamental diffracted beams, i.e., those that appear for the substrate as well as the overlayer. When the submonolayer grows with a superlattice [(nxm)] structure, superlattice reflections will appear in the diffraction pattern. An analysis of the superlattice reflections will provide only partial information on the surface order, as the shape of these spots, which is independent of beam energy, contains information on the overlayer domains only (at least in the kinematic limit [59.3]). Furthermore, many systems order in p(1 × 1) structures, which produce only fundamental reflections. In the fundamental reflections for both superlattice and $p(1\times1)$ ordering, substrate and overlayer scattering interfere and these reflections therefore contain information on both the substrate and overlayer domain size distribution. A kinematic model is developed to evaluate the major contributions to the profiles. The calculations are completely general in the specification of size distributions and are applicable to all overlayer systems. The kinematic calculations provide estimates, at any given beam energy, of the proportion of the intensity profile that is sensitive to the domain size distribution. Multiple-scattering results are also utilized to investigate the same question. Specific application is made to the (01) beam measured from LEED from GaAs(110) $p(1\times1)$-Sb.

The determination of the shape of the angular profile of a diffracted beam requires evaluating the correlation function of the repeating elements that comprise the crystal. To calculate the intensity distribution in a diffracted beam we divide the crystal into columns perpendicular to the surface, that have the repeat distance a of the substrate unit cell. We denote be F_k the column scattering factor for column type k. For clean surfaces the columns will be identical. When overlayer atoms are deposited on the surface a number of different configurations can be realized, some of which are shown in Fig. 59.1. In homoepitaxy the overlayer unit cell A has a scattering factor identical to that of the substrate unit cell B, and occupies the bulk lattice site, i.e., $f_A = f_B$ and $t_{AB} = t_{BB}$. In heteroepitaxy either or both of these equalities may not pertain.

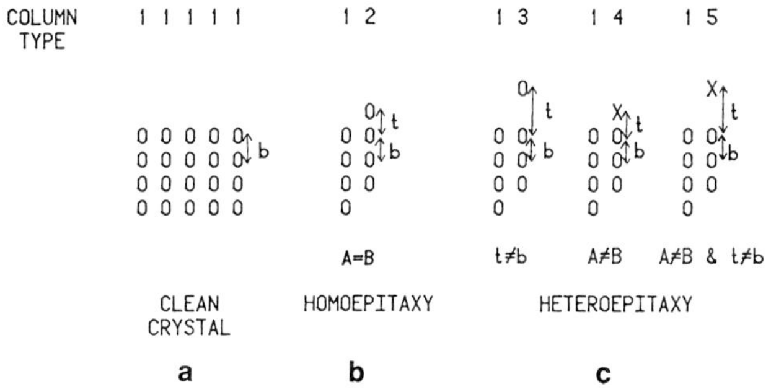

COLUMN TYPE 1 1 1 1 1 | 1 2 | 1 3 | 1 4 | 1 5

A=B | t≠b | A≠B | A≠B & t≠b

CLEAN CRYSTAL | HOMOEPITAXY | HETEROEPITAXY

a **b** **c**

Fig.59.1a-c. Possible configurations of adatoms deposited on a substrate: (**a**) clean surface, (**b**) homoepitaxy, (**c**) heteroepitaxy. In homoepitaxy the overlayer atom of type A is located at a bulk lattice site on a substrate of type A. In heteroepitaxy either the overlayer atom type is different from that of the substrate or the overlayer atom is not at a bulk lattice site

Lent and *Cohen* [59.4] considered the case where only the outermost atoms in each layer exposed to the vacuum contribute to the measured profile. They assumed that these atoms have identical scattering factors with no energy or angle dependence. By dividing the correlation function into a term independent of separation M, and another containing the remainder of the correlation, they show that the angular profile can be decomposed into two components, one broad or "diffuse", reflecting the finite-size broadening, and the other narrow, reflecting the long-range order in the substrate. The narrowest width of the latter is determined by the instrument response. In this model the intensity has a form in which the broadening is simply related to the perpendicular component of momentum transfer $S_\perp$, the step height t, and the coverage θ. The intensity scattered by the outermost atoms only, I_{outer}, is given by

$$I_{outer}(S_\perp,\mathbf{S}_\parallel) = G_0(S_\perp,\mathbf{S}_\parallel)\{\theta + (1-\theta) - [2 - 2\cos(S_\perp t)]\theta(1-\theta)\}$$

$$+ G_1(S_\perp,\mathbf{S}_\parallel)\{[2 - 2\cos(S_\perp t)]\theta(1-\theta)\} \quad , \qquad (59.1)$$

where G_0 is the profile corresponding to diffraction from a perfect surface as broadened by instrument response and G_1 is the profile of the instrument-response-convoluted ordered-domain size broadening.

For systems that order homoepitaxially, this result can be simply extended to take into account all atoms in a column, because the unit magnitude of the atomic scattering factors in the model of [59.4] can be replaced by the magnitude of the column scattering factor $|F_1|$. The phase relations are unchanged. The diffracted intensity is then

$$I(S_\perp,\mathbf{S}_\parallel) = |F_1|^2\, I_{outer}(S_\perp,\mathbf{S}_\parallel) \quad . \qquad (59.2)$$

For heteroepitaxial systems there is no simple relationship between the scattering factors of columns that contain occupied overlayer sites and those that do not. This is true both for deposition of type A atoms onto a type A substrate where the overlayer atom is not located at the bulk lattice site and for cases where the overlayer atom is of type B on a substrate of type A (for which, in general $|F_1| \neq |F_2|$ and $\eta_2 - \eta_1 \neq S_\perp t$, independent of the site that is occupied).

For the case of one possible adatom per column there are two column types and the intensity distribution is

$$I(S_\perp,\mathbf{S}_\parallel) = G_1(S_\perp,\mathbf{S}_\parallel)[(|F_1|^2 + |F_2|^2 - 2|F_1||F_2|\cos\phi)\theta(1-\theta)]$$

$$+ G_0(S_\perp,\mathbf{S}_\parallel)[\theta|F_1|^2 + (1-\theta)|F_2|^2 - (|F_1|^2 + |F_2|^2$$

$$- 2|F_1||F_2|\cos\phi)\theta(1-\theta)] \quad , \qquad (59.3)$$

where $\phi = \eta_2 - \eta_1$. To determine the distribution of domain sizes it is necessary to evaluate what proportion of the measured profile consists of the G_1 component and to estimate the variation of the column scattering factors across the profile. For homoepitaxial systems this analysis is relatively straightforward. Here the relative proportions of narrow and broad components in the profile are simply related to the coverage. In particular, at $\theta = 1/2$ and at beam energies such that $S_\perp t = (2n+1)\pi$, the beam appears maximally broadened. These are the easiest conditions at which to infer a distribution of domain sizes because here $I(S_\perp,\mathbf{S}_\parallel) \sim G_1(S_\perp,\mathbf{S}_\parallel)$. Similarly, for systems where the resolving power of the instrument is high compared to the finite-size broadening one is trying to resolve, it is possible to remove the narrow component by inspecting the profile [59.5]. In practice, this depends on the instrument, but for all instruments there is an ordered-domain size limit beyond which it is not possible to separate the components by inspection. Then determining the column scattering factors would enable a decomposition of the profile into broad and narrow components. Calculating the magnitudes and phases of column scattering factors therefore has a twofold goal. First, knowing these quantities (in particular at the Bragg condition in $\mathbf{S}_\parallel$) will allow an evaluation of the relative proportions of the G_0 and G_1 components at any given energy. Second, estimating the variation of the column scattering factors across a profile will improve the evaluation of the angular variation of G_1 alone.

In general a calculation of the column scattering factors should include contributions from all atoms in the crystal [59.6]. Each atom should have the appropriate energy- and angle-dependent scattering factor and the correct position coordinate. In the kinematic approximation the amplitude scattered from any column is given by

$$F_k(\mathbf{S}) = \sum_{m=0} \alpha^m \sum_{n=1}^{N_m} f^k_{n,m}(\mathbf{S}) \exp[i(\mathbf{r}_{n,m} \cdot \mathbf{S})] \quad . \tag{59.4}$$

The index n runs over all N_m atoms distributed among the m layers that contribute to the diffracted intensity. Then $f^k_{n,m}(\mathbf{S})$ denotes the atomic scattering factor for an atom at site n of layer m in column type k. Further, $f^k_{n,m}(\mathbf{S})$ can be written as a scattering power times a phase factor that depends on **S**. Here α is a measure of the attenuation and is given by the ratio of amplitudes scattered by any two succeeding layers. The proportion of the total diffracted intensity distributed in the broad component of the profile at a given beam energy is given by the ratio W:

$$W = \frac{\int G_1(S_\perp, S_\parallel)\,(|F_1|^2 + |F_2|^2 - 2|F_1||F_2| \cos\theta)\theta(1-\theta)\, dS_\parallel}{\int I(S_\perp, S_\parallel)\, dS_\parallel} \quad . \tag{59.5}$$

For homoepitaxial systems this ratio shows a sinusoidal variation as a function of the perpendicular component of momentum transfer $S_\perp$ between the limits of 0 and 1. The ratio W and the full profiles of the (01) beam are plotted in Fig.59.2 for two cases. The first case is calculated for homoepitaxial deposition assuming column scattering factors $F_1 = 1$ and $F_2 = \exp(iS_\perp t)$ and a geometric distribution of domain sizes. The second case shows the same plots for one half of a monolayer of Sb deposited on GaAs(110) with column scattering factors calculated in a kinematic calculation according to (59.4). We took calculated, fully dynamical values for the atomic scattering factors [59.7] that have been shown to be quite accurate for LEED applications [59.8]. The relative magnitudes and phases of the Sb, Ga, and As scattering factors vary with angle and energy and modulate the dependence of the LEED angular profile on $S_\perp$ accordingly. For α we used reasonable values based on the attenuation of a low-energy electron beam [59.9] and the [110] interplanar distance in GaAs. We adopted the best estimates from the dynamical LEED calculations for the GaAs(110) structure [59.10] and for the structure of a monolayer of Sb deposited on GaAs(110) [59.8] to model respectively those columns with vacant overlayer sites and those with overlayer sites occupied by Sb atoms. The assumption that the atomic coordinates of the different column types can be represented by those obtained from either clean GaAs (110) or GaAs(110) with a full monolayer of Sb has greater validity when the ordered patches are large and therefore approximate those conditions. Step edge relaxation would modify the quantitative results.

Multiple scattering of the electrons may be important in affecting the magnitude and phase of the column structure factors. To date, multiple-scattering calculations of structure factors have concentrated on evaluating the intensity only at the Bragg condition in $S_\parallel$ (I-V profiles). (The calculations of *Jagodzinski* and co-workers [59.12] are a notable exception.) Multiple-scattering I-V calculations do, however, predict values for the magnitude and phase of the column scattering factors, although the phase information is generally not presented. We used values for the GaAs/Sb system calculated for us by *Tong* and *Mei* [59.13] to determine the ratio W. In Fig.

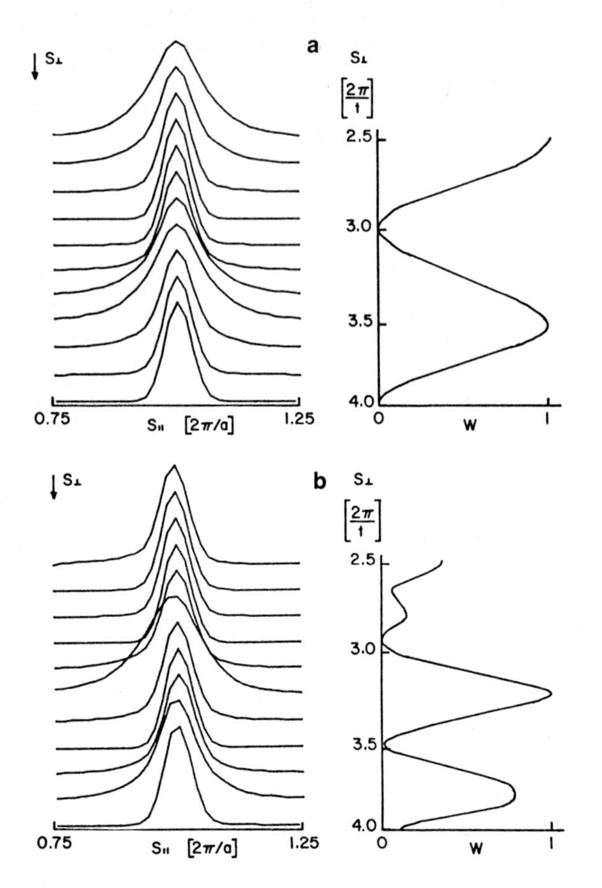

Fig.59.2a,b. Comparisons of kinematic calculations: (**a**) homoepitaxial systems, (**b**) 1/2 monolayer of Sb on GaAs(110). Angular profiles and the ratio W from (59.5) are plotted for various values of $S_\perp$

59.3 this ratio is compared with that calculated using the kinematic approximation, as described above, and with the experimentally measured FWHM of the diffracted beam. For experimentally measured profiles in which it is difficult to separate the broad component of the profile from the instrument response function, values of the FWHM of the profile that are larger than the instrument response function at the same energy will occur as the proportion of the G_0 component in the profile decreases. The magnitude of the measured FWHM beyond that due to the instrument response therefore indicates the proportion of intensity in the broad component G_1 relative to that in the narrow component G_0. Its energy dependence should therefore be directly comparable to that of W. The experimentally measured variation of the FWHM of the (01) beam for a 1/2 monolayer of Sb deposited on GaAs(110) is shown along with the calculated variation of W with energy for a 10 eV energy interval around an energy where the beam shows locally maximal broadening [59.11]. The corresponding angular profiles at the condition of maximum broadening are also shown for each case. To determine these the variation of the column scattering factors across the profile is estimated in the kinematic approximation for both calculations. The calculations in this as yet very restricted effort do not predict well the absolute value of the broadening at a given energy. The sensitivity of the relative weighting of the G_0 and G_1 components to the absolute

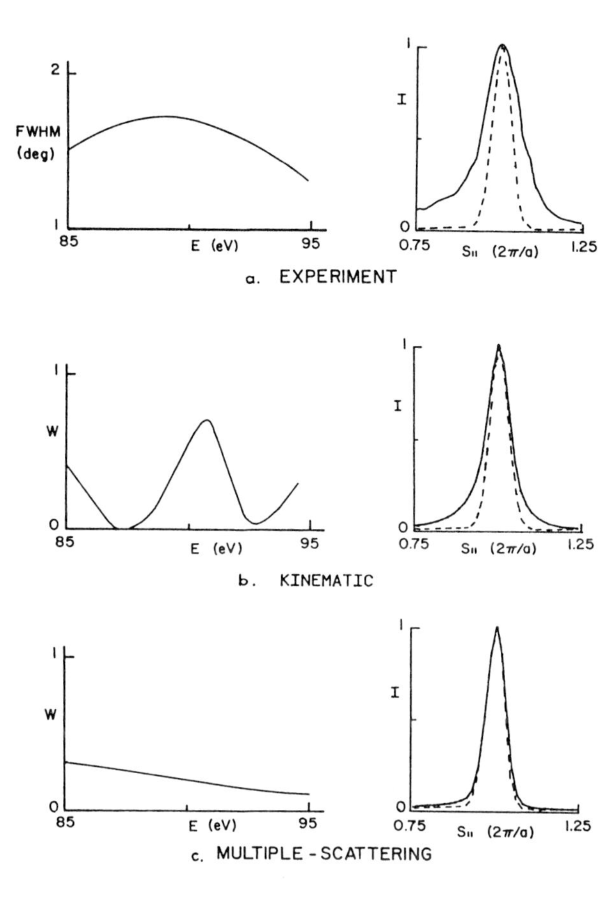

Fig.59.3a-c. Comparison of the measured FWHM and angular profile for 1/2 monolayer of Sb on GaAs(110) with calculated values of W and angular profiles. (**a**) Energy dependence of the FWHM and the angular profile measured at 90 eV. (---) in the profile is the instrument response function at 90 eV. (**b**) Energy dependence of W, the ratio of diffuse to narrow components of a profile, and the angular profile calculated at 91 eV, where W has a local maximum. (---) is the instrument response function at 91 eV. W is determined from a kinematic calculation. (**c**) Energy dependence of W using column scattering calculated using multiple-scattering theory [59.13]. The angular profile at 90 eV is shown for comparison. (---) is the instrument response function

values of the column scattering factors, in particular to the values around the Bragg condition, where G_0 is sharply peaked, makes these calculations very sensitive to correct atomic positions, scattering powers, and phase factors. This has both negative and positive aspects. When other parameters are not well known this method determines accurate domain size with difficulty. On the other hand, an analysis of the profile shape may more sensitively determine atomic coordinates than analysis of I-V profiles does by itself.

The dependence of the shape of the broadened profile on the variation of the column scattering factors across the spot is sensitive to the relative values of the column scattering factors, which vary slowly across a typical profile.

An upper limit to the average domain size can be set by assuming that the maximally broadened profile measured experimentally contains no component of the narrow function G_0. (It will always be possible to satisfy this condition at certain points in energy-coverage space.) An inspection of the angular distribution of intensity near the Bragg condition provides an indication of how good is the approximation that there is no narrow component in the angular

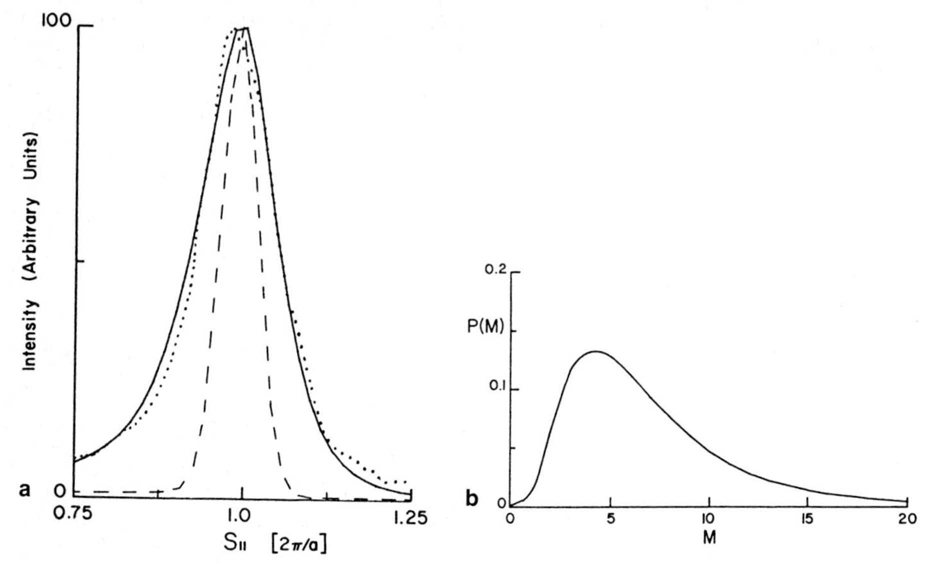

Fig.59.4a,b. Comparison of experimental results for GaAs(110) p(1 × 1)-Sb with model calculation. (**a**) Calculated (——), experimental (····) and instrument response (----) profiles. The solid curve is calculated with the island size distribution shown in (*b*). (**b**) Log-normal island size distribution. P(N) is the probability of having an island of width N. The mean island width for the fit in (*a*) is $\langle N \rangle = 8.3$

profile. For profiles with a smaller curvature through the Bragg peak than the instrument response the G_0 component will clearly be small. The above approximation is then valid, and the experimental profile can be fitted directly using correlation functions generated for different distributions of domain sizes.

Figure 59.4 shows profiles obtained from a clean cleaved GaAs(110) crystal (a measure of the instrument response), and from a cleaved GaAs(110) crystal on which 1/2 monolayer of Sb was deposited [59.11] along with a calculated fit. The profiles were measured at an energy of 90 eV, one of the maximally out of phase conditions.

The calculated fit was performed as follows. A size distribution of overlayer islands was introduced into the model by generating a series of numbers, each denoting whether the corresponding overlayer site was vacant or occupied. The correlation functions of the different column types were calculated from this distribution of occupied sites. The correlation functions can be calculated for a completely arbitrary arrangement of occupied overlayer sites and thus encompass the special-case functions used in other models [59.4,14]. The angular profiles are evaluated by Fourier transforming the correlation functions, multiplying the result by the appropriate column scattering factors, (59.4), and convoluting the resulting expression with the instrument response function. The instrument response function, obtained from measurements on the clean surface [59.11] can, to a good approximation, be represented by a Gaussian function. The island size distribution is then varied and the resulting angular profiles are fitted to the experimental profiles. A log-normal island size distribution gave the best fit to the data. The angular profile calculated with a log-normal island size distribution with a mean island width of 8.3 unit cells (corresponding to an island width of 44 Å) was compared with the measured angular profile in Fig. 59.4a.

In general, the analysis of surface domains requires a two-dimensional treatment. In [59.15] we have shown that for a slit detector, an "effective" one-dimensional correlation function is appropriate. Therefore the one-dimensional model used here gives a good indication of the type of fit that can be obtained, although the accuracy of the values determined for the average island size should be viewed with discretion.

In summary, we have calculated angular profiles of diffracted fundamental beams from substrates containing an overlayer with an arbitrary distribution of ordered-domain sizes and reiterated that these profiles consist of diffuse and narrow components. We have shown that at any given energy and coverage, the relative proportion of the broad component resulting from the distribution of finite-size overlayer islands and of the narrow component resulting from the constant part of the correlation function is determined by the magnitude and phase of the column scattering factors. In homoepitaxy an energy can be found where the narrow component in the profile is absent. In heteroepitaxy this is not the case and there is a complicated dependence of broadening on energy. The limited calculations we have performed so far for the latter case do not agree well with experiment. We believe that the absolute values of the column scattering factors that we could generate from available information did not have sufficient accuracy to predict the amount of broadening in a profile at a given energy. The kinematic calculation is able to predict the modulation of the broad profile by the relative variation of the column scattering factor across the profile. Nevertheless, by properly interpreting the variation of the experimental profile with energy a size distribution can be generated to fit the data and an upper limit to the mean size of the distribution can be established.

Acknowledgment. This research was supported by NSF under grant No. DMR 83-18601.

References

59.1 M. Henzler: Surf. Sci. **73**, 240 (1978);
M. Henzler: In *Electron Spectroscopy for Surface Analysis*, ed. by H. Ibach, Topics Curr. Phys., Vol.4 (Springer, Berlin, Heidelberg 1977)
59.2 M.G. Lagally: In *Chemistry and Physics of Solid Surfaces IV*, ed. by R. Vanselow, R. Howe, Springer Ser. Chem. Phys., Vol.20 (Springer, Berlin, Heidelberg 1982)
59.3 T.-M. Lu, L.H. Zhao, M.G. Lagally, G.-C. Wang, J.E. Houston: Surf. Sci. **122**, 519 (1982)
59.4 C.S. Lent, P.I. Cohen: Surf. Sci. **139**, 121 (1984)
59.5 K.D. Gronwald, M. Henzler: Surf. Sci. **117**, 180 (1982)
59.6 D. Saloner, M.G. Lagally: J. Vac. Sci. Technol. **A2**, 935 (1984)
59.7 M. Fink, J. Ingram: Atomic Data **4**, 129 (1972);
D. Gregory, M. Fink: Atomic and Nuclear Data Tables **14**, 39 (1974)
59.8 A. Kahn, J. Carelli, C.B. Duke, A. Paton, W.K. Ford: J. Vac. Sci. Technol. **20**, 775 (1982)

59.9 M.B. Webb, M.G. Lagally: Solid State Physics **28**, 301 (1973)
59.10 C.B. Duke, S.L. Richardson, A. Paton, A. Kahn: Surf. Sci. **127**, L135 (1983)
59.11 H.M. Clearfield: Ph.D. dissertation, in preparation;
H.M. Clearfield, M.G. Lagally: In preparation
59.12 H. Jagodzinski: Acta Cryst. **1**, 201, 208, 298 (1949);
H. Jagodzinski, W. Moritz, D. Wolf: Surf. Sci. **77**, 233, 249, 265, 283 (1978);
W. Moritz, D. Wolf: Preprint, Munich University
59.13 S.Y. Tong, W.N. Mei: Private communication
59.14 J.M. Pimbley, T.-M. Lu: J. Appl. Phys. **55**, 182 (1984)
59.15 D. Saloner, M. Henzler, M.G. Lagally: In preparation

VI.2 Experimental Studies

60. Diffusion and Interaction of Adatoms[1]

G. Ehrlich

Coordinated Science Laboratory, University of Illinois at Urbana-Champaign, 1101 W. Springfield Ave., Urbana, IL 61801, USA

Compared to the wealth of information now available about atomic arrangements in overlayers and surfaces, rather little is known about the energetics governing the atomic interactions which underlie the structures formed. This imbalance has been redressed somewhat, as recently considerable information has been derived about atomic interactions from Monte Carlo simulations [60.1-3]. In these studies, interactions between atoms in a lattice gas are varied until the simulations match the experimentally observed phase diagram for the adsorbed layer. This macroscopic and rather indirect approach has also been complemented by insights concerning interactions at surfaces derived by direct examination on the atomic level.

Several techniques now appear capable of providing information about individual adatoms – electron microscopy [60,4,5], scanning tunneling microscopy [60.6], and field ion microscopy [60.7] – but only the last has so far generated a significant amount of data dealing with atomic phenomena. Field ion microscopy has some limitations. Observations of chemisorbed gases, for example, induce significant perturbations in the surface [60.8]. Layers of metal and semiconductor atoms, however, have been successfully examined on the atomic level, and so we discuss such studies here. Actually, a fair body of work relying upon the field ion microscope to derive information about atomic events has already grown up. Instead of attempting to survey these efforts [60.9], we shall present a few simple examples to demonstrate the type of information successfully obtained and the techniques of analysis useful in quantitative studies of surface interactions on the atomic level.

60.1 The Spatial Distribution of Adatoms

Binding of atoms at different sites on a crystal has been of interest for some time in surface studies. Consider a surface in an equilibrium state; the probability P_i of finding the adatom at a specified site i, as compared to some reference site 0, is given by

$$P_i/P_0 = \exp(-F_i/kT) \quad ; \tag{60.1}$$

here P_0 is the probability of the atom being at the reference site, and F_i is the relative free energy of the system with the atom at site i as compared

1 Supported by the National Science Foundation under Grant DMR 82-01884.

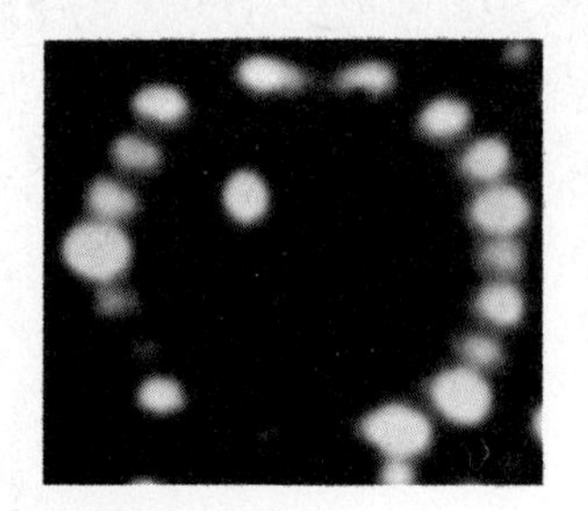

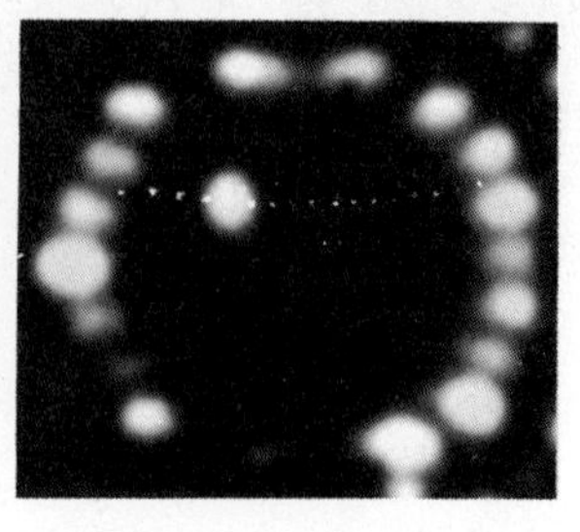

Fig.60.1 Fig.60.2

Fig.60.1. Field ion image of a single rhenium adatom on a W(211) plane, *at left*, is compared with a hard-sphere model of the same surface, *at right*

Fig.60.2. Binding sites in one channel of W(211), mapped out by determining the position of an adatom after repeated diffusion steps [60.10]. White dots superposed on ion image mark locations of adatom center

to being at site 0. To predict the distribution of adsorbate over a surface, we must have information about the free energy F_i —a considerable challenge to theoretical chemists. The problem can be turned around, however. With the field ion microscope it is possible to locate individual atoms on a surface; the probability P_i and therefore also the relative free energy F_i can now be measured.

The field ion image of a single rhenium atom on the (211) plane of tungsten is shown in Fig.60.1. It is in a sense disappointing, as only the adatom and the substrate atoms forming the periphery of the (211) plane are revealed. That, it turns out, is not a significant drawback: the adatom itself can be used to map out its binding sites on the surface. The surface is warmed up until the adatom can move about and then is imaged again. By repeating this procedure often enough, all the sites accessible to the adatom can be located, as in the map in Fig.60.2 [60.10]. For tungsten the (211) plane is made up of rows of close-packed lattice atoms [60.11] which form channels along which the adatoms are constrained to move. Thus both examination and analysis of interactions on W(211) are especially simple, and much of our presentation will be concentrated on this particular plane. In Fig.60.2 the atom moving over the surface has settled at the same site several times, and there is some scatter in the positions actually measured. This dispersion is indicative of the errors in individual observations; however, the scatter is small in comparison with the spacing between sites, and causes no problem.

With the sites pinpointed, it now remains to establish the frequency with which they are occupied by an adatom. Here it is all-important to establish equilibrium conditions. To insure this the surface must be brought to a temperature at which the adatom can diffuse over a distance comparable to the channel length during each equilibration interval [60.12]. At equilibrium the atom can select sites for occupation based on the strength of binding rather than on the kinetics of diffusion. It is also important to note that the surface is heated in the absence of *any* applied voltages, so that equilibration occurs in a field-free environment, just as in any ordinary experiment. High voltages are applied only for imaging, which is done at low temperatures, $T \approx 20$ K, under conditions such that the applied field has little effect on adatom location.

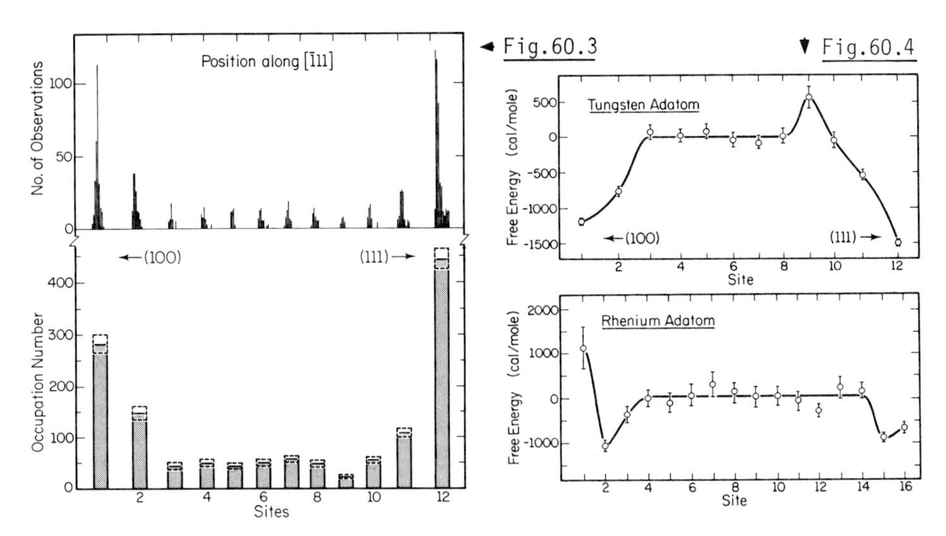

Fig.60.3. Equilibrium distribution of a single tungsten adatom on W(211) at 363 K [60.10]. *Top*: frequency with which adatom is observed at different locations, showing spread in position measurements. *Bottom*: distribution over the 12 binding sites established by measurements at top. Total observations: 1340

Fig.60.4. Free energy of a tungsten as well as a rhenium atom on W(211), relative to free energy of the atom at a central site [60.10], all at 363 K

The results of distribution measurements [60.10] for a tungsten atom on W(211), Fig.60.3, are surprising. Contrary to what might have been expected by considering the coordination of an adatom with lattice atoms on a hard-sphere model, the tungsten atom prefers to occupy sites close to the ends of a channel rather than at the center. The effect of the channel ends extends over quite a long range – on the side of the (211) facing the (111) plane, sites as far as four spacings from the end differ significantly in their properties from center sites. It is also apparent that the two sides of the (211) plane are not equivalent in their binding properties, but that is expected from the absence of a plane of symmetry perpendicular to the close-packed channels. The difference in the free energy of an adatom close to the ends and in the center of a channel is quite significant for a tungsten adatom; as shown in Fig.60.4, it amounts to ≈1.6 kcal/mole.

Measurements of this type are still limited; nevertheless they have already revealed a clear chemical specificity. Rhenium atoms do not favor the ends as markedly as tungsten atoms. Recent observations by *Wrigley* [60.13] suggest that for silicon, binding is much stronger close to the ends, with the free energy ≈2 kcal/mole lower than at the center. Also, for silicon the preference is strongly in factor of the channel end facing (100) rather than (111), as observed for tungsten atoms. It should be noted that there is a practical limit to the scope of such distribution measurements. During equilibration the adatom must not be lost from the plane by diffusion over the edges. A number of systems satisfy this requirement, and it will be of interest to acquire distribution data for a variety of adatoms and substrates, in order to delineate structural and chemical trends. It is already clear, even from the limited data available, that simple notions based on pairwise interactions are not adequate to describe atomic binding on the surface.

The studies described so far have been done with one adatom on the plane of interest and explore the binding of the adatom at different sites. Suppose now that a second adatom is placed on the (211) in a separate channel. The equilibrium distribution of the two adatoms should now be responsive to the interactions between the atoms. If we confine ourselves to adatoms at the L central sites of a channel unaffected by the ends, then the probability P_x of observing two adatoms at a separation x from each other, where x is measured along the channel direction, is given by

$$P_x = C(2 - \delta_{x0})(L - x)\exp[-F_x/kT] \quad . \tag{60.2}$$

Now F_x is the free energy of interaction for two adatoms in channels a specified distance from each other, δ_{x0} is Kronecker's delta, and C just insures normalization of the probabilities. Again, all that is necessary in order to explore this free energy is to observe the frequency with which adatom pairs occur at the separation x in a thoroughly equilibrated system. Such measurements have been made for adatoms on a (211) plane of tungsten [60.14], with the atoms separated by an intervening empty channel, as in Fig.60.5. The results of Graham for tungsten adatoms, Fig.60.6, suggest that there are no significant interactions at these rather long ranges: the observed distribution conforms quite well to that expected for two adatoms without interaction

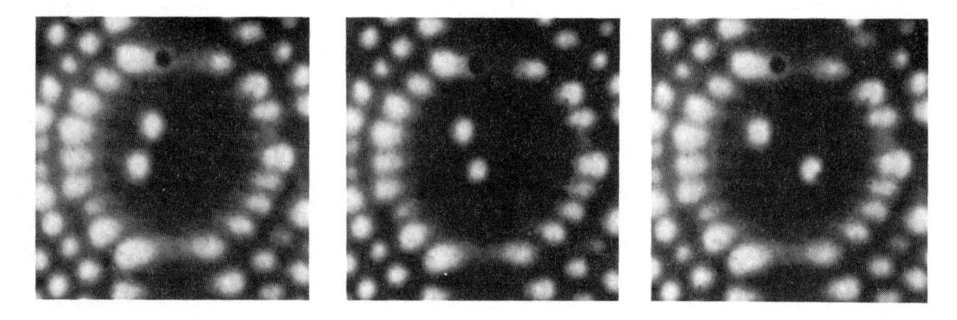

Fig.60.5. Tungsten (211) plane with two rhenium adatoms separated by an empty channel. Equilibration at 289 K for 90 s. Photographs by courtesy of K. Stolt

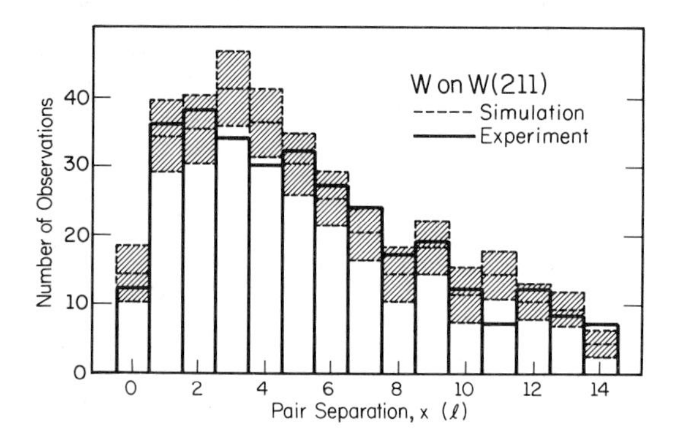

Fig.60.6. Distribution of lateral distances between two tungsten adatoms on W(211) separated by an empty channel [60.14]. System equilibrated at room temperature. Simulation is for two noninteracting adatoms

distributed at random over sites in separate channels. When two adatoms are placed in adjacent channels, then stronger interactions come into play, as evidenced by the fact that dimers form which are stable [60.15] even at 400 K. Distance distribution measurements for such a system are feasible and should eventually afford a detailed map of the effective interactions between adatoms.

60.2 Dissociation Equilibrium of Clusters

The complete potential between two atoms on a channeled surface has not yet been satisfactorily worked out by the method just described. More limited information about the energetics of the interactions has been obtained, however, by examining the equilibrium between bound and dissociated dimers [60.15]. Suppose just for simplicity that the range of interatomic forces is limited, so that when $x > 2$, that is when the lateral separation between atoms in adjacent channels is 2 or more spacings, interactions have become negligible. Bound dimers can then exist in two states, labeled by their lateral separation as 0 (straight) and 1 (staggered). If there are L equivalent sites available for occupancy in a channel, the ratio of dimers in state 1 to dissociated dimers is just

$$P_1/P_D = \frac{2}{L-2} \exp(-F_1/kT) \quad . \tag{60.3}$$

An estimate of the free energy difference F_1 between bound dimers in state 1 and dissociated dimers, both illustrated in Fig.60.7, can be obtained from observing the frequency with which dissociated species occur at a fixed temperature, as in Fig.60.8. The temperature dependence of the fraction of dimers dissociated then yields the energy and entropy difference between bound and dissociated dimers. In practice, implementation of such estimates is made difficult by several technical problems. The most significant of these arises from the fact that after equilibration the sample does not immediately drop to 20 K. Cooling occurs at a finite rate, which depends upon the structure of the sample support. Corrections must therefore be made to allow for the possibility that during cooling the distribution may adjust itself to some intermediate temperature before finally being frozen in by the loss of mobility.

Despite these problems, observations of the adatom-dimer equilibrium have been made by *Stolt* et al. [60.15] for Re_2 on W(211) and are shown in Fig.60.9. From the measured temperature dependence, we infer an energy difference E_1 of -3.7 ± 1.1 kcal/mole between dimers in state 1 and dissociated species. Unfortunately the statistical scatter is too large to allow a meaningful estimate of the entropy.

Thermodynamic information about different configurations of a cluster can also be deduced in much the same way, by measuring how frequently different states occur in an equilibrium ensemble. The ratio of dimers in state 1 to state 0 should, at equilibrium, be given by

$$\frac{P_1}{P_0} = \frac{2(L-1)}{L} \exp[-(F_1 - F_0)/kT] \quad . \tag{60.4}$$

Provided proper account is taken of possible annealing effects, energy and entropy differences can be obtained as usual from the temperature dependence. In Fig.60.10, this is done for rhenium dimers on W(211). For these cross-channel clusters, the thermodynamic data so derived are

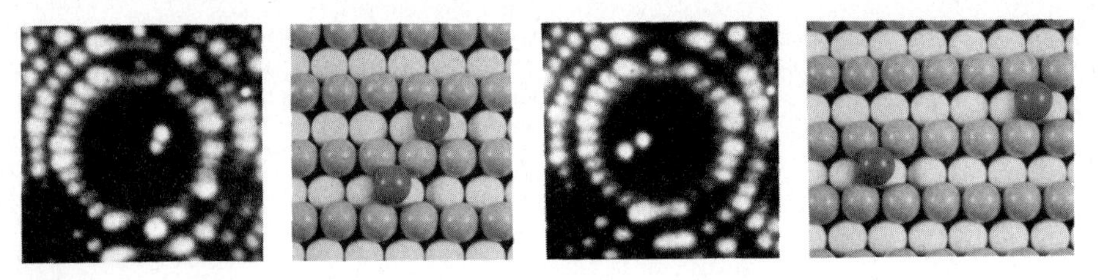

Fig.60.7. Rhenium cross-channel dimer on W(211) [60.15]. Dimer in configuration 1 is shown both in field ion image and on model *at left*; dissociated dimer is *at right*

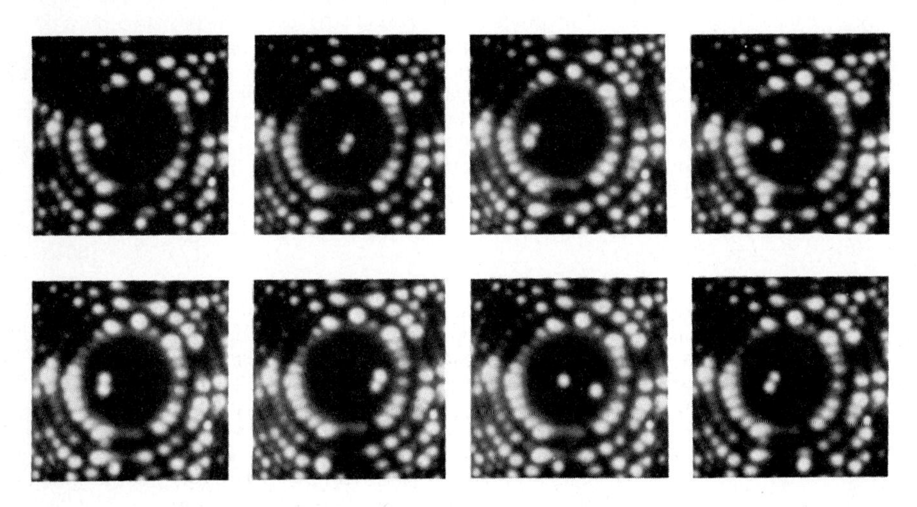

Fig.60.8. Dissociation sequence [60.15] for a rhenium cross-channel dimer on W(211). Equilibration between images is for 3 s at 392 K

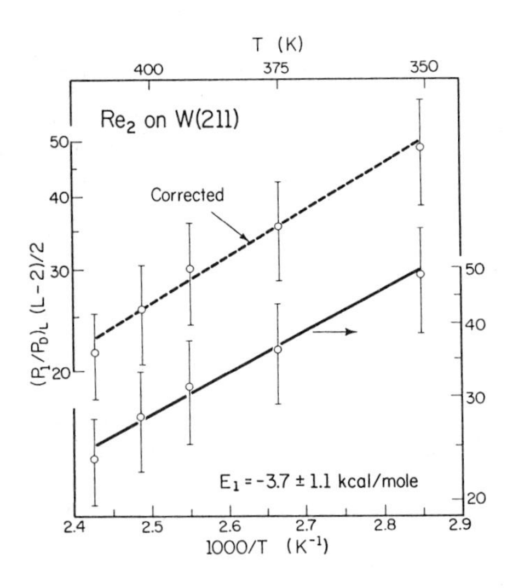

Fig.60.9. Equilibrium [60.15] between dissociated cross-channel rhenium dimers and dimers in state 1 on W(211). *Corrected curve* obtained by allowing for changes in the distribution during the slow temperature decay after equilibration

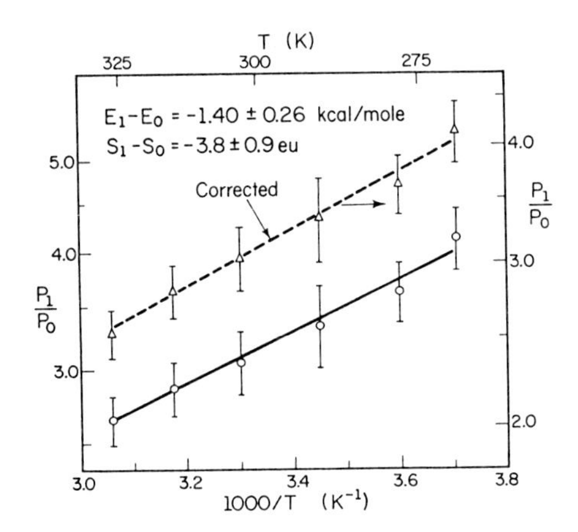

Fig.60.10. Equilibrium between different bound states of rhenium [60.15] cross-channel dimers on W(211)

$$E_1 = -3.7 \pm 1.1 \text{ kcal/mole}, \; E_0 = -2.3 \pm 1.2 \text{ kcal/mode},$$

$$S_1 - S_0 = -3.8 \pm 0.9 \text{ eu} \quad .$$

The energy difference between bound and dissociated dimers is surprising. It is small, on the order of van der Waals interactions in rare gas crystals, but despite that the dimers are quite stable to dissociation at 400 K. This stability arises from the rather limited configurational entropy of the dissociated dimer, which can assume only a small number of configurations on channels with a finite number of binding sites. Most important of all, however, is a general conclusion: the ability to discern individual adatoms on a crystal makes it possible to apply standard thermodynamic relations and procedures in examining the energetics of well-defined clusters.

60.3 Cluster Motion and Atomic Interactions

Considerable additional information about atomic interactions may be derived from quantitative studies of cluster diffusion, provided the mechanism of cluster motion is known or can be guessed. As an example of this idea, consider once more cross-channel dimers on W(211). For these it is well established that diffusion occurs by jumps of the individual atoms making up the dimer [60.16]. If the conversion of dimers from state 0 to 1 occurs at a rate a and the reverse process occurs at a rate b, as illustrated in Fig.60.11, then it can be shown, by analogy with an ordinary random walk, that the mean-square displacement of the center of mass of the dimer amounts to [60.17]

$$\langle \Delta R^2 \rangle = 2Dt = 2[aP_0 + bP_1](\ell/2)^2 = P_0 a \ell^2 t \quad ; \qquad (60.5)$$

ℓ is the spacing between atom sites, t the time of the diffusion interval, and D the diffusivity. The probability P_0 is known from equilibrium studies, which also provide the ratio $P_0/P_1 = b/a$. Measurements of the mean-square displacement $\langle \Delta R^2 \rangle$ therefore can be made to yield the rate a at which an atom jumps away from its partner in the dimer, as well as the reverse rate b.

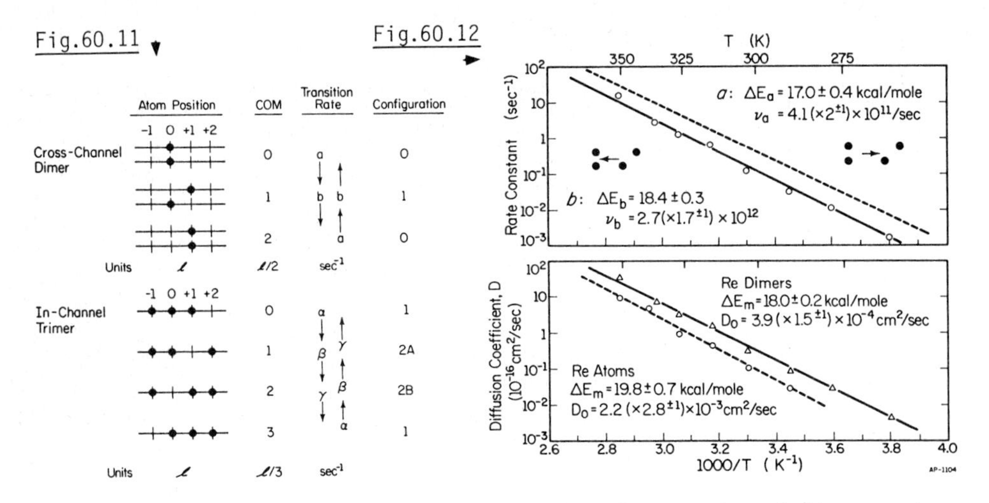

Fig.60.11. Schematic of jump rates in diffusion of cross-channel dimers and in-channel trimers on W(211); ℓ gives the spacing between nearest-neighbor sites

Fig.60.12. Diffusion of rhenium cross-channel dimers [60.16] on W(211). *Bottom*: Comparison of adatom and dimer diffusivities. *Top*: Jump rates important in dimer diffusion. ΔE gives the relevant activation energy, ν the frequency factor

The diffusion of rhenium dimers on W(211) has been studied [60.16], and turns out to be quite unusual. As evident from Fig.60.12, the diffusivity is greater than that of a single adatom. A rate analysis has been made, and yields the values $a = 8 \times 10^{11} \exp(-17.0 \times 10^3/R_G T)$ and $b = 3 \times 10^{12} \exp(-18.4 \times 10^3/R_G T)$, where R_G represents the gas constant. Most interesting in the present context is the fact that for a dimer the barrier to jumps out of state 0 and into state 1 is smaller than the barrier of 19.8 kcal/mole opposing the motion of an isolated adatom on the (211) plane. This reduction arises from interactions between the moving atom and its stationary partner, and we can therefore infer how the effective potential varies with the lateral distance separating rhenium atoms in a cross-channel dimer.

Such a plot for a rhenium adatom moving away from another rhenium adatom in an adjacent channel is shown in Fig.60.13. The effective potential between the two is nonmonotonic: the minimum in the interaction curve occurs at the saddle position that separates dimers in the straight and staggered configurations. At smaller as well as larger distances, the potential rises. The location of this minimum is responsible for the fact that rhenium dimers move more rapidly than a single rhenium adatom. This type of behavior is not general, however. As an example, for iridium adatoms interactions diminish uniformly with separation, and the barrier to dimer diffusion exceeds that for motion of single atoms [60.18]. The chemical nature of the adatoms obviously plays an important role in affecting interactions.

Observations have also been made on two rhenium adatoms in the same channel [60.19], shown in Fig.60.14. These form in-channel dimers which have quite different properties than cross-channel dimers of rhenium, which should not

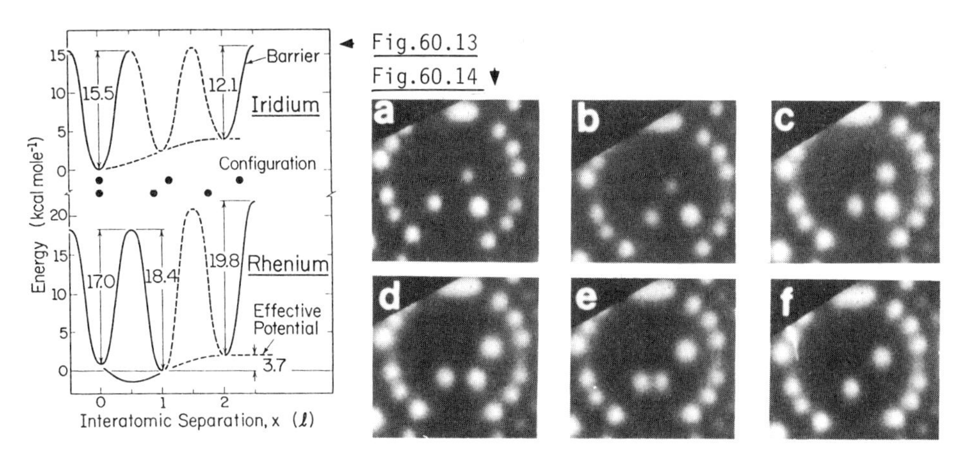

Fig.60.13. Effective potentials for rhenium and iridium cross-channel dimers on W(211) deduced from equilibrium and diffusion measurements

Fig.60.14a-f. Formation of in-channel dimer [60.19] on W(211). Three Re atoms are present on the plane, 2 in one channel and another separated from these by an empty channel. On repeated heating to 286 K for 30 s, the lower two combine into a dimer. Prior to coalescing (**e**), distance between the two atoms is 3 spacings. After dimer formation (**f**), individual atoms are not resolved

be too surprising: adatoms in the same channel can appraoch each other much more closely than when in adjacent rows of the (211) plane. In-channel rhenium dimers diffuse much more slowly than single rhenium adatoms. The activation energy for dimer motion amounts to 23.3 ± 1.2 kcal/mole compared to 19.8 kcal/mole for singles. If this motion again occurs by single atom jumps, then we can infer that interactions between rhenium adatoms in the same channel have their maximum strength at close to one nearest-neighbor spacing. However, a short-range potential with a minimum at the nearest-neighbor separation would not account for the observations. Clusters bound by such a potential would be able to dissociate without further activation when a jump occurs. This has not been observed. Furthermore, in-channel trimers do not diffuse or dissociate even at $T > 430$ K. Diffusion of the trimer center of mass is governed by the relation [60.17,19]

$$\langle \Delta R^2 \rangle = \frac{2\alpha\beta\gamma\ell^2 t}{(2\alpha + \gamma)(2\beta + \gamma)} \quad , \tag{60.6}$$

where the rate constants α, β and γ are illustrated in Fig.60.11. If atomic interactions are additive, then the activation energy for trimer motion can be inferred from the adatom potentials which describe dimer behavior and this imposes further limits on the course of the potential. To account for the absence of dissociation in either dimers or trimers, we must postulate a potential roughly of the form in Fig.60.15. This is in the nature of a minimal curve, inasmuch as dissociation has not been observed; the actual potential could be rather deeper and of longer range. Even the curve in Fig.60.15 already shows the expected effects: interactions between adatoms in the same row are considerably stronger than between atoms in adjacent channels. The impor-

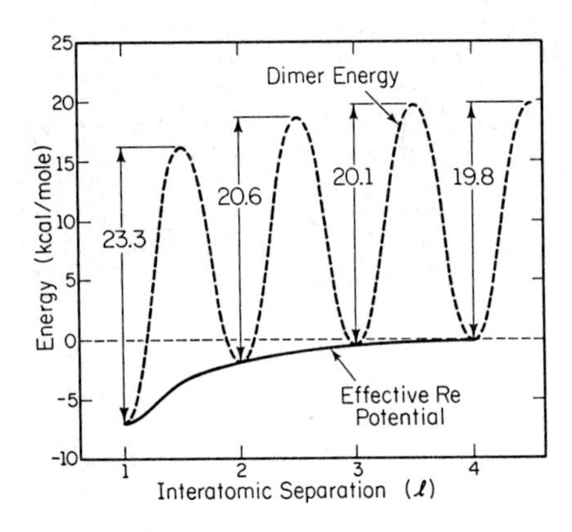

Fig.60.15. Effective potentials for rhenium in-channel dimers [60.19] on W(211)

tant conclusion is that forces between adatoms will be dominated by the atomic arrangement of the substrate.

A note of caution is appropriate here: if the mechanism has been properly identified, the kinetic methods just described give us information about effective potentials, that is, they yield the difference in the energy when the adatom is in the saddle position compared to being at a normal position. However, in the saddle there may be considerable displacements of lattice atoms around the moving atom, and these effects may be more pronounced than with an adatom at a normal binding site. The quantitative reliability of such measurements also has to be scrutinized carefully. The effective potentials are derived as the difference of two large quantities, the activation energy for diffusion of clusters compared to that of adatoms, which are themselves subject to considerable error. The uncertainty in the potentials derived by kinetic methods is therefore likely to be significant. Nevertheless, these techniques do yield important information.

60.4 Dissociation Rates and the Additivity of Interactions

If the information on atomic interactions derived by these techniques is to be useful for predictions about more complicated structures, then interactions will have to be pairwise additive; that is, many-atom effects can make only a small contribution to the total. It should be noted, however, that in studies of phase diagrams it has become customary [60.1-3] to invoke trio interactions and the question therefore arises: can many-atom contributions to binding be measured directly? In principle the answer is, of course, yes. Pair interactions can be derived from observations on two adatoms, trio effects should manifest themselves in binding studies of trimers and so. So far only one system [60.20,21] has been examined this way—rhenium on W(110).

Pair potentials on a two-dimensional plane can be derived, just as outlined above for a one-dimensional system, by repeated observation of two adatoms, such as the rhenium adatoms in Fig.60.16. The distance distribution function must now be written as

$$P(R) = CP_0(R) \exp[-F(R)/kT] \quad . \qquad (60.7)$$

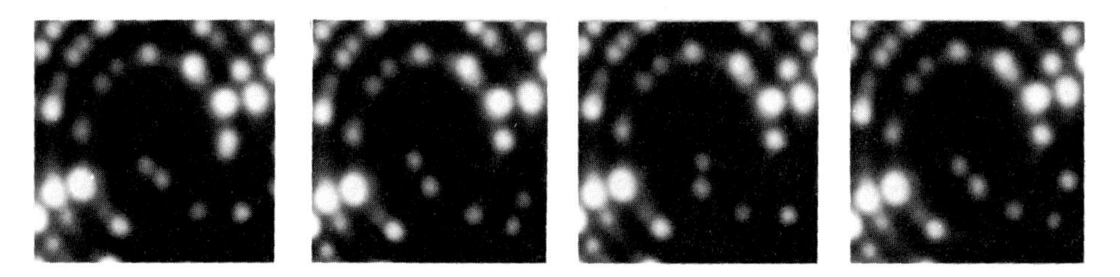

Fig.60.16. Two rhenium adatoms on W(110), equilibrated by heating at 400 K for 30 s. Photographs courtesy of H.-W. Fink

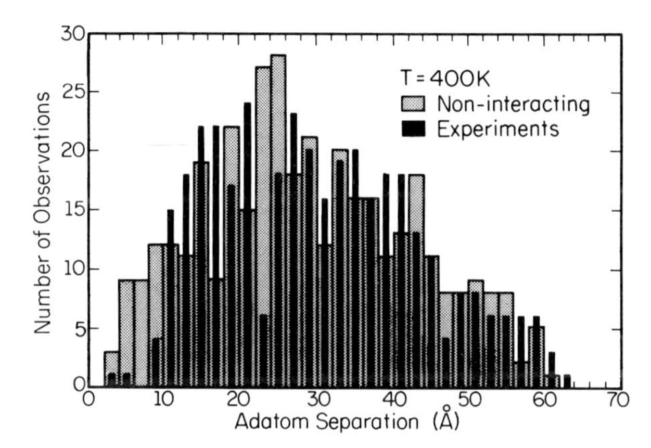

Fig.60.17. Equilibrium distribution of separations between two rhenium adatoms [60.21] on W(110). For comparison, a distribution for noninteracting adatoms, obtained by Monte Carlo simulation, is shown in light gray

Here $P_0(R)$ gives the probability of finding pairs of noninteracting atoms at a distance R, C is the usual normalization term and F(R) is the free energy of interaction between adatoms at a distance R from each other. Several measurements [60.9,22-24] have been made of the distance distribution for rhenium on W(110). In Fig.60.17 are shown recent data taken by *Fink* [60.20,21] under much the same conditions under which trimers were subsequently studied. The available data are not extensive enough to deduce F(R) in its entirely; however, from the distribution in Fig.60.17 it is evident that rhenium dimers with a close interatomic separation are not found. The number of rhenium pairs observed at separations $R < 10$ Å is much smaller than the number for noninteracting atoms obtained by Monte Carlo simulations, which is also shown in Fig.60.17. Evidently, at close separations there are repulsive interactions between two rhenium adatoms on W(110). On this point, there is good agreement in all recent measurements. We estimate an average repulsion of ≈50 meV and that should make it difficult to form a rhenium trimer when the number of adatoms present on W(110) is increased. This is not the case, however.

In Fig.60.18 are shown three rhenium adatoms deposited on W(110). After warming to $T > 420$ K they combine into a trimer despite the repulsions between pairs. The trimer is quite stable to dissociation; it maintains its identity while diffusing over the surface at $T \approx 400$ K. Only at higher temperatures is dissociation into three adatoms observed.

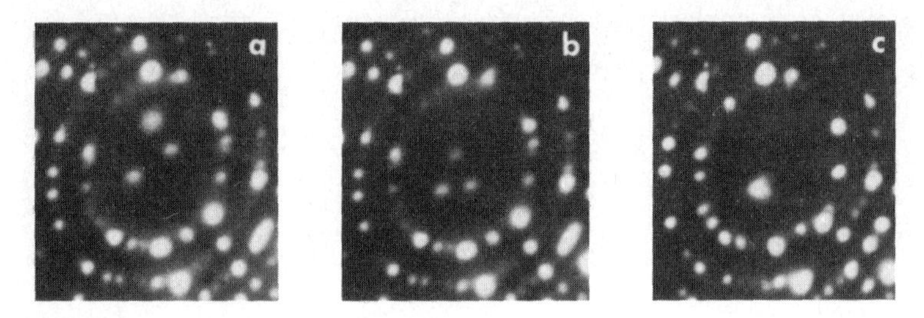

Fig.60.18a-c. Formation of rhenium trimer [60.21] on W(110). (**a**) Three rhenium atoms after deposition; (**b**) trimer is apparent after heating to 450 K for 100 s

The stability of the trimer can be assessed in two different ways: observations of the number of bound vs dissociated trimers should yield the free energy difference between these species. The idea is just the same as in the thermodynamic measurements on dimers vs adatoms on W(211), described above, but the detailed relations must obviously be modified to account for the additional dimension of the (110) plane. There is an alternative procedure which we have not described; that is to measure the dissociation rate, or more accurately, to determine the mean lifetime for dissociation, $\langle\tau\rangle$. This quantity can be written as

$$\langle\tau\rangle = \tau_0 \exp(\Delta E_D/kT) \quad . \tag{60.8}$$

The activation energy for dissociation ΔE_D can in turn be represented by the sum of the binding energy of an atom in the trimer E_B, plus the activation energy ΔE_m for moving the atom away from the remnant, so that

$$\Delta E_D = \Delta E_m + E_B \quad . \tag{60.9}$$

Measurements of the time at $T = 420$ K required to dissociate a rhenium trimer on W(110) are plotted in Fig.60.19. The large scatter in the times is typical of a Poisson process [60.25]. In fact, an alternative to just averaging the observed lifetimes in order to find $\langle\tau\rangle$ is to fit the observed probability $P(\tau)$ that dissociation occurs within a time τ to the expression expected for Poisson statistics,

$$P(\tau) = 1 - \exp[-\tau/\langle\tau\rangle] \quad . \tag{60.10}$$

The value of $\langle\tau\rangle = 304$ s obtained from the dissociation data plotted following (60.10) agrees very well with the straight average of 323 s. Ideally, the activation energy for dissociation ΔE_D should be derived from the temperature dependence of the lifetime $\langle\tau\rangle$ represented by an Arrhenius plot. No reliable values of dissociation energies have been obtained in this way by field ion microscopic observations because the experimentally accessible temperature range is generally too small [60.23,26]. As a less desirable alternative, the dissociation energy may be estimated by guessing the prefactor τ_0. Using a standard value of $\tau = 10^{-12}$s and assuming that the activation energy for migration of rhenium adatoms [60.27] on W(110), 1.04 eV, accounts for ΔE_m, we arrive at a value of the binding energy $E_B = 0.25$ eV.

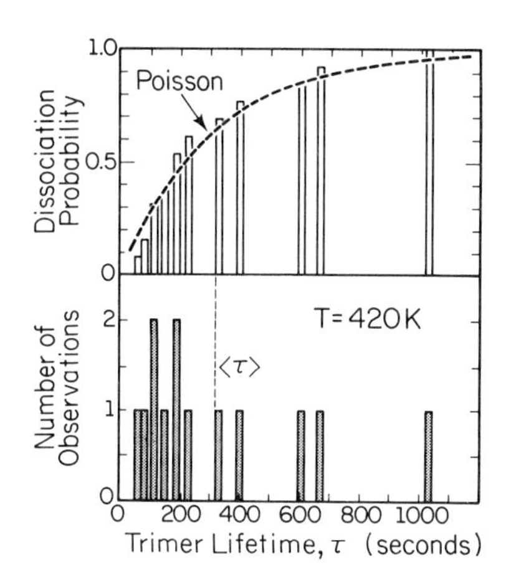

Fig.60.19. Dissociation of rhenium trimer [60.21] on W(110). *Bottom*: direct determination of $\langle\tau\rangle = 323$ s by averaging individual lifetimes. *Top*: fit to Poisson distribution (60.10) gives $\langle\tau\rangle = 304$ s

As a general procedure, this way of deriving activation energies represents a shortcut which can introduce large and unknown errors. In this particular example, the results obtained for the binding energy of rhenium in a trimer from dissociation rate measurements agree nicely with separate equilibrium studies [60.21]. And the results are quite surprising. Interactions between two rhenium adatoms are repulsive to a distance ≈10 Å. Interactions between three rhenium adatoms are attractive, with three-atom interactions providing all the cohesion. In all likelihood this constitutes a special case, but this one example should serve as a warning that in metal layers at least many-atom interactions are likely to make quite a significant contribution.

60.5 Summary

Some of the detailed results presented in this survey will undoubtedly be modified as more experimental studies accumulate. It is already clear, however, that the forces operating on adatoms are quite long range, and strongly affected by the atomic arrangement of the substrate. Most important at the moment is the realization that we command observational techniques, such as field ion microscopy, which provide a view of individual atoms and clusters on well-defined surfaces. These techniques make it possible to apply thermodynamic and kinetic methods, which are standard for developing reliable information about chemical interactions in more macroscopic experiments, to the examination of surface interactions and binding on the atomic level.

Acknowledgments. I am much indebted to Drs. H.-W. Fink, D.A. Reed, K. Stolt, and J.D. Wrigley for help in preparing this review, and to R.T. Gladin for his photographic work.

References

60.1 L.D. Roelofs: In *Chemistry and Physics of Solid Surfaces IV*, ed. by R. Vanselow, R. Howe, Springer Ser. Chem. Phys., Vol.20 (Springer, Berlin, Heidelberg 1982) Chap.10
60.2 W. Selke, K. Binder, W. Kinzel: Surf. Sci. **125**, 74 (1983)
60.3 L.C.A. Stoop: Thin Solid Films **103**, 375 (1983)

60.4 A.V. Crewe: Science **221**, 325 (1983);
M. Utlaut: Phys. Rev. B**22**, 4650 (1980);
M. Isaacson, D. Kopf, M. Utlaut, N.W. Parker, A.V. Crewe: Proc. Nat. Acad. Sci. USA **74**, 1802 (1977);
M.S. Isaacson, J. Langmore, N.W. Parker, D. Kopf, M. Utlaut: Ultramicroscopy **1**, 359 (1976)
60.5 K. Takayanagi: Jpn. J. Appl. Phys. **22**, 24 (1983); Surf. Sci. **104**, 527 (1981); Ultramicroscopy **8**, 145 (1982);
G. Honjo, K. Yagi: In *Current Topics in Materials Science*, Vol.6, ed. by E. Kaldis (North-Holland, Amsterdam 1980) p.195
60.6 A.M. Bard, G. Binnig, H. Rohrer, C. Gerber, E. Stoll, A. Baratoff, F. Salvan: Phys. Rev. Lett. **52**, 1304 (1984);
G. Binnig, H. Rohrer: In *Scanning Electron Microscopy* 1983, pt. 3, 1079 (1983);
G. Binnig, H. Rohrer, C. Gerber, E. Weibel: Surf. Sci. **131**, L379 (1983);
G. Binnig, H. Rohrer: Surf. Sci. **126**, 236 (1983); Phys. Bl. **39**, 16 (1983); Helv. Phys. Acta **55**, 726 (1982)
60.7 J.A. Panitz: J. Phys. E**15**, 1281 (1982);
K.M. Bowkett, D.A. Smith: *Field Ion Microscopy* (North-Holland, Amsterdam 1970);
E.W. Müller, T.T. Tsong: *Field Ion Microscopy Principles and Applications* (American Elsevier, New York 1969)
60.8 R.T. Lewis, R. Gomer: Surf. Sci. **26**, 197 (1971);
A.E. Bell, L.W. Swanson, D. Reed: Surf. Sci. **17**, 418 (1969);
D.W. Bassett: Brit. J. Appl. Phys. **18**, 1753 (1967);
G. Ehrlich, F. Hudda: Philos. Mag. **8**, 1587 (1963)
60.9 For reviews of field ion microscopic studies, see D.W. Basset: In *Surface Mobilities on Solid Material*, ed. by V.T. Binh (Plenum, New York 1983) p.63, 83; and
G. Ehrlich: In *Proc. 9th Int. Vacuum Congress and 5th Int. Conf. Solid Surfaces*, ed. by J.L. de Segovia (ASEVA, Madrid 1983) p.3
60.10 H.-W. Fink, G. Ehrlich: Surf. Sci. **143** (1984)
60.11 For a recent structural study, see H.L. Davis, G.-C. Wang: Bull. Am. Phys. Soc. **29**, 221 (1984)
60.12 D.A. Reed, G. Ehrlich: Surf. Sci. **120**, 179 (1982)
60.13 J.D. Wrigley, G. Ehrlich: An Atomic View of Metal Silicon Interactions, 44th Ann. Conf. Phys. Electronics, Princeton, NJ, June 1984
60.14 W.R. Graham, G. Ehrlich: Phys. Rev. Lett. **32**, 1309 (1984)
60.15 K. Stolt, J.D. Wrigley, G. Ehrlich: J. Chem. Phys. **69**, 1151 (1978)
60.16 K. Stolt, W.R. Graham, G. Ehrlich: J. Chem. Phys. **65**, 3206 (1976)
60.17 D.A. Reed, G. Ehrlich: J. Chem. Phys. **64**, 4616 (1976)
60.18 D.A. Reed, G. Ehrlich: Philos. Mag. **32**, 1095 (1975)
60.19 D.A. Reed, G. Ehrlich: Surf. Sci. (in press)
60.20 H.-W. Fink, G. Ehrlich: Phys. Rev. Lett. **52**, 1532 (1984)
60.21 H.-W. Fink, G. Ehrlich: J. Chem. Phys. (in press)
60.22 T.T. Tsong: Phys. Rev. Lett. **31**, 1206 (1973)
60.23 D.W. Bassett, D.R. Tice: In *The Physical Basis for Heterogeneous Catalysis*, ed. by E. Drauglis, R.I. Jaffee (Plenum, New York 1975) p. 231
60.24 T.T. Tsong, R. Casanova: Phys. Rev. B**24**, 3063 (1981)
60.25 D.R. Cox, P.A. Lewis: *The Statistical Analysis of Series of Events* (Methuen, London 1966) Chap.2
60.26 D.W. Bassett, D.R. Tice: Surf. Sci. **40**, 499 (1973)
60.27 D.W. Bassett, M.J. Parsley: J. Phys. D**3**, 707 (1970)

61. Atom-Probe and Field Ion Microscope Studies of the Atomic Structure and Composition of Overlayers on Metal Surfaces

T.T. Tsong and M. Ahmad

Physics Department, The Pennsylvania State University
University Park, PA 16802, USA

The field ion microscope can give a two-dimensional view of the atomic structure of a solid surface. The composition of the top and deeper surface layers can be analyzed atomic layer by atomic layer with a time-of-flight atom probe. If an adsorption layer structure is commensurate with the substrate, then its structure can be derived reliably from the field ion image. We have observed an adsorption layer superstructure, silicon on W(110), which can be directly correlated to the experimentally measured pair energies of silicon adatom-adatom interactions. The true atomic layer by atomic layer compositional analysis of surfaces can be best illustrated by a study of surface segregation of platinum base binary alloys, where the number of surface layers enriched with the segregants is found to vary from one system to another, ranging from 1 to 4 atomic layers. For some alloys, the second atomic layer is depleted with segregants. This result is consistent with an oscillatory structure in the single-ion potential in the near surface layers.

61.1 Introduction

The field ion microscope (FIM) can give a two-dimensional view of the atomic structure of a solid surface with a lateral resolution of better then 3 Å [61.1a]. This resolution refers to the shortest separation between two atoms when they can still be seen in the FIM image as separate image spots. The position of an adsorbed atom, however, can be determined to a precision better than 0.5 Å using a mapping technique, and the lattice structure of a surface can be revealed even when it is not seen in the image [61.1b]. To achieve the best resolution, the nearly hemispherical emitter surface must be well developed to the degree of atomic perfection since the magnified FIM image is produced by a radial projection of field ions formed near the surface. When the radius of the hemispherical surface is only slightly over 100 atomic diameters, the atomically well developed surface is not really a smooth surface, but has structures such as lattice steps and flat facets. The radial projection is therefore not uniform, and the image magnification varies from one region to another. Even within a plane, the image magnification varies from the plane center to the plane edge. Field ion images thus cannot be used to determine lattice parameters with good accuracy. They can, however, show a two-dimensional view of the structure of an overlayer from which its structure can be derived provided the overlayer structure is commensurate with the substrate and no atomic rearrangement occurs for the substrate surface. Of course the overlayer atoms must be able to resist field evaporation by the high electric field needed for forming a field ion image, which ranges from 2.0 to 4.5 V/Å. One will have to consider the effect of the high field on the atomic structure also. Another difficulty of the FIM is that electronegative gases such as oxygen, nitrogen, and hydrogen cannot be imaged in the field ion microscope.

There are some unique capabilities of the FIM and the atom probe in surface studies which include the availability of atomically perfect surface planes prepared by field evaporation, the possibility of studying the behavior of single atoms [61.2-4], and the capability of identifying surface atoms one by one and atomic layer by atomic layer with the time-of-flight atom probe. Due to length limitation, only two subjects will be discussed here, namely a successful correlation between adatom-adatom interaction and adsorption layer superstructure formation [61.5], and an atomic layer by atomic layer compositional depth profiling of alloy surfaces in surface segregation [61.6]. The composition depth profiles can shed some light on the atomic interaction in the near surface layers, which may cause surface reconstruction.

61.2 FIM Observations of the Atomic Structure of Metal Surfaces

In the FIM, an atomically well-developed surface is usually prepared by low-temperature field evaporation. Figure 61.1 shows the atomic structure of a few planes of a WSi_2 crystal where only W atoms are imaged. In most cases the atomic structures of surface planes revealed in the field ion images agree with those expected from the bulk structure of the emitter crystal. In a few cases, disagreements have been found. An example is the p(2 × 2) and c(2 × 2) image structures observed for W(100) especially when the field evaporation is done at high temperature [61.7b]. This surface is known to be reconstructed from LEED studies [61.8]. The c(2 × 2) LEED pattern observed turns out to be much more complicated to interpret than the vertical shift model originally proposed by *Felter* et al. [61.8]. Although LEED studies now generally favor a lateral shift model, FIM results have been interpreted to favor the vertical shift model.

In field ion microscopy, three types of surfaces enter into consideration: a thermally equilibrated surface, a low-temperature field evaporated (LTFE) surface, and a high-temperature field evaporated (HTFE) surface. A LTFE surface is not a thermodynamic equilibrated surface since no atomic jumps can occur to minimize the surface free energy after field evaporation. A HTFE surface will reveal a surface structure of minimum free energy of the surface under the influence of the applied field which may range from 2 V/Å to 5 V/Å. That an applied field can change the atomic structure of a surface [61.9] is vividly seen for the Rh (001). When a (1 × 1) Rh(100) plane is subjected to pulsed-laser heating in an applied field of 2 to 3 V/Å, field evaporation

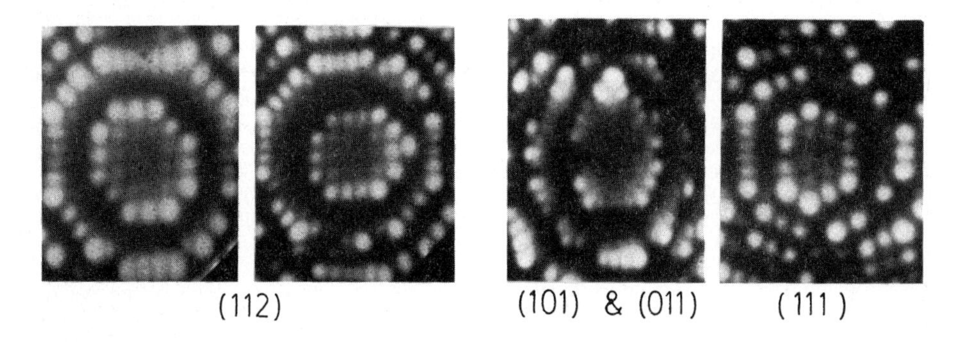

Fig.61.1. FIM images showing the atomic structure of a few planes of a WSi_2 crystal where only W atoms are imaged. For details, see [61.7a]

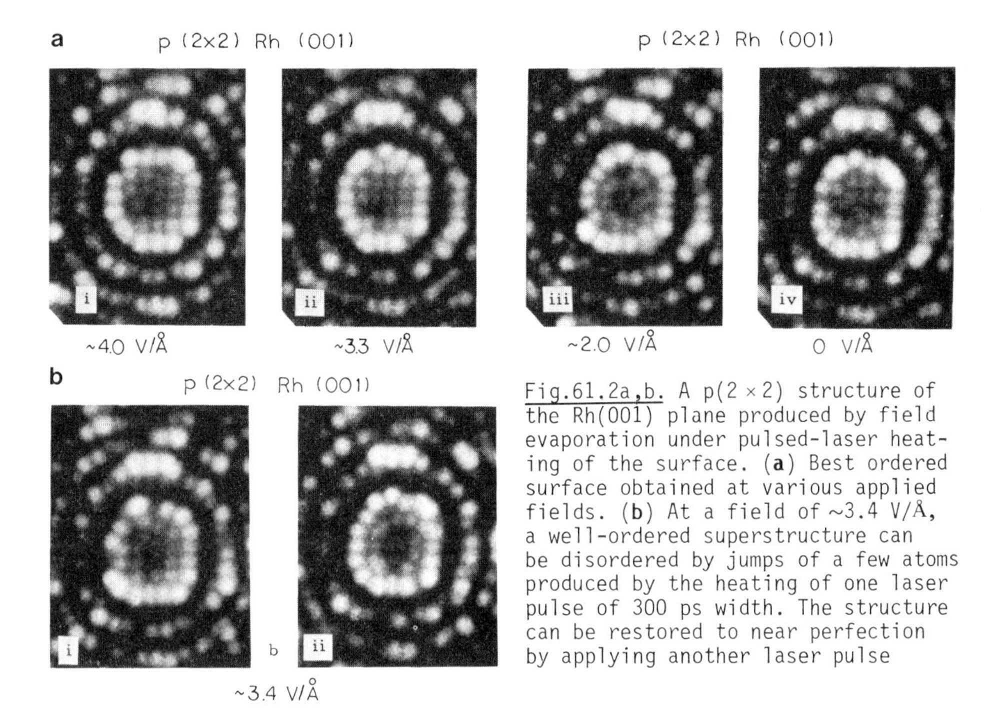

Fig.61.2a,b. A p(2 × 2) structure of the Rh(001) plane produced by field evaporation under pulsed-laser heating of the surface. (**a**) Best ordered surface obtained at various applied fields. (**b**) At a field of ~3.4 V/Å, a well-ordered superstructure can be disordered by jumps of a few atoms produced by the heating of one laser pulse of 300 ps width. The structure can be restored to near perfection by applying another laser pulse

and atomic jumps can be observed to occur and eventually a p(2 × 2) surface is formed. This well-ordered surface can be disordered by heating the surface by pulsed laser in the absence of an applied field, Fig.61.2.

Although effects of the applied field have to be carefully considered in interpreting field evaporated surfaces, there is little question that the energetics involved in surface reconstructions can be revealed in the field ion image. For example, it has been known for some time that the field evaporation behavior of the W(001) plane is quite different from other planes, but a detailed study was reported only after LEED experiments found a reconstruction of this surface. It is surprising that only a few FIM studies of surface reconstruction have been reported, and there are still no detailed FIM analyses of the atomic structures of thermally annealed surfaces.

61.3 Adatom-Adatom Interactions and Adsorption Layer Superstructure Formation

In thermodynamic equilibrium, adatoms on a surface assume a superstructure of minimum free energy and maximum entropy of this particular adatom coverage. Assuming that the pair potential energies in adatom-adatom interactions are additive, then one would expect the lattice parameters of the superstructure to correspond well to bonds of relative minimum in the pair potential energy. At least in one system, Si adatoms on the W(110), such correlation has been found [61.5].

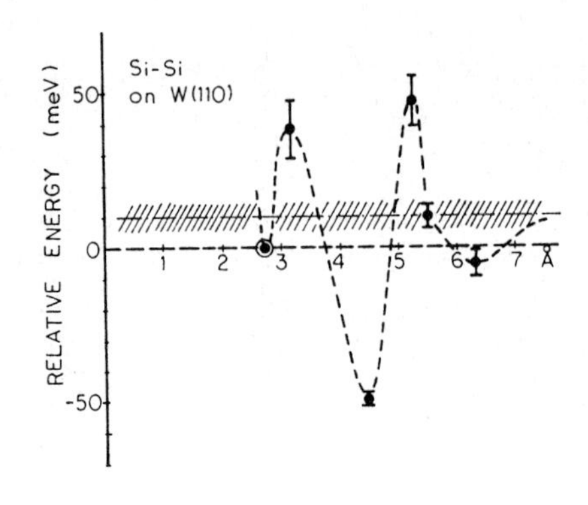

Fig.61.3. The relative pair potential of Si-Si interaction of the W(110) plane. See text for explanations

Silicon adatom-adatom interaction on a metal surface is most interesting since Si atoms form covalent bonds in solid state. It is interesting to see if Si adatoms on a metal surface interact very differently from metal adatoms. When measuring pair distributions with metal adatoms, two atoms are deposited onto a plane. For silicon adatoms, because of a peculiar imaging property, it was difficult to obtain a pair distribution with only two adatoms on a plane. Instead, three to six adatoms were deposited onto a plane. The relative frequencies of observing a diatomic cluster of different bond lengths were measured after heating the surface at 300 K for 50 s in field free conditions. From these frequencies, the relative pair energies at 6 bond distances, from the first to sixth nearest-neighbor distances of the W(110) lattice, were derived, Fig.61.3.

In deriving these pair energies, we assume that all the adatoms sit in equivalent sites, although this is by no means certain. As the adatom-adatom interaction is very much weaker than the adatom-substrate atom interaction, this is a reasonable assumption. We have to point out here that the pair energies are derived only for six discrete points. A detailed form of the pair potential cannot be determined from these data alone. The dashed curve merely connects these points smoothly; it is not intended to show the pair potential function. Also the asymptotic line of the pair potential is not known; the figure uses the pair energy at the nearest-neighbor distance as a reference.

Obviously when many Si adatoms are deposited on a plane, they interact with one another to form a superlattice. Although many-body effects may play an important role in adsorption-layer superstructure formation [61.10], if the adatom coverage is low then these effects may be less important, and the superstructure may correlate well with the pair potential. The pair potential shown in Fig.61.3 indicates that a stable adatom superstructure of $(2\sqrt{2}/\sqrt{3} \times 4/\sqrt{3})$ R35.26° structure will be formed if the adatom coverage is about 1/4 of a monolayer, and if the asymptotic energy is slightly higher than the reference line, or the pair energy at the nearest-neighbor distance. In fact this adsorption layer superstructure of silicon has been subsequently found, Fig.61.4. A Monte Carlo simulation indicates that the asymptotic value of the pair potential should be at least 10 meV above the pair energy at the nearest-neighbor distance.

To form an adlayer superstructure, the adatom-adatom interaction has to have at least two bonds of relative minimum in the pair potential (these two bonds may have the same length but in different directions). Also the poten-

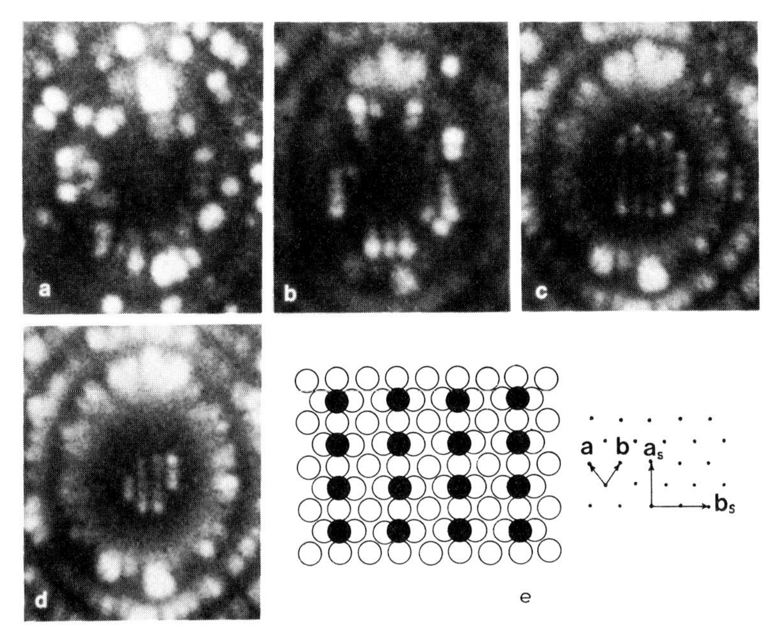

Fig.61.4a-e. From the pair potential shown in Fig.61.3, an adsorption layer of structure shown in (**e**) can be expected as in (**a-d**). About 1/4 monolayer of Si atoms are deposited with the substrate kept at 80 K. Heating the surface to 300 K for 50 s suffices to form a superstructure of the adsorption layer which is clearly revealed by some field evaporation of the layer. **a** and **b** are lattice vectors of the substrate, and $\mathbf{a}_S$ and $\mathbf{b}_S$ are those of the adsorption layer superstructure

tial well depths have to be considerably larger than the thermal energy (kT_d) of the diffusion temperature when adatoms can hop from one site to the other. We want to point out here that the interaction responsible for the adsorption layer superstructure formation is not the electronic indirect interaction alone, as proposed in [61.11], but is the combined effect of all the possible adatom-adatom interactions. The FIM experiments measure pair energies of this combined effect.

61.4 Atomic Layer by Atomic Layer Compositional Analysis of Solid Surface

The capability of the atom probe in the compositional analysis of single atomic layers was first utilized in a study of an ordered Pt_3Co alloy where the composition of superlattice and fundamental layers were analyzed one by one [61.12]. Atomic layer by atomic layer compositional analysis is now a common practice in field ion microscopy. Unfortunately the number of atoms which can be detected from one atomic layer is very small and few studies can be considered truly quantitative. An exception is the study of surface segregation of dilute alloys where quantitatively reliable data can be obtained by repeated measurements under an identical heat treatment of the samples [61.13].

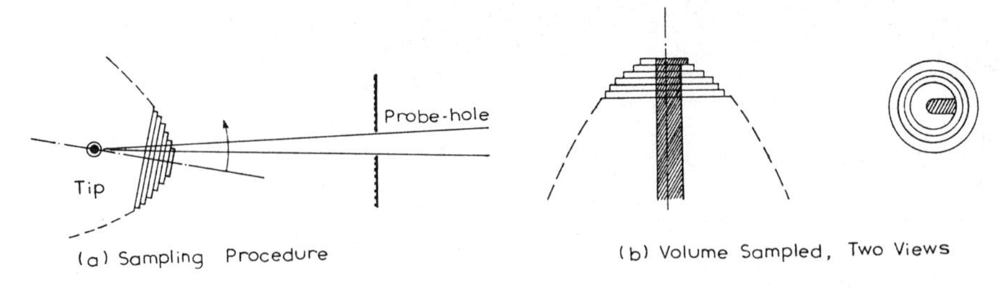

Fig.61.5a,b. The sampling procedure and the volume sampled in an atom-probe analysis of the composition of surface layers in surface segregation

Surface segregation in alloys has important implications for catalysis, metallurgy and material properties. Although *Gibbs* predicted already in the 19th century that it would occur and many techniques have already been applied to study this phenomenon, only recently can quantitatively reliable composition depth profiles be derived with the atom probe. Detailed experimental procedures can be found elsewhere [61.13]. The alloy tip is first degassed and field evaporated to develop a clean and atomically perfect surface. The tip is then annealed to a high temperature, 600 to 800°C, for 3 to 5 minutes to equilibrate the distribution of alloy species in the near-surface layers. It is then quenched to ~80 K at an estimated rate of $\sim 10^4$K/s which freezes the distribution of the alloy species in the sample. The tip is then imaged again. The thermal end form shows enlarged facets of low index planes. A controlled pulsed field evaporation is then carried out with the probe hole aimed at the edge of the top surface layer as in Fig.61.5. The evaporated ions are detected one by one and mass analyzed from their flight times. The field evaporation reduces the size of the layer, and the tip orientation is readjusted so that the receding edge of the top surface layer always remains aimed at the probe hole. This process is continued until the top layer is completely removed, and is then repeated for subsequent layers. As there is no intermixing of mass signals from different atomic layers, the composition depth profile obtained has a true single atomic layer depth resolution. For statistical reliability, the same procedure is repeated many times for many samples heat-treated under identical conditions. The number of ions detected for each atomic layer ranges from two hundred to over one thousand ions which gives a statistical uncertainty of less than 2%. To make sure that a thermodynamic equilibrium distribution of the alloy species has been reached, a composition depth profile is derived at a higher annealing temperature. Figure 61.6 gives two composition depth profiles of a Pt-Rh alloy showing enrichment of Rh at the top surface layer and a significant depletion of Rh in the second layer. From the third layer on, the composition returns to the bulk value. This nonmonotonic depth profile persists for alloys of different composition, Fig.61.7.

As far as we are aware the atom probe is the only instrument capable of providing reliable composition depth profiles. Ever since we found the enrichment of Cu in NiCu to be confined to the top surface layer, most theories are now based on the assumption that the enrichment of segregants in alloys is confined to the top layer. Further atom-probe measurements, however, show a wide range of chemical specificity of the segregation behavior as listed in Table 61.1. The enrichment may extend to a few atomic layers in depth and the compositional variation may decay monotonically or nonmonotonically into

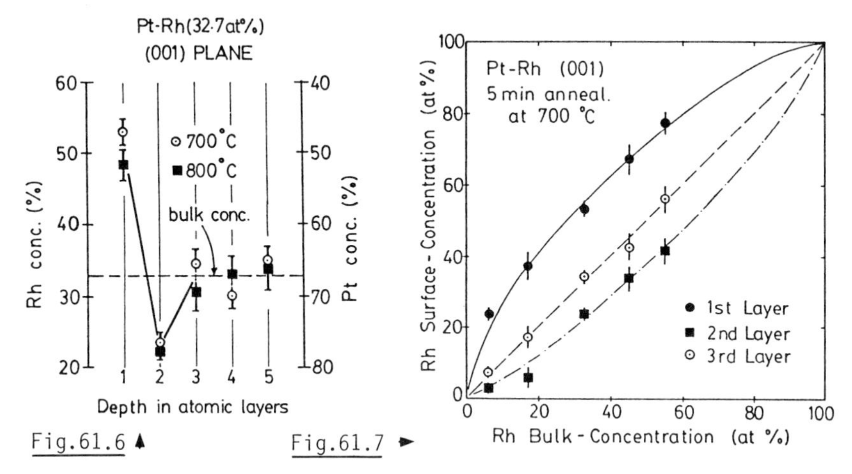

Fig.61.6. A composition depth profile of Pt-32.7 at %Rh equilibrated at 700°C and at 800°C. Both show an Rh enrichment in the top layer and a depletion in the second layer. From the third layer the composition returns to the bulk value

Fig.61.7. Similar segregation behavior persists for five alloys of different composition

Table 61.1. Binary alloy surface segregation behavior observed in atom-probe studies

Alloy system	Segregating element	Extent of segregation (No. of atomic layers)	Nature of convergence to bulk value
Pt-Rh	Rh	1	Oscillatory
Pt-Ir	Pt	2	Monotonic
Pt-Au	Au	4	Monotonic
Ni-Cu	Cu	1	Oscillatory

the bulk. The nonmonotonic segregation behavior is especially interesting since a pseudopotential calculation by *Barnett* et al. shows that the single-ion potential at the surface oscillates about a constant bulk value [61.14]. This oscillation may induce a surface atomic rearrangement. Since the segregant atoms should equilibrate under the influence of such a potential, our nonmonotonic composition depth profile may in fact reflect this oscillation also.

Acknowledgments. The author acknowledges the support of NSF, and the contribution of R. Casanova and C.F. Ai to some of the results presented in this paper.

References

61.1a E.W. Müller, T.T. Tsong: *Field Ion Microscopy, Principles and Applications* (Elsevier, New York 1969); and in Prog. Surf. Sci. **4**, 1 (1978)
61.1b T.T. Tsong: Phys. Rev. B**6**, 417 (1972);
P.W. Cowan, T.T. Tsong: Surf. Sci. **67**, 158 (1977)
61.2 G. Ehrlich, F.G. Hudda: J. Chem. Phys. **44**, 1039 (1966); see also G. Ehrlich: This Volume, p.375
61.3 D.W. Bassett, M.J. Parsely: Nature (London) **221**, 1046 (1969)
61.4 T.T. Tsong: J. Chem. Phys. **55**, 4658 (1971); Phys. Rev. B**6**, 417 (1972)
61.5 T.T. Tsong, R. Casanova: Phys. Rev. Lett. **47**, 113 (1981);
R. Casanova, T.T. Tsong: Thin Solid Films **93**, 41 (1982)
61.6 T.T. Tsong, Y.S. Ng, S.V. Krishnaswamy: Appl. Phys. Lett. **32**, 778 (1978);
Y.S. Ng, T.T. Tsong, S.B. McLane: Phys. Rev. Lett. **42**, 588 (1979);
T.T. Tsong, Y.S. Ng, S.B. McLane: J. Appl. Phys. **73**, 1464 (1980)
61.7a T.T. Tsong et al.: J. Vac. Sci. Technol. B**1**, 915 (1983)
61.7b A.J. Melmed, R.T. Tung, W.R. Graham, G.D.W. Smith: Phys. Rev. Lett. **43**, 1521 (1979);
R.T. Tung, W.R. Graham, A.J. Melmed: Surf. Sci. **115**, 576 (1982).
Tsong and Sweeny found a (1×1) structure by field evaporating the W(001) plane at 21 K, T.T. Tsong, J. Sweeny: Solid State Commun. **30**, 767 (1979)
61.8 T.E. Felter, R.A. Barker, P.J. Estrup: Phys. Rev. Lett. **38**, 1138 (1977);
M.K. Debe, D.A. King: Phys. Rev. Lett. **39**, 708 (1977)
61.9 C.F. Ai, T.T. Tsong: Surf. Sci. **127**, L165 (1983)
61.10 T.L. Einstein: CRT Crit. Rev. Solid State Mat. Sci. **7**, 261 (1978) and references therein
61.11 T.B. Grimley: Proc. Phys. Soc. (London) B**90**, 751; B**92**, 776 (1973);
N.R. Burke: Surf. Sci. **58**, 349 (1976);
K.H. Lau, W. Kohn: Surf. Sci. **75**, 69 (1978);
A.M. Gavovich, L.G. Il'Chenko, E.A. Pashitskii, Y.A. Romanov: Surf. Sci. **94**, 179 (1980)
61.12 T.T. Tsong, S.V. Krishnaswamy, S.B. McLane, E.W. Müller: Appl. Phys. Lett. **23**, 1 (1973)
61.13 M. Ahmad, T.T. Tsong: Appl. Phys. Lett. **44**, 40 (1984); also to be published
61.14 R.N. Barnett, R.G. Barrera, C.L. Cleveland, U. Landman: Phys. Rev. B**28**, 1667 (1983)

62. LEED Studies of Physisorbed Noble Gases on Metals and Interadatom Interactions*

M.B. Webb and E.R. Moog

University of Wisconsin-Madison, Madison, WI 53706, USA

Low-energy electron diffraction studies of noble gas adsorption provide opportunities to study a number of interesting phenomena. There are also a number of experimental problems. Approaches to these opportunities and problems are discussed and illustrated with selected results.

62.1 Introduction

Physisorption refers to adsorption where the binding energies are small — of the order of tenths of an eV/atom and primarily due to dispersion forces — in contrast to chemisorption involving ordinary chemical bonds with energies of perhaps several eV/atom. The apparently simple system of physisorbed noble gases allows the study of an amazingly large number of interesting phenomena, many of which have been discussed in these proceedings. There are 2-dimensional gas and fluid phases and commensurate and incommensurate monolayer solids. On many metal surfaces, where registry forces can be unimportant, the system provides a good approximation to 2d physics regarding the nature of long-range order, melting, and phase transitions and critical phenomena. Going somewhat away from 2d, to substrates where registry forces due to the atomic structure are important, studying commensurate-incommensurate transitions and orientational epitaxy is possible. Finally, yet further from 2d, observing thicker films allows studying wetting and growth modes [62.1].

The weak interactions in physisorption make possible or convenient a number of experiments which are difficult with chemisorbed systems. First, because of small adsorption energy, there is an appreciable and experimentally manageable equilibrium vapor pressure above the adsorbed film. Thus experiments can be done in equilibrium with the 3d gas phase and the equilibrium vapor pressure gives a simple and direct measure of the adsorbate chemical potential. This allows determining various thermodynamic properties like heats and entropies of adsorption without the necessity of model-dependent interpretations of kinetic experiments. In some systems like reconstructed semiconductor surfaces, equilibrium within the film can be achieved only through the gas phase because diffusion barriers can be appreciable compared to desorption energies [62.2]. Secondly, the adatom-adatom interactions within the adsorbed film are due to the same type of weak interactions; thus they are a larger fraction of the total energy and easier to study experimentally.

*Work supported by NSF Grant No. DMR 8114843 and in part by the University of Wisconsin Graduate Research Committee.

Most of the phenomena listed above depend on these interadatom interactions and their strength relative to the registry forces. Much of our work has been aimed at a detailed understanding of these interadatom forces. Finally, because of the weak interactions, the adsorbate is much less likely to disturb the substrate structure. This makes the interpretation simpler and also leads to the possibility of using the noble gases as a probe of the substrate surface properties. There have been a number of experiments exploiting this possibility [62.2,3].

Besides these opportunities, the weak interactions also make certain experimental problems more severe in physisorption studies. Vacuum requirements are particularly stringent because the substrate surface at cryogenic temperature is itself a cryopump for residual gases. Beam-stimulated desorption, especially when using electrons as the probe, can be important [62.4]. Since much of the interesting physics manifests itself in small structural and thermodynamic effects, the experiments often require more accuracy than other surface-science studies. Finally, precise coverage measurements are required, especially for incommensurate layers where there is no superlattice for reference. One desires a coverage measurement which does not include extrinsically adsorbed gas on thermal facets or other heterogeneities.

X-ray [62.5], neutron [62.6], atom [62.7] and electron diffraction have all been used for structural investigations. For some time our group has been involved in studies of noble gas adsorption on metal and semiconductor surfaces using primarily low-energy electron diffraction. In the space available, rather than review these results, many of which have already appeared [62.2,8-10], it seems more useful to discuss the approaches to some of these opportunities and problems, which can have broader applications, and then illustrate them briefly with selected results.

62.2 Diffraction

X-ray, neutron, atom and electron diffraction have all been used as structural probes of physisorbed films and all the techniques complement one another. X-ray diffraction using synchroton radiation gives extreme lateral resolution so that, except for the very best substrates, the information is limited by the substrate perfection. X-ray and neutron diffraction are most easily used with high specific area substrates and have the important advantage of being compatible with high 3d gas pressures. Atoms and neutrons allow energy and momentum resolution to map phonon dispersion relations.

Low-energy electron diffraction is the simplest and most natural technique for studying single-crystal substrates. It allows a large range of both parallel and perpendicular momentum transfers. While interpretation of diffracted intensities can be complicated, this is not a problem in determining lateral structures. Whereas conventional display apparatus has limited lateral resolution, instruments capable of very high resolution are available or are being developed [62.11]. Importantly, LEED also gives considerable information characterizing the perfection of the substrate [62.11]. We feel that its most serious limitation in physisorption studies is the restriction to pressures compatible with the electron mean-free path, making interesting parts of the phase diagram inaccessible. This can be partially overcome by the gas-handling technique discussed below.

Let us first consider measurement of the lateral spacing in an adsorbed film. The low-energy electron trajectories are soft and very sensitive to small fields. This usually limits the accuracy of absolute measurements of the lateral momentum transfer to a percent or so which is insufficient for our purpose. However, careful angular profiles of diffracted beams taken with a Faraday collector allow measurements of small differences in $S_{\parallel}$ with a precision corresponding to differences in lateral spacings of a few thousandths of an Ångstrom. Making such a relative measurement between the adsorbed film and a subsequently grown bulk noble gas crystal and using the known bulk lattice parameter allows the absolute film lattice parameter to be determined with the same precision [62.8].

One is also interested in studying 2d fluid phases. Diffuse scattering from disordered adsorbates has often been observed visually in LEED. Again, careful Faraday collector measurements allow detailed study of diffuse scattering and its evolution with coverage, temperature, etc., giving information about the lateral pair distribution function [62.10].

Finally, one is interested in the vertical structure of the film which is available from an analysis of the LEED I-V data. For incommensurate overlayers, this analysis is simplified since a number of multiple-scattering processes are then unimportant. In fact, the adsorbate-substrate separation can be simply determined from the ratio of the specular I-V profiles from the clean and monolayer-covered surface [62.8,9]. This ratio shows a prominent and periodic modulation corresponding to the interference of the scattered amplitudes from the adsorbate and the substrate. Such a determination has been confirmed by full dynamic calculations [62.12]. The temperature dependence of this ratio gives an estimate of the thermal vibrations normal to the surface, but this estimate is considerably less accurate than information from atom diffraction.

62.3 Vacuum, Gas Handling and Coverage Measurements

The vacuum requirements are severe because at low temperatures the surface adsorbs any residual gas. This problem is more serious because at the low temperatures and pressures of the experiments the approach to equilibrium can be slow and the surface must be exposed to the residual gas for long times — perhaps several hours. One convenient approach is to nearly surround the sample with a shroud which is independently cooled to He temperatures while the sample is still hot. This serves both as a thermal shield and a cryopump and the surface can be kept clean for the required time. However, the cryoshield would then pump the sample gas if the chamber were backfilled. Therefore, instead of backfilling, the sample gas is introduced as a well-collimated beam directed at the place on the sample surface where the experiment is performed. Thermally desorbed gas or beam atoms not sticking have negligible probability of returning to the sample surface before being cryopumped. The beam may be valved to deliver a sequence of equal gas doses or, for "equilibrium" experiments, a constant flux impinging on the sample surface plays the role of the arrival of atoms from a 3d gas phase. The effective pressure represented by a given flux is determined by subsequently cooling the surface and noting the temperature where the bulk noble gas solid forms. The effective pressure is then read from the well-known bulk phase diagram. In this way thermodynamic experiments can be done without ever measuring the pressure directly, which is an inaccurate measurement particularly in the presence of thermal transpiration. It is much easier to maintain a constant flux than a constant pressure in a UHV system. This method of handling the sample gas has other advantages. Most importantly for experiments using electrons as a probe, since the electrons need move only short distances through the gas beam, the effective pressure at the surface can be some orders of magnitude higher than pressures consistent with the electron mean-free path in backfilling experiments. Secondly, since the bulk solid-gas coexistence is used as the fiducial for the effective pressure, it turns out that determining various thermodynamic properties like heats of adsorption depends only on measurements of small temperature differences rather than on absolute temperatures [62.8].

With electron probes, stimulated desorption is always a potential problem. In dosing experiments, i.e., experiments in the absence of the 3d gas phase, it is usual to use very small incident currents to minimize the stimulated desorption. The problem is much less serious in equilibrium experiments since most often the stimulated desorption is negligible compared to the thermal desorption in the dynamic equilibrium [62.4,8].

In all adsorption experiments, it is necessary to know the coverage. In many experiments the coverage is inferred from the exposure; in others it is measured by Auger or core level spectroscopy or by integrating the thermal desorption spectra. In some investigations a secondary coverage measurement is made from changes in the work function. All of these have some drawbacks: they are not direct and require some auxiliary calibration, there are often line-shape changes and signals are not linear in coverage. Most coverage measurements include the total adsorbate—that extrinsically adsorbed on steps, facets, and other heterogeneities as well as that intrinsically adsorbed on the ideal areas of the surface. It is the extrinsically adsorbed gas that is responsible for the commonly observed low coverage and high binding energy tail on isotherms which precludes the study of the intrinsic 2d gas phase.

In the special case of incommensurate overlayers, like the noble cases on many substrates, there is a convenient alternative coverage measurement [62.8]. Since the overlayer is incommensurate, its diffracted amplitude interferes with that from the substrate only for the specular beam. The effect of the overlayer on the nonspecular substrate diffracted beams is only to attenuate them. (There are small corrections for the overlayer thermal diffuse scattering. One also can check that noble gas overlayers do not affect the substrate thermal vibrations.) Thus a measurement of a substrate nonspecular diffracted beam intensity is easily converted into a coverage measurement [62.8]. The measurement is calibrated, for example, by measuring the attenuation by a complete monolayer for which the lattice parameter has been precisely measured. Since the substrate diffracted beam originates from the coherently scattering regions of the surface, the attenuation is due to that gas adsorbed only on the coherently scattering regions.

62.4 Selected Results

We now use selected results from studies of noble gas adsorption on Ag(111) [62.8,9] and Pd(100) [62.10] surfaces to illustrate some of the capabilities of LEED for investigating physisorbed films. Xenon, Kr and Ar adsorbed on Ag(111) form incommensurate triangular lattices in the 2d solid monolayer phase. The Xe interatomic spacing in the monolayer solid, determined from measurements relative to the bulk, is shown as a function of temperature in Fig.62.1. The closed symbols denote an unconstrained layer, i.e., the surface is partially filled with 2d solid islands, there is no 3d gas present and the gas remains on the surface at the low temperatures because of its long lifetime. Under these conditions the spreading pressure is essentially zero. The interatomic spacing is about 2% larger than in bulk Xe. In contrast, with a constant flux of gas corresponding to an effective pressure of 7×10^{-9} Torr, the interatomic spacing follows the open symbols. This much

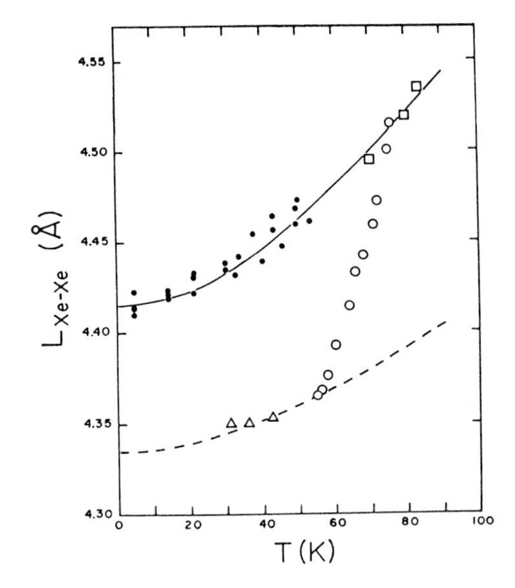

Fig.62.1. The Xe-Xe interatomic spacing as a function of temperature for Xe adsorbed on Ag(111). (●) denote the unconstrained layer, (○) the monolayer solid in equilibrium with a 3d gas of effective pressure 7×10^{-9} Torr, (□) are spacings at monolayer condensation for other effective pressures, and (----) is the bulk Xe spacing. (Δ) denote a 20 layer thick film

larger thermal expansion is the result of the equilibrium spreading pressure. More pictorially, with the 3d gas present it is energetically favorable for more atoms to squeeze into the film to take advantage of the substrate holding potential at the expense of the energy to compress the film. This compression terminates at the bulk lattice parameter where the bilayer forms. By the second layer, the film has healed to its bulk structure and subsequent growth is layer by layer.

An example of an isobar, i.e., the coverage vs temperature at a fixed effective pressure, for Xe on Ag(111) is shown in Fig.62.2. The coverage has been measured from the attenuation of the Ag(10) beam, using a calibration from the lattice parameter measured at monolayer condensation. The vertical risers signal first-order transitions: the first indicates the coexistence of the 3d gas with 2d gas and 2d monolayer solid; the second is the coexistence of the monolayer and bilayer, etc. The small tail at temperatures above monolayer condensation is the intrinsically adsorbed 2d gas [62.13]. The slope of the plateaus and the different riser heights are due to the compression we saw in the lattice parameter data. This is demonstrated in Fig.62.2b where we have plotted the area which would be occupied by the measured coverage at the measured lattice parameter; here the plateaus are flat and the steps are of equal height. This, incidentally, sets a limit on the number of vacancies and adatoms in the monolayer. From a series of such isobars the phase diagram can be determined and the latent and isosteric heats can be derived. The lateral isothermal compressibility of the film can also be extracted [62.8]. All these properties and the monolayer-bilayer transition have been understood quantitatively from both statistical mechanics and computer simulations using a model of the interadatom interactions consisting of the known isolated pair potential with important corrections due to the presence of the substrate, mainly the Sinanoglu-Pitzer-McLachlan interaction [62.14].

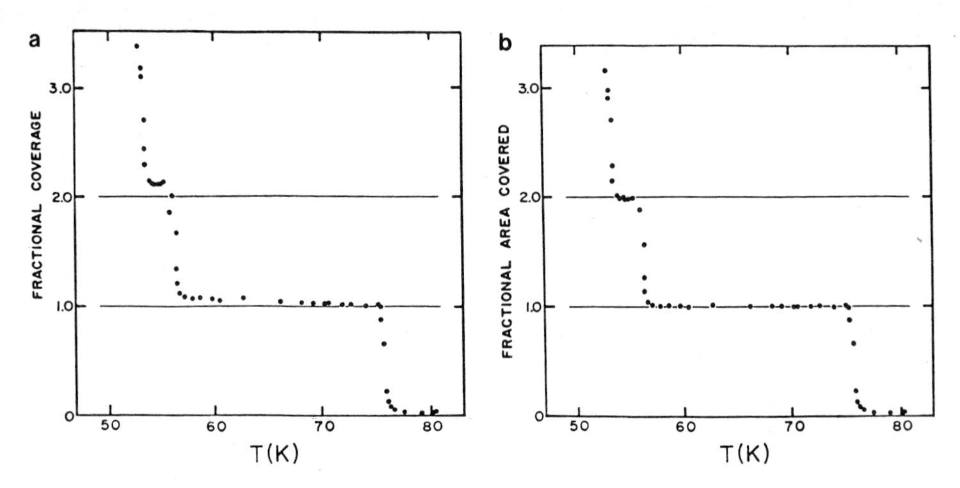

Fig.62.2. (**a**) A Xe/Ag(111) isobar. The fractional Xe coverage as a function of temperature for an effective pressure of 2×10^{-8} Torr, determined from the attenuation of the Ag(10) diffracted beam. (**b**) The fractional area covered by the Xe in (*a*), found by using the lattice constant from Fig.62.1

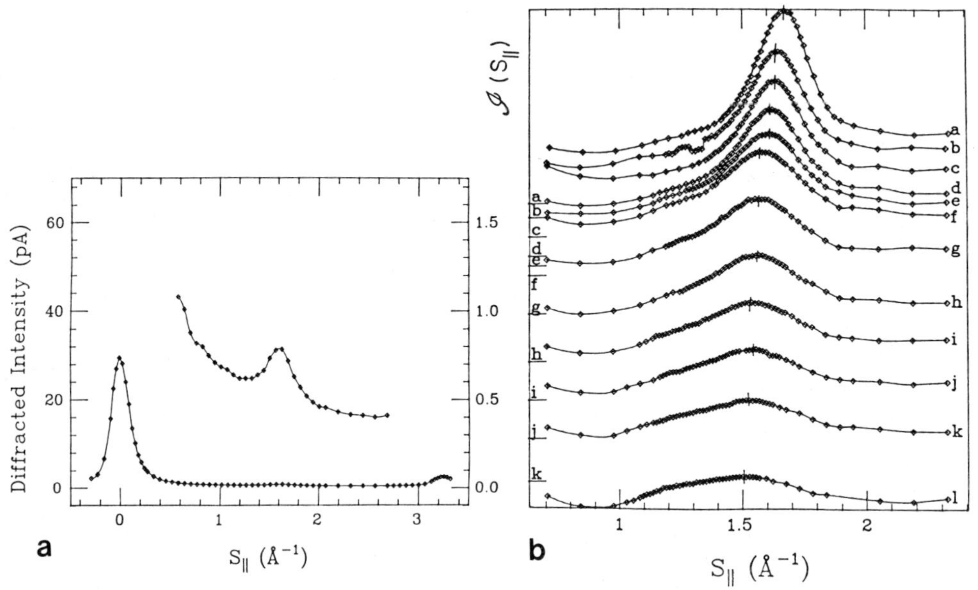

Fig.62.3. (**a**) An angular profile of the diffracted intensity at 326 eV and normal incidence for Pd(100) plus a monolayer of Xe. The upper curve has been enlarged 40 ×. (**b**) A series of interference functions for different coverages of a well-annealed Xe layer. The Xe coverages are: *a*, 0.97; *b*, 0.95; *c*, 0.93; *d*, 0.91; *e*, 0.90; *f*, 0.89; *g*, 0.85; *h*, 0.80; *i*, 0.76; *j*, 0.72; *k*, 0.68; *l*, 0.61. The zeroes are indicated by the dashes at the left and have been displaced by amounts proportional to the coverage

When the substrate is Pd(100) instead of Ag(111), the situation changes drastically [62.10]. We observe no 2d crystallization of the monolayer even down to 10 K, and at all temperatures and coverages the diffraction is a continuous diffuse ring like that from a fluid. Figure 62.3a shows an angular profile taken radially across the ring, and Fig.62.3b shows a family of interference functions for various coverages. As the coverage increases the ring expands, becoming sharper and more intense, as expected if the interadatom interactions were repulsive everywhere. The repulsive nature of the interactions is also shown by the decreasing isosteric heats derived from the isobars. At a coverage of 0.8 there is a break in the apparent Xe-Xe spacing as a function of coverage which possibly indicates the crystallization to a poorly ordered solid as expected for the hard disc system [62.15]. The physics included in the model interaction used to understand the noble gases on Ag(111) is insufficient to understand the behavior on Pd(100). The direct overlap interactions between the electronic structure of the adatom/substrate complexes, which could be ignored for adsorption on Ag(111), are apparently important for Pd(100).

References

62.1 For reviews see J.G. Dash: *Films on Solid Surfaces* (Academic, New York 1975);
S.K. Sinha (ed.): *Ordering in Two Dimensions* (North-Holland, Amsterdam 1980);
B. Mutaftschiev (ed.): *Interfacial Aspects of Phase Transformations* (Reidel, 1982);
Symposium on Statistical Mechanics of Adsorption, Surf. Sci. **125**, 1-325 (1983)

62.2 E. Conrad, M.B. Webb: Surf. Sci. **129** (1983)

62.3 K. Wandelt: J. Vac. Sci. Technol. A**2**, 802 (1984);
J.E. Demuth, A.J. Schell-Sorokin: J. Vac. Sci. Technol. A**2**, 808 (1984);
N.D. Lang, A.R. Williams: Phys. Rev. B**25**, 2940 (1982)

62.4 E.R. Moog, J. Unguris, M.B. Webb: Surf. Sci. **134**, 849 (1983)

62.5 R.J. Birgeneau, P.M. Horn, D.E. Moncton: This Volume, p.404

62.6 P. Dutta, S.K. Sinha, P. Vora, M. Nielsen, L. Passell, M. Bretz: *Ordering in Two Dimensions*, ed. by S.K. Sinha (North-Holland, Amsterdam 1980) p.169

62.7 S.J. Sibener, K.D. Gibson: Submitted to J. Chem. Phys.

62.8 J. Unguris, L.W. Bruch, E.R. Moog, M.B. Webb: Surf. Sci. **87**, 415 (1979); **109**, 522 (1981)

62.9 P.I. Cohen, J. Unguris, M.B. Webb: Surf. Sci. **58**, 429 (1976)

62.10 E.R. Moog, M.B. Webb: Submitted to Surf. Sci.

62.11 M. Henzler: This Volume, p.351

62.12 N. Stoner, M.A. Van Hove, S.Y. Tong, M.B. Webb: Phys. Rev. Lett. **40**, 243 (1978)

62.13 J. Unguris, L.W. Bruch, M.B. Webb, J.M. Phillips: Surf. Sci. **114**, 219 (1982)

62.14 L.W. Bruch, P.I. Cohen, M.B. Webb: Surf. Sci. **59**, 1 (1976);
L.W. Bruch, J. Unguris, M.B. Webb: Surf. Sci. **87**, 437 (1979);
L.W. Bruch, J.M. Phillips: Surf. Sci. **91**, 1 (1980);
L.W. Bruch: Surf. Sci. **125**, 12 (1983);
L.W. Bruch, M.S. Wei: Surf. Sci. **100**, 481 (1980)

62.15 D.G. Chae, F.H. Ree, T. Ree: J. Chem. Phys. **50**, 1581 (1969);
F.H. Ree, W.G. Hoover: J. Chem. Phys. **40**, 939 (1964)

63. Phases and Phase Transitions in Two Dimensional Systems with Competing Interactions

R.J. Birgeneau

Department of Physics, Massachusetts Institute of Technology
Cambridge, MA 02139, USA

P.M. Horn

IBM Thomas J. Watson Research Center, Yorktown Heights, NY 10598, USA

D.E. Moncton

Brookhaven National Laboratory, Upton, NW 11973, USA

We summarize the results of synchrotron X-ray studies of the phases and phase transitions in two prototypical two-dimensional systems where competing interactions play an essential role. The first, stage-4 bromine intercalated graphite, $C_{28}Br_2$, exhibits a transition from a centered $\sqrt{3} \times 7$ commensurate phase to a stripe domain incommensurate phase where technically there are no Bragg peaks but only power law singularities with exponents $\eta_n = 2n^2/49$ where n is the harmonic number. The second, monolayer krypton on graphite, is one of the most extensively studied surface overlayer systems. This system exhibits a $\sqrt{3} \times \sqrt{3}$ commensurate phase, a modulated hexatic reentrant fluid phase and both nonrotated and rotated incommensurate triangular solid phases. These phases and the transitions between them have been studied in detail. Finally, we discuss briefly perspectives for the future.

63.1 Introduction

Much of the current activity in the field of phase transitions is centered on systems with competing interactions. It turns out that the physics of such systems is both very rich and quite subtle. Not surprisingly, particularly interesting behavior occurs in two dimensions where enhanced fluctuation effects lead to entirely new and novel phases of matter. Surfaces and surface overlayers provide model laboratories for the realization of many idealized two-dimensional phase transition problems.

In the last several years, experiment and theory for 2D surface phase transitions have advanced symbiotically. On the theoretical side, real space renormalization group techniques have facilitated calculations using realistic potentials for real surface systems. From our vantage point, the experimental advances have been particularly dramatic. It became clear after the first few experiments that study of the important issues in two-dimensional (2D) phase transitions would require a drastic improvement in resolution over that provided by conventional LEED, X-ray or neutron diffraction techniques. For X-ray and neutron techniques the angular resolution is limited primarily by the brilliance of the source, since perfect Si crystals can provide a resolution of order 0.0001 $Å^{-1}$. As is by now well known, electron storage rings operating at a few GeV provide high intensity, highly directional, polychromatic beams of X-rays which are ideal for such studies. Pioneering experiments have been carried out at SSRL [63.1-3] and at DESY [63.4]. Over the next few years we expect these initial activities to evolve into routine techniques for surface scientists.

The number of X-ray scattering studies of surface phase transitions is still quite limited. The first moderate resolution experiments utilized traditional X-ray sources and exfoliated graphite as a substrate [63.2].

Next came high-resolution synchrotron studies utilizing similar substrates [63.2,4] as well as experiments on single-crystal free-standing liquid crystal films [63.3]. These experiments were soon followed by studies of monolayer Pb on single-crystal Cu(110) [63.1]. As we shall discuss in this paper, the graphite experiments have now been extended to single-crystal substrates [63.5]. A number of X-ray studies of surface reconstruction, albeit not from a phase transition point of view, have also been reported [63.1,6].

In this paper we review recent results by the authors and their co-workers (G.S. Brown, K. D'Amico, P. Dimon, P.A. Heiney, S.E. Nagler, E. Specht, P.W. Stephens, and M. Sutton) on two separate problems: studies of the weakly incommensurate phase in bromine-intercalated graphite [63.7] and studies of the phases and phase transitions of krypton on graphite in the monolayer coverage range [63.2,5]. These represent prototypes of the behavior of 1D and 2D incommensurabilities in 2 dimensions.

In Sect.63.2 we discuss work on $C_{28}Br_2$; Sect.63.3 reviews recent results on krypton on graphite; in the final section we discuss briefly future perspectives.

63.2 Bromine-Intercalated Graphite

Bromine-intercalated graphite is manifestly not a surface system. However, for stage 4 material, $C_{28}Br_2$, the Br_2 layers are effectively decoupled from one another so that one observes ideal 2D behavior [63.2]. There is substantial charge transfer between the carbon and Br_2 layers; thus this system illustrates the behavior also expected for *chemisorbed* surface systems.

At room temperature, the Br_2 has a centered $\sqrt{3}\times 7$ structure, commensurate with the graphite (001) basal planes. At 342.20 ± 0.05 K, the Br_2 layers undergo a novel commensurate-incommensurate transition (CIT) in which the Br_2 lattice expands relative to the graphite in the 7-fold direction while staying commensurate in the $\sqrt{3}$ direction [63.7]. Initially, the diffraction results seemed quite puzzling since some peaks moved up in Q while others moved down. Hoever, it was quickly realized that the results could be explained by assuming that the thermal expansion was effected by the creation of a lattice of 1D domain walls. Specifically, assuming true long-range order (LRO), if there is a net displacement of τ across each domain wall with the walls occurring every N unit cells, then the structure factor in the 7-fold (y) direction is given by

$$S(q_y)=\left\{\frac{\sin\dfrac{Nbq_y}{2}}{\sin\dfrac{bq_y}{2}}\right\}^2 \delta\left(q_y - n\,\frac{2\pi}{Nb+\tau}\right) \quad , \tag{63.1}$$

where b is the lattice constant and n is an integer. High-resolution longitudinal scans in the commensurate and incommensurate phases are shown in Fig.63.1. Note that the linewidths indicate that the Br_2 coherence distance is of the order of 1 micron or more. In the commensurate phases the positions can be explained using (63.1) with a displacement $\tau=(2/7)b$ at each domain wall [63.7]. The intensities can be approximately explained by (63.1). However, 2D continuous symmetry systems cannot have true LRO so the delta-function Bragg condition in (63.1) cannot be correct [63.8]. As discussed

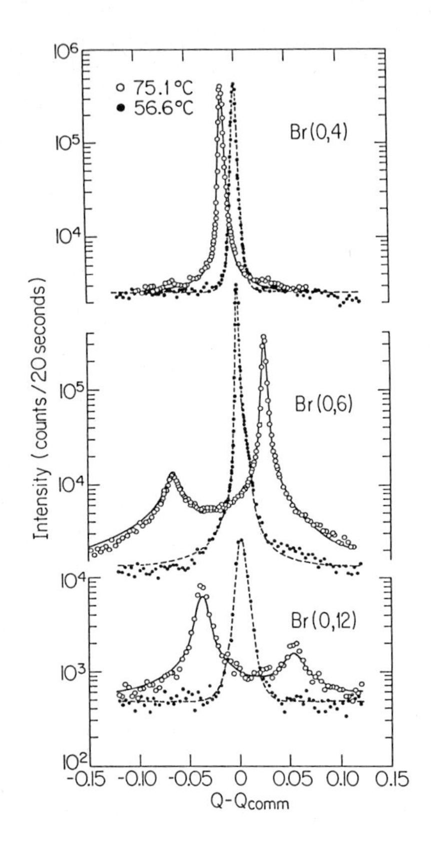

Fig.63.1. Diffraction profiles [63.7b] of various peaks in the commensurate phase (●), at 56.6°C, and in the incommensurate phase (○), at 75.1°C. The transition temperature is 69.5°C. (----) are model commensurate line shapes; (——) are model power-law line shapes. To improve clarity not every datum point is shown

below, this lack of LRO manifests itself in the unusual line shapes evident in Fig.63.1.

The idealized model of an array of 1D domain walls in two dimensions has been extensively studied theoretically [63.9]. A number of authors have developed an elegant description of the commensurate-incommensurate transition in this system; in the weakly incommensurate phase it is assumed that the walls wander and collide, leading to an entropic repulsion which varies as the inverse square of the wall separation. This situation leads to several quantitative predictions. First, the wall separation (ℓ) should obey

$$\ell = \ell_0 t^{-1/2} \quad \text{with} \quad t = (T - T_c)/T_c \quad . \tag{63.2}$$

This form has been confirmed [63.7] for 5 different harmonics in $C_{28}Br_2$. Second, the Bragg peaks should actually be power law singularities

$$S(Q) \sim (Q - Q_n)^{-2+\eta_n} \quad \text{with} \quad \eta_n = \frac{2n^2}{p^2} \quad , \tag{63.3}$$

where p is the periodicity of the commensurate structure relative to the reference lattice and n is the harmonic number. Thus, in the weakly incommensurate phase, η_n is predicted to be independent of temperature, wall density and the microscopic elastic constants. It is this latter remarkable prediction which motivated the experiments of *Mochrie* et al. [63.7]. The solid lines in Fig.63.1 are calculated assuming the above result and with only the relative intensities as adjustable parameters. Clearly the fits are excellent. We consider this result to be an outstanding success for the theory, illustrating dramatically the consequences of the lack of true LRO in 2D continuous symmetry systems. Such effects will occur in all incommensurate chemisorbed surface phases including those conventionally described in an antiphase domain wall language.

63.3 Krypton on Graphite

Krypton on graphite is one of the most extensively explored surface systems with diffraction [63.2,4,10], heat capacity [63.11] and vapor pressure isotherm [63.12] information available. For submonolayer coverages the Kr forms a $\sqrt{3}\times\sqrt{3}$ R30° structure. In their seminal LEED experiments [63.10], *Fain* and co-workers discovered that as the coverage is increased above one $\sqrt{3}\times\sqrt{3}$ monolayer, the Kr exhibits a CIT to a triangular structure incommensurate with respect to the graphite; at saturation, the incommensurability is about 5%. The hierarchy of interactions for this physisorbed system is (i) the Kr-Gr binding of ~1500 K, (ii) the Kr-Kr Lennard-Jones potential with a well depth ~180 K, (iii) the corrugation of the Kr-Gr potential with a difference in binding energy between the Gr hexagon center and corner sites estimated to be between 40 K and 80 K. In domain wall language, this means that incommensurate Kr is in the broad domain wall regime with wall full widths of order 6 lattice constants.

Early work focused on the nature of the commensurate melting transition which was predicted to be in the universality class of the 2D 3-component Potts models [63.13]. For technical reasons which are too detailed to discuss here, the nature of this melting transition and indeed whether it is first or second order has not yet been satisfactorily resolved. The first experiment which we performed where the high resolution provided by synchrotron X-ray techniques gave qualitatively new information was our study of the Kr CIT [63.2]. Here, to our amazement, we discovered that even at quite low temperatures, weakly incommensurate Kr was disordered; the Kr overlayer, however, recovered order with increasing incommensurability. We speculated that the weakly incommensurate phase should be considered a *domain wall fluid*. It was anticipated further that this fluid must connect smoothly onto the normal high-temperature fluid. These speculations were later put on firm theoretical ground by a number of different groups, albeit based on somewhat different mechanisms for the disorder [63.14]. One of the most interesting results of these experiments is that, as shown in Fig.63.2, the incommensurability for temperatures below 100 K follows a simple universal law [63.2,10]; the exponent of 1/3 still has not been explained other than heuristically.

Recent work [63.2] has focused on the behavior at high temperatures in the coverage range between one and two layers. Figure 63.3 shows a series of scans taken at a fixed pressure of 310 Torr as a function of decreasing temperature. We note that an important advantage of X-ray surface techniques is that they may be used in the presence of a dense vapor or liquid overlayer. These data were taken using as a substrate exfoliated ZYX graphite, which is an azimuthal powder but well aligned in the vertical direction; the surface correlation length is about 2000 Å. The scans in Fig.63.2 center around the $\sqrt{3}\times\sqrt{3}$ primary wave vector $Q = 1.70$ Å^{-1}. At 129.5 K the Kr forms a weakly correlated fluid, $\xi \approx 18$ Å, with a mean spacing close to the commensurate solid

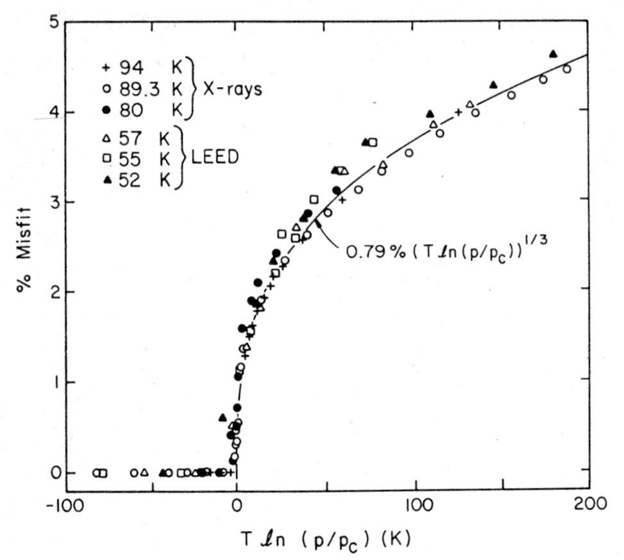

Fig.63.2. Peak shift in reduced units vs chemical potential difference at a series of temperatures for Kr on graphite from Stephans et al. [63.2]. (□) and (△,▲) are LEED data from [63.10], (○,●) and (+) are X-ray data discussed in [63.2]

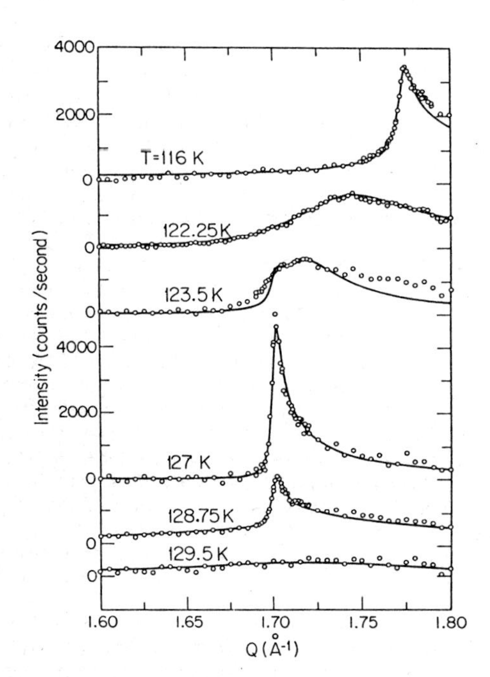

Fig.63.3. Kr(10) diffraction profiles at P = 310 Torr. (——) are model line shapes from *Specht* et al. [63.2]; they are discussed briefly in the text and more extensively in [63.2]

value. At ~128.75 K a commensurate solid peak appears on top of the liquid response, indicating a first-order fluid-commensurate solid transition. We note that the solid lines all represent fits to model liquid and solid line shapes, as discussed in [63.2]. With further decrease in temperature, at ~123.5 K the solid melts into a reentrant incommensurate fluid; finally at

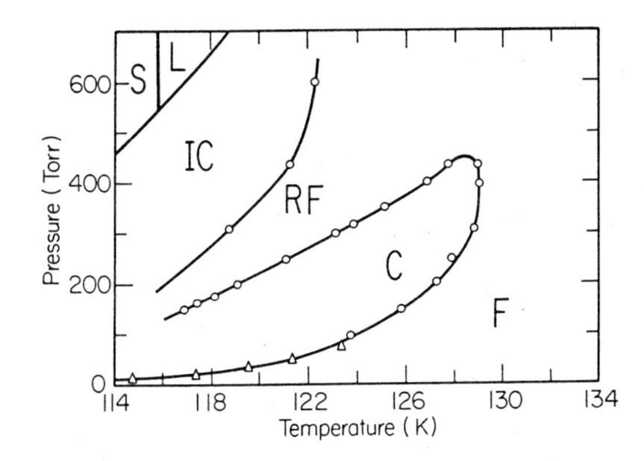

Fig.63.4. Phase diagram of near-monolayer krypton: *F*: fluid, *C*: commensurate solid, *RF*: reentrant fluid, *IC*: incommensurate solid, *S*: bulk solid, *L*: bulk liquid. (Δ) are from [63.12], (○) from Specht et al. [63.2]

~118 K the fluid freezes into an incommensurate solid. The phase diagram is shown in Fig.63.4. At the tip of the commensurate peninsula the Kr coverage is about 1.5 monolayers [63.11]. The reentrant fluid observed here connects continuously onto the modulated fluid observed at lower temperatures ($T \leqslant 94$ K). This unusual phase diagram illustrates clearly the novel behavior which may occur in surface systems with competing interactions.

These experiments have now been extended to a single-crystal graphite substrate by *D'Amico* et al. [63.5]. Such experiments, of course, provide orientational information which is unavailable from powdered substrates. The single-crystal experiments confirm the principal features deduced from the ZYX measurements. They show, in addition, that the reentrant fluid is orientationally ordered, that is, it is a *hexatic fluid* with axes coinciding with the $\sqrt{3} \times \sqrt{3}$ R30° Gr axes; further, the transverse scans are quite sharp, indicating that the angular fluctuations have a small net amplitude.

The most important new result is on the nature of the Novaco-McTague effect [63.15]. *Novaco* and *McTague* first predicted that in the linear response regime the axes of an incommensurate overlayer would rotate relative to those of the substrate; this rotation is a consequence of the fact that in a Lennard-Jones solid the shear elastic constant is $\sim 1/\sqrt{3}$ times the compressional elastic constant so that the overlayer prefers to take up the stress due to its interaction with the substrate by transverse displacement of the atoms. This prediction was first confirmed by *Fain* and co-workers for Ar on Gr [63.10]; it has since been studied for metallic overlayers [63.17]. *Shiba* [63.18] noted that with decreasing incommensurability one must cross over from the sine-modulation to the domain wall regime. Shiba has shown that a second-order transition should occur in which the rotation angle goes to zero at some finite incommensurability. The threshold value provides a measure of the domain wall width in his theory.

This transition was observed by *Fain* and co-workers [63.19] in their early LEED work in Kr on graphite; however, it could not be studied quantitatively. Synchrotron X-ray techniques provide an angular resolution of better than 0.002 degrees [63.20] so that the substrate inevitably provides the ultimate limit in the accuracy of such experiments. We show in Fig.63.5 results by *D'Amico* et al. [63.5] on the Kr rotation angles as a function of lattice constant at ~85 K, where a second-order transition with the mean field exponent $\beta = 1/2$ is indeed observed. From the threshold incommensurability value of

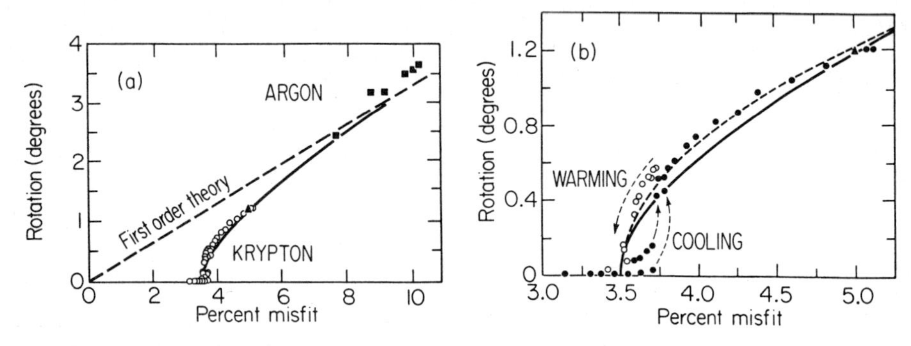

Fig.63.5. (**a**) Misfit vs rotation angle for Kr from [63.5] (circles) and [63.10,19] (Δ), and for Ar (■) from [63.16]. (----) is the Novaco-McTague prediction for a Lennard-Jones solid with $C_L/C_T = \sqrt{3}$. (——) is Shiba's prediction [63.18] scaled for a critical misfit of 3.5%. (**b**) Kr data in (*a*) shown on an expanded scale to illustrate the hysteresis. (----) is the result of a fit to a square root law. Figure from [63.5]

3.5% one deduces a domain wall full width of $\lambda \sim 6$ lattice constants, consistent with the work of *Stephens* et al. [63.2], who estimated λ from the modulation peak intensities. As is evident from Fig.63.5, *Shiba*'s theory [63.18] gives a good account of the overall variation of rotation angle versus misfit. The success may reflect partly the fact that because of the large lengths involved this is a mean-field transition and Shiba's calculations neglect fluctuations. The small hysteresis observed near the transition is probably connected with kinetics and may not indicate a first-order transition; further study of this aspect of the problem is required.

63.4 Discussion: Perspectives for the Future

In this paper we have limited our discussion to two prototypical systems where the physics is dominated by competing interaction effects. Synchrotron X-ray experiments have also been reported on 2D continuous symmetry melting of argon, krypton and xenon on graphite [63.2,4], on the wetting behavior of ethylene on graphite [63.21] and on the structures and transitions of simple molecules such as O_2 on graphite [63.22]. A beautiful family of experiments has been reported on free-standing liquid crystal films with thicknesses varying from 2 molecules to macroscopic values [63.3,23]. All of these have been studied with resolutions varying between 10^{-3} and 10^{-4} Å^{-1}.

Clearly, the next important step in the field is to wed synchrotron X-ray scattering with conventional surface techniques, all carried out in situ. Initial versions of such apparati have been successful [63.1] and new facilities are now being constructed by a number of groups. It should be emphasized that X-rays will not replace electrons as probes of surface structures. Rather, high-resolution X-ray techniques are complementary to LEED which provides a rapid, low-resolution survey. As illustrated by the two examples discussed here, an important advantage of X-rays beyond the evident high resolution is that peak profiles and intensities may be analyzed quantitatively to yield precise information about the underlying corre-

tance and the position of the second atom in the unit cell on the 0.5 (formerly the 0.25) position along the [1$\bar{1}$0].

65.4 Melting Studies

To study the melting characteristics of these Pb layers we chose to monitor the strongest peak along the [1$\bar{1}$0] for both phases, the (0.8,0) and the (0.777,0) reflections. By repeated radial scans as a function of sample temperature the progress of the melting could be followed through the peak intensity, the peak position and the line shape of the reflection. At crucial points in the transition transverse curves were also obtained. After the sample was dosed with Pb the initial Auger ratio was determined. The sample was heated slowly while monitoring the diffraction peak and the Pb/Cu Auger ratio at periodic intervals. The sample was then cooled to roughly 200°C, well below the melting temperatures observed for these coverages, and the heating process was repeated. We studied melting transitions for ten Pb surface coverages. Ordered diffraction peaks were not observed for Pb/Cu ratios below 0.3.

This work brought out some startling new results and explained some of the puzzling data which had previously been obtained. The first surprise is that the as-deposited Pb layer is not a metastable phase as postulated by MFE but is rather an equilibrium high-density phase which can be reversibly melted. This was discovered by careful monitoring of the Auger signal from overlayer and substrate during the annealing and X-ray scattering scans, a unique capability of the chamber used for the experiment [65.5]. The commensurate, high-density Pb layer exists over a range of coverages which are characterized by Auger intensity ratios from 0.35 to 0.55. Below ~0.34 the surface structure changes to the incommensurate phase first seen by MFE. The second interesting feature of the data is that the incommensurability increases as the coverage decreases below the critical ratio of ~0.34. The variation of melting temperature with coverage is shown in Fig.65.3. The area of the plot denoted by 1 is the incommensurate regime with decreasing melting temperature and increasing incommensurability as the coverage is reduced. Region 2 is the commensurate overlayer, with melting temperature decreasing as the coverage increases.

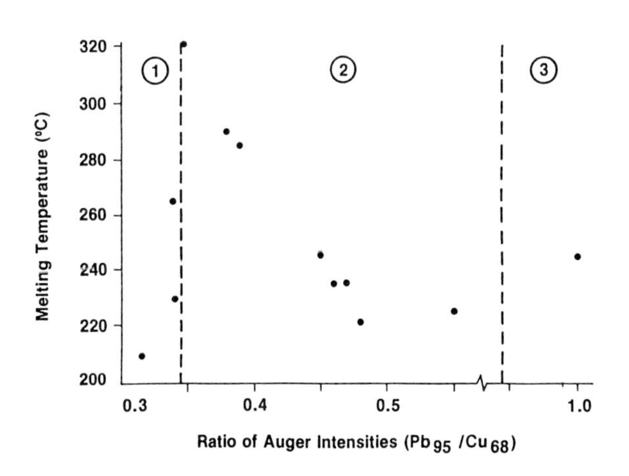

Fig.65.3. The melting temperature of Pb layers on a Cu(110) surface as a function of coverage

At very high Pb coverages (Pb/Cu Auger ratio of 1.0 or larger) a third Pb structure was discovered which had diffraction peaks with d spacings very similar to bulk fcc Pb, namely Pb(220) reflections. This structure had a constant melting temperature of 246°C over a range of coverages up to a Pb/Cu ratio of 5.0. In this third regime the melting transition seems to be first order, being sharper than the 1°C resolution of our controller but 80°C lower than the known bulk Pb melting temperature. Alloying with Cu is unlikely to be the cause of this melting temperature depression as the minimum Pb-Cu alloy melting temperature (the eutectic very near the Pb-rich side of the Pb-Cu phase diagram) is 326°C, only 1°C below the pure Pb melting temperature [65.6]. Also note that the temperature was monitored by two thermocouples which agreed within 5°C and which consistently measured slightly higher temperatures than the surface temperature measured by an infrared pyrometer. This result is consistent with radiative losses from the sample surface; a lower surface temperature would imply a lower melting temperature than the 246°C cited. Thus, we believe that this drop in the melting temperatures is a real phenomenon and further experiments are in progress.

65.5 Conclusions

The results presented here give a picture of the growth of Pb on the Cu(110) substrate. At coverage below one monolayer the structure is incommensurate along the [1$\bar{1}$0], commensurate along the [001] and has a two-atom unit cell. Above one monolayer the structure has a higher density four-atom unit cell and is commensurate in both directions. The melting temperature as a function of composition exhibits a sharp maximum at a coverage of approximately one monolayer and at the commensurate-incommensurate transition.

At very high coverages (over ~5 monolayers effective coverage) the emergence of Pb bulk-like crystallites occurs. These grow as 1 μ islands on top of the several complete layers of Pb grown directly on the Cu substrate. These crystallites exhibit an extremely low melting temperature (80°C below bulk Pb). Further studies are being performed to identify the exact mechanism of this transition but we hypothesize the low melting temperature is caused by a transition from spherical crystallites to a smooth continuous liquid layer on the Cu surface.

Acknowledgments. We should like to thank G. Hughes, J.B. Stark, R. Hewitt, A. Dayem and E. Westerwick for their assistance. This work was partially done at SSRL which is supported by the NSF; the DOE, Office of Basic Energy Sciences; and the NIH, Biotechnology Resource Program, Division of Research Resources.

References

65.1 D.R. Nelson, B.I. Halperin: Phys. Rev. B**19**, 2457 (1979;
S. Ostlund, B.I. Halperin: Phys. Rev. B**23**, 335 (1981)
65.2 W.C. Marra, P.H. Fuoss, P. Eisenberger: Phys. Rev. Lett. **49**, 1169 (1982)
65.3 J. Henrion, G.E. Rhead: Surf. Sci. **29**, 20 (1972)
65.4 W.C. Marra: Ph.D. Thesis, Stevens Institute of Technology (1981)
65.5 S. Brennan, P. Eisenberger: Nucl. Instr. Meth. **222**, 164 (1984)
65.6 M. Hansen: *Constitution of Binary Alloys* (McGraw-Hill, New York 1958) p.609

Index of Contributors

Abraham, F.F. 334
Ahmad, M. 389
Allan, D.C. 298
Als-Nielsen, J. 313
Aono, M. 187

Bachrach, R.Z. 308
Bahr, C.C. 191
Bartelt, N.C. 357
Barton, J.J. 191
Bartoš, I. 113
Batra, I.P. 251,285
Bean, J.C. 163
Behm, R.J. 257
Bennemann, K.H. 226
Birgeneau, R.J. 404
Bohr, J. 313
Bonapace, C. 317
Brennan, S. 421
Bringans, R.D. 308
Brocksch, H.-J. 226
Broughton, J.Q. 163
Burdick, S. 347

Chabal, Y.J. 70
Chen, C.D. 170
Chiaradia, P. 66
Chiarotti, G. 66
Christmann, K. 257
Ciccacci, F. 66
Citrin, P.H. 149
Cohen, M.L. 4,321
Cook, M.R. 285
Cowley, J.M. 55
Cricenti, A. 66

Daw, M.S. 41
Duke, C.B. 317

Eckert, J. 219
Egelhoff, Jr., W.F. 199
Ehrlich, G. 375
Einstein, T.L. 357
Eisenberger, P. 246,421
El-Batanouny, M. 347
Engel, T. 251
Ertl, G. 48,257

Fain, Jr., S.C. 413
Falicov, L.M. 12
Farrell, H.H. 163
Feidenhans'l, R. 313
Ferrer, S. 204
Fock, J.H. 313
Foiles, S.M. 41
Fuoss, P.H. 246,421

Garcia, N. 204
Gay, J.G. 35
Gorse, D. 176
Güntherodt, H.J. 48

Halicioğlu, T. 231
Hall, B.M. 156
Heinz, K. 105
Henzler, M. 351
Hidber, H.R. 48
Himpsel, F.J. 285
Holloway, S. 18
Horn, P.M. 404
Howard, J. 219
Hsu, T. 55
Hu, G.Y. 341
Hughes, G.J. 246

Ibach, H. 156
Ignatiev, A. 326
Ishizawa, Y. 187

Johnson, R.L. 313
Jona, F. 92,285,293

Kahn, A. 317
Kirczenow, G. 347
Koukal, J. 113

Lackey, C.D. 269
Lagally, M.G. 366
Lapujoulade, J. 176
Lehwald, S. 156
Li, K. 317
Liang, K.S. 246
Liu, H. 285
Louie, S.G. 29
Lu, T.-M. 361

Madey, T.E. 264
Mailhiot, C. 317
Marcus, P.M. 92,285,293
Martini, K.M. 347
Mayol, R. 117
McRae, E.G. 278
Mei, W.N. 303
Mele, E.J. 298
Moncton, D.E. 404
Moog, E.R. 397
Müller, K. 105

Nicol, J.M. 219
Nielsen, M. 313
Nørskov, J.K. 18
Northrup, J.E. 321

Ocal, C. 204
Oelhafen, P. 48
Oshima, C. 187

Parellada-Sabata, J. 117
Paton, A. 317
Pendry, J.B. 124,131,135
Penka, V. 257
Pimbley, J.M. 361

Richardson, N.V. 269
Richter, R. 35
Rieder, K.H. 251
Ringger, M. 48
Robey, S.W. 191
Robinson, I.K. 60,313
Rocca, M. 156
Roelofs, L.D. 357

Salanon, B. 176
Saldin, D.K. 131,135
Saloner, D. 366
Salvat, F. 117
Schaefer, J.A. 163
Schlögl, R. 48
Schöbinger, M. 334
Schwankner, R. 257
Selci, S. 66
Selloni, A. 170
Shen, Y.R. 77
Shirley, D.A. 191
Shu, Y.S. 293
Slichter, C.P. 84
Smith, J.R. 35
Souda, R. 187
Stöhr, J. 140
Strozier, Jr., J.A. 92
Surman, M. 269

Takai, T. 231
Tersoff, J. 12,54
Tiller, W.A. 231
Toney, M. 313
Tong, S.Y. 156,303
Tosatti, E. 170
Tromp, R.M. 285
Tsong, T.T. 389

Victora, R.H. 12
Vvedensky, D.D. 131,135
Van Hove, M.A. 100
Vanderbilt, D. 29

Wandelt, K. 48
Webb, M.B. 397
Willis, R.F. 237
Wright, C.J. 210
Wu, N.J. 326

Xu, G. 303
Xu, M.-L. 156

Yang, W.S. 293
Ying, S.C. 341
You, H. 413

Zhan, S.B. 321

Subject Index

Acetylene *see* C_2H_2
Adatom clusters 379ff.
Adatom-adatom interactions 375ff., 389ff.,397ff.
Adsorbate binding 18ff.
Adsorption isotherms 78,281
AES *see* Auger electron spectroscopy
Ag on Si(111) 187ff.
Ag(100) 38
Ag(110)
  clean 93,179
  with H_2O 266
  with H_2O and Br 266
  with O_2 and O 267
Ag(111)
  with Ar 400ff.
  with Kr 400ff.
  with Xe 400ff.
Alloys 14,93,389ff.
Al on GaAs(110) 9
Alumina 84,205
Al(110) 93
Al(111)
  with NH_3 266
  with NH_3 and O 266
Amorphized Si 151ff.
Angle-resolved photoemission extended fine structure 191ff.
Angle-resolved X-ray photoelectron spectroscopy 199ff.
Ar
  on Ag(111) 400ff.
  on Pd(100) 400ff.
ARPEFS *see* Angle-resolved photoemission extended fine structure
ARXPS *see* Angle-resolved X-ray photoelectron spectroscopy
Atom diffraction 176ff.,251ff.
Atom probe 389ff.
Atom-surface scattering 178ff., 251ff.
Attenuation 114,117ff.
Au diffusion in aluminum oxide 204ff.
Auger electron spectroscopy 117
Autocorrelation 352
Au(110)-(1 × 2) 61ff.,226ff.,231ff.

Beam set neglect method 101,131
Benzene *see* C_6H_6
Bond lengths 23,140ff.
Bonzel-Ferrer model for fcc(110)-(1 × 2) 226ff.,231ff.
Bound-state transition 143
Br
  in graphite 405ff.
  on Ag(110) with H_2O 266
Buckling model for Si(111) 7,66,171

C *see* Diamond, graphite
Cabrera-Mott theory of oxidation 204ff.
Carbon monoxide *see* CO
$C_{28}Br_2$ 405ff.
CEMS *see* Conversion electron Mössbauer spectroscopy
CH_3O (methoxy)
  on Cu(100) 143
CH_3OH (methanol)
  on Cu(100) 143
C_2H_2 (acetylene)
  on Ni(100) 102
  on Ni(111) 102
  on Pt(111) 102,146
  on Pt particles 88ff.
C_2H_3 (ethylidyne)
  on Pt(111) 102,146
  on Rh(111) 102
C_2H_4 (ethylene)
  on Pt(111) 146
  on ZnNaA zeolite 219ff.
C_3H_4 (methylacetylene)
  on Rh(111) 102
C_3H_5 (propylidyne)
  on Rh(111) 102
C_4H_4S (thiophene)
  on PT(111) 146

C_5H_5N (pyridine)
  on Pt(110) 270ff.
  on Pt(111) 146
$C_6H_5CH_3$ (toluene)
  on Pt(111) 146
C_6H_6 (benzene)
  on Ni(111) 133
  on Pt(111) 102,146
  on Rh(111) 102
$C_{10}H_8$ (naphthalene)
  on Rh(111) 102
Chemical shift 87
Cluster approach to LEED 96ff.
Cluster calculations 74
Coadsorption 146,266ff.,273ff.
CO (carbon monoxide)
  on Cu(100) 101,143
  on Cu(110) with K 273ff.
  on graphite 413ff.
  on Ni(100) 101
  on Pd(100) 101
  on Pt(111) with Na 146
  on Pt particles 87ff.
  on Rh(111) 79,101
  on Ru(0001) 101,266ff.
Co magnetization 13ff.
Co on Cu(111) 15
Co(0001)
  with H 22
  with O 22
Combined space method 101
Commensurate-incommensurate transition 334,341ff.,347ff.,426
Conversion electron Mössbauer spectroscopy 117
Coordination number 210ff.
Correlation length 358
Corrugation function 179
Critical exponent 357ff.
Critical phenomena 357ff.
  see also Phase transitions
Cr magnetization 13ff.
Cr(110) with H 22
Cs on Rh(111) 80
Cu on Ni(100) 199ff.
Cu_3Au 93
Cu-Ni interface magnetization 14
Cu(100)
  clean 38,93,115
  with HCO_2 144ff.
  with CH_3OH, CH_3O 143
  with CO 101,143
  with Ni 15
  with S 191ff.
Cu(110)
  clean 93,179ff.
  with O 246ff.
  with CO and K 273ff.
  with Pb 421ff.
Cu(111)
  with Co 15
  with H 22
  with O 22
  with Ni 15
Cu(311) 179ff.
Cu(511) 179ff.
Cu(711) 179ff.

D 72ff.,270ff.
Damping 114,117ff.
Dangling bonds 171
Dead magnetic layers 16
Density of states 114
Deuteration of pyridyne 270ff.
Deuterium *see* D
Diamond (111) 8,29ff.
Dielectric function 67
Diffuse LEED 124ff.,131ff.
Diffusion
  along surface 24ff.,88,375ff.
  of clusters 381ff.
  through layers 204ff.
Dimer model for Si(100)-(2 × 1) 7,72,93,293ff.
Dipole interactions 70
Dipole moments for H and D on Si(100) 75
Dipole scattering 269
Dipole selection rules 269
Dislocations 56,347ff.
Disordered adsorbates 101,124ff.,131ff.,154
Dispersion of particles 84
Domain size 352,366ff.
Domain wall 61,334ff.,341ff.,362,405
  crossing 334
  fluid 407
  interaction 61
Dynamic dipole moment 70,271
Dynamical LEED *see* Multiple scattering

EELS *see* Electron energy loss spectroscopy
Effective medium theory 18ff.,42,178
Electron
  acceptor 80
  beam damage 103
  channeling 199
  damping 113ff.
  donor 80
  energy loss spectroscopy *see* HREELS
Electron microscopy 55ff.
Electron spectroscopy for chemical analysis *see* X-ray photoemission electron spectroscopy

Electron-hole pairs 113
Electron-phonon interactions 113, 156ff.,170ff.
Electron-stimulated desorption 264ff.
Electronic band structure 4ff.,29ff., 67ff.
  ion angular distributions 264ff.
Embedded atom method 41ff.
Encapsulation 205
Epitaxy 199ff.
Esbjerg-Nørskov potential 20ff., 178ff.,252
ESCA *see* Electron spectroscopy for chemical analysis
ESD *see* Electron-stimulated desorption
ESDIAD *see* Electron-stimulated desorption ion angular distributions
Ethylene *see* C_2H_4
Ethylidyne *see* C_2H_3
EXAFS *see* Extended X-ray absorption fine structure
Extended Hückel theory 102
  X-ray absorption fine structure 141

FeCo alloy magnetization 14
FeCo(110) with Fe 16
Fe magnetization 13ff.
Fe(100) 38
Fe(110)
  with H 22
  with O 22
Fe(210) 93
Fe(211) 93
Fe(310) 93
Field ion microscopy 375ff.,389ff.
FIM *see* Field ion microscopy
Fluctuations in long- and short-range order 237ff.
Force-field calculations 102
Formate *see* HCO_2

GaAs(110)
  clean 9
  with Al 9
  with Sb 317ff.,366ff.
GaAs(111)-(2 × 2) 93,303ff.,308ff.
GaAs($\bar{1}\bar{1}\bar{1}$)-(2 × 2) 308ff.
GaP(110) with Sb 317ff.
GaSb(110) with Sb 317ff.
Ge on Si(111) 321ff.
Ge(111)-(2 × 1) 8,68ff.
Ge-Si(100) alloys with OH 163ff.
Graphite
  clean 56
  with Br 405ff.
  with CO 413ff.
  with K 326ff.
  with Kr 334ff.,407ff.
  with N_2 413ff.
Grazing-incidence X-ray scattering 61,246ff.,421ff.
Green's function 113

H
  on Co(0001) 22
  on Cr(110) 22
  on Cu(111) 22
  on Fe(110) 22
  on Mn 22
  on MoS_2 215ff.
  on Ni(110) 41ff.,257ff.
  on Ni(111) 22
  on Pd black 213
  on Pd(111) 41ff.
  on Pt black 214ff.
  on Pt particles 84ff.
  on Raney Ni 210ff.
  on Sc(0001) 22
  on Si(100)-(2 × 1) 70ff.
  on Si(111)-(7 × 7) 280ff.
  on Ti(0001) 22
  on V(110) 22
  on W(100) 237ff.,341ff.
  on WS_2 215ff.
HCO_2 (formate)
  on Cu(100) 144ff.
Healing length 14
Helium diffraction 175ff.,251ff.
Hellmann-Feynman forces 7,299
Herringbone structure 413ff.
Heterojunction 5
Hexatic fluid 409
High-resolution electron energy loss spectroscopy 70,100,156ff.,163ff., 269ff.
H_2O
  on Ag(110) 266
  on Ag(110) with Br 266
  on Ni(110) 267
  on Ni(111) 266
  on Ni(111) with O 266
  on Ru(0001) 266ff.
  on Ru(0001) with Na 266ff.
Hopping integral 171
HREELS *see* High-resolution electron energy loss spectroscopy
Hydroxyl *see* OH

ICISS *see* Impact collision ion scattering spectroscopy
Impact collision ion scattering spectroscopy 187ff.
Impact scattering 269

InAs(110) with Sb 317ff.
Incommensurate phases 334ff.
Inelastic neutron scattering 210ff., 219ff.
Infrared spectroscopy 70ff.,219ff., 281
InP(110) with Sb 317ff.
InSb(110) with Sb 317ff.
InSb(111)-(3 × 3) 313ff.
Instrument response function 352, 358,361
Instrument transfer function 352
Intercalation 326ff.
Interface magnetization 14ff.
Internal conversion 117
Ion scattering spectroscopy 204ff.
IR *see* Infrared spectroscopy
Ir(100)
  clean 107ff.
  with K 109ff.
Ir(110)-(1 × 2) 226ff.
Islands 70,366
Isotopic substitution 70
ISS *see* Ion scattering spectroscopy
I-V curves
  in LEED 95
  in HREELS 158

K
  on graphite 326ff.
  on Ir(100) 109ff.
  on Rh(111) 80
  on Cu(110) with CO 273ff.
Kikuchi beam 199
Knight shift 85
Kr
  on Ag(111) 400ff.
  on graphite 334ff.,407ff.
  on Pd(100) 400ff.

Lattice distortion 170
Lattice dynamics 237ff.
Lattice parameter determination 398ff.
Lattice-gas disorder 131
LCAO method 29
LEED *see* Low-energy electron diffraction
Local density functional 6,19,29, 42,321
Louse 49
Low-energy electron diffraction 92ff.,100ff.,278,285ff.,293ff., 303ff.,326ff.
  diffuse LEED 124ff.,131ff.
  patterns 397ff.,413ff.
  spot profiles 351ff.,361ff., 366ff.,397ff.

Magnetization 12
Methanol *see* CH_3OH
Methoxy *see* CH_3O
Methylacetylene *see* C_3H_4
Microchannel plates 103
Missing-row model for fcc(110)-(1 × 2) 61,226ff.,231ff.
Mixed basis 6
Mn
  with H 22
  with O 22
Molecular adsorbates 100ff.,131ff., 140ff.,264ff.
Molecular dynamics 41ff.,335,347ff.
Molecular orientation 140ff.
Monte Carlo technique 42,118,232
MoS_2 with H 215ff.
Multilayer relaxation 93,226ff.
Multiple internal reflection in IR 71
Multiple quantum coherence 89
Multiple scattering
  in LEED 92,100ff.,124ff.,131ff.
  in NEXAFS/XANES 124ff.,135ff., 140ff.

Na
  on Pt(111) with CO 146
  on Rh(111) 80
Naphthalene *see* $C_{10}H_8$
Near-edge X-ray absorption fine structure 124ff.,135ff.,140ff. (*see also* XANES)
Near-neighbor multiple scattering 101
Nearly free electron model 113
Negative-ion resonance 163
Neutron scattering 210ff.,219ff., 416
NEXAFS *see* Near-edge X-ray absorption fine structure
NH_3
  on Al(111) 266
  on Al(111) with O 266
  on Ni(110) 266
  on Ni(110) with Na 266
  on Ni(111) 266
  on Ni(111) with O 266
  on Ru(0001) 266
  on Ru(0001) with Na 266
  on Ru(0001) with O 266
Ni_3Al 93
Nickel silicide 150
Ni-Cu interface 14
Ni magnetization 13ff.
Ni
  on Cu(100), Cu(111) 15
  on Si(111) 150

$NiSi_2(111)$ 150
Ni(100)
clean 38,156ff.
with C_2H_2 102
with CO 101
with Cu 199ff.
with O 129,138,251ff.
with S 191ff.
Ni(110)
clean 179
with H 41ff.,257ff.
with H_2O 267
with NH_3 266ff.
with NH_3 and Na 266ff.
with O 22
with OH and O 267
with S 191ff.
Ni(111)
with C_2H_2 102
with C_6H_6 133
with H 22
with H_2O 266
with H_2O and O 266
with NH_3 266
with NH_3 and O 266
with O 22
NMR *see* Nuclear magnetic resonance
Noble gases 334ff.,397ff.
N_2 on graphite 413ff.
Nuclear magnetic resonance 84ff.

O
on Ag(110) 267
on Co(0001) 22
on Cr(110) 22
on Cu(110) 246ff.
on Cu(111) 22
on Fe(110) 22
on Mn 22
on Ni(100) 129,138,251ff.
on Ni(110) 22
on Ni(110) with OH 22
on Ni(111) 22
on Sc(0001) 22
on Si(111) 81
on Ti(0001) 22
on V(110) 22
O_2
on Ag(110) 267
on Rh(111) 78
OH
on Ge-Si(100) alloys 163ff.
on Ni(110) 22
on Ni(110) with O 22
Optical potential 113
Optical spectroscopy 66ff.
Optical transitions 66ff.
Order parameter 357
Order-disorder transitions 41ff.
Organometallic clusters 100
Overlayer magnetization 15ff.
Overtone 163
Oxidation 18ff.,204ff.

Pair correlation function 352,357, 361ff.,366ff.,376ff.
Parallel relaxation 93
Particles 84ff.
Pb on Cu(110) 421ff.
Pd black with H 213
Pd(100)
clean 48ff.
with Ar 400ff.
with CO 101
with Kr 400ff.
with Xe 400ff.
Pd(110) 179
Pd(111) with H 41ff.
$Pd_{81}Si_{19}$ alloy 48ff.
Phase transitions 41ff.,109ff., 181ff.,237ff.,257ff.,341ff.,357ff., 397ff.,404ff.,421ff.
Phonons 113,156ff.,170ff.,300ff.
soft modes 237ff.
Photon-stimulated desorption 264ff.
Photothermal displacement spectroscopy 67
Physisorption 397
Pinwheel structure 413ff.
Plasmon excitation 113
Polarons 170ff.
Position-sensitive detector 103
Propylidyne *see* C_3H_5
PSD *see* Photon-stimulated desorption
Pseudopotential 4,321
Pt(110)
(1×2) 226ff.
with C_5H_5N 270ff.
Pt(111)
clean 56
with C_2H_2 102,146
with C_2H_3 102,146
with C_2H_4 146
with C_4H_4S 146
with C_5H_5N 146
with $C_6H_5CH_3$ 146
with C_6H_6 102,146
with CO and Na 146
Pt-Au alloy 395
Pt black with H 214ff.
Pt-Cu alloy 395
Pt-Ir alloy 395
Pt particles
with C_2D_2, C_2H_2 88ff.

Pt particles (cont.)
with CO 87ff.
with H 84ff.
Pt-Rh alloy 394ff.
Pyridine *see* C_5H_5N

Quadrupole-quadrupole interactions (electrostatic) 413ff.

Ramsauer-Townsend resonance 195, 197
Raney nickel with H 210ff.
Rayleigh surface waves 300
Reactions at surfaces 146ff.,149ff., 269ff.
Re
on W(110) 384ff.
on W(211) 376ff.
Re(10$\bar{1}$0) 93
Reconstruction 4ff.,29ff.,61,63, 183ff.,226ff.,285ff.,298ff.
adsorbate-induced 257ff.
Recursion method 228
Reentrant fluid 408ff.
Reflection electron microscopy 55ff.
REM *see* Reflection electron microscopy
Renormalized forward scattering 131,135
Rh(100) 390ff.
Rh(111)
with C_2H_3 102
with C_3H_4 102
with C_3H_5 102
with C_6H_6 102
with $C_{10}H_8$ 102
with CO 79,101
with Cs 80
with K 80
with Na 80
with O_2 78
Roughening transition 181ff.
Rutherford ion backscattering 278, 288
Ru(0001)
with CO 101,266ff.
with NH_3 266
with NH_3 and O 266
with NH_3 and Na 266ff.
with H_2O 266ff.
with H_2O and Na 266ff.

S
on Cu(100) 191ff.
on Ni(100) 191ff.
on Ni(110) 191ff.
Sb
on GaAs(110) 317ff.,366ff.
on GaP(110) 317ff.
on GaSb(110) 317ff.
on InAs(110) 317ff.
on InP(110) 317ff.
on InSb(110) 317ff.
Scanning electron microscopy 48ff.
Scanning tunneling microscopy 48ff., 54,280
theory 54
Sc(0001)
with H 22
with O 22
Schottky barrier 5
Second harmonic generation 77ff.
SEDOR *see* Spin-echo double resonance
Self-consistent local orbital transitions 67
Seiwatz model for Si(111) 8
Selection rules in optical method 35ff.
SEM *see* Scanning electron microscopy
SEXAFS *see* Surface extended X-ray absorption fine structure
Shape resonances in HREELS 163ff.
SHG *see* Second harmonic generation
Short-range order 101,361ff.
Si(100)
(2×1) 7,93,293ff.,298ff.
(2×1)-H 70ff.
Si(111)
(1×1) 7,94
(2×1) 7ff.,67ff.,93,173ff., 285ff.
(7×7) 7,63ff.,96,278ff.
(7×7) with H 280ff.
with Ni 150
with O 81
amorphized 151ff.
with Ag 187ff.
with Ge 321ff.
Si on W(110) 389ff.
Soliton 237ff.,341ff.
superlattice 237ff.
Specific heat 357ff.
Spin-echo double resonance 86
Spin-lattice relaxation 87
Spot profile in LEED 95,105ff., 351ff.,361ff.,366ff.,397ff.
Stacking faults 55,278ff.
Steps 9,56,61,72,179ff.,362
STM *see* Scanning tunneling microscopy
Strain energy minimization 288
Strong metal-support interaction 205
Superlattice 5,391ff.
Surface extended X-ray absorption fine structure 140ff.,149ff.

Surface extended X-ray absorption fine structure (cont.)
and clean surfaces 151ff.
Surface magnetization 13ff.
phonon 300ff.
segregation 389ff.
state 5,170ff.
state polaron 170
Synchrotron radiation 60,246ff., 404ff.

TED *see* Transmission electron diffraction
Terrace size 352
Thiophene *see* C_4H_4S
Three-body interactions 232
Ti(0001)
with H 22
with O 22
Tight binding 6,12,102,227,298, 317ff.
Toluene *see* $C_6H_5CH_3$
Total energy 4ff.,18ff.,29ff.,35ff., 41ff.,226ff.,231ff.,298ff.,317ff., 321ff.
Total reflection in XRD 313
Transfer width 353
Transmission electron diffraction 281ff.

Ultraviolet photoelectron emission spectroscopy 48ff.,309ff.
UPS *see* Ultraviolet photoelectron emission spectroscopy

Vacancy-buckling model for GaAs(111)-(2 × 2) 304ff.
Van der Waals radii 102
V(100) 93
V(110)
with H 22
with O 22
Vibrations
atomic adsorbates 24,70ff.,210ff.
clean surface 156ff.,298
deexcitation 70
Video LEED 94ff.,103,105ff.

W(100)
clean 183ff.,341ff.
with H 237ff.,341ff.
W(110)
with Re 384ff.
with Si 389ff.
W(211) with Re 376ff.
Wave matching 5
Wiggler 246
WS_2 with H 215ff.
WSi_2 390

XANES *see* X-ray absorption near-edge structure
Xe
on Ag(111) 400ff.
on Pd(100) 400ff.
XPS *see* X-ray photoelectron emission spectroscopy
X-ray absorption near-edge structure (*see also* NEXAFS) 124ff.,135ff., 140ff.
XRD *see* X-ray diffraction
X-ray diffraction 60ff.,246ff., 313ff.,404ff.,416,421ff.
X-ray photoelectron emission spectroscopy 48ff.,204ff.,281

Zeolites 219ff.
ZnNaA zeolite 219ff.

π-bonded chain model for Si(111)-(2 × 1) 8,29ff.,66,93,173,285ff.
π-bonded molecule model for Si(111)-(2 × 1) 8

EXAFS and Near Edge Structure III

Proceedings of an International Conference, Stanford, CA, July 16–20. 1984

Editors: **K. O. Hodgson, B. Hedman, J. E. Penner-Hahn**
1984. 392 figures. XV, 533 pages. (Springer Proceedings in Physics, Volume 2). ISBN 3-540-15013-7

Contents: Fundamental Aspects of EXAFS and XANES. – EXAFS Data Analysis. – Biological Systems. – Calatytic Systems and Small Metal Clusters. – Surface Structure. – Amorphous Materials and Glasses. – Geology and Geochemistry, and High Pressure. – Other Applications. – Related Techniques and Instrumentation. – Index of Contributors.

V. F. Kiselev, O. V. Krylov

Adsorption Processes on Semiconductor and Dielectric Surfaces I

1985. 70 figures. VIII, 287 pages. (Springer Series in Chemical Physics, Volume 32). ISBN 3-540-12416-0

The monograph is the first in a series of three written by a physicist and a chemist which explore the relationship between adsorption and catalysis on the one hand, and the electronic processes taking place on the semiconductor-dielectric interface on the other. This volume summarizes the vast knowledge on electron-acceptor and protondonor sites in adsorption and catalysis, and investigates ways of controlling their activities. Also discussed are the hybridization of surface atoms and plausible mechanisms of elementary processes in adsorption and catalysis.

Springer-Verlag
Berlin
Heidelberg
New York
Tokyo

Thin-Film and Depth-Profile Analysis

Editor: **H. Oechsner**
With contributions by numerous experts

1984. 99 figures. XI, 205 pages. (Topics in Current Physics, Volume 37). ISBN 3-540-13320-8

Contents: Introduction. – The Application of Beam and Diffraction Techniques to Thin-Film and Surface Micro Analysis. – Depth-Profile and Interface Analysis of Thin Films by AES and XPS. – Secondary Neutral Mass Spectronometry (SNMS) and Its Application to Depth-Profile and Interface Analysis. – In-Situ Laser Measurements of Sputter Rates During SIMS/AES In-Depth Profiling. – Physical Limitations to Sputter Profiling at Interfaces – Model Experiments with Ge/Si Using KARMA. – Depth Resolution and Quantitative Evaluation of AES Sputtering Profiles. – The Theory of Recoil Mixing in Solids. – Additional References with Titles. – Subject Index.

I. Pockrand

Surface Enhanced Raman Vibrational Studies at Solid/Gas Interfaces

1984. 60 figures. IX, 164 pages. (Springer Tracts in Modern Physics, Volume 104). ISBN 3-540-13416-6

Contents: Introduction. – Fundamentals of Surface Enhanced Raman Scattering. – Experimental. – Pyridine Adsorption. – Hydrocarbon Adsorption. – Carbon Monoxide Exposure and Carbonaceous Deposits. – Oxygen Exposure. – Water Adsorption. – Other Adsorbates. – Selected Applications and Related Surface Enhanced Phenomena. – Summary and Outlook. – Appendix: Recent Developments and Results. – References. – Subject Index. – Material Index.

Springer-Verlag
Berlin
Heidelberg
New York
Tokyo

Structural Studies of Surfaces

With contributions by **K. Heinz, K. Müller, T. Engel, K.-H. Rieder**

1982. 120 figures. VII, 180 pages. (Springer Tracts in Modern Physics, Volume 91). ISBN 3-540-10964-1